Methods in Enzymology

Volume 338
NUCLEAR MAGNETIC RESONANCE
OF BIOLOGICAL MACROMOLECULES
Part A

METHODS IN ENZYMOLOGY

EDITORS-IN-CHIEF

John N. Abelson   Melvin I. Simon

DIVISION OF BIOLOGY
CALIFORNIA INSTITUTE OF TECHNOLOGY
PASADENA, CALIFORNIA

FOUNDING EDITORS

Sidney P. Colowick and Nathan O. Kaplan

Methods in Enzymology

Volume 338

Nuclear Magnetic Resonance of Biological Macromolecules

Part A

EDITED BY

Thomas L. James

SCHOOL OF PHARMACY
UNIVERSITY OF CALIFORNIA
SAN FRANCISCO, CALIFORNIA

Volker Dötsch

SCHOOL OF PHARMACY
UNIVERSITY OF CALIFORNIA
SAN FRANCISCO, CALIFORNIA

Uli Schmitz

GENELABS TECHNOLOGIES, INC.
REDWOOD CITY, CALIFORNIA

ACADEMIC PRESS

San Diego London Boston New York Sydney Tokyo Toronto

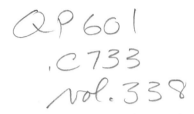

This book is printed on acid-free paper. ∞

Academic Press
A Harcourt Science and Technology Company
525 B Street, Suite 1900, San Diego, California 92101-4495, USA
http://www.academicpress.com

Academic Press
Harcourt Place, 32 Jamestown Road, London NW1 7BY, UK
http://www.academicpress.com

International Standard Book Number: 0-12-182239-7

PRINTED IN THE UNITED STATES OF AMERICA
01 02 03 04 05 06 07 SB 9 8 7 6 5 4 3 2 1

Table of Contents

Section I. General Techniques

v

Section II. Nucleic Acids and Carbohydrates

Contributors to Volume 338

Article numbers are in parentheses following the names of contributors.
Affiliations listed are current.

NORZEHAN ABDUL-MANAN (8), *Vertex Pharmaceuticals Incorporated, Cambridge, Massachusetts 02139*

KAZUYUKI AKASAKA (5), *Department of Molecular Science, Kobe University, Graduate School of Science and Technology, Kobe 657-8501, Japan*

DANEEN T. ANGWIN (9), *Structural, Analytical, and Medicinal Chemistry, Pharmacia Corporation, Kalamazoo, Michigan 49001*

DAINA AVIZONIS (10), *NMR Systems, Varian Incorporated, Palo Alto, California 94301*

SAMUEL E. BUTCHER (17), *Department of Chemistry and Biochemistry, University of California, Los Angeles, California 90077*

TERESA CARLOMAGNO (2), *Department of Molecular Biology and Skaggs Institute of Chemical Biology, MB 33, The Scripps Research Institute, La Jolla, California 92037*

DAVID A. CASE (1), *Department of Molecular Biology, The Scripps Research Institute, La Jolla, California 92037*

FLORENCE CORDIER (4), *Department of Structural Biology, Biozentrum, University of Basel, CH-4056 Basel, Switzerland*

JENNY CROMSIGT (16), *Department of Medical Biosciences—Medical Biophysics, Umeå University, S-901 87 Umeå, Sweden*

VLADIMIR P. DENISOV (7), *Department of Physical Chemistry 2, Condensed Matter Magnetic Resonance Group, Lund University, S-221 00 Lund, Sweden*

ANDREW J. DINGLEY (4), *Institute of Physical Biology, Heinrich-Heine-Universität, D-40225 Düsseldorf, Germany*

KATHLEEN A. FARLEY (9), *Structural, Analytical, and Medicinal Chemistry, Pharmacia Corporation, Kalamazoo, Michigan 49001*

SHAUNA FARR-JONES (10), *Department of Pharmaceutical Chemistry, University of California, San Francisco, California 94143*

JULI FEIGON (17), *Department of Chemistry and Biochemistry, University of California, Los Angeles, California 90077*

JASNA FEJZO (8), *Vertex Pharmaceuticals Incorporated, Cambridge, Massachusetts 02139*

L. DAVID FINGER (17), *Department of Chemistry and Biochemistry, University of California, Los Angeles, California 90077*

RUBEN L. GONZALEZ, JR. (18), *Department of Structural Biology, Physical Biosciences Division, University of California, Lawrence Berkeley National Laboratory, Berkeley, California 94720*

CHRISTIAN GRIESINGER (2), *Institut für Organische Chemie, Universität Frankfurt, Frankfurt D-60439, Germany, and Max-Planck Institut für Biophysikalische Chemie, Göttingen D-37077, Germany*

STEPHAN GRZESIEK (4), *Department of Structural Biology, Biozentrum, University of Basel, CH-4056 Basel, Switzerland*

MAURICE GUÉRON (15), *Groupe de Biophysique, L'Ecole Polytechnique et de l'UMR 7643 du CNRS, Palaiseau 91128, France*

VINEET GUPTA (12), *Laboratory of Molecular Biophysics, The Rockefeller University, New York, New York 10021*

BERTIL HALLE (7), *Department of Physical Chemistry 2, Condensed Matter Magnetic Resonance Group, Lund University, S-221 00 Lund, Sweden*

MIRKO HENNIG (2), *Department of Molecular Biology and Skaggs Institute of Chemical Biology, MB 33, The Scripps Research Institute, La Jolla, California 92037*

JEFFREY C. HOCH (6), *Rowland Institute for Science, Cambridge, Massachusetts 02142*

NICHOLAS V. HUD (17), *School of Chemistry and Biochemistry, Georgia Institute of Technology, Atlanta, Georgia 30332-0400*

JOCHEN JUNKER (2), *Institut für Organische Chemie, Universität Frankfurt, Frankfurt D-60439, Germany*

MASATSUNE KAINOSHO (11), *Department of Chemistry, Japan Science and Technology Corporation & Graduate School of Science, Tokyo Metropolitan University, Hachioji, Tokyo 192-0397, Japan*

CHOJIRO KOJIMA (11), *Department of Chemistry, Japan Science and Technology Corporation & Graduate School of Science, Tokyo Metropolitan University, Hachioji, Tokyo 192-0397, Japan*

ĒRIKS KUPČE (3), *Varian, Incorporated, Walton-on-Thames, Surrey KT12 2QF, United Kingdom*

LESTER J. LAMBERT (12), *Laboratory of Molecular Biophysics, The Rockefeller University, New York, New York 10021*

CHRISTOPHER A. LEPRE (8), *Vertex Pharmaceuticals Incorporated, Cambridge, Massachusetts 02139*

JEAN-LOUIS LEROY (15), *Groupe de Biophysique, L'Ecole Polytechnique et de l'UMR 7643 du CNRS, Palaiseau 91128, France*

DAVID LIVE (13), *Department of Biochemistry, Molecular Biology and Biophysics, Medical School and College of Biological Sciences, University of Minnesota, Minneapolis, Minnesota 55455*

JONATHAN M. MOORE (8), *Vertex Pharmaceuticals Incorporated, Cambridge, Massachusetts 02139*

TAKASHI NAGATA (12), *Laboratory of Molecular Biophysics, The Rockefeller University, New York, New York 10021*

EDWARD P. NIKONOWICZ (14), *Department of Biochemistry and Cell Biology, Rice University, Houston, Texas 77251*

AKIRA ONO (11), *Department of Chemistry, Japan Science and Technology Corporation & Graduate School of Science, Tokyo Metropolitan University, Hachioji, Tokyo 192-0397, Japan*

AKIRA "MEI" ONO (11), *Department of Chemistry, Japan Science and Technology Corporation & Graduate School of Science, Tokyo Metropolitan University, Hachioji, Tokyo 192-0397, Japan*

JEFFREY W. PENG (8), *Vertex Pharmaceuticals Incorporated, Cambridge, Massachusetts 02139*

ANH-TUÂN PHAN (15), *Groupe de Biophysique, L'Ecole Polytechnique et de l'UMR 7643 du CNRS, Palaiseau 91128, France*

BERND REIF (2), *Institut für Organische Chemie, Universität Frankfurt, Frankfurt D-60439, Germany*

CHRISTIAN RICHTER (2), *Bruker AG, Fällanden CH-8117, Switzerland*

JÜRGEN SCHLEUCHER (16), *Department of Medical Biosciences–Medical Biophysics, Umea University, S-901 87 Umea, Sweden*

JURGEN SCHMIDT (13), *National Stable Isotope Resource, Bioscience Division, B1-S, Los Alamos National Laboratory, Los Alamos, New Mexico 87545*

HARALD SCHWALBE (2), *Center for Magnetic Resonance, Francis Bitter Magnet Laboratory and Department of Chemistry, Massachusetts Institute of Technology, Cambridge, Massachusetts 02139*

LOUIS A. "PETE" SILKS III (13), *National Stable Isotope Resource, Bioscience Division, B1-S, Los Alamos National Laboratory, Los Alamos, New Mexico 87545*

ALAN S. STERN (6), *Rowland Institute for Science, Cambridge, Massachusetts 02142*

BRIAN J. STOCKMAN (9), *Structural, Analytical, and Medicinal Chemistry, Pharmacia Corporation, Kalamazoo, Michigan 49001*

IGNACIO TINOCO, JR. (18), *Department of Chemistry, University of California, Lawrence Berkeley National Laboratory, Berkeley, California 94720-1460*

BERND VAN BUUREN (16), *Department of Medical Biosciences–Medical Biophysics, Umea University, S-901 87 Umea, Sweden*

MILTON H. WERNER (12), *Laboratory of Molecular Biophysics, The Rockefeller University, New York, New York 10021*

SYBREN WIJMENGA (16), *Department of Medical Biosciences–Medical Biophysics, Umea University, S-901 87 Umea, Sweden*

DAVID S. WISHART (1), *Faculty of Pharmacy and Pharmaceutical Sciences, University of Alberta, Edmonton, Alberta T6G 2N8, Canada*

HIROAKI YAMADA (5), *Department of Molecular Science, Kobe University, Graduate School of Science and Technology, Kobe 657-8501, Japan*

Preface

It has now been 12 years since the initial volumes (176 and 177) of *Methods in Enzymology* on the application of NMR spectroscopy to the biological sciences were published. These were followed by Volumes 239 and 261 which were published in 1994 and 1995, respectively.

Exciting new aspects of NMR continue to be developed, prompting the development of Volumes 338 and 339 which contain articles describing the new methodologies and how they are being applied to elicit important biochemical information. Clearly, basic research into the complicated phenomenon of NMR still yields fruit, sustaining further advances in the applications of NMR to the study of important biological and pathological systems. Especially prominent have been developments in using dipolar coupling and TROSY techniques. While these have broad applicability, they have especially increased the size of proteins and nucleic acids amenable to study by NMR. There have been further developments in solid state NMR. These, in particular, have set the stage for structure determination of membrane proteins—incredibly important, but so far relatively intractable using X-ray crystallography or solution NMR techniques. Methods for data analysis and for structure and dynamics elucidation continue apace. There have been advances in preparing samples of proteins and nucleic acids in a manner such that previously unassailable systems can now be studied. For the most part, this means that methods of introducing isotopic labels in clever new ways have been developed.

To function, biopolymers must interact with other molecules. Studies of complexes formed by biopolymers are covered extensively in these latest volumes. NMR has also developed as an important tool in drug discovery; chapters are included to describe this.

As always, the orientation of the chapters is aimed at a "hands-on" approach. The chapter authors have emphasized those aspects of their topic that have greatest utility to other researchers.

THOMAS L. JAMES
VOLKER DÖTSCH
ULI SCHMITZ

METHODS IN ENZYMOLOGY

VOLUME XXXVI. Hormone Action (Part A: Steroid Hormones)
Edited by BERT W. O'MALLEY AND JOEL G. HARDMAN

VOLUME XXXVII. Hormone Action (Part B: Peptide Hormones)
Edited by BERT W. O'MALLEY AND JOEL G. HARDMAN

VOLUME XXXVIII. Hormone Action (Part C: Cyclic Nucleotides)
Edited by JOEL G. HARDMAN AND BERT W. O'MALLEY

VOLUME XXXIX. Hormone Action (Part D: Isolated Cells, Tissues, and Organ Systems)
Edited by JOEL G. HARDMAN AND BERT W. O'MALLEY

VOLUME XL. Hormone Action (Part E: Nuclear Structure and Function)
Edited by BERT W. O'MALLEY AND JOEL G. HARDMAN

VOLUME XLI. Carbohydrate Metabolism (Part B)
Edited by W. A. WOOD

VOLUME XLII. Carbohydrate Metabolism (Part C)
Edited by W. A. WOOD

VOLUME XLIII. Antibiotics
Edited by JOHN H. HASH

VOLUME XLIV. Immobilized Enzymes
Edited by KLAUS MOSBACH

VOLUME XLV. Proteolytic Enzymes (Part B)
Edited by LASZLO LORAND

VOLUME XLVI. Affinity Labeling
Edited by WILLIAM B. JAKOBY AND MEIR WILCHEK

VOLUME XLVII. Enzyme Structure (Part E)
Edited by C. H. W. HIRS AND SERGE N. TIMASHEFF

VOLUME XLVIII. Enzyme Structure (Part F)
Edited by C. H. W. HIRS AND SERGE N. TIMASHEFF

VOLUME XLIX. Enzyme Structure (Part G)
Edited by C. H. W. HIRS AND SERGE N. TIMASHEFF

VOLUME L. Complex Carbohydrates (Part C)
Edited by VICTOR GINSBURG

VOLUME LI. Purine and Pyrimidine Nucleotide Metabolism
Edited by PATRICIA A. HOFFEE AND MARY ELLEN JONES

VOLUME LII. Biomembranes (Part C: Biological Oxidations)
Edited by SIDNEY FLEISCHER AND LESTER PACKER

VOLUME LIII. Biomembranes (Part D: Biological Oxidations)
Edited by SIDNEY FLEISCHER AND LESTER PACKER

VOLUME LIV. Biomembranes (Part E: Biological Oxidations)
Edited by SIDNEY FLEISCHER AND LESTER PACKER

Section I

General Techniques

[1] Use of Chemical Shifts in Macromolecular Structure Determination

By DAVID S. WISHART and DAVID A. CASE

Introduction

Chemical shifts are the universal language of nuclear magnetic resonance (NMR). They communicate in a simple way detailed molecular information that almost every chemist can understand. Although chemical shifts have long been used as tools for covalent structural analysis, it is important to remember that they can also provide detailed information about noncovalent structure, solvent interactions, ionization constants, ring orientations, hydrogen bond interactions, and other phenomena. One major force driving a growing interest in chemical shifts has been the rapid growth in the number of macromolecular NMR assignments in repositories such as the BioMagResBank (BMRB). Another important factor has been improvements in computer hardware and software that now make accurate chemical shift calculations on fragments of biological macromolecules almost routine.[1-3]

The purpose of this article is to provide readers with practical advice on how to use chemical shifts as an aid in understanding, generating or refining macromolecular structures. Some attempt will be made to update material presented an earlier volume of this series.[4,5] However, no attempt will be made to extensively survey the theory and history of macromolecular chemical shifts, as several excellent reviews already exist.[3,6,7]

Chemical Shift Referencing

Chemical shifts are among the most precisely measurable but least accurately measured spectral parameters in all of NMR spectroscopy. The issue of accuracy, as opposed to precision, lies at the heart of how reliably chemical shifts can be used for structural interpretation and analysis. Systematic errors in chemical shift measurements as small as 0.05 ppm for ^1H shifts, 0.3 ppm for ^{13}C shifts, or 0.5 ppm for ^{15}N shifts can make a significant difference in the

[1] E. Oldfield, *J. Biomol. NMR* **5**, 217 (1995).
[2] D. A. Case, *Curr. Opin. Struct. Biol.* **8**, 624 (1998).
[3] D. A. Case, *Curr. Opin. Struct. Biol.* **10**, 197 (2000).
[4] D. S. Wishart and B. D. Sykes, *Methods Enzymol.* **239**, 363 (1994).
[5] D. A. Case, H. J. Dyson, and P. E. Wright, *Methods Enzymol.* **239**, 392 (1994).
[6] L. Szilagyi, *Prog. Nucl. Magn. Reson. Spectrosc.* **27**, 325 (1995).
[7] M. P. Williamson and T. Asakura, *Meth. Mol. Biol.* **60**, 53 (1997).

identification of secondary structure,[4,8-11] the measurement of hydrogen bond lengths,[12,13] the determination of dihedral angle restraints,[14-16] the convergence of a chemical shift refinement[17-19] or the development of empirical chemical shift "surfaces."[14,15,20-22]

It is important to remember that chemical shifts are relative frequency measurements, not absolute measurements. In organic chemistry and for compounds dissolved in organic solvents, TMS (tetramethylsilane) has been the *de facto* [1]H and ^{13}C chemical shift standard since the 1970s.[23] However, it has only been recently that a set of universal standards for chemical shift referencing in aqueous solutions has been proposed[24,25] and adopted[26] by the IUPAC and IUBMB. Prior to that decision, more than a dozen different chemical shift standards had been in use at different times by different labs around the world.[4]

The specific IUPAC recommendations for biological molecules are that internal DSS (2,2-dimethyl-2-silapentane-5-sulfonic acid), a water-soluble, pH-insensitive form of TMS, should be the standard used for ^{1}H and ^{13}C referencing. In addition, external anhydrous liquid ammonia should be used for ^{15}N referencing, external 100% trifluoroacetic acid should be used ^{19}F referencing,[25] and internal 10% trimethylphosphate is recommended for ^{31}P referencing.[26] Because of the difficulties in working with some of these reference compounds, an alternative indirect referencing procedure has been strongly advocated.[24,26,27] In particular, by using predetermined nucleus-specific frequency ratios (called Ξ or

[8] D. S. Wishart, B. D. Sykes, and F. M. Richards, *Biochemistry* **31**, 1647 (1992).

[9] W. J. Metzler, K. L. Constantine, M. S. Friedrichs, A. J. Bell, and E. G. Ernst, *Biochemistry* **32**, 13818 (1993).

[10] A. M. Gronenborn and G. M. Clore, *J. Biomol. NMR* **4**, 455 (1994).

[11] K. Ösapay and D. A. Case, *J. Biomol. NMR* **4**, 215 (1994).

[12] G. Wagner, A. Pardi, and K. Wuthrich, *J. Am. Chem. Soc.* **105**, 5948 (1983).

[13] D. S. Wishart, B. D. Sykes, and F. M. Richards, *J. Mol. Biol.* **222**, 311 (1991).

[14] D. S. Wishart and A. M. Nip, *Biochem. Cell Biol.* **76**, 153 (1998).

[15] M. Iwadate, T. Asakura, and M. P. Williamson, *J. Biomol. NMR* **13**, 199 (1999).

[16] G. Cornilescu, F. Delaglio, and A. Bax, *J. Biomol. NMR* **13**, 289 (1999).

[17] J. Kuszewski, A. M. Gronenborn, and G. M. Clore, *J. Magn. Reson. Ser. B* **107**, 293 (1995).

[18] J. Kuszewski, J. Qin, A. M. Gronenborn, and G. M. Clore, *J. Magn. Reson. Ser. B* **106**, 92 (1995).

[19] G. M. Clore and A. M. Gronenborn, *Proc. Natl. Acad. Sci. U.S.A.* **95**, 5891 (1998).

[20] S. Spera and A. Bax, *J. Am. Chem. Soc.* **113**, 5490 (1991).

[21] H. Le and E. Oldfield, *J. Biomol. NMR* **4**, 341 (1994).

[22] R. D. Beger and P. H. Bolton, *J. Biomol. NMR* **10**, 129 (1997).

[23] IUPAC, *Pure Appl. Chem.* **29**, 627 (1972).

[24] D. S. Wishart, C. G. Bigam, J. Yao, F. Abildgaard, H. J. Dyson, E. Oldfield, J. L. Markley, and B. D. Sykes, *J. Biomol. NMR* **6**, 135 (1995).

[25] T. Maurer and H. R. Kalbitzer, *J. Magn. Reson. B* **113**, 177 (1996).

[26] J. L. Markley, A. Bax, Y. Arata, C. W. Hilbers, R. Kaptein, B. D. Sykes, P. E. Wright, and K. Wüthrich, *J. Biomol. NMR* **12**, 1 (1998).

[27] A. Bax and J. Subramanian, *J. Magn. Reson.* **67**, 565 (1986).

TABLE I
IUPAC/IUBMB RECOMMENDED Ξ (XI) RATIOS FOR INDIRECT
CHEMICAL SHIFT REFERENCING IN BIOMOLECULAR NMR[a]

Nucleus	Compound	Ξ Ratio
^1H	DSS	1.000 000 000
^{13}C	DSS	0.251 449 530
^{15}N	Liquid NH_3	0.101 329 118
^{19}F	CF_3COOH	0.940 867 196
^{31}P	$(CH_3)_3PO_4$	0.404 808 636

[a] Relative to DSS.

the Greek letter xi) derived for DSS (^{13}C), liquid ammonia (^{15}N), trifluoroacetic acid (^{19}F), and trimethylphosphate (^{31}P), it is possible to determine the zero point reference for these compounds (and hence these nuclei) using the absolute ^1H frequency of internal DSS. Some of the more commonly used Ξ values are presented in Table I. A more extensive list is available at the BioMagResBank (www.bmrb.wisc.edu/bmrb).

As an example, let us assume one wished to reference the ^{15}N dimension of an $^{15}N-^1H$ HSQC experiment. First, the sample must contain a detectable amount of dissolved DSS (say 100 μM). Prior to collecting the spectrum, determine the ^1H carrier frequency (say 500,000,087.2 Hz) of the spectrometer. Second, determine the ^1H DSS frequency relative to the carrier frequency (assume it is 2521.2 Hz upfield of the carrier). This implies the absolute DSS frequency is $500,000,087.2 - 2521.2 = 499,997,566.0$ Hz. Third, multiply this DSS ^1H frequency by the ^{15}N Ξ ratio found in Table I (the result is 50,664,312.4 Hz). This value corresponds to the hypothetical ^{15}N resonance frequency of external liquid ammonia, which by definition is 0 ppm. If the ^{15}N carrier (decoupler) frequency is also known or measured (say it is 50,670,450.8 Hz), then the ^{15}N chemical shift scale, in ppm, can be fully determined. Because magnetic fields drift and spectrometer frequencies vary over time, this indirect referencing procedure must generally be repeated each time a new sample is placed in a spectrometer. For most spectrometers it is possible to write a simple computer program to perform this referencing task routinely. A macro called XREF for indirect referencing is available over the Web (see Table V for more information). Regardless of whether one chooses to use the direct or indirect referencing procedures, *properly referenced spectra are absolutely key to obtaining meaningful chemical shift information.*

Given the long-standing problems in the biomolecular NMR community concerning chemical shift referencing, one might ask: "How accurate are the protein or nucleic acid chemical shifts that have been published over the past 20 years?" The answer to this question is important because it affects our ability to compare, extract, and reproduce data from other laboratories. One of us (DSW) has undertaken a long-term retrospective survey of the protein chemical shift data deposited in the BMRB to attempt to identify and correct potentially misreferenced data sets. The process involves predicting protein ^{1}H, ^{13}C, and ^{15}N chemical shifts using X-ray or NMR coordinate data and then comparing those predictions to the measured values of corresponding proteins found in the BMRB. The program, called SHIFT-COR (Table V), was originally developed and tested on a series of proteins known to be correctly referenced via IUPAC conventions. Although the residue specific shift predictions for this program are often less than perfect, the global averages for ^{1}H, ^{13}Cα, ^{13}Cβ, ^{13}CO, and ^{15}N shifts (calculated over all residues in a given protein) have been found to be quite sensitive to chemical shift referencing errors. Of the more than 60 proteins surveyed to date it appears that (1) there are essentially no detectable problems with ^{1}H shift referencing; (2) approximately 20% of the proteins deposited in the BMRB have ^{13}Cα ^{13}CO and ^{13}Cβ shifts that are misreferenced (> 0.5 ppm); and (3) approximately 30% of the proteins deposited in the BMRB have ^{15}N shifts that are misreferenced (> 1 ppm). Other investigators have reported similar problems with ^{15}N and ^{13}C shifts over smaller protein data sets.[15,16] This result is quite worrisome—particularly for those who would like to use experimental chemical shift data in structural analysis and comparison.

Databases of "re-referenced" protein chemical shifts are available in the TALOS program[16] and in RefDB, which is composed of those proteins that have already gone through the SHIFTCOR filter, and is available over the Web (Table V). Interestingly, ^{13}CO data seem to have their own particular chemical shift referencing problems, which are distinct from ^{13}Cα and ^{13}Cβ referencing issues. Indeed, it is not uncommon to find a protein with correctly referenced ^{13}Cα and ^{13}Cβ shifts but with ^{13}CO shifts that are systematically off by 1 or 2 ppm. Evidently this problem may be related to the tendency of some spectroscopists to use offset-synthesized pulses to excite ^{13}CO resonances while still keeping their carrier at the ^{13}Cα frequency. When this is done, the resulting spectrum will be folded many times and may often show resonances at reasonable (but incorrect) ^{13}CO ppm values. Essentially, the referencing error arises when one uses the ppm frequency of the offset synthesized pulse as the carrier frequency in the ^{13}CO dimension (A. Bax, personal communication).

Random Coil Shifts

Random coil shifts are defined as the characteristic chemical shifts of amino acid residues or nucleic acid bases in short, disordered polymers. They are generally used in spin assignment and residue identification,[28–31] but they are also

[28] K. Wüthrich, "NMR of Proteins and Nucleic Acids." John Wiley & Sons, New York, 1986.
[29] S. Grzesiek and A. Bax, *J. Biomol. NMR* **3**, 185 (1993).

useful in determining the so-called secondary chemical shift ($\delta\Delta$), which is the difference between the observed shift and the corresponding random coil value. Secondary chemical shifts primarily contain noncovalent structural or dynamic information, as opposed to simple covalent information. They can be valuable in identifying secondary structure, determining ring pucker, delineating flexible regions, locating hydrogen bonds, setting dihedral restraints, and detecting aromatic stacking interactions.[6,14,30]

For polypeptides, a considerable body of work on random coil shifts exists.[32–39] Because many of the early chemical shift measurements used nonstandard referencing protocols, those chemical shift tables published prior to 1994 should be used with caution. Probably the best sets of random coil shifts to use are those proposed by Braun et al.,[36] Merutka et al.,[37] and Wishart et al.[38] Random coil phosphoamino acid shifts have also been measured.[40] A complete set of IUPAC referenced amino and phosphoamino acid backbone (including $^{13}C\beta$ and $^{1}H\beta$) random coil chemical shifts are listed in Tables II and III.

Based on the large body of existing chemical shift assignments[41] and the random coil values presented here, we suggest the following limits on allowable deviations from random coil shifts should be used: ± 1.3 ppm for $^{1}H\alpha$ shifts, ± 5 ppm for $^{13}C\alpha$ and $^{13}C\beta$, ± 4 ppm for ^{13}CO, and ± 10 ppm for ^{15}N. Exceptions should be checked for assignment or chemical shift referencing errors.

Defining random coil shifts for nucleic acids is a more difficult problem than for amino acids.[28,30,31] This is because nucleic acid shifts in polynucleotides (particularly ^{1}H shifts) are much more sensitive to sequence effects than is the case for amino acids in polypeptides. The use of chemical shifts from isolated nucleic acids or nucleic acid analogs instead of polynucleotides is also problematic because complications arising from intermolecular interactions make these measurements somewhat questionable.[28] Nevertheless, as has been previously shown with amino acids,[13] it is possible to generate a reasonably good set of random coil nucleic

[30] S. S. Wijmenga, M. Kruithof, and C. W. Hilbers, *J. Biomol. NMR* **10**, 337 (1997).

[31] A. Dejaegere, R. A. Bryce, and D. A. Case, *in* "Modeling NMR Chemical Shifts. Gaining Insights into Structure and Environment" (J. C. Facelli and A. C. de Dios, eds.), p. 194. American Chemical Society, Washington, D.C., 1999.

[32] R. Richarz and K. Wüthrich, *Biopolymers* **17**, 2133 (1978).

[33] A. Bundi and K. Wüthrich, *Biopolymers* **18**, 285 (1979).

[34] J. Glushka, M. Lee, S. Coffin, and D. Cowburn, *J. Am. Chem. Soc.* **111**, 7716 (1989).

[35] V. Thanabal, D. O. Omichinsky, M. D. Reily, and W. L. Cody, *J. Biomol. NMR* **4**, 47 (1994).

[36] D. Braun, G. Wider, and K. Wüthrich, *J. Am. Chem. Soc.* **116**, 8466 (1994).

[37] G. Merutka, H. J. Dyson, and P. E. Wright, *J. Biomol. NMR* **5**, 14 (1995).

[38] D. S. Wishart, C. G. Bigam, A. Holm, R. S. Hodges, and B. D. Sykes, *J. Biomol. NMR* **5**, 67 (1995).

[39] K. W. Plaxco, C. J. Morton, S. B. Grimshaw, J. A. Jones, M. Pitkeathly, I. D. Campbell, and C. M. Dobson, *J. Biomol. NMR* **10**, 221 (1997).

[40] E. A. Bienkiewicz and K. J. Lumb, *J. Biomol. NMR* **15**, 203 (1999).

[41] B. R. Seavey, E. A. Farr, W. M. Westler, and J. L. Markley, *J. Biomol. NMR* **1**, 217 (1991).

TABLE II
"RANDOM COIL" CHEMICAL SHIFTS FOR COMMON AMINO
AND PHOSPHOAMINO ACIDS[a]

Amino acid	^{1}HN	^{15}N	^{1}Hα	^{13}Cα	^{1}Hβ	^{13}Cβ	^{13}CO
Ala	8.24	123.8	4.32	52.5	1.39	19.1	177.8
Cys(r)	8.32	118.8	4.55	58.2	2.93/2.93	28.0	174.6
Cys(o)	8.43	118.6	4.71	55.4	3.25/2.99	41.1	174.6
Asp	8.34	120.4	4.64	54.2	2.72/2.65	41.1	176.3
Glu	8.42	120.2	4.35	56.6	2.06/1.96	29.9	176.6
Phe	8.30	120.3	4.62	57.7	3.14/3.04	39.6	175.8
Gly	8.33	108.8	3.96	45.1	—	—	174.9
His	8.42	118.2	4.73	55.0	3.29/3.16	29.0	174.1
				56.3 (pH 9)		30.8 (pH 9)	
Ile	8.00	119.9	4.17	61.1	1.87	38.8	176.4
Lys	8.29	120.4	4.32	56.2	1.84/1.75	33.1	176.6
Leu	8.16	121.8	4.34	55.1	1.62/1.62	42.4	177.6
Met	8.28	119.6	4.48	55.4	2.11/2.01	32.9	176.3
Asn	8.40	118.7	4.74	53.1	2.83/2.75	38.9	175.2
Pro	—	—	4.42	63.3 (trans)	2.29/2.94	32.1 (trans)	177.3
				62.8 (cis)		34.5 (cis)	
Gln	8.32	119.8	4.34	55.7	2.12/1.99	29.4	176.0
Arg	8.23	120.5	4.34	56.0	1.86/1.76	30.9	176.3
Ser	8.31	115.7	4.47	58.3	3.89/3.87	63.8	174.6
pSer	8.65	115.5	4.60	57.4	4.22/4.13	66.8	174.9
Thr	8.15	113.6	4.35	61.8	4.24	69.8	174.7
pThr	8.43	113.4	4.43	61.8	4.66	74.0	175.1
Val	8.03	119.2	4.12	62.2	2.08	32.9	176.3
Trp	8.25	121.3	4.66	57.5	3.29/3.27	29.6	176.1
Tyr	8.12	120.3	4.55	57.9	3.03/2.98	37.8	175.9
pTyr	8.21	120.1	4.61	57.0	3.15/3.02	38.7	176.7

[a] Measured at 25°C, pH 5. Data given in ppm. Note that Cys(r) refers to cysteine and Cys(o) refers to cystine. Data compiled from Refs. 32, 38, and 40.

acid chemical shifts by taking the average values for each ^{1}H type from previously assigned polynucleotides with known three-dimensional (3D) structures.[30] It has been suggested that these values should be slightly modified so that sequence dependent effects and variations in sugar pucker could be more appropriately accommodated.[31] Based on these and other considerations we have attempted to produce a "consensus" set of ^{1}H, ^{13}C, and ^{15}N random coil shifts for RNA and DNA. These are given in Table III.

Calculating Chemical Shifts

Interpreting chemical shifts in structural terms is based on fundamental knowledge of how electrons shield nuclei in various environments. Here we briefly review

TABLE III
"Random Coil" Chemical Shifts Relative to DSS for Nucleic Acid Bases[a]

Base	DNA			RNA		
	^1H	^{15}N	^{13}C	^1H	^{15}N	^{13}C
A/C/G/T (sugar) 2′	2.24		40.5	4.02		76.3
A/C/G/T (sugar) 2″	2.17					
A/C/G/T (sugar) 3′	4.36		78.4	4.14		73.9
A/C/G/T (sugar) 4′	4.00		86.7	4.00		83.1
A/C/G/T (sugar) 5′	3.72		67.6	3.78		66.3
A/C/G/T (sugar) 5″	3.60			3.58		
A (sugar) 1′	5.23		85.1	5.03		93.1
A (base) 2	8.68		154.3	8.68		154.3
A (base) 8	8.60		141.2	8.60		141.2
A (base) 6		80.5			80.5	
C (sugar) 1′	5.48		85.1	5.28		93.1
C (base) 4		97.8	158.7		97.8	158.7
C (base) 5	6.20		98.2	6.20		98.2
C (base) 6	7.80		142.2	7.80		142.2
G (sugar) 1′	5.25		85.1	5.05		93.1
G (base) 1		146.3			146.3	
G (base) 2		73.8			73.8	
G (base) 8	7.69		137.9	7.69		137.9
T (sugar) 1′	5.80		85.1			
T (base) 3		158.8				
T (base) 6	8.68		138.9			
T (base) 7	8.60		14.4			
U (sugar) 1′				5.60		93.1
U (base) 2					160.5	
U (base) 5				5.95		103.5
U (base) 6				7.85		142.0

[a] Data given in ppm at 25°. ^1H random coil shifts were taken from Ref. 31. ^{13}C and ^{15}N random coil shifts were determined from averages of DNA (#4104, #4103, #4165) and RNA (#4235, #4345, #4346) shifts deposited in the BioMagResBank.

quantum mechanical and more classical approaches to calculating chemical shifts from structure.

Quantum Mechanical Approaches

The chemical shielding tensor (σ) relates the effective magnetic field ($\mathbf{H}_{eff}$) felt at a probe nucleus to the applied spectrometer field ($\mathbf{H}_0$):

$$\mathbf{H}_{eff} = (1 - \sigma)\mathbf{H}_0 \tag{1}$$

The effective field need not be aligned with the spectrometer field, so that the shielding is a tensor quantity. The fundamental connections between chemical shielding and molecular electronic structure have been presented in many places[42,43] and will not be repeated here. The basic procedure for determining the wavefunction of a molecule in the presence of an external magnetic field involves replacing the momentum operator $\mathbf{p}$ with $\mathbf{p} - (e/c)\mathbf{A}$, where $\mathbf{A}$ is the vector potential associated with the external field $\mathbf{H}_0$.[42] This leads to a theory that appears to depend on the "gauge," since a constant field can be generated by many vector potentials. In the most common gauge,

$$\mathbf{A} = \tfrac{1}{2}(\mathbf{H}_0 \times \mathbf{r}) \tag{2}$$

so that $\mathbf{A}$ has a dependence on the origin of the coordinate system used to define the position vector $\mathbf{r}$. Results for exact wavefunctions are invariant to changes in the gauge origin, but many approximate solutions are not. Most current practical calculations tackle this problem by placing a gauge dependence in the expansion functions themselves, either at the atomic orbital or at a localized molecular orbital level. This can have the effect of greatly accelerating convergence for finite basis sets that are necessarily incomplete. The development and efficient implementation of approaches to surmounting the so-called "gauge invariance" problem has greatly accelerated the convergence of results with basis set size and made high-quality chemical shielding calculations a realistic possibility.[44,45]

The general accuracy of shielding calculations has been enhanced as well by the development of density functional theory (DFT), which provides a computationally straightforward way to incorporate some important effects beyond the Hartree–Fock approximation, in a manner efficient enough to be applied to molecules or fragments containing dozens of atoms. Although it is certainly not a "magic bullet," this theory appears to offer systematic improvement over Hartree–Fock theory both for energies and for chemical shieldings.[46–52] Some DFT implementations use a sum-over-states perturbation formula and modify the energy denominators to partially account for the change in the exchange-correlation energy densities in

[42] R. McWeeny, "Methods of Molecular Quantum Mechanics," 2nd ed. Academic Press, San Diego 1989.

[43] T. Helgaker, M. Jaszunski, and K. Ruud, *Chem. Rev.* **99**, 293 (1999).

[44] K. Wolinski, J. F. Hinton, and P. Pulay, *J. Am. Chem. Soc.* **112**, 8324 (1990).

[45] W. Kutzelnigg, U. Fleischer, and M. Schindler, *NMR, Basic Principles and Progress* **23**, 165 (1990).

[46] M. Bühl, M. Kaupp, O. L. Malkina, and V. G. Malkin, *J. Computat. Chem.* **20**, 91 (1998).

[47] P. J. Wilson, R. D. Amos, and N. C. Handy, *Mol. Phys.* **97**, 757 (1999).

[48] P. Wilson, R. Amos, and N. Handy, *Chem. Phys. Lett.* **312**, 475 (1999).

[49] C. Adamo and V. Barone, *Chem. Phys. Lett.* **298**, 113 (1998).

[50] M. Bienati, C. Adamo, and V. Barone, *Chem. Phys. Lett.* **311**, 69 (1999).

[51] K. B. Wiberg, *J. Computat. Chem.* **20**, 1299 (1999).

[52] K. K. Baldridge and J. S. Siegel, *J. Phys. Chem. A* **103**, 4038 (1999).

the excited states. This appears to be a valuable correction, especially when correlation effects are large.[47,53,54] Also of note has been the development of methods to compute shifts from more conventional correlated wavefunctions, such as those arising from perturbation theory or CI or MCSCF expansions[43,55]; these methods allow benchmark calculations on small molecules that can provide important guidance for interpreting calculations on larger systems.

It is not straightforward to give general rules about the expected accuracy of chemical shielding calculations. It is important to distinguish computations of absolute or relative shieldings for atoms in widely different chemical environments from studies of trends in a single type of shift as conformational parameters are varied. The latter are generally more reliable and converge at a lower level of theory than the former. We consider first test sets that span a wide range of chemical environments. Baldridge and Siegel[52] have shown that density functional calculations for proton shifts in simple hydrocarbons can be computed with an average error of about 0.1 ppm, but that extrapolations to more complicated molecules increases the expected error to about 0.3 ppm. Rablen et al.[56] report an rms (root mean square) error versus experiment of 0.15 ppm for proton shifts in 80 organic molecules, using a slight linear scaling of density functional results. Errors for carbon and nitrogen shifts are significantly larger than for protons. Carbon shifts for a wide set environments are found to be within about 3–5 ppm of experiment in a number of studies.[51,54,57,58] Computed nitrogen shifts are generally in somewhat worse agreement with experiment. This is probably because electron correlation effects can be both important and variable in a range of different chemical environments.

Comparisons that look at the same type of atom in different conformations (generally arising from rotations about single bonds) are expected to fare better. For example, Pearson et al.[59] looked at DFT simulations of the $C\alpha$ and $C\beta$ shifts in valine residues from three proteins. They found a mean deviation of less than 1 ppm for relative shifts (over a range of of 12 ppm for $C\alpha$ and 10 ppm for $C\beta$). On a related note, it is common to find that conformation-dependent shifts are often insensitive to basis-set effects or to the method of geometry optimization. For example, studies of trends in peptides or sugars give nearly the same results for geometries optimized at the Hartree–Fock level versus those optimized using a molecular mechanics

[53] V. G. Malkin, O. L. Malkina, M. E. Casida, and D. R. Salahub, J. Am. Chem. Soc. **116,** 5898 (1994).

[54] L. Olsson and D. Cremer, J. Chem. Phys. **105,** 8995 (1996).

[55] J. Gauss, Ber. Bunsen-Ges. Phys. Chem. **99,** 1001 (1995).

[56] P. R. Rablen, S. A. Pearlman, and J. Finkbiner, J. Phys. Chem. A **103,** 7357 (1999).

[57] J. R. Cheeseman, G. W. Trucks, T. A. Keith, and M. J. Frisch, J. Chem. Phys. **104,** 5497 (1996).

[58] L. Olsson and D. Cremer, J. Phys. Chem. **100,** 16881 (1996).

[59] J. G. Pearson, H. Le, L. K. Sanders, N. Godbout, R. H. Havlin, and E. Oldfield, J. Am. Chem. Soc. **119,** 11941 (1997).

force field.[60,61] Nevertheless, the general level of error cited here is large enough to suggest that empirical corrections or calibrations will continue to be used in combination with quantum chemical studies to extract the greatest amount of structural information from protein and nucleic acid chemical shifts.

Over the past decade, there have been a variety of systematic quantum chemistry studies of chemical shifts in peptides and nucleic acid fragments,[59-69] and it is not possible to give any comprehensive review of these here. Some ways in which these calculations have been used will be cited below.

Classical and Semiclassical Approaches

There is a long history of classical approaches to understanding environmental effects on chemical shifts, and the basic ideas have not significantly changed since the early writings on this subject 40 years ago.[70] In addition to "local" contributions that depend upon substituents and geometries very close to the nucleus of interest, there are long-range or "environmental" effects that arise from groups not chemically bonded to the probe nucleus. The latter are usually divided into three categories: (a) effects arising from the magnetic susceptibilities of neighboring groups; (b) electric field effects that polarize the bond containing the probe nucleus; and (c) close-contact effects that arise from the Pauli or van der Waals repulsion of charge clouds when there are close nonbonded contacts, as in hydrogen bonds. It is generally easiest to discern long-range interactions for ^{1}H and ^{19}F, since these atoms are bonded to only one other atom, so that local geometric effects tend to be reasonably constant for a particular type of nucleus. On the other hand, ^{13}C shifts generally report local geometric features, since they are bonded to several atoms and (for aliphatic carbons) can be in the center of torsion angles that differ from one residue to another. The next few paragraphs provide an overview of these contributions; more detailed discussions can be found elsewhere.[60]

[60] D. Sitkoff and D. A. Case, *J. Am. Chem. Soc.* **119**, 12262 (1997).

[61] A. P. Dejaegere and D. A. Case, *J. Phys. Chem. A* **102**, 5280 (1998).

[62] A. C. de Dios, J. G. Pearson, and E. Oldfield, *Science* **260**, 1491 (1993).

[63] A. C. de Dios and E. Oldfield, *Solid State Nucl. Magn. Reson.* **6**, 101 (1996).

[64] H. Le and E. Oldfield, *J. Phys. Chem.* **100**, 16423 (1996).

[65] A. C. de Dios, *Prog. NMR Spectrosc.* **29**, 229 (1996).

[66] A. C. de Dios, J. L. Roach, and A. E. Walling, *in* "Modeling NMR Chemical Shifts. Gaining Insights into Structure and Environment" (J. C. Facelli and A. C. de Dios, eds.), p. 220. American Chemical Society, Washington, D.C., 1999.

[67] C. Scheurer, N. Skrynnikov, S. F. Lienin, S. K. Straus, R. Brüschweiler, and R. R. Ernst, *J. Am. Chem. Soc.* **121**, 4242 (1999).

[68] X.-P. Xu, W.-L. A. K. Chiu, and S. C. F. Au-Yeung, *J. Am. Chem. Soc.* **120**, 4230 (1998).

[69] D. Sitkoff and D. A. Case, *Prog. Nucl. Magn. Reson. Spectrosc.* **32**, 165 (1998).

[70] A. D. Buckingham, T. Schaefer, and W. G. Schneider, *J. Chem Phys.* **32**, 1227 (1960).

Magnetic Susceptibility Contributions. If a chemical group is far from the probe nucleus, its contribution to the induced magnetic field can be approximated by a magnetic dipole. The shielding at a probe nucleus then depends in a relatively simple way on group magnetic susceptibilities[71]:

$$\sigma = \frac{\chi}{r^3} - \frac{3\chi(\mathbf{rr}^T)}{r^5} \tag{3}$$

In this "McConnell" equation, χ is the susceptibility tensor, $\mathbf{r}$ is the vector from the center of the remote group to the probe nucleus, and $\mathbf{rr}^T$ is the outer product of $\mathbf{r}$ with itself. It is easy to show that contributions to the isotropic shift arise only when χ is anisotropic. This has led to a long interest in group suscepti-bility anisotropies and their influence on liquid state spectra.[6] Aromatic groups have large anisotropies that can be modeled as "ring currents" as well as by point magnetic dipoles, and it has long been recognized that a major contri-bution to shift dispersion of protons in biomolecules arises from the effects of nearby aromatic groups.[72,73] These susceptibility anisotropies can be calculated with good reliability by quantum chemical methods, offering an alternative to the empirical fits that are usually used[74,75]; this may be especially helpful in estimat-ing ring-current effects for novel groups where there is no established empirical database. For the common aromatic groups appearing in proteins and nucleic acids, various functional forms for ring-current shifts have been tested and calibrated against observed data.[72,73,76] The basic ideas may be found in textbooks[77] and reviews.[11]

It should be noted that peptide groups also have a nonnegligible susceptibility anisotropy in solution. It is only about $\frac{1}{10}$ of that of a benzene ring, but the large number of peptide groups in proteins can magnify their effects. In particular, secondary structural preferences for Hα and Hβ shifts appear to have a large contribution from this mechanism.[11,78]

An especially attractive case for structure determination arises in some para-magnetic proteins, where unpaired electrons at transition metal sites can make dom-inant contributions to susceptibility anisotropies. This contribution is completely analogous to that of aromatic groups and is generally called the pseudo-contact

[71] H. M. McConnell, *J. Chem. Phys.* **27**, 226 (1957).
[72] C. Giessner-Prettre and B. Pullman, *Q. Rev. Biophys.* **20**, 113 (1987).
[73] K. Ösapay and D. A. Case, *J. Am. Chem. Soc.* **113**, 9436 (1991).
[74] K. Ruud, H. Skaane, T. Helgaker, K. L. Bak, and P. Jorgensen, *J. Am. Chem. Soc.* **116**, 10135 (1994).
[75] D. A. Case, *J. Biomol. NMR* **6**, 341 (1995).
[76] C. W. Haigh and R. B. Mallion, *Prog. NMR Spectrosc.* **13**, 303 (1980).
[77] R. K. Harris, "Nuclear Magnetic Resonance Spectroscopy, A Physicochemical View." Longman Scientific & Technical, Essex, U.K., 1986.
[78] T. Asakura, K. Taoka, M. Demura, and M. P. Williamson, *J. Biomol. NMR* **6**, 227 (1995).

term. For the case of an axially symmetric χ, one has

$$\delta_{pc} = \frac{1}{3}r^{-3}\Delta\chi(3\cos^2\theta - 1) \tag{4}$$

where θ is the angle between the metal-nucleus vector $\mathbf{r}$ and the unique axis of the susceptibility tensor, and $\Delta\chi$ is the difference between magnetic susceptibilities along the unique axis and perpendicular to it. These shifts can provide an important source of structural information in paramagnetic systems, as discussed below.[79-85]

Electric Field Effects. Magnetic susceptibilities in neighboring groups affect chemical shifts by directly contributing to the effective magnetic field at the probe nucleus. Nearby charges or dipoles can have a more indirect effect by polarizing chemical bonds containing the probe nucleus, which in turn affects the shielding. There have been a number of attempts to calibrate the magnitude of this effect, either through quantum chemistry calculations[69] or through empirical analyses of shielding changes when polar substituents are incorporated at various locations in organic molecules.[86-88] In principle, this mechanism could provide a valuable probe for electric fields in biomolecules, but practical questions of disentangling the field effect from other contributions to shift dispersion have limited its use so far. Fluorine may be an attractive nucleus here, since its shifts have a high sensitivity to local fields.[89-91]

Close Contact Interactions. A third general environmental effect, especially important in solvation shifts, is a deshielding that occurs when the electron cloud of a nonbonded partner overlaps that of the probe nucleus. This can be most clearly seen for rare gases (which have no polarity or susceptibility anisotropy),[60] but is also an important aspect of the deshielding arising from hydrogen bond interactions, where contact distances can be quite short. This effect is generally

[79] S. D. Emerson and G. N. La Mar, *Biochemistry* **29**, 1545 (1990).

[80] M. Gochin and H. Roder, *Protein Sci.* **4**, 296 (1995).

[81] L. Banci, I. Bertini, K. L. Bren, M. A. Cremonini, H. B. Gray, C. Luchinat, and P. Turano, *J. Biol. Inorg. Chem.* **1**, 117 (1996).

[82] P. X. Qi, R. A. Beckman, and A. J. Wand, *Biochemistry* **35**, 12275 (1996).

[83] L. Banci, I. Bertini, G. Gori Savellini, A. Romagnoli, P. Turano, M. A. Cremonini, C. Luchinat, and H. B. Gray, *Proteins* **29**, 68 (1997).

[84] L. Banci, I. Bertini, J. G. Huber, C. Luchinat, and A. Rosato, *J. Am. Chem. Soc.* **120**, 12903 (1998).

[85] K. Tu and M. Gochin, *J. Am. Chem. Soc.* **121**, 9276 (1999).

[86] M. Grayson and W. T. Raynes, *Magn. Reson. Chem.* **33**, 138 (1995).

[87] R. J. Abraham, M. A. Warne, and L. Griffiths, *J. Chem. Soc., Perkin Trans.* **2**, 203 (1997).

[88] R. J. Abraham, M. A. Warne, and L. Griffiths, *J. Chem. Soc., Perkin Trans.* **2**, 881 (1997).

[89] J. G. Pearson, E. Oldfield, F. S. Lee, and A. Warshel, *J. Am. Chem. Soc.* **115**, 6851 (1993).

[90] S. E. Chambers, E. Y. Lau, and J. T. Gerig, *J. Am. Chem. Soc.* **116**, 3603 (1994).

[91] J. Feeney, J. E. McCormick, C. J. Bauer, B. Birdsall, C. M. Moody, B. A. Starkmann, D. W. Young, P. Francis, R. H. Havlin, W. D. Arnold, and E. Oldfield, *J. Am. Chem. Soc.* **118**, 8700 (1996).

modeled as an r^{-6} distance dependence, but is probably more closely related to exchange–repulsion interactions than it is to dispersion interactions.[60]

The combination of susceptibility, electric field, and close-contact terms is known to give a reasonably good account of proton chemical shift dispersion in proteins.[8,11,17,73,78,83,92] As would also be expected, calculated proton shifts from quantum chemistry calculations can also often be interpreted using these ideas. Comparison of density functional results to their "classical" analogs for a wide variety of peptide conformations[19] support the notion that quantum calculations can in many places be replaced by much simpler classical models.

Interpreting Chemical Shifts in Proteins and Nucleic Acids

In interpreting macromolecular chemical shifts it is usually assumed that the various contributions discussed above are approximately independent and additive, so that the chemical shift "equation" can be written as[15,73,93]:

$$\Delta\delta = \delta_{total} - \delta_{rc} = \delta_{tor} + \delta_{ring} + \delta_{HB} + \delta_e + \delta_{side} + \delta_{misc} \tag{5}$$

where δ_{rc} is the random coil or "intrinsic" chemical shift of an amino acid residue, δ_{tor} is the backbone torsional contribution, δ_{ring} is the ring current contribution, δ_{HB} is the contribution arising from hydrogen bonding or close contacts, δ_e is the electric field or local charge contribution, δ_{side} is the side chain torsional contribution, and δ_{misc} encompasses other chemical shift contributions including solvent, temperature, motional averaging, and covalent bond geometry effects. In an effort to quantify these contributions we have assembled in Table IV rough estimates of the contributions each effect has on ^1H, ^{13}C, and ^{15}N chemical shifts. This information was derived from literature analyses[14,15,62,64,78] and preliminary results from a semiempirical chemical shift prediction program called SHIFTX (see Table V). Table IV shows that each of the six major nuclei have their chemical shifts determined by different contributions. The lack of overlap or "orthogonality" in chemical shift contributions suggests that shifts from different nuclei reveal complementary information about the local environment.

$^1H\alpha$ Shifts

Efforts to predict and understanding these shifts date from the late 1960s.[94,95] Because of their sensitivity to backbone torsional angles, $^1H\alpha$ shifts have long

[92] M. P. Williamson, J. Kikuchi, and T. Asakura, *J. Mol. Biol.* **247**, 541 (1995).
[93] M. P. Williamson, T. Asakura, E. Nakamura, and M. Demura, *J. Biomol. NMR* **2**, 93 (1992).
[94] J. L. Markley, D. H. Meadows, and O. Jardetzky, *J. Mol. Biol.* **27**, 25 (1967).
[95] H. Sternlicht and D. Wilson, *Biochemistry* **6**, 2881 (1967).

TABLE IV
DETERMINANTS OF ^{1}H, ^{13}C, AND ^{13}N CHEMICAL SHIFTS IN PROTEINS[a]

Attribute	^{1}HN	^{15}N	^{1}Hα	^{13}Cα	^{13}Cβ	^{13}CO
Random coil	0	50	25	50	75	25
Torsions (Φ/Ψ)	0	0	50	25	10	50
Torsions (Φ/Ψ_{i-1})	25	25	0	0	0	0
Side chain (χ)	5	0	0	5	5	5
Side chain (χ_{i-1})	5	5	0	0	0	0
Hydrogen bonds	25	5	5	5	0	5
Ring currents	10	0	10	0	5	5
Local charges	10	0	0	0	0	0
Miscellaneous	20	15	10	5	5	10

[a] Data given as %.

been used as quick and reliable indicators of secondary structure.[6,96–98] As a general rule, residues in stable helices typically exhibit upfield secondary chemical shifts (0.38 ppm on average) and residues in β sheets exhibit downfield shifts (0.38 ppm on average). However, ^{1}Hα shifts are also affected by many other factors as well.[11]

A semiquantitative breakdown of these effects and their respective contributions to the total (δ_{total}) measurable ^{1}Hα chemical shift is given in the third column of Table IV. In making these estimates we assume that the range of measured ^{1}Hα shifts is between 3.0 and 6.0 ppm—or about 3.0 ppm. The range in random coil or intrinsic shifts is about 0.75 ppm (see Table II), or about 25% of the total observed ^{1}Hα variation. With maximal upfield shifts of $\sim$0.6 ppm and maximal downfield shifts of $\sim$0.7 ppm,[14,20] backbone ϕ/ψ torsional effects typically contribute to 1.3 ppm, or about 50% of the observed ^{1}Hα range. Ring currents cause upfield or downfield shifts of up to 1.5 ppm, but because they affect only 10–15% of all residues, ring current effects contribute an average of about 0.3 ppm to the observed ^{1}Hα range. Hydrogen bonding,[12] solvent interactions, and proximal charges may account for about 0.1 ppm (or 5%) of the observable ^{1}Hα chemical shift variation. With about 90% of the ^{1}Hα chemical shift variation accounted for by these effects, the best semiempirical ^{1}Hα shift predictions typically attain correlation coefficients of about 0.90 with observed chemical shifts.[73,92] The remaining 10% uncertainty in ^{1}Hα chemical shift contributions probably arises in part from inaccuracies in 3D protein structures, dynamic averaging, or differences between solution and solid state conformations.[7]

[96] D. C. Delgarno, B. A. Levine, and R. J. P. Williams, *Biosci. Rep.* **3**, 443 (1983).
[97] A. Pastore and V. Saudek, *J. Magn. Reson.* **90**, 165 (1990).
[98] D. S. Wishart and B. D. Sykes, *J. Biomol. NMR* **4**, 171 (1994).

If one ignores the intrinsic or random coil contribution to ^{1}Hα shifts and looks only at secondary shift effects ($\Delta\delta$), it is evident that the backbone ϕ/ψ angles account for nearly 70% of the observed variation in ^{1}Hα shifts. This makes the ^{1}Hα nucleus perhaps the most sensitive to ϕ/ψ angles of all NMR active nuclei. Given this sensitivity, it is no wonder that ^{1}Hα shifts tend to be the most reliable predictors of backbone dihedral angles and secondary structure.[14,16] Because there is an upper limit of about ±0.6 ppm to ^{1}Hα torsional contributions, it is often possible to identify (without prior knowledge of the 3D structure) those that are affected by ring currents. This can be done simply by looking for ^{1}Hα secondary shifts ($\Delta\delta$) that exceed ±0.7 ppm. This ring current information can sometimes be used in setting up crude distance constraints or in verifying the correctness of an initial fold. Similarly, variations in ^{1}Hβ, ^{1}Hγ, or ^{1}Hδ secondary chemical shifts that exceed ±0.3 ppm can also be used as good indicators of proximal aromatic rings.

^{1}HN and ^{15}N Shifts

^{1}HN shifts are very sensitive to the protein environment, but our understanding of these shifts is weak. A synopsis of the ^{1}HN contributions is provided in Table IV. Here we can see that, unlike ^{1}Hα shifts, ^{1}HN shifts are affected by a different set of backbone dihedral angles (ϕ/ψ_{i-1}). Furthermore, the contribution of random coil or intrinsic shifts to ^{1}HN chemical shifts is essentially negligible. Hydrogen bonds play a fivefold more significant role in ^{1}HN shifts than in ^{1}Hα shifts. In making the calculations presented in Table IV we assumed the total ^{1}HN shift range is 4.0 ppm (6.50–10.50 ppm), the torsional contribution is 1.2 ppm (±0.60 ppm), the hydrogen bond contribution is 2.0 ppm with 60% of all amides being hydrogen bonded (total $= 1.2$ ppm), and ring current effects contribute about 0.5 ppm, while local charge effects average to 0.3 ppm. These effects account for about 75% of the ^{1}HN chemical shift variation. This coincides nicely with the correlation coefficients of 0.75 that the best semiempirical prediction techniques can offer.[92,99,100] The remaining 25% uncertainty in ^{1}HN chemical shift contributions probably arises from solvent and temperature effects, side chain effects, variations in backbone ω angles, inaccuracies in 3D coordinates, dynamic averaging, or differences between solution and solid state conformations.[78] As a rule, ^{1}HN shifts are not very good predictors of dihedral angles or indicators of secondary structure.[4,16] However, when the dihedral angles are largely invariant or known (as in a helix or in β sheets), ^{1}HN shifts are exquisitely sensitive indicators of hydrogen bond lengths and hydrogen bond energies.[12,13,78]

[99] J. Herranz, C. Gonzalez, M. Rico, J. L. Nieto, J. Santoro, M. A. Jimenez, M. Bruix, J. L. Neira, and F. J. Blanco, *Magn. Reson. Chem.* **30**, 1012 (1992).

[100] S. Neal and D. S. Wishart (unpublished data).

Nitrogen shifts are also sensitive measures of protein structure and dynamics. Like ^{1}HN shifts, they are primarily affected by ϕ/ψ_{i-1} backbone dihedral angles. Unlike ^{1}HN shifts, however, ^{15}N shifts are also extremely sensitive to the nature of attached side chains (i.e., the amino acid residue type). Approximately 15 ppm of the chemical shift variation associated with ^{15}N shifts arises from these covalent effects. Interestingly, this variation can also be used to distinguish Gly, Ser, and Thr from the other 17 amino acids (Table II)—a fact that can be helpful during the sequential assignment process. The sensitivity of ^{15}N shifts to attached side chains is also manifested in the so-called "nearest neighbor effects" which are particularly exaggerated with the ^{15}N nucleus.[36,38] In particular, ^{15}N shifts for amino acid residues in random coil polypeptides can vary by up to 3 ppm from their listed random coil values depending on the residue that precedes them. The nearest neighbor effect actually arises from the interaction or proximity of the side chain of the $i-1$ residue with the ^{15}N nucleus of the following residue.[64] This interaction is most evident when β-branched amino acids (Ile, Val, and Thr) or aromatic amino acids precede the residue of interest. The extent of this interaction appears to be closely correlated to the χ angle of the preceding residue (χ_{i-1}).

^{15}N shifts are also sensitive to hydrogen bonding in a manner not unlike ^{1}HN shifts. In general the same trends (upfield for long hydrogen bonds, downfield for short hydrogen bonds) are seen in ^{1}H shifts as in ^{15}N shifts. Indeed, a common positive slope (δ^{15}N/δ^1H) of about 4 is typically seen among ^{15}N/^{1}H cross peaks that have moved when hydrogen bonds are shortened or lengthened during some perturbation. If ^{15}N/^{1}H cross peak movements are seen to have substantially different slopes (0 or negative), this is usually a good indicator that either ring-current effects or side-chain rotations are influencing one or the other nucleus. As with ^{1}HN shifts, ^{15}N shifts are neither good predictors of backbone dihedral angles nor very good indicators of secondary structure.[16,21]

In making the calculations for ^{15}N presented in Table IV we assumed the total ^{15}N shift range is 15.0 ppm (109–124 ppm), the torsional contribution is 7 ppm (±3.5 ppm), the preceding side chain effect contributes up to 6 ppm with about 25% of all residues being affected (total = 1.5 ppm) and the hydrogen bond contribution averages about 3.0 ppm with 60% of all amides being hydrogen bonded (total = 1.8 ppm). These effects account for about 85% of the ^{1}HN chemical shift variation, which coincides nicely with the correlation coefficients of 0.85 claimed by the best ^{15}N prediction methods.[62,100] The remaining 15% uncertainty in ^{15}N chemical shift contributions probably arises from solvent and temperature effects, local side-chain effects, variations in backbone ω angles, inaccuracies in 3D coordinates, or differences between solution and solid state conformations.[78]

^{13}C Shifts

As can be seen in Table IV, all three types of carbon nuclei in proteins (^{13}Cα, ^{13}Cβ, and ^{13}CO) are highly sensitive to backbone ϕ/ψ torsion angles. Indeed, if the intrinsic or random coil contributions to the different ^{13}C shifts are ignored and only the secondary shift effects ($\Delta\delta$) are calculated, it is evident that the backbone ϕ/ψ angles account for half or more of the observed variation in ^{13}C shifts. This is why ^{13}C chemical shifts play a useful role not only in identifying secondary structure but also in predicting backbone dihedral angles.[14,16,20,98] As a general rule, ^{13}Cα shifts experience a downfield shift of about 2.5 ppm in helices and an upfield shift of 2.0 ppm in beta sheets. ^{13}CO shifts are affected in a similar way with downfield shifts being characteristic of helices and upfield shifts being typical of β sheets. In contrast, ^{13}Cβ shifts are shifted downfield by about 2.5 ppm in β sheets, but assume a near random coil value in helices. ^{13}Cβ shifts are only strongly upfield shifted when a residue has positive ϕ/ψ angles, making the ^{13}Cβ shift a particularly good indicator of this unusual ϕ/ψ combination.[14] Side-chain orientation, particularly for β-branched (Ile, Val, Thr) and aromatic (Phe, Trp, Tyr, His) amino acids, also plays a role in determining ^{13}C shifts.[15,59] However, differences between solution and crystal structures combined with the inherent mobility of side chains probably mask the true significance of side chain (χ_1 angle) effects on ^{13}C shifts. In fact, it is likely that ^{13}C shifts will play an increasingly important role in determining side chain orientation once these side chain trends are more clearly delineated.

In contrast to ^{1}H and ^{15}N shifts, ring currents have a relatively weak effect on ^{13}C shifts.[15,101] This may simply be a reflection of the fact that most ^{13}C shifts are "shielded" by attached proton or oxygen atoms, so close interactions with aromatic rings are not sterically allowed. Not only are ring current contributions relatively small for ^{13}C shifts, there also appears to be little contribution from hydrogen bond or electrostatic effects (at least for ^{13}Cα and ^{13}Cβ shifts). Hence ^{13}C secondary shifts are primarily determined by local geometric or covalent factors ($\phi/\psi/\chi$ angles) while ^{1}H and ^{15}N secondary shifts are primarily determined by "through-space" or dipolar factors.

Whereas ^{13}Cα and ^{13}Cβ shifts are generally quite predictable using empirical methods (correlation coeffients between observed and calculated of 0.98 and 0.99, respectively), the same cannot be said of ^{13}CO shifts. Indeed, there is still some question over what features contribute most significantly to ^{13}CO shifts. Some evidence suggests that nearest neighbor effects may play a large role[102]; other data suggest that hydrogen bond interactions outweigh dihedral contributions.[63,103] Empirical studies on a large number of "re-referenced" protein ^{13}CO shifts suggest that neither hydrogen bond nor nearest neighbor effects make significant contributions[14] and that ^{13}CO shifts can be calculated with a correlation coefficient of $\sim$0.90[100] using the weighting scheme outlined in Table IV.

[101] L. Blanchard, C. N. Hunter, and M. P. Williamson, *J. Biomol. NMR* **9**, 389 (1997).

[102] J. Yao, H. J. Dyson, and P. E. Wright., *FEBS Lett.* **419**, 285 (1997).

[103] I. Ando, T. Kameda, N. Asakawa, S. Kuroki, and H. Kurosu, *J. Mol. Struct.* **441**, 213 (1998).

Proton Shifts in Nucleic Acids

The use of chemical shifts in nucleic acid structure determination is much less advanced than for proteins. In large measure, this is due to concern that crystal structures for small nucleic acids may be more influenced by crystal packing forces than is the case for proteins. This means that the empirical methods used for proteins (which rely on crystal structures during the parameter development stage) are less useful for nucleic acids. This situation may change as more reliable NMR structures become available, and as a larger database of assigned structures is used, which should allow errors arising from local crystal packing effects to be averaged out.

Two groups have recently considered the ability of ring-current, susceptibility, and electrostatic models to predict proton shifts in nucleic acids.[30,31,75] If the reference shift for each type of proton is taken as the average value found in a database of molecules for which X-ray or NMR structures are available, the root-mean-square error of the predictions is 0.17 ppm. Assigning reference shifts from fragments (as discussed above) increases the overall error to about 0.25 ppm, but does allow discrimination between protons at different chemical positions. The general trend that sugar protons in duplex DNA and RNA are downfield from their random-coil positions, and that base protons are upfield of their positions in isolated bases, are readily explained by ring-current effects. Base protons are generally more reliably predicted than are sugar protons. Nevertheless, the overall correlation of predicted to experimental results is significantly worse than for proteins. This may reflect inadequacies in the structures that were used, or it may reflect limitations of the computational models, particularly in their treatment of the effects of the charges on the sugar–phosphate backbone.

Chemical Shift Refinement

Dihedral Angle Restraints

Given their sensitivity to backbone torsions, it is not surprising that chemical shifts are particularly useful in setting dihedral angle restraints.[14,16,22,104] The first study on the use of $^{13}C\alpha$ chemical shifts for deriving backbone dihedral constraints suggested that very generous limits ($\pm 100°$) and very stringent secondary chemical shift cutoffs (> 1.5 ppm) had to be used to avoid potential errors.[104] Based on comparative measures of "extent" and "uniqueness" it was suggested that $^{13}C\alpha$ shifts were probably comparable to $^3J_{HNHa}$ coupling constants in terms of their ability to define secondary structure. Luginbuhl *et al.*[104] also suggested that because of their relatively modest predictive power, $^{13}C\alpha$ shift-derived constraints should be used only in the early phases of protein structure refinement. Although this early

[104] P. Luginbuhl, T. Szyperski, and K. Wüthrich, *J. Magn. Reson. B* **109,** 229 (1995).

study was quite cautious in its assessment of chemical shifts, much more has since been learned about how chemical shifts can be used in setting dihedral restraints. It now appears that if multiple backbone (^{1}H, ^{13}C, and ^{15}N) chemical shifts are appropriately combined, backbone dihedral angle constraints can be determined with surprisingly good precision.

For example, secondary structures determined through heteronuclear chemical shift index (CSI) methods are now frequently used to set backbone dihedral constraints. Typically the dihedral limits are $\phi = -60 \pm 40°$; $\psi = -40 \pm 30°$ for CSI-defined helices and $\phi = -120 \pm 60°$; $\psi = 100 \pm 80°$ for CSI-defined β strands.[14] These limits are typically tighter than those commonly derived from J-coupling data. As always, care must be taken to avoid overinterpreting "sparse" or conflicting chemical shift or CSI data. Although CSI methods can provide dihedral constraints for regular secondary structures (about 60% of the residues in a typical protein), more precise angular constraints for all residues can often be derived from either chemical shift hypersurfaces[20,22,105] or database comparisons.[16] Early efforts by Oldfield and co-workers[1,105] demonstrated that reasonably accurate dihedral angle constraints could be derived by superimposing chemical shift probability plots derived from measured ^{1}Hα, ^{13}Cα, ^{13}Cβ, and ^{15}N chemical shifts and their corresponding hypersurfaces. Although limited to a few amino acids (Gly, Ala, Val) and a relatively small number of proteins, this ingenious approach demonstrated that dihedral angles could be determined to remarkably good precision ($\pm 12°$) using chemical shift data alone. This approach has been extended to all 20 amino acids using empirically derived chemical shift hypersurfaces,[14,22] but the precision of the dihedral angle predictions is not maintained for all residue types.

More recently Cornilescu et al.[16] have demonstrated impressive results using the TALOS program (Table V). This innovative approach to dihedral angle prediction is based on the simple observation that similar amino acid sequences with similar chemical shifts have similar backbone dihedral angles. In particular, TALOS breaks a query sequence into overlapping amino acid triplets, Each triplet and its corresponding heteronuclear chemical shifts (^{1}Hα, ^{13}Cα, ^{13}CO, ^{13}Cβ, and ^{15}N) are compared to a database of known sequences, shifts, and dihedral angles. The dihedral angles from the closest sets of database triplets are used to "predict" the dihedral angle for the central residue of the query amino acid triplet. TALOS is able to predict about 70% of protein backbone dihedral angles to within 30° (summed difference of $\Delta\phi + \Delta\psi$). A particular strength of the TALOS approach is that it accounts for nearest neighbor effects in a way that cannot easily be done with chemical shift hypersurfaces. Furthermore, as the database of known chemical shifts and dihedral angles grows, the accuracy of TALOS' predictions is expected to grow progressively as well. A similar dihedral angle prediction program called

[105] H. Le, J. G. Pearson, A. C. de Dios, and E. Oldfield, J. Am. Chem. Soc. 117, 3800 (1995).

TABLE V

PROGRAMS AND DATABASES FOR MACROMOLECULAR CHEMICAL SHIFT ANALYSIS

Program	Purpose	Access	Ref.[a]
AMBER	^{1}H chemical shift refinement of peptides, proteins, and nucleic acids	www.amber.ucsf.edu/amber	(1)
BMRB	Chemical shift repository	www.bmrb.wisc.edu/bmrb	(2)
Dalton	HF, MCSCF shift calculations	www.kjemi.uio.no/software/dalton	(3)
DeMon	DFT shift calculations	www.cerca.umontreal.ca/deMon	(4)
Gaussian98	HF, DFT, MP2 shift calculations	www.gaussian.com	—
PQS	HF, DFT shift calculations	www.pqs-chem.com	(5)
RefDB	Database of reference-corrected protein chemical shifts	http://redpoll.pharmacy.ualberta.ca	(6)
SHIFTCOR	For verifying, correcting, and referencing protein chemical shifts using PDB files	http://redpoll.pharmacy.ualberta.ca	(6)
SHIFTS	Calculation of ^{1}H chemical shifts for peptides and proteins from PDB files	www.scripps.edu/case	(7)
SHIFTOR	Prediction of dihedral angles from ^{1}H, ^{13}C, and ^{13}N chemical shifts	http://redpoll.pharmacy.ualberta.ca	(6)
SHIFTX	Calculation of ^{1}H, ^{13}C, ^{15}N shifts for peptides and proteins from PDB files	http://redpoll.pharmacy.ualberta.ca	(8)
SHIFTY	Prediction of ^{1}H, ^{13}C, and ^{15}N shifts from sequence homology	www.bmrb.wisc.edu/bmrb http://redpoll.pharmacy.ualberta.ca	(9)
SIMPRED	Prediction of ^{1}H, ^{13}C, and ^{15}N shifts from sequence and secondary structure	http://redpoll.pharmacy.ualberta.ca	(6)
TALOS	Prediction of dihedral angles from ^{1}H, ^{13}C, and ^{15}N chemical shifts	http://spin.niddk.nih.gov/bax/software/TALOS/info.html	(10)
THRIFTY	For rapid structure generation via chemical shift threading	http://redpoll.pharmacy.ualberta.ca	(11)
TOTAL	Calculation of ^{1}H and ^{13}C chemical shifts for peptides and proteins from PDB files	www.shef.ac.uk/uni/projects/nmr/resources.html	(12)
XPLOR/CNS	^{1}H and ^{13}C chemical shift refinement of peptides and proteins	http://xplor.csb.yale.edu/xplor-info http://cns.csb.yale.edu/v1.0	(13)
XREF	Varian macros for indirect referencing	http://redpoll.pharmacy.ualberta.ca	(14)

Key to References: (1) D. A. Pearlman, D. A. Case, J. W. Caldwell, W. R. Ross, T. E. Cheatham III, S. DeBolt, D. Ferguson, G. Seibel, and P. Kollman, *Comput. Phys. Comm.* **91**, 1 (1995); (2) B. R. Seavey, E. A. Farr, W. M. Westler, and J. L. Markley, *J. Biomol. NMR* **1**, 217 (1991); (3) T. Helgaker, M. Jaszunski, and K. Ruud, *Chem. Rev.* **99**, 293 (1999); (4) V. G. Malkin, O. L. Malkina, M. E. Casida, and D. R. Salahub, *J. Am. Chem. Soc.* **116**, 5898 (1994); (5) K. Wolinski, J. F. Hinton, and P. Pulay, *J. Am. Chem. Soc.* **112**, 8324 (1990); (6) H. Zhang M.Sc. Thesis, University of Alberta (2001); (7) K. Ösapay and D. A. Case, *J. Am. Chem. Soc.* **113**, 9436 (1991); (8) S. Neal, A. M. Nip, and D. S. Wishart (in preparation); (9) D. S. Wishart, M. S. Watson, R. F. Boyko, and B. D. Sykes, *J. Biomol. NMR* **10**, 329 (1997); (10) G. Cornilescu, F. Delaglio, and A. Bax, *J. Biomol. NMR* **13**, 289 (1999); (11) D. S. Wishart (in preparation); (12) M. P. Williamson, J. Kikuchi, and T. Asakura, *J. Mol. Biol.* **247**, 541 (1995); (13) A. T. Brunger, P. D. Adams, G. M. Clore, W. L. DeLano, P. Gros, R. W. Grosse-Kunstleve, J. S. Jiang, J. Kuszewski, M. Nilges, N. S. Pannu, R. J. Read, L. M. Rice, T. Simonson, and G. L. Warren, *Acta Cryst. D* **54**, 905 (1998); and (14) H. Monzavi (unpublished).

SHIFTOR (Table V) also makes use of the TALOS concept. SHIFTOR uses a different sequence comparison algorithm so that the running time is greatly reduced and the predictions are slightly improved.

Dihedral angles predicted with high confidence (rating > 6) by either TALOS or SHIFTOR are typically within $\pm 15°$ of the observed value. For helices, it is usually safe to set the error to less than $\pm 10°$. Dihedral angles predicted with moderate confidence (rating 5–6) can usually be given a variance of $\pm 20°$. Typically, this means that $\sim 70\%$ of the residues in a protein can be assigned dihedral angle restraint limits of $\pm 20°$ or less. If the protein is mostly helical, the number of residues that can be assigned is closer to 80% and the average limits can be lowered to $\pm 10°$. For comparison, $J(\text{NH-N}\alpha)$ couplings measured with a precision of ± 0.8 Hz corresponds to an ϕ error of $\pm 15°$.[106,107] Cross-correlated relaxation techniques[108,109] are typically able to provide ψ angle restraints with an error of $\pm 10°$. Average backbone dihedral angle uncertainties in well-refined X-ray structures are typically $\pm 7°$.[106] The central advantage to using chemical shift-derived dihedral angle restraints over J-couplings and cross-correlated relaxation data is that chemical shifts provide data for two backbone angles (ϕ and ψ) as opposed to just one. Furthermore, these dihedral restraints can be determined without additional NMR experiments, and the precision of chemical shift measurements is not compromised by increases in molecular weight. Despite these advantages, most NMR spectroscopists still use very qualitative methods for assigning dihedral angle restraints.

Distance Restraints

Although chemical shifts are most commonly used for dihedral restraint generation, there is also a long history of using dipolar or through-space ^{1}H chemical shifts to define approximate distance restraints.[12,110,111] For instance, hydrogen bond or electrostatic effects on amide protons are detectable up to 2.5 Å away, ring current effects on ^{1}H shifts typically extend up to 6 Å, while paramagnetic effects have been shown to extend more than 20 Å. In theory, if one knows that a certain ^{1}H shift is caused by a ring current, then it should be possible to assign an approximate distance between that proton and the aromatic ring based on the direction and extent of the ring current shift. The difficulty with this approach is that without prior knowledge of the 3D structure one cannot be certain which protons are truly ring current shifted, or which rings are involved.

[106] Y. Wang, A. M. Nip, and D. S. Wishart, *J. Biomol. NMR* **10**, 373 (1997).
[107] A. C. Wang and A. Bax, *J. Am. Chem. Soc.* **118**, 2483 (1996).
[108] D. Yang, R. Konrat, and L. E. Kay, *J. Am. Chem. Soc.* **119**, 11938 (1997).
[109] B. Reif, M. Hennig, and C. Griesinger, *Science* **276**, 1230 (1997).
[110] R. M. Keller and K. Wüthrich, *Biochim. Biophys. Acta* **285**, 326 (1972).
[111] C. C. McDonald and W. D. Phillips, *Biochemistry* **12**, 3170 (1973).

In simple situations, where there is only one known aromatic ring, or one paramagnetic center, or one possible hydrogen bond, the problem of identifying the source of the shift is trivial. In these situations, fairly precise distance restraints can often be derived using simple equations.[12] Similarly, if an approximate 3D structure is known, then dipolar shifts can also be used to obtain more precise distance restraints.[80] One little-known application of this concept concerns the precise determination of hydrogen bond lengths in helices. It has been consistently shown that ^{1}HN shifts in helices are highly sensitive to the O–H distances in hydrogen bonds. Several inverse distance relationships ($1/r$ and $1/r^3$) have been proposed.[12,13,78] What is striking about this relationship is that the length-dependent hydrogen bond effects are so precise, it is only possible to detect them in the most highly resolved X-ray structures (resolution <1.5 Å). Because ^{1}HN shifts are sensitive to O–H perturbations as small as 0.1 Å, distance restraints with errors as small as ±0.1 Å can be assigned purely on the basis of distances derived from ^{1}HN chemical shifts. The same kind of tight distance restraints can also be applied to β hairpins where a comparable well-defined, hydrogen-bonded structure exists.[112]

Direct Chemical Shift Refinement

"Direct" refinements against NMR data involve target functions that depend upon the difference between observed and calculated parameters, rather than using the more indirect information that is present in distance or dihedral angle restraints. Chemical shift refinement only became a possibility for peptides and proteins after reasonably accurate semiempirical theories and sufficiently fast computer programs became widely available.[20,73,92,113] Chemical shift refinement options are now available in AMBER,[114,115] XPLOR, and CNS,[17,18,116] with XPLOR/CNS offering both ^{1}H shifts and ^{13}C shift refinement. In general, ^{1}H chemical shift refinement of nonexchangeable protons has been shown to improve the quality of most structures and increase the definition of previously ill-defined segments.[19] Typically the most significant improvements are seen in regions where there is close spatial proximity to aromatic groups.[17,114,117] Despite the potential improvements available through ^{1}H chemical shift refinement, this approach appears not to be

[112] A. C. Gibbs, L. H. Kondejewski, W. Gronwald, A. M. Nip, R. S. Hodges, B. D. Sykes, and D. S. Wishart, *Nature Struct. Biol.* **5,** 284 (1998).

[113] D. D. Laws, A. C. deDios, and E. Oldfield, *J. Biomol. NMR* **3,** 607 (1993).

[114] K. Ösapay, Y. Theriault, P. E. Wright, and D. A. Case, *J. Mol. Biol.* **244,** 183 (1994).

[115] B. Celda, C. Biamonti, M. J. Arnau, R. Tejero, and G. T. Montelione, *J. Biomol. NMR* **5,** 161 (1995).

[116] A. T. Brünger, P. D. Adams, G. M. Clore, W. L. DeLano, P. Gros, R. W. Grosse-Kunstleve, J. S. Jiang, J. Kuszewski, M. Nilges, N. S. Pannu, R. J. Read, L. M. Rice, T. Simonson, and G. L. Warren, *Acta Cryst. D* **54,** 905 (1998).

[117] P. J. Lodi, J. A. Ernst, J. Kuszewski, A. B. Hickman, A. Engelman, R. Craigie, G. M. Clore, and A. M. Gronenborn, *Biochemistry* **34,** 9826 (1995).

widely used in the NMR community. This may be due in part to a small numerical error in CNS/XPLOR that leads to a persistent 0.5 ppm oversestimate of $^1H\alpha$ shifts in aromatic amino acids (His, Phe, Trp, Tyr). ^{13}C shift refinement has been successfully applied to human thioredoxin,[17,18] epidermal growth factor, [115] transcription factor GAGA,[118] and valine side chain refinement in interleukin β, calmodulin, and staph nuclease.[59] This process appears to be most useful in defining regions that are ordered, but possess no regular secondary structure.[19] The sensitivity of ^{13}C shifts to side-chain conformations[59,119] also makes ^{13}C chemical shift refinement a particularly attractive alternative to using NOE and $J\alpha\beta$ coupling constant measurements to estimate χ angles.

Perhaps the most impressive application of direct chemical shift refinement can be found in paramagnetic proteins.[80,83,120] Electronuclear interactions involving paramagnetic metals such as iron, cobalt, and various lanthanides can, under the right conditions, give rise to substantial (>10 ppm) chemical shifts.[121] These shifts can be divided into two forms, a contact and a pseudo-contact form. The contact or Fermi contact shift is a scalar or through-bond phenomenon that fades rapidly after four or five bonds. The pseudo-contact shift, Eq. (1), is a through-space phenomenon having a $1/r^3$ dependence that can extend up to 20 Å away from the paramagnetic center.

In practical terms, paramagnetic shift refinement requires at least twice as much work as diamagnetic shift refinement, since it requires spectra for two forms of the protein, one with a diamagnetic metal and the other with a suitable paramagnetic metal. The observed chemical shift differences between the diamagnetic and paramagnetic forms are assumed to arise entirely from the pseudo-contact shifts. Obviously if structural or solution conditions change, this is not a valid assumption. Care must be taken to reproduce the exact solution conditions of the first measurements and to ensure that one is looking at either a fully diamagnetic or fully paramagnetic form. It is possible to refine both the susceptibility tensor and the protein coordinates during the minimization process. Computer programs and protocols to orient the g tensor have been written and are available (FANTASIAN, PSEUDOREM).[83,122] AMBER[83] and XPLOR[80] have also been modified to explicitly perform pseudo-contact ^1H shift refinement.

Because of the long-range effects of pseudo-contact shifts, precise information about atomic positions, global shape, and domain orientation can be derived by including this information in the refinement process. Indeed, in some cases it is

[118] J. G. Omichinski, P. B. Pedone, G. Felsenfeld, A. M. Gronenborn, and G. M. Clore, *Nature Struct. Biol.* **4**, 657 (1997).

[119] K. R. MacKenzie, J. H. Prestegard, and D. M. Engelman, *J. Biomol. NMR* **7**, 256 (1996).

[120] M. Assfalg, L. Banci, I. Bertini, M. Bruschci, and P. Turano, *Eur. J. Biochem.* **256**, 261 (1998).

[121] L. Lee and B. D. Sykes, *Biochemistry* **22**, 4373 (1983).

[122] G. Williams, N. J. Claydon, G. R. Moore, and R. J. P. Williams, *J. Mol. Biol.* **183**, 447 (1985).

possible to determine atomic positions with a precision of less than 0.1 Å.[80] Being able to position ^{1}H atoms with this kind of precision also aids in the assignment process. In particular, stereospecific ^{1}H assignments can often be made because the shift differences between different diastereomers can be so precisely determined. In addition, this technique can provide independent evidence about the location of the paramagnetic metal atom.[123] Given that many proteins either have or could accommodate a paramagnetic center, it is clear that paramagnetic shift refinement will play an increasingly important role in many future structural studies.

Using Chemical Shift Anisotropy

In the liquid state, molecular tumbling averages out the anisotropic portions of the shielding tensor, so that only the isotropic portion is reflected in resonance positions. Chemical shift anisotropy (CSA) information can be uncovered either by moving to oriented or partially oriented samples, or by studying relaxation effects in liquids, where the transient effects of CSA interactions have important consequences. Since chemical shielding creates effective fields that scale with the spectrometer field [*cf.* Eq. (1)], relaxation by this mechanism becomes more important as NMR spectrometer fields increase.

Interpreting CSA relaxation in biomolecules can be a difficult task since the magnitudes and orientations of the shielding tensors are generally not known and since CSA relaxation competes with dipolar relaxation. One way to generate spectra with a fairly straightforward interpretation is to consider cross-correlated relaxation between CSA and dipolar terms.[124] Several groups have extracted CSA-related information for amide groups in proteins in liquids, both for ^{15}N[67,125–129] and for ^{1}H.[130–132] Interestingly, proton CSA appears to vary in a predictable way with hydrogen bond strength.[69,132] A key point that is not yet fully resolved is the extent to which the ^{15}N CSA in amides varies from one residue to another. The origins of this variability (as with the isotropic ^{15}N shift itself) have so far eluded generally applicable structural interpretations.[67]

[123] S. J. Wilkens, B. Xia, F. Weinhold, J. L. Markley, and W. M. Westler, *J. Am. Chem. Soc.* **120**, 4806 (1998).

[124] L. G. Werbelow, *in* "Encyclopedia of Nuclear Magnetic Resonance" (D. M. Grant, R. K. Harris, eds.), p. 4072. John Wiley, London, 1996.

[125] N. Tjandra, A. Szabo, and A. Bax, *J. Am. Chem. Soc.* **118**, 6986 (1996).

[126] D. Fushman, N. Tjandra, and D. Cowburn, *J. Am. Chem. Soc.* **120**, 10947 (1998).

[127] D. Fushman, N. Tjandra, and D. Cowburn, *J. Am. Chem. Soc.* **121**, 8577 (1999).

[128] C. D. Kroenke, M. Rance, and A. G. Palmer III, *J. Am. Chem. Soc.* **121**, 10119 (1999).

[129] J. Boyd and C. Redfield, *J. Am. Chem. Soc.* **121**, 7441 (1999).

[130] M. Tessari, H. Vis, R. Boelens, R. Kaptein, and G. W. Vuister, *J. Am. Chem. Soc.* **119**, 8985 (1997).

[131] N. Tjandra and A. Bax, *J. Am. Chem. Soc.* **119**, 8076 (1997).

[132] G. Wu, C. J. Freure, and E. Verdurand, *J. Am. Chem. Soc.* **120**, 13187 (1998).

Carbon CSA effects are also receiving increasing attention in protein NMR, probing CSA-dipolar cross-correlated relaxation along the $C\alpha$–$H\alpha$ bond,[133] or along $C\alpha$–C' or H–C', where C' is the carbonyl carbon.[134–136] In measuring the $C\alpha$ CSA one uses an (HA)CA(C)NH experiment to detect intensity differences in $^{13}C\alpha$–$^1H\alpha$ doublets caused by relaxation interference between $^{13}C\alpha$ dipolar and CSA effects. The anisotropy is defined as the difference between the shielding parallel to the C–H bond and the shielding orthogonal to this bond. In helices, this anisotropy is typically between 0 and 10 ppm, with an average value of 6 ppm. In β sheets it is between 20 and 33 ppm, with an average value of 27 ppm. There is essentially no overlap between helical and β sheet values. This suggests that ^{13}C CSA values could be used to both unambiguously determine secondary structure, and to define backbone dihedral angles.[69,133] Since the orientation of the carbonyl carbon shielding tensor is closely related to the direction of the C=O bond vector, it can also provide valuable information about the structure and dynamics of the peptide plane which can be difficult to obtain by other means.[137,138]

Similar CSA-related experiments are being pursued in nucleic acids as well. These experiments promise to provide additional insights into the detailed nature of local nucleic acid structure and dynamics.[139–141] In particular, the relative CSA (projected along the C–H bond direction) for C1′ and C3′ appears to offer a sensitive way to determine sugar puckers.[61,142]

CSA effects in paramagnetic systems are easiest to understand in the so-called "Curie limit,"[143–146] where electron spin relaxation is fast enough that nuclear spins interact only with the thermal average of the electron spin (the "Curie spin"): hence, the interaction between a remote nucleus and the metal center is modulated by molecular motion but not by the electronic spin relaxation itself. In this limit, the effects of the paramagnetic center on remote nuclei are the same as (but often

[133] N. Tjandra and A. Bax, *J. Am. Chem. Soc.* **119,** 9576 (1997).

[134] M. W. Fischer, L. Zeng, A. Majumdar, and E. R. Zuiderweg, *Proc. Natl. Acad. Sci. USA* **95,** 8016 (1998).

[135] Y. Pang, L. Wang, M. Pellecchia, A. V. Kurochkin, and E. R. P. Zuiderweg, *J. Biomol. NMR* **14,** 297 (1999).

[136] T. Norwood, M. Tillett, and L. Lian, *Chem. Phys. Lett.* **300,** 429 (1999).

[137] G. Cornilescu, J. L. Marquardt, M. Ottiger, and A. Bax, *J. Am. Chem. Soc.* **120,** 6836 (1998).

[138] M. Pellecchia, Y. Pang, L. Wang, A. V. Kurochkin, A. Kumar, and E. R. P. Zuiderweg, *J. Am. Chem. Soc.* **121,** 9165 (1999).

[139] C. Kojima, A. Ono, M. Kainosho, and T. L. James, *J. Magn. Reson.* **136,** 169 (1999).

[140] J. Boisbouvier, B. Brutscher, J.-P. Simorre, and D. Marion, *J. Biomol. NMR* **14,** 241 (1999).

[141] P. Damberg, J. Jarvet, P. Allard, and A. Grslund, *J. Biomol. NMR* **15,** 27 (1999).

[142] J. Boisbouvier, B. Brutscher, A. Pardi, D. Marion, and J.-P. Simorre, *J. Am. Chem. Soc.* **122,** 6779 (2000).

[143] I. Bertini, P. Turano, and A. J. Vila, *Chem. Rev.* **93,** 2833 (1993).

[144] M. Gueron, *J. Magn. Reson.* **19,** 58 (1975).

[145] I. Bertini, C. Luchinat, and D. Tarchi, *Chem. Phys. Lett.* **203,** 445 (1993).

[146] R. Ghose and J. H. Prestegard, *J. Magn. Reson.* **128,** 138 (1997).

larger than) that of other types of localized susceptibilities (such as aromatic rings), and the fact that the system is "paramagnetic" i.e., has unpaired electrons) becomes irrelevant. Consider the projection of the shielding tensor along and perpendicular to some particular direction **d** (such as N–H bond direction). If the susceptibility is isotropic, it is straightforward to show that[3]:

$$\Delta\sigma = -r^{-3}\chi\, P_2(\cos\theta) \tag{6}$$

where $\Delta\sigma$ is the difference in shielding along and perpendicular to **d**, and θ is the angle between **d** and the metal-nucleus vector. The large effect of metal ions on CSA makes this a potentially powerful technique for analyzing paramagnetic proteins.[147] Boisbouvier et al.[148] examined the effect of a metal ion on ^{1}H CSA/^{15}N–^{1}H dipolar cross correlation in cytochrome c', seeing effects out to 25 Å, with enough precision to obtain useful geometric restraints out to 15 Å. A similar approach is likely to be useful for other metalloenzymes.

Chemical Shifts in Protein Structure Identification and Generation

Here we consider the use of chemical shifts as "lower resolution" probes that can be used either for rapid structural characterization or on conformationally flexible systems where the conventional refinement techniques discussed above are not applicable.

Secondary Structure

Perhaps the most widespread application of chemical shifts in protein structure generation has been in the area of secondary structure identification. A number of elegant and simple methods have been described, including $\Delta\delta$ plots,[96] smoothed ^{1}Hα chemical shift plots for helix identification,[97] spectral integration for secondary structure quantification,[149] ^{1}H chemical shift indices,[8] smoothed ^{13}Cβ–^{13}Cα plots,[9] helix capping box identification,[10] and heteronuclear chemical shift indices or CSI plots.[98]

The most popular of these techniques are the $\Delta\delta$ and CSI plots. The $\Delta\delta$ plots are particularly useful because they present the data in a raw format. On the other hand, CSI plots are simpler to interpret, easier to delineate, and can be automatically generated via several freely available computer programs (Table V). As might be expected, the more chemical shift data one has, the more reliable the

[147] I. C. Felli, H. Desvaux, and G. Bodenhausen, *J. Biomol. NMR* **12**, 509 (1998).
[148] J. Boisbouvier, P. Gans, M. Blackledge, B. Brutscher, and D. Marion, *J. Am. Chem. Soc.* **121**, 7700 (1999).
[149] D. S. Wishart, B. D. Sykes, and F. M. Richards, *FEBS Lett.* **293**, 72 (1991).

secondary structure prediction. Heteronuclear CSI methods, which make simultaneous use of $^1H\alpha$, $^{13}C\alpha$, $^{13}C\beta$ and ^{13}CO chemical shift information, are generally the most accurate (93% agreement between X-ray secondary structure designations). Homonuclear methods ($^1H\alpha$ alone, $^{13}C\alpha$ alone, etc.) typically have a slightly lower level of agreement (80–85%). Incomplete chemical shift data or data collected under unusual conditions or with unusual amino acids can further lower the expected level of agreement with X-ray or NOE data.

In general, chemical shift methods are better at identifying helices than β sheets. However, this differential accuracy appears to be largely a function of the widely varying definitions of what constitutes a β sheet.[150,151] Although X-ray or NOE-derived secondary structures are often considered to be the most correct or accurate, it appears that secondary structures derived from heteronuclear CSI methods are likely to be just as correct.[18,98] This suggests that secondary structure identification in NMR (via heteronuclear CSI) can be reliably performed without the need to measure NOEs. This is particularly useful for peptides or flexible proteins.[99,152,153]

In addition to these simple graphical methods for identifying secondary structure, several newer chemical shift approaches have emerged, including ^1HN chemical shift analysis,[78] neural network analyses,[154] $^{13}C\alpha$ chemical shift anisotropy measurements (discussed above), and $^{13}C\alpha$ deuterium isotope shifts.[155]

Deuterium isotope shifts are small (parts per billion) shifts that arise from vibrational differences between ^1H and ^2H when attached to heavier ^{13}C or ^{15}N atoms. The magnitude of these shifts is governed by many effects, including the number of bonds separating the ^2H from the observed nucleus, the extent of isotopic substitution, the level of hydrogen bonding, and the local conformation. Interestingly, the three-bond isotope shift $^3\Delta C\alpha(ND)$ (i.e., the ^{13}C shift change brought about from having a deuterium on the $i + 1$ amide nitrogen) has been shown be highly dependent on the backbone ψ angle, and hence the local secondary structure.[155] By using an HCA(CO)N pulse sequence and separating the minute chemical shift changes using an E. COSY technique, these small shifts can be quite accurately measured on $^{13}C/^{15}N$ labeled proteins. Typically $^3\Delta C\alpha(ND)$ values greater than 30 ppb are indicative of β sheets, while $^3\Delta C\alpha(ND)$ values less than 25 ppb are indicative of α helices. There is almost no overlap in these values. The fact that $^3\Delta C\alpha(ND)$ values are reasonably well correlated with the ψ angle ($\pm 20°$) suggests that this approach could be used to extract ψ dihedral angle restraints.

[150] W. Kabsch and C. Sander, *Biopolymers* **22,** 2577 (1983).

[151] F. M. Richards and C. E. Kundrot, *Proteins: Struct. Funct. Genet.* **3,** 71 (1998).

[152] M. D. Reily, V. Thanabal, and D. O. Omichinsky, *J. Am. Chem. Soc.* **114,** 6251 (1992).

[153] D. Eliezer, J. D. Chung, H. J. Dyson, and P. E. Wright, *Biochemistry* **39,** 2984 (2000).

[154] K. Huang, M. Andrec, S. Heald, P. Blake, and J. H. Prestegard, *J. Biomol. NMR* **10,** 45 (1997).

[155] M. Ottiger and A. Bax, *J. Am. Chem. Soc.* **119,** 8070 (1997).

Chemical Shift Threading

Chemical shift threading is a new and potentially powerful approach for rapidly generating 3D protein structures using only chemical shift data (Wishart, in preparation). This is a modified form of protein threading, where a query protein sequence is "threaded" incrementally through a large database of known 3D protein structures. As the sequence is pushed through each structure (much as a pipe cleaner is pushed through a pipe) the stability of the new fold is evaluated using a quickly calculated heuristic potential.[156] Protein threading is an effective method for identifying the compatibility of a new protein sequence with a 3D fold that might not be obvious from normal sequence comparisons. It is also a way of "predicting" the approximate 3D structure of novel protein sequences.[156] In chemical shift threading, both the protein sequence and the experimentally observed chemical shifts are used in the threading process. Specifically, the protein sequence of interest is incrementally threaded through a structure database, and then at each step the ^{1}H, ^{13}C, and/or ^{15}N chemical shifts for that query sequence are calculated and compared to the observed chemical shifts. When a structural alignment is found where both the observed and predicted chemical shifts agree (to some preset tolerance), one can infer that the correct 3D fold has been found. Unlike classical threading techniques, which use no experimental information, chemical shift threading uses experimental data (chemical shifts) in evaluating the fit between the sequence and the putative fold. The use of experimental data gives chemical shift threading a significant edge over other threading techniques in terms of both efficacy and accuracy.

An example of how effective chemical shift threading can be is shown in Fig. 1, where a 3D structure of ubiquitin has been generated using only chemical shift data as input. Here the backbone ^{13}C and ^{1}H chemical shifts of ubiquitin[157] were used to conduct a search against a nonredundant version of the Protein Data Bank that explicitly excluded known ubiquitin homologs. The best "chemical shift" match was found with the G chain of β-elongin (1VCB), a protein sharing less than 25% sequence identity with ubiquitin. From this chemical shift alignment a 3D structure was built (Fig. 1A), which was found to be within 1.4 Å RMSD of the actual structure of ubiquitin (Fig. 1B). Chemical shift refinement[17] further reduced the pairwise RMSD to 1.1 Å. The correlations between the initial ubiquitin model's calculated secondary chemical shifts and the actual ubiquitin secondary shifts are quite high (^{1}Hα = 0.70, ^{13}Cα = 0.81), whereas for nonmatching or incorrect folds the correlations are all quite low (average ^{1}Hα ∼ 0.2, ^{13}Cα ∼ 0.01). This difference is substantial and illustrates the discriminatory

---

[156] D. T. Jones and J. M. Thornton, *Curr. Opin. Struct. Biol.* **6**, 210 (1996).

[157] A. J. Wand, J. L. Urgauer, R. P. McEvoy, and R. J. Bieber, *Biochemistry* **35**, 6116 (1996).

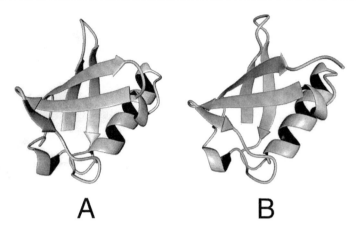

A   B

FIG. 1. Structure of human ubiquitin generated by chemical shift threading (A) compared to the actual X-ray structure (B) of ubiquitin (PDB: 1UBQ). The backbone RMSD between the two structures is 1.4 Å. Structure A was generated using published ^{1}H and ^{13}C chemical shifts for ubiquitin. The entire structure generation process took only a few minutes on a 250 MHz processor.

power that chemical shifts have in identifying the correct fold from the "noise" of other biologically compatible folds. Interestingly, the entire structure generation process (from initial chemical shifts to final 3D structure) took only a few minutes.

A key limitation of chemical shift threading is that it is only effective if the approximate fold for the query protein is already contained in the Protein Data Bank.[156] Consequently, chemical shift threading is not likely to be particularly effective in identifying or generating novel protein folds. Nevertheless, an analysis of 30 randomly chosen chemical shift sets recently deposited in the BMRB indicates that more than 80% (25/30) could have had their 3D structures generated (within 3 Å RMSD) using this chemical shift threading technique. This suggests that chemical shift threading could be applicable to a field such as structural proteomics, where rapid and approximate structure generation is of paramount importance. It is also worth noting that the concept of chemical shift threading need not be restricted to chemical shifts, so that other NMR data could be included. The chemical shift threading program called THRIFTY is available over the Web (Table V).

Solution Structure Generation

It is of considerable interest to explore the extent to which 3D structure determination of proteins could be carried out directly from chemical shift data alone. Key to the success of shift-based structure generation will be the ability

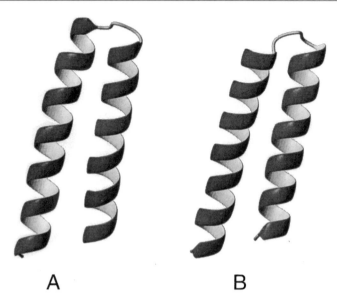

A   B

FIG. 2. X-ray structure of ColE1 ROP protein (A) compared to the ROP structure (B) generated by SHIFTOR and PEPMAKE using published ^{1}H and ^{13}C chemical shifts (BMRB #4072). Again the only pieces of information used to generate structure B were the chemical shifts and the amino acid sequence of ROP.

to generate accurate backbone dihedral angles and distance restraints from raw chemical shift data.[1,105] As we have already seen, significant strides have been made with the introduction of TALOS and SHIFTOR for backbone dihedral angle prediction. Although these two approaches still lack the precision and overall reliability needed for large-scale (> 60 residue) structure generation, they offer a promising lead for generating initial models of smaller peptides and proteins. Good success has already been achieved with the generation of a series of small cyclic peptides,[112] a calmodulin fragment,[14] and the ColE1 Rop protein (Fig. 2) using little more than chemical shift data. All three efforts relied on the combined use of approximate dihedral angles generated from chemical shifts, genetic algorithms for conformational sampling, and heuristic folding potentials[158] to assess the correctness of predicted folds. Based on these promising results it is likely that small protein structures (<100 residues) may soon be generated by combining the latest techniques in *ab initio* structure prediction[159] with the latest techniques in chemical shift-based dihedral angle prediction.

[158] S. H. Bryant and C. E. Lawrence, *Proteins: Struct. Funct. Genet.* **16,** 92 (1993).
[159] M. J. E. Sternberg, P. A. Bates, L. A. Kelley, and R. M. MacCallum, *Curr. Opin. Struct. Biol.* **9,** 368 (1999).

Although 3D structure generation using chemical shifts alone is clearly a desirable goal, it is likely that the most immediate successes in generating structures via chemical shifts will be attained using a combination of other NMR parameters. Certainly the combination of chemical shifts (as restraints) and NOE data is an obvious choice and, as might be expected, some success has already been reported.[22,160] However, a host of other NMR parameters, including J-coupling constants, residual dipolar couplings, T_1/T_2 ratios, cross-relaxation rates, and other spectral values, may be more readily measured and more reliably used in conjunction with chemical shift data.[19] Indeed, efforts along these lines are already underway in a number of laboratories. Delaglio et al.[161] have demonstrated how chemical shift information in conjunction with residual dipolar coupling data could be used to generate a 3D structure of ubiquitin. The principal of this technique is to use a TALOS-like program to confirm whether the dihedral angles estimated through dipolar coupling measurements for seven-residue molecular fragments were consistent with the chemical shifts measured in solution. This "inverse" TALOS approach permitted backbone dihedral angles to be accurately selected from several possible angles compatible with dipolar coupling measurements. From this information a 3D model of ubiquitin could be generated independently of NOE data. This particular application actually demonstrates how solution-state NMR has moved toward using the methods long used in solid-state NMR to help solve some outstanding problems in protein structure generation.

Solid-State NMR Structure Generation

Advances in molecular biology, spectrometer technology and pulse sequence methodology have shown that solid state NMR may be especially useful in studying the structure and dynamics of integral membrane proteins.[162–165] The fact that most conventional parameters (NOEs and J-coupling constants) relevant to solution structure determination are not readily obtainable from solid-state NMR spectra has forced solid-state spectroscopists to rely on the things they can measure: chemical shifts, chemical shift anisotropies, and dipolar couplings. Techniques such as CP-MAS,[166,] DIPSHIFT,[167] REDOR,[168] and heteronuclear PISEMA[169]

[160] P. J. Kraulis, J. Mol. Biol. 243, 696 (1994).
[161] F. Delaglio, G. Kontaxis, and A. Bax, J. Am. Chem. Soc. 122, 2142 (2000).
[162] R. Fu and T. A. Cross, Ann. Rev. Biophys. Biomol. Struct. 28, 235 (1999).
[163] F. M. Marassi and S. J. Opella, Curr. Opin. Struct. Biol. 8, 640 (1998).
[164] F. M. Marassi, A. Ramamoorthy, and S. J. Opella, Proc. Natl. Acad. Sci. U.S.A. 94, 8551 (1997).
[165] R. R. Ketchem, W. Hu, and T. A. Cross, Science 261, 1457 (1993).
[166] S. O. Smith, K. Ascheim, and M. Groesbeek, Q. Rev. Biophys. 29, 395 (1996).
[167] M. Hong, J. D. Gross, C. M. Rienstra, and R. G. Griffin, J. Magn. Reson. 129, 85 (1997).
[168] T. Guillon and J. Schaefer, J. Magn. Reson. 81, 196 (1989).
[169] C. H. Wu, A. Ramamoorthy, and S. J. Opella, J. Magn. Reson. 109, 270 (1994).

have allowed chemical shifts, dipolar couplings, or combinations of the two to be resolved and measured in proteins as large as 190 residues. By combining theoretical knowledge about the orientational dependence of ^{15}N chemical shifts, ^{1}H chemical shifts, and ^{1}H–^{15}N dipolar couplings, it is actually possible to obtain precise orientational constraints (restriction plots) that can be used to generate 3D protein structures.[163,164] Interestingly, these same structure generation concepts may also be applied to weakly oriented proteins in dilute bicelle, lanthanide, or phage containing solutions.[133,161] Among the more useful observations from this solid-state work are that ^{15}N chemical shifts can be used as very direct indicators about the orientation and structure of membrane proteins relative to the membrane bilayer. For strongly oriented systems, ^{15}N shifts between 180 and 230 ppm have been shown to be indicative of transmembane helices oriented parallel to the bilayer surface, whereas ^{15}N shifts between 60 and 130 ppm are indicative of amphipathic helices floating on top or oriented perpendicular to the bilayer surface. These chemical shift differences are fundamentally related to the differences in the σ_{11} and σ_{33} components of the ^{15}N shielding tensor relative to the direction of the external magnetic field. Similar orientational differences in ^{15}N chemical shifts can also be measured for weakly oriented systems in solution.[170]

Acknowledgments

DSW thanks H. Zhang, A. Gibbs, and D. Waldman for advice and assistance in preparing this manuscript. DSW acknowledges financial support by the Natural Sciences and Engineering Research Council (Canada), the Protein Engineering Network of Centres of Excellence (Canada), and Bristol Myers-Squibb (Canada). DAC was supported by NIH grant GM45811.

[170] M. Ottiger, N. Tjandra, and A. Bax, *J. Am Chem. Soc.* **119**, 9825 (1997).

[2] Cross-Correlated Relaxation for Measurement of Angles between Tensorial Interactions

By Harald Schwalbe, Teresa Carlomagno, Mirko Hennig,
Jochen Junker, Bernd Reif, Christian Richter, and
Christian Griesinger

1. Introduction

Structure calculations of proteins, oligonucleotides, and other molecules of biological interest rely on the measurement of proton–proton distances from nuclear Overhauser effect spectroscopy (NOESY) data, on the determination of scalar coupling constants to obtain torsion angle restraints, and on the measurement of dipolar coupling constants to obtain projection restraints. Although for the first class of parameters, increasing molecular weight does not put a limitation on the precision and accuracy of the extracted parameters for structure determination, the size of scalar coupling constants, especially when measured between two protons, is systematically affected for larger molecules. However, in many cases, local conformational analysis of torsion angles is of importance in gaining information on the dynamic conformation of the biomolecule. Flexible parts of the biological macromolecule are frequently involved in interaction with other molecules, and the conformational averaging often cannot be described by analysis of NOE data alone. The structural interpretation of cross-correlated relaxation[1] between different relaxation mechanisms has been introduced and has become an additional tool for obtaining torsion angle restraints in liquid state nuclear magnetic resonance (NMR) spectroscopy. Since the introduction of the technique, a number of experiments have been developed to determine protein backbone angles ϕ and ψ, pseudorotation phases in RNA and phosphodiester backbone angles, and glycosidic torsion angles. Furthermore, cross-correlated relaxation has been used to obtain dynamic information on protein backbone motions. Structural information can also be obtained for ligands weakly bound to biological macromolecules in the so-called transfer-NOE regime, in which the bound ligand is in fast exchange with the free ligand. There are a number of interesting features that make the analysis of cross-correlated relaxation a technique with considerable potential for the study of larger molecules. The size of the effect scales linearly with the overall correlation time τ_c of the biological macromolecule. As opposed to scalar coupling constants, the structural interpretation of cross-correlated relaxation does not require any parameterization. In favorable cases, known reference distances and geometries can be used to overcome the problem of the determination of effective correlation

[1] B. Reif, M. Hennig, and C. Griesinger, *Science* **276**, 1230 (1997).

METHODS IN ENZYMOLOGY, VOL. 338

times as discussed below. Finally, the effect of cross-correlated relaxation can in principle be a long-range effect. Provided efficient ways for the excitation of the desired coherences can be developed, there is no distance limitation for the effect of cross-correlated relaxation. Cross-correlated relaxation transverse relaxation optimized spectroscopy (TROSY) is exploited in the TROSY experiments[2] introduced by Wüthrich and co-workers. It relies on the same physical principles as dipole–chemical shift anisotropy (CSA) cross-correlated relaxation. However, TROSY experiments select the slow-relaxing components of a multiplet to increase the sensitivity of the experiment.

Cross-correlated relaxation arises from the interference of two anisotropic spin interactions that are averaged in a correlated way by the reorientation of the molecule in solution. The anisotropic interactions can be quantitatively described by a second rank tensor. The rate of this cross-correlated relaxation reflects the orientation of the two anisotropic interactions with respect to each other as well as their mutual motions and the overall correlation time of the molecule.[1] As opposed to NMR spectroscopy of liquids, such tensor projections have been determined and interpreted in structural terms in solid state local field separated,[3] spin diffusion,[4] or multiple quantum NMR spectroscopy.[5]

In spectra of liquids, anisotropic spin interactions between two second rank tensors can only be detected through relaxation. It is not required that the two interaction tensors have spins in common, but cross-correlated relaxation can be observed for remote spins separated by an arbitrary number of covalent bonds. The main tensors exploited in solution NMR are the tensors of dipolar coupling and of chemical shift anisotropy. The interactions between these two tensors can be classified as summarized in Table I. In Table I, the operator $\hat{\Gamma}_{I_k,I_m}^{CSA,CSA}$ represents the cross-correlated relaxation of the chemical shift anisotropy of spin $\hat{I}_k$ and of $\hat{I}_m$, while the operator $\hat{\Gamma}_{I_lI_k,I_mI_n}^{DD,DD}$ describes cross-correlated relaxation between the dipolar tensors connecting spins $\hat{I}_k$ and $\hat{I}_l$ as well as $\hat{I}_m$ and $\hat{I}_n$, respectively. It is interesting to note that the spins $\hat{I}_k$, $\hat{I}_l$, $\hat{I}_m$, and $\hat{I}_n$ can be nuclear or electron spins. There may also be mutually identical spins among the four spins. In order to calculate the effect of cross-correlated relaxation and how the effect reflects the molecular

[2] K. Pervushin, R. Riek, G. Wider, and K. Wüthrich, Proc. Natl. Acad. Sci. USA 94, 12366 (1997).

[3] R. K. Hester, J. L. Ackermann, B. L. Neff, and J. S. Waugh, Phys. Rev. Lett. 36, 1081 (1976); M. Linder, A. Höhener, and R. R. Ernst, J. Chem. Phys. 73, 4959 (1980).

[4] G. Dabbagh, D. P. Weli, and R. Tycko, Macromolecules 27, 6183 (1994).

[5] K. Schmidt-Rohr, Macromolecules 29, 3975 (1996); K. Schmidt-Rohr, J. Am. Chem. Soc. 118, 7601 (1996); X. Feng, Y. K. Lee, D. Sandström, M. Edén, H. Maisel, A. Sebald, and M. H. Levitt, Chem. Phys. Lett. 257, 314 (1996); X. Feng, P. J. E. Verdegem, Y. K. Lee, D. Sandström, M. Edén, P. Bovee-Geurts, J. W. de Grip, J. Lugtenburg, H. J. M. de Groot, and M. H. Levitt, J. Am. Chem. Soc. 119, 6853 (1997); M. Hong, J. D. Gross, and R. G. Griffin, J. Phys. Chem. B 101, 5869 (1997).

TABLE I

POSSIBLE TENSOR INTERACTIONS GIVING RISE
TO CROSS-CORRELATED RELAXATION

	CSA	Dipolar coupling
CSA	$\hat{\Gamma}^{CSA,CSA}_{I_k,I_m}$	$\hat{\Gamma}^{DD,CSA}_{I_l I_k,I_m}$
Dipolar coupling	$\hat{\Gamma}^{CSA,DD}_{I_k,I_n I_m}$	$\hat{\Gamma}^{DD,DD}_{I_l I_k,I_n I_m}$

structure, we need to calculate how the relaxation operators given in Table I affect specific coherences. A formulation of the relaxation theory is given in the following to provide the mathematical tools for the calculation of cross-correlated relaxation effects.

2. Relaxation Theory

2.1. Hamiltonian for Dipolar Coupling and Chemical Shift Anisotropy

The main magnetic interactions that cause cross-correlated relaxation in liquid state NMR and that are exploited to obtain structural or dynamic information are dipolar couplings and chemical shift anisotropy.

The Hamiltonian for the dipolar coupling is given by:

$$\hat{H}^{DD}_{kl} = b_{kl} \left\{ 3\frac{1}{r^2_{kl}}(\hat{I}_k \cdot \mathbf{r}_{kl})(\hat{I}_l \cdot \mathbf{r}_{kl}) - \hat{I}_k \hat{I}_l \right\}$$

$$= b_{kl} \sum_{q=-2}^{+2} F^{(-q)}_{kl}(\theta_{kl}, \phi_{kl}) \hat{A}^{(q)}_{kl}(\hat{I}_k, \hat{I}_l) \tag{1}$$

$$b_{kl} = -\mu_0 \frac{\gamma_k \gamma_l \hbar}{4\pi r^3_{kl}}$$

In Eq. (1), γ_k and γ_l denote the gyromagnetic ratio of the nuclei k and l, $\hbar$ the Planck constant divided by 2π, and r_{kl} the distance between the two nuclei. The angles θ_{kl}, ϕ_{kl} refer to the orientation of the vector $\mathbf{r}_{kl}$ with respect to the static magnetic field. The exact expressions for the second rank tensor operators $\hat{A}^{(q)}_{kl}(\hat{I}_k, \hat{I}_l)$ and the time dependent modified spherical harmonics $F^{(q)}_{kl}(\theta_{kl}, \phi_{kl})$ are given in Table II. The contribution to the Hamiltonian due to CSA in the principal

TABLE II

TENSOR OPERATORS IN ROTATING FRAME AND MODIFIED SPHERICAL HARMONICS
FOR DIPOLAR COUPLING AND CSA INTERACTION[a]

Tensor operators for dipolar interaction $b_{kl} = -\mu_0 \frac{\gamma_K \gamma_l \hbar}{4\pi r_{kl}^3}$	Tensor operators for CSA interaction $b_k = \frac{1}{3}(\sigma_\parallel - \sigma_\perp)\gamma_k B_0$	Modified spherical harmonics	Frequency
q	$\hat{A}_{kl}^{(q)}(\hat{I}_k, \hat{I}_l)$	$\hat{A}_k^{(q)}(\hat{I}_k)$	$F_k^{(q)}(\theta, \phi), F_{kl}^{(q)}(\theta, \phi)$	ω_q
-2	$\sqrt{\frac{3}{8}} \hat{I}_k^- \hat{I}_l^-$	—	$\sqrt{\frac{3}{2}} \sin^2\theta \exp(+2i\phi)$	$\omega(\hat{I}_k) + \omega(\hat{I}_l)$
-1	$\sqrt{\frac{3}{8}} \hat{I}_{k,z} \hat{I}_l^-$	—	$\sqrt{6} \sin\theta \cos\theta \exp(+i\phi)$	$\omega(\hat{I}_l)$
-1	$\sqrt{\frac{3}{8}} \hat{I}_k^- \hat{I}_{l,z}$	$\sqrt{\frac{3}{8}} \hat{I}_k^-$	$\sqrt{6} \sin\theta \cos\theta \exp(+i\phi)$	$\omega(\hat{I}_k)$
0	$\hat{I}_{k,z} \hat{I}_{l,z}$	$\hat{I}_{k,z}$	$3\cos^2\theta - 1$	0
0	$\frac{1}{4}(\hat{I}_k^+ \hat{I}_l^- + \hat{I}_k^- \hat{I}_l^+)$	—	$3\cos^2\theta - 1$	$\omega(\hat{I}_k) - \omega(\hat{I}_l)$
$+1$	$\sqrt{\frac{3}{8}} \hat{I}_{k,z} \hat{I}_l^+$	$\sqrt{\frac{3}{8}} \hat{I}_k^+$	$\sqrt{6} \sin\theta \cos\theta \exp(-i\phi)$	$-\omega(\hat{I}_l)$
$+1$	$\sqrt{\frac{3}{8}} \hat{I}_k^+ \hat{I}_{l,z}$	—	$\sqrt{6} \sin\theta \cos\theta \exp(-i\phi)$	$-\omega(\hat{I}_k)$
$+2$	$\sqrt{\frac{3}{8}} \hat{I}_k^+ \hat{I}_l^+$	—	$\sqrt{\frac{3}{2}} \sin^2\theta \exp(-2i\phi)$	$-\omega(\hat{I}_k) - \omega(\hat{I}_l)$

[a] The calibration has been chosen such that $\frac{1}{4\pi} \int F^{(q)}(\theta, \phi) F^{(-q)}(\theta, \phi) d\cos\theta d\phi = \frac{4}{5}$ is independent of q.

axis frame (PAS) of the chemical shift tensor of the nucleus k can be written as[6]

$$\hat{H}_k^{CSA,PAS} = \gamma_k \sum_{i=x,y,z} B_i \hat{\sigma}_{ii}^k \hat{I}_{k,i} \tag{2}$$

$$= \gamma_k \frac{\sigma_{xx}^k + \sigma_{yy}^k + \sigma_{zz}^k}{3} \mathbf{B}\hat{I}_k + \frac{1}{3}\gamma_k \left(\sigma_{xx}^k - \sigma_{zz}^k\right)$$
$$\times [2\hat{I}_{k,x} B_x - \hat{I}_{k,y} B_y - \hat{I}_{k,z} B_z] + \frac{1}{3}\gamma_k \left(\sigma_{yy}^k - \sigma_{zz}^k\right)$$
$$\times [2\hat{I}_{k,y} B_y - \hat{I}_{k,x} B_x - \hat{I}_{k,z} B_z]$$

The B_i denote the components of the applied field B_0 in the PAS. In the laboratory frame (LF), in analogy to Eq. (1), the CSA Hamiltonian can be separated into the time-dependent, orientation-dependent functions $F_k^{(q)}(\theta_k, \phi_k)$ and the time-independent spin operator terms $\hat{A}_k^{(q)}(\hat{I}_k)$. The expressions for the second rank tensor operators $\hat{A}_k^{(q)}(\hat{I}_k)$ and the time-dependent modified spherical harmonics

[6] M. Goldman, "Quantum Description of High-Resolution NMR in Liquids." Clarendon Press, Oxford, 1988; R. R. Ernst, G. Bodenhausen, and A. Wokaun, "Principles of Nuclear Magnetic Resonance in One and Two Dimensions." Clarendon Press, Oxford, 1989.

$F_k^{(q)}(\theta_k, \phi_k)$ are summarized in Table II.

$$\hat{H}_k^{CSA,LF} = b_{k,x} \sum_{q=-1}^{1} F_k^{(q)}(\theta_{k,x}, \phi_{k,x})\hat{A}_k^{(q)}(\hat{I}_k) + b_{k,y} \sum_{q=-1}^{1} F_k^{(q)}(\theta_{k,y}, \phi_{k,y})\hat{A}_k^{(q)}(\hat{I}_k)$$

(3)

with

$$b_{k,x} = \frac{1}{3}(\sigma_{xx}^k - \sigma_{zz}^k)\gamma_k B_0 \quad \text{and} \quad b_{k,y} = \frac{1}{3}(\sigma_{yy}^k - \sigma_{zz}^k)\gamma_k B_0$$

and $\theta_{k,x}$, $\phi_{k,x}$ and $\theta_{k,y}$, $\phi_{k,y}$ are the polar angles of the x and y principal axes of the CSA tensor with respect to the main field.

For an axially symmetric CSA tensor $\sigma_{xx}^k = \sigma_{yy}^k$, this equation can be rewritten into:

$$\hat{H}_k^{CSA,PAS} = \gamma_k \frac{2\sigma_{xx}^k + \sigma_{zz}^k}{3} \mathbf{B}\hat{I}_k + \frac{1}{3}\gamma_k (\sigma_{zz}^k - \sigma_{xx}^k) [2\hat{I}_{k,z}B_z - \hat{I}_{k,x}B_x - \hat{I}_{k,y}B_y]$$

$$\hat{H}_k^{CSA,LF} = b_k \sum_{q=-1}^{1} F_k^{(q)}(\theta_k, \phi_k)\hat{A}_k^{(q)}(\hat{I}_k)$$

(4)

with

$$b_k = \frac{1}{3}(\sigma_{zz}^k - \sigma_{xx}^k)\gamma_k B_0$$

for axially symmetric CSA tensors. θ_k, ϕ_k are the polar angles of the z principal axis of the CSA tensor in the laboratory frame.

2.2. Relaxation Superoperator and Spectral Density Functions

In the Liouville–von Neumann equation

$$\frac{d}{dt}\hat{\sigma} = -i[\hat{H}_0, \hat{\sigma}(t)] - \sum_{V,W} \hat{\hat{\Gamma}}_{V,W}(\hat{\sigma}(t) - \hat{\sigma}(0))$$

(5)

the relaxation superoperator has the general form

$$\hat{\hat{\Gamma}}_{VW}\hat{\sigma} = b_V b_W \sum_q [\hat{A}_V^{(-q)}, [\hat{A}_W^{(q)}, \hat{\sigma}]]j_{VW}^q(\omega_q)$$

(6)

The indices V and W refer to the magnetic interactions that are the source for relaxation. The term $j_{VW}^q(\omega_q)$ denotes the spectral density function and is obtained by evaluating the correlation function of the spherical harmonics

$$j_{VW}^q(\omega_q) = \int_0^\infty d\tau \overline{F_V^{(q)}(t)F_W^{(-q)}(t+\tau)} \exp(-i\omega_q \tau)$$

(7)

The bar indicates time average over t. In the autocorrelation case ($V = W$), both relaxation sources V and W are originating from the same interaction, e.g., dipolar coupling between the same pair of nuclei or the chemical shift anisotropy of the same nucleus. In the cross correlation case, the time-dependent spherical harmonics refer to different kinds of tensorial interactions ($V \neq W$).

In the following, we concentrate on the cross correlation case. The theory for the description of intramolecular dipolar interaction has been developed.[7-11] The spectral density functions introduced in Eq. (6) depend on the reorientation of the molecule in solution. Formulae have been derived for the three cases of isotropic, axially symmetric, and generally anisotropic reorientation of the molecule and are given in sections 2.3.1, 2.3.2, and 2.3.3. A review on this topic is found in Ref. 40.

2.3. Models for Rotational Reorientation

2.3.1. Spherical Top Molecules. The spectral density function for iso-tropic rotational diffusion has been derived by Hubbard, Kuhlmann, and Baldeschweiler[36a,37]:

$$
\begin{aligned}
J^q_{VW}(\omega_q) &= \frac{2}{5}(3\cos^2\theta_{V,W} - 1)\left[\frac{6D}{(6D)^2 + \omega_q^2}\right] \\
&= \frac{1}{5}(3\cos^2\theta_{V,W} - 1)\left[\frac{2\tau_c}{1 + (\omega_q\tau_c)^2}\right]
\end{aligned}
\tag{8}
$$

D denotes the diffusion constant, with $1/\tau_c = 6D$. V and W denote the different interactions. $\theta_{V,W}$ describes the projection angle between the principal axes of the two interactions V and W. If V denotes, e.g., a dipolar interaction between two spins A_1 and A_2 and W a dipolar interaction between a second set of two spins B_1 and B_2, then θ is the included angle between the two bond vectors connecting A_1 and A_2 and B_1 and B_2, respectively.

2.3.2. Asymmetric Top Molecules. The spectral density function for the asym-metric top assumes the form

$$
\begin{aligned}
J^q_{VW}(\omega_q) = \frac{1}{10}\Big\{ & 12\cos\theta_V \cos\theta_W \sin\theta_V \sin\theta_W \sin\phi_V \sin\phi_W \frac{b_1}{b_1^2 + \omega_q^2} \\
& + 12\cos\theta_V \cos\theta_W \sin\theta_V \sin\theta_W \cos\phi_V \cos\phi_W \frac{b_2}{b_2^2 + \omega_q^2} \\
& + 3\sin^2\theta_V \sin^2\theta_W \sin 2\phi_V \sin 2\phi_W \frac{b_3}{b_3^2 + \omega_q^2} \\
& + \Big[3\cos^2\left(\tfrac{\zeta}{2}\right)\sin^2\theta_V \sin^2\theta_W \cos 2\phi_V \cos 2\phi_W
\end{aligned}
$$

[7] P. S. Hubbard, *Phys. Rev.* **109**, 1153 (1958); P. S. Hubbard, *J. Chem. Phys.* **51**, 1647 (1969); P. S. Hubbard, *J. Chem. Phys.* **52**, 563 (1970).

[8] K. F. Kuhlmann and J. D. Baldeschweiler, *J. Chem. Phys.* **43**, 572 (1965).

[9] H. Shimizu, *J. Chem. Phys.* **37**, 765 (1969).

[10] D. M. Grant and L. G. Werbelow, *J. Magn. Reson.* **21**, 369 (1976).

[11] L. G. Werbelow and D. M. Grant, *Adv. Magn. Reson.* **9**, 189 (1977).

$$+ \sin^2\left(\tfrac{\zeta}{2}\right)(3\cos^2\theta_V - 1)(3\cos^2\theta_W - 1)$$

$$+ \sqrt{3}\cos\left(\tfrac{\zeta}{2}\right)\sin\left(\tfrac{\zeta}{2}\right)((3\cos^2\theta_V - 1)\sin^2\theta_W\cos 2\phi_W$$

$$+ (3\cos^2\theta_W - 1)\sin^2\theta_V\cos 2\phi_V)\Big]\frac{b_4}{b_4^2 + \omega_q^2}$$

$$+ \left[3\sin^2\left(\tfrac{\zeta}{2}\right)\sin^2\theta_V\sin^2\theta_W\cos 2\phi_V\cos 2\phi_W\right.$$

$$+ \cos^2\left(\tfrac{\zeta}{2}\right)(3\cos^2\theta_V - 1)(3\cos^2\theta_W - 1)$$

$$- \sqrt{3}\cos\left(\tfrac{\zeta}{2}\right)\sin\left(\tfrac{\zeta}{2}\right)((3\cos^2\theta_V - 1)\sin^2\theta_W\cos 2\phi_W$$

$$+ (3\cos^2\theta_W - 1)\sin^2\theta_V\cos 2\phi_V\Big]\frac{b_5}{b_5^2 + \omega_q^2}\Bigg\} \tag{9}$$

in which the notation of Woessner[12] has been used:

$$D = \frac{1}{3}(D_{xx} + D_{yy} + D_{zz})$$

$$L^2 = \frac{1}{3}(D_{xx}D_{yy} + D_{xx}D_{zz} + D_{yy}D_{zz})$$

$$\tan\zeta = \sqrt{3}\left[\frac{D_{xx} - D_{yy}}{2D_{zz} - D_{xx} - D_{yy}}\right]$$

$$b_1 = 4D_{xx} + D_{yy} + D_{zz} \tag{10}$$

$$b_2 = D_{xx} + 4D_{yy} + D_{zz}$$

$$b_3 = D_{xx} + D_{yy} + 4D_{zz}$$

$$b_4 = 6D + 6\sqrt{D^2 - L^2}$$

$$b_5 = 6D - 6\sqrt{D^2 - L^2}$$

2.3.3. *Axially Symmetric Top Molecules.* For the case that $D_{xx} = D_{yy} = D_\perp$, Eq. (9) can be simplified yielding the spectral density function of the symmetric top rotator ($D_{zz} = D_\parallel$).

$$J_{VW}^q(\omega_q) = \frac{1}{20}\{(3\cos^2\theta_V - 1)(3\cos^2\theta_W - 1)J_{VW}^{q,0}$$

$$+ 12\cos\theta_V\cos\theta_W\sin\theta_V\sin\theta_W\cos(\phi_V - \phi_W)J_{VW}^{q,1}$$

$$+ 3\sin^2\theta_V\sin^2\theta_W\cos(2\phi_V - 2\phi_W)J_{VW}^{q,2}\} \tag{11}$$

where the reduced spectral density functions ($-2 \le m \le +2$)

$$J_{VW}^{q,m} = \frac{2\tau_{c,m}}{1 + (\omega_q\tau_{c,m})^2} \tag{12}$$

[12] D. E. Woessner, *J. Chem. Phys.* **36**, 1 (1962); D. E. Woessner, *J. Chem. Phys.* **37**, 647 (1962).

have been used. The correlation times $\tau_{c,m}$ can be rewritten as diffusion constants $D_{\parallel}$ and $D_{\perp}$ according to

$$1/\tau_{c,m} = 6D_{\perp} + m^2(D_{\parallel} - D_{\perp}) \tag{13}$$

2.3.4. Inclusion of Internal Motion. Internal motion can be incorporated into the spectral density either by the Lipari and Szabo[13] approach or by explicit calculation of the motion from, e.g., motional models of molecular dynamics trajectories. In Eqs. (8), (9), and (11), the spectral densities are Fourier transformations of the motion of the molecule with respect to the external magnetic field. In these equations it is assumed that internal motion is absent. Rewriting the spectral densities as a convolution ($\times$) of a Fourier transformation of the global motion and the Fourier transformation of the local motion, the first term $J_{V,W}^q(\omega_q)$ in Eq. (11) for axially symmetric diffusion is given by:

$$J_{VW}^{q,\text{local motion}}(\omega_q) = \frac{1}{5} FT\left\{ \overline{P_2[\cos\theta_V(t)]P_2[\cos\theta_W(t+\tau)]} \times J_{VW}^{q,0} \right\}\Big|_{\omega_q} + \cdots \tag{14}$$

The Fourier transformation concerns τ; the average is taken with respect to t. $P_2[\cos\theta]$ denotes the Legendre polynomial $(3\cos^2\theta - 1)/2$. If the internal motion is uncorrelated and fast with respect to the global motion, Eq. (14) can be directly used to compare field-dependent experimental relaxation rates with predicted rates and to analyze molecular dynamics trajectories or other motional models.

Application of the Lipari and Szabo approach assumes in addition an exponential decay of the correlation function with the characteristic rate τ_i from time 0 to the time τ according to

$$\overline{P_2[\cos\theta_V(t)]P_2[\cos\theta_W(t+\tau)]}$$

$$= \exp\left(-\frac{\tau}{\tau_i}\right)\{\overline{P_2[\cos\theta_V(t)]P_2[\cos\theta_W(t)]} - \overline{P_2[\cos\theta_V(t)]P_2[\cos\theta_W(t+\infty)]}\}$$

$$+ \overline{P_2[\cos\theta_V(t)]P_2[\cos\theta_W(t+\infty)]}$$

$$= \exp\left(-\frac{\tau}{\tau_i}\right)\{[1 - (S_{VW}^{q,0})^2] + (S_{VW}^{q,0})^2\}\overline{P_2[\cos\theta_V(t)]P_2[\cos\theta_W(t+\infty)]} \tag{15}$$

where $(S_{VW}^{q,0})^2$ is the order parameter for the respective interaction and τ_i corresponds to the internal correlation time.[14,15]

[13] G. Lipari and A. J. Szabo, *J. Am. Chem. Soc.* **104**, 4546 (1982).
[14] R. Brüschweiler and D. A. Case, *Prog. NMR Spectr.* **26**, 27 (1994).
[15] B. Brutscher, N. R. Skrynnikov, T. Bremi, R. Brüschweiler, and R. R. Ernst, *J. Magn. Reson.* **130**, 346 (1998).

TABLE III
TENSOR OPERATORS THAT GIVE RISE TO CROSS-CORRELATED RELAXATION[a]

Tensor operators	$\sqrt{\frac{3}{8}}\hat{I}^+_m\hat{I}^+_n$	$\sqrt{\frac{3}{8}}\hat{I}^+_m\hat{I}_{n,z}$	$\hat{I}_{m,z}\hat{I}_{n,z}$	$\frac{1}{4}(\hat{I}^+_m\hat{I}^-_n + \hat{I}^-_m\hat{I}^+_n)$	$\hat{I}_{m,z}$	$\sqrt{\frac{3}{8}}\hat{I}^+_m$
$\sqrt{\frac{3}{8}}\hat{I}^-_k\hat{I}^-_l$	$\omega_k + \omega_l$ $= \omega_m + \omega_n$	—	—	—	—	—
$\sqrt{\frac{3}{8}}\hat{I}^-_k\hat{I}_{l,z}$	—	$\omega_k = \omega_m$	—	—	—	$\omega_k = \omega_m$
$\hat{I}_{k,z}\hat{I}_{l,z}$	—	—	0	$0 = \omega_m - \omega_n$	0	—
$\frac{1}{4}(\hat{I}^+_k\hat{I}^-_l + \hat{I}^-_k\hat{I}^+_l)$	—	—	$0 = \omega_l - \omega_k$	$\omega_k - \omega_l = \omega_m - \omega_n$	$0 = \omega_k - \omega_l$	—
$\hat{I}_{k,z}$	—	—	0	$0 = \omega_m - \omega_n$	0	—
$\sqrt{\frac{3}{8}}\hat{I}^-_k$	—	$\omega_k = \omega_m$	—	—	—	$\omega_k = \omega_m$

[a] Frequencies of spectral densities.

2.4. Spectral Densities of Cross-Correlated Relaxation

Not all anisotropic interactions that are cross-correlated by motion give rise to cross-correlated relaxation. This is due to the fact that to be cross-correlated, the eigenfrequencies of the anisotropic interactions in the laboratory frame must be identical. Table III collects the spectral densities expected for cross correlation of anisotropic interactions. There are tight requirements for the frequencies of the involved spins. For example, the cross-correlated relaxation between the dipolar couplings of spins I_l and $I_k(I^-_k I_{l,z})$ and the dipolar couplings of spins I_m and $I_n(I^+_m I_{n,z})$ that depend on spectral densities sampled at a frequency different from 0 such as ω_k requires that $\omega_k = \omega_m$. Operators connected with spectral densities of 0 are boxed in Table III. These are the most important operators for macromolecular applications.

2.5. Calculation of Cross-Correlated Relaxation using Double Commutator Approach

In the following, the evolution of coherences, for which cross-correlated relaxation is a secular interaction, will be calculated. Nonsecular terms will not be discussed here.[16] Cross-correlated relaxation under spin lock conditions for which nonsecular terms become secular again will also not be discussed.[17] The operators of importance are given in Table IV. It is obvious from Table IV that for macromolecular applications only operators of the type $I_{k,z}$ or $I_{k,z}I_{l,z}$ contribute.

[16] B. Reif, A. Diener, M. Hennig, M. Maurer, and C. Griesinger, *J. Magn. Reson.* **143**, 45 (2000).
[17] R. Brüschweiler, C. Griesinger, and R. R. Ernst, *J. Am. Chem. Soc.* **111**, 8034 (1989).

TABLE IV
DENSITY MATRICES YIELDING SECULAR CROSS-CORRELATED RELAXATION WITH TENSOR OPERATORS[a]

Tensor operators	$\sqrt{\frac{3}{8}}\hat{I}_m^+\hat{I}_n^+$	$\sqrt{\frac{3}{8}}\hat{I}_m^+\hat{I}_{n,z}$	$\hat{I}_{m,z}\hat{I}_{n,z}$	$\frac{1}{4}(\hat{I}_m^+\hat{I}_n^- + \hat{I}_m^-\hat{I}_n^+)$	$\hat{I}_{m,z}$	$\sqrt{\frac{3}{8}}\hat{I}_n^-$
$\sqrt{\frac{3}{8}}\hat{I}_k^-\hat{I}_l^-$	ac	—	—	—	—	—
$\sqrt{\frac{3}{8}}\hat{I}_k^-\hat{I}_{l,z}$	—	$\hat{I}_{k,\gamma}[\hat{I}_{l,z}]$ with $\hat{I}_k = \hat{I}_m$	—	—	—	$\hat{I}_{k,\gamma}[\hat{I}_{l}$ with $\hat{I}_k = \hat{I}$
$\hat{I}_{k,z}\hat{I}_{l,z}$	—	—	$\hat{I}_{k,\delta}[\hat{I}_{l,z}]\hat{I}_{m,\delta}[\hat{I}_{n,z}]$	$0 = \omega_m - \omega_n,\ \hat{I}_{m,\gamma}\hat{I}_{l,\delta}$	$\hat{I}_{k,\delta}[\hat{I}_{l,z}]\hat{I}_{m,\delta}$	—
$\frac{1}{4}(\hat{I}_k^+\hat{I}_l^- + \hat{I}_k^-\hat{I}_l^+)$	—	—		$[\hat{I}_{k,z}\hat{I}_{n,\gamma'}]$		—
			ns	ac	ns	
$\hat{I}_{k,z}$	—	—	$\hat{I}_{k,\delta}\hat{I}_{m,\delta}[\hat{I}_{n,z}]$	$0 = \omega_m - \omega_n,\ \hat{I}_{m,\gamma}\hat{I}_{l,\delta}$	$\hat{I}_{k,\delta}\hat{I}_{m,\delta}$	—
				$[I_{n,\gamma'}]$		
$\sqrt{\frac{3}{8}}\hat{I}_k^-$	—	$\hat{I}_{k,\gamma}[\hat{I}_{n,z}]$ with $\hat{I}_k = \hat{I}_m$	—	—	—	ac

[a] Chemical shifts of different spins have been assumed to be nondegenerate. ns, Nonsecular; ac, autocorrelat... relaxation only; $\gamma = x$ or y or z, $\gamma' = x$ or y; z only if $\gamma \neq z$. $\delta = x$ or y, [ ] means that this operator may be the... Any operators of spins not occurring in the tensor operators may be added.

When considering operators that are secular and contribute to the spectral density at zero frequency, only the following three cases remain.

2.5.1. Cross-Correlated Relaxation between Two Dipoles. The single quantum coherence of spin $\hat{I}_k$ that is scalar coupled to spins $\hat{I}_l$ and $\hat{I}_n$ will exhibit different line widths for the multiplet components. Using the commutator rules we find:

$$\left(\hat{I}_k^+\hat{I}_l^\alpha\hat{I}_n^\alpha\right)^\bullet = b_{kl}b_{kn}\left\{\left[\hat{I}_{k,z}\hat{I}_{l,z}, \left[\hat{I}_{k,z}\hat{I}_{n,z}, \hat{I}_k^+\hat{I}_l^\alpha\hat{I}_z^\alpha\right]\right]\right.$$
$$\left. + \left[\hat{I}_{k,z}\hat{I}_{n,z}, \left[\hat{I}_{k,z}\hat{I}_{l,z}, \hat{I}_k^+\hat{I}_l^\alpha\hat{I}_z^\alpha\right]\right]\right\}j_{kl,kn}^0(0)$$
$$= \frac{1}{2}b_{kl}b_{kn}\hat{I}_k^+\hat{I}_l^\alpha\hat{I}_n^\alpha j_{kl,kn}^0(0) \tag{16}$$

The sign changes with replacement of one α by β. It stays the same when replacing $\hat{I}_k^+$ by $\hat{I}_k^-$.

For double or zero quantum coherences of spins $\hat{I}_k$ and $\hat{I}_m$ one finds:

$$\left(\hat{I}_k^+\hat{I}_l^\alpha\hat{I}_m^+\hat{I}_n^\alpha\right)^\bullet = b_{kl}b_{mn}\left\{\left[\hat{I}_{k,z}\hat{I}_{l,z}, \left[\hat{I}_{m,z}\hat{I}_{n,z}, \hat{I}_k^+\hat{I}_l^\alpha\hat{I}_m^+\hat{I}_z^\alpha\right]\right]\right.$$
$$\left. + \left[\hat{I}_{m,z}\hat{I}_{n,z}, \left[\hat{I}_{k,z}\hat{I}_{l,z}, \hat{I}_k^+\hat{I}_l^\alpha\hat{I}_m^+\hat{I}_z^\alpha\right]\right]\right\}j_{kl,mn}^0(0)$$
$$= \frac{1}{2}b_{kl}b_{mn}\hat{I}_k^+\hat{I}_l^\alpha\hat{I}_m^+\hat{I}_n^\alpha j_{kl,mn}^0(0) \tag{17}$$

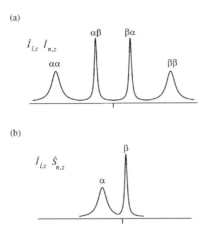

FIG. 1. (a) Schematic representation of the spectrum of single quantum or double or zero quantum coherences where the active spins are scalar coupled to $\hat{I}_l$ and $\hat{I}_n$, respectively. Spin states (α or β) for spins $\hat{I}_l$ and $\hat{I}_n$ are marked in the figure. The multiplet components show differential line broadening due to cross-correlated relaxation. (b) Schematic representation of the spectrum of the single quantum coherence of spin $\hat{I}_k$ scalar coupled to $\hat{I}_l$ and $\hat{S}_n$, respectively. The hyperfine coupling to the electron spin $\hat{S}_n$ is not resolved. The multiplet components show differential line broadening due to cross-correlated relaxation.

The sign changes with replacement of one $+$ by $-$ as well as with replacement of one α by β. The resulting multiplet is shown in Fig. 1a.

It should be noted that one of the spins, e.g., $\hat{S}_n$, might be an electron. In this case, the relaxation mechanism is called cross correlation between Curie relaxation and dipolar coupling. Because of the high gyromagnetic ratio of the electron we need to take into account that the populations of the α and β states are not equal. In fact, the average over $\langle\hat{S}_{n,z}\rangle$ is given by $\frac{\gamma_e\hbar(S+1)S}{6kT}$, thus yielding

$$
\begin{aligned}
\left(\hat{I}_k^+\hat{I}_l^\alpha\right)^\bullet &= b_{kl}b_{kn}\hat{I}_k^+\hat{I}_l^\alpha\hat{I}_{n,z}\,j_{kl,kn}^0(0)\\
&= \frac{\gamma_e\hbar(S+1)S}{6kT}b_{kl}b_{kn}\hat{I}_k^+\hat{I}_l^\alpha\,j_{kl,kn}^0(0)
\end{aligned}
\tag{18}
$$

in which S is the spin of the electron and γ_e is its gyromagnetic ratio. The resulting multiplet is shown in Fig. 1b.

2.5.2. *Cross-Correlated Relaxation between One Dipole and One CSA.* The commutators from Table II can be applied to obtain the differential line widths of multiplets of single or double and zero quantum coherences.

$$
\begin{aligned}
\left(\hat{I}_k^+\hat{I}_l^\alpha\right)^\bullet &= b_{kl}b_k\big\{\big[\hat{I}_{k,z}\hat{I}_{l,z},\big[\hat{I}_{k,z},\hat{I}_k^+\hat{I}_l^\alpha\big]\big]+\big[\hat{I}_{k,z},\big[\hat{I}_{k,z}\hat{I}_{l,z},\hat{I}_k^+\hat{I}_l^\alpha\big]\big]\big\}j_{kl,k}^0(0)\\
&= b_{kl}b_{kn}\hat{I}_k^+\hat{I}_l^\alpha\,j_{kl,k}^0(0)
\end{aligned}
\tag{19}
$$

The sign changes with replacement of one α by β.

$$\left(\hat{I}_k^+ \hat{I}_l^\alpha \hat{I}_m^+\right)^{\cdot} = b_{kl}b_{mn}\{[\hat{I}_{k,z}\hat{I}_{l,z}, [\hat{I}_{m,z}, \hat{I}_k^+ \hat{I}_l^\alpha \hat{I}_m^+]]$$
$$+ [\hat{I}_{m,z}, [\hat{I}_{k,z}\hat{I}_{l,z}, \hat{I}_k^+ \hat{I}_l^\alpha \hat{I}_m^+]]\} j_{kl,m}^0(0) = b_{kl}b_m \hat{I}_k^+ \hat{I}_l^\alpha \hat{I}_m^+ j_{kl,m}^0(0)$$
$$(20)$$

The sign changes with replacement of one $+$ by $-$ as well as α by β. The resulting multiplet is shown for both cases in Fig. 1b.

2.5.3. Cross-Correlated Relaxation between Two CSA Interactions. The commutators from Table II yield in this case a uniform broadening of all lines of the multiplet.

$$\left(\hat{I}_k^+ \hat{I}_m^+\right)^{\cdot} = b_k b_m \{[\hat{I}_{k,z}, [\hat{I}_{m,z}, \hat{I}_k^+ \hat{I}_m^+]] + [\hat{I}_{m,z}, [\hat{I}_{k,z}, \hat{I}_k^+ \hat{I}_m^+]]\} j_{k,m}^0(0)$$
$$= 2b_k b_m \hat{I}_k^+ \hat{I}_m^+ j_{k,m}^0(0) \qquad (21)$$

The sign changes with replacement of one $+$ by $-$. It should be noted that the same equal line broadening is observed for cross-correlated relaxation of $\hat{I}_k$, $\hat{I}_m$ double quantum coherence because of their dipolar interactions to one spin $\hat{I}_l = \hat{I}_n$. The formula reads:

$$\left(\hat{I}_k^+ \hat{I}_m^+\right)^{\cdot}$$
$$= b_{kl}b_{ml}\{[\hat{I}_{k,z}\hat{I}_{l,z}, [\hat{I}_{m,z}\hat{I}_{l,z}, \hat{I}_k^+ \hat{I}_m^+]] + [\hat{I}_{m,z}\hat{I}_{l,z}, [\hat{I}_{k,z}\hat{I}_{l,z}, \hat{I}_k^+ \hat{I}_m^+]]\} j_{kl,mn}^0(0)$$
$$= \frac{1}{2}b_{kl}b_{mn}\hat{I}_k^+ \hat{I}_m^+ j_{kl,ml}^0(0) \qquad (22)$$

3. Experimental Procedures

3.1. *Measurement of Cross-Correlated Relaxation in Evolution Periods*

Cross-correlated relaxation rates can be extracted from the analysis of differential line intensities in experiments, in which SQ, DQ, or ZQ coherence have evolved in a *constant time* manner. The intensities and the integrals directly reflect the

TABLE V

SIZE OF INTERACTIONS RELEVANT FOR
CROSS-CORRELATED RELAXATION

	$b_{kl} = -\mu_0 \frac{\gamma_k \gamma_l \hbar}{4\pi r_{kl}^3}$
CH (1.10 Å)	2.0391×10^4
NH (1.04 Å)	1.0825×10^4
H,H (1.78 Å)	2.1298×10^4
C,C (1.55 Å)	0.2040×10^4
H,e (10 Å)	0.310×10^4

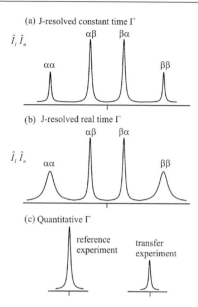

FIG. 2. Categorization of different experimental approaches to quantify cross-correlated relaxation.

cross-correlated relaxation rates. For their calculation, the size of the relevant interactions are summarized in Table V. We call this way of measuring cross-correlated relaxation rates the J-resolved constant time Γ experiment, since the multiplet is resolved because of the J couplings to the coupling spins I_l and I_n, but the cross-correlated relaxation evolves during a constant time. Each line intensity and integral is affected by the cross-correlated relaxation rate as shown in Fig. 2a and expressed in Eqs. (16) and (17). This measurement scheme was first introduced in Ref. 1.

If the cross-correlated relaxation rate evolves during a *real time* evolution, line integrals and intensities do not directly reflect the relaxation rates. Instead, relaxation rates have to be deconvoluted from a potentially complicated line shape that is often not known since it includes, e.g., small long-range coupling constants. However, deconvolution procedures have been developed that account for different and potentially unknown line shapes and are described in Refs. 16 and 18. This experiment is the J-resolved real time Γ experiment. The corresponding multiplets and line shapes obtained are depicted in Fig. 2b.

3.2. Measurement of Cross-Correlated Relaxation in Nonevolution Periods

Cross-correlated relaxation can transfer a given operator A uniquely to a new operator B that could not be created because of other interactions in the sequence.

[18] B. Reif, H. Steinhagen, B. Junker, M. Reggelin, and C. Griesinger, *Angew. Chem.* **110,** 2006 (1998); *Angew. Chem. Int. Ed. Engl.* **37,** 1903 (1998).

Normally, this operator transformation takes place in a constant time period.

$$
\begin{aligned}
\left(\hat{I}_k^+ \hat{I}_l^\alpha \hat{I}_m^+ \hat{I}_n^\alpha\right)' &= b_{kl}b_{mn}\Big\{\big[\hat{I}_{k,z}\hat{I}_{l,z}, \big[\hat{I}_{m,z}\hat{I}_{n,z}, \hat{I}_k^+ \hat{I}_l^\alpha \hat{I}_m^+ \hat{I}_z^\alpha\big]\big] \\
&\quad + \big[\hat{I}_{m,z}\hat{I}_{n,z}, \big[\hat{I}_{k,z}\hat{I}_{l,z}, \hat{I}_k^+ \hat{I}_l^\alpha \hat{I}_m^+ \hat{I}_z^\alpha\big]\big]\Big\} j_{kl,mn}^0(0) \\
&= \frac{1}{2} b_{kl}b_{mn}\hat{I}_k^+ \hat{I}_l^\alpha \hat{I}_m^+ \hat{I}_n^\alpha j_{kl,mn}^0(0)
\end{aligned}
$$

For example, the operator $2\hat{I}_{k,x}\hat{I}_{l,z}\hat{I}_{m,x}$ can be understood as the following linear combination:

$$
2\hat{I}_{k,x}\hat{I}_{l,z}\hat{I}_{m,y} = \frac{1}{4i}(\hat{I}_k^+ \hat{I}_m^+ - \hat{I}_k^- \hat{I}_m^- - \hat{I}_k^+ \hat{I}_m^- + \hat{I}_k^- \hat{I}_m^+)(\hat{I}_l^\alpha - \hat{I}_l^\beta)(\hat{I}_n^\alpha + \hat{I}_n^\beta)
$$

This operator is transferred into:

$$
2\hat{I}_{k,y}\hat{I}_{m,x}\hat{I}_{n,z} = \frac{1}{4i}(\hat{I}_k^+ \hat{I}_m^+ - \hat{I}_k^- \hat{I}_m^- + \hat{I}_k^+ \hat{I}_m^- - \hat{I}_k^- \hat{I}_m^+)(\hat{I}_l^\alpha + \hat{I}_l^\beta)(\hat{I}_n^\alpha - \hat{I}_n^\beta)
$$

by dipole–dipole cross-correlated relaxation. The transfer efficiency can be taken from Eq. (17) by noting that each change of $+$ to $-$ or α to β inverts the sign of the cross-correlated relaxation rate: $\frac{1}{2}b_{kl}b_{mn}j_{kl,mn}^0(0)$. This means that the first term of the start operator: $\frac{1}{4i}\hat{I}_k^+ \hat{I}_m^+ \hat{I}_l^\alpha \hat{I}_n^\alpha$ is transformed into the first operator of the target operator $\frac{1}{4i}\hat{I}_k^+ \hat{I}_m^+ \hat{I}_l^\alpha \hat{I}_n^\alpha$ with the attenuation of $e^{-\frac{1}{2}b_{kl}b_{mn}j_{kl,mn}^0(0)T}$. Another component of the start operator, $-\frac{1}{4i}\hat{I}_k^+ \hat{I}_m^+ \hat{I}_l^\beta \hat{I}_n^\alpha$, is transformed into this operator in the target operator: $\frac{1}{4i}\hat{I}_k^+ \hat{I}_m^+ \hat{I}_l^\beta \hat{I}_n^\alpha$ with $-e^{+\frac{1}{2}b_{kl}b_{mn}j_{kl,mn}^0(0)T}$. Looking at all 16 start and target operators, it is obvious that there are eight terms $e^{-\frac{1}{2}b_{kl}b_{mn}j_{kl,mn}^0(0)T}$ and eight terms $-e^{+\frac{1}{2}b_{kl}b_{mn}j_{kl,mn}^0(0)T}$, amounting to a total transfer amplitude from $2\hat{I}_{k,x}\hat{I}_{l,z}\hat{I}_{m,y}$ to $2\hat{I}_{k,y}\hat{I}_{m,x}\hat{I}_{n,z}$ of $\frac{1}{2}(e^{-\frac{1}{2}b_{kl}b_{mn}j_{kl,mn}^0(0)T} - e^{+\frac{1}{2}b_{kl}b_{mn}j_{kl,mn}^0(0)}T) = -\sinh(\frac{1}{2}b_{kl}b_{mn}j_{kl,mn}^0(0)T)$. The intensity of the signal derived from the new operator depends in a quantitative way on the size of the cross-correlated relaxation rate. By application of sets of refocusing pulses, specific cross-correlated relaxation pathways can be suppressed as discussed in Section 3.3. We call this type of experiment a quantitative Γ experiment. This scheme depicted in Fig. 2c was first introduced for the measurement of HCCH[19] and HCNH cross-correlated relaxation rates.[20]

3.3. Evolution of Cross-Correlated Relaxation under Influence of π Pulses

For the design of pulse sequences that measure cross-correlated relaxation rates, pulse sequence elements affect the evolution of coherence under cross-correlated relaxation in a certain predictable way.[21] This is different from autocorrelated relaxation for which pulse sequence elements normally do not affect

[19] I. C. Felli, C. Richter, C. Griesinger, and H. Schwalbe, *J. Am. Chem. Soc.* **121**, 1956 (1999).
[20] P. Pelupessy, E. Chiarparin, R. Ghose, and G. Bodenhausen, *J. Biomol. NMR* **13**, 375 (1999).
[21] M. H. Levitt and L. Di Bari, *Bull. Magn. Reson.* **16**, 94 (1994).

relaxation. This short section introduces some rules to calculate an effective Liouvillian in a given pulse sequence.

π pulses allow us to manipulate the evolution of heteronuclear cross-correlated relaxation in a similar way as they allow us to manipulate the evolution of couplings and chemical shifts in heteronuclear spin systems. The time evolution of the density matrix under the Hamilton operator and the relaxation superoperator is given by

$$\hat{\sigma}^{\bullet} = i[\hat{H}, \hat{\sigma}] - \hat{\hat{\Gamma}}(\hat{\sigma} - \hat{\sigma}_0) \tag{23}$$

Ignoring the inhomogeneous part, assuming weak coupling and the secular approximation for the relaxation superoperator, we can treat the evolution under $\hat{\hat{\Gamma}}$ independently for each eigencoherence of the Hamiltonian $\hat{H}$:

$$\hat{\sigma}^{\bullet} = -\hat{\hat{\Gamma}}\hat{\sigma} \tag{24}$$

Incorporation of a π pulse in the pulse sequence leads to the equation of motion

$$(\hat{\hat{P}}\hat{\sigma})^{\bullet} = -\hat{\hat{\Gamma}}(\hat{\hat{P}}\hat{\sigma}) \tag{25}$$

π pulses have the property that for many relaxation mechanisms either the commutator or the anticommutator vanishes:

$$[\hat{\hat{\Gamma}}, \hat{\hat{P}}] = \hat{\hat{\Gamma}}\hat{\hat{P}} - \hat{\hat{P}}\hat{\hat{\Gamma}} = 0 \quad \text{or} \quad [\hat{\hat{\Gamma}}, \hat{\hat{P}}]_+ = \hat{\hat{\Gamma}}\hat{\hat{P}} + \hat{\hat{P}}\hat{\hat{\Gamma}} = 0 \tag{26}$$

In the first case, evolution of coherence under a certain relaxation is not affected by the π pulse:

$$\hat{\hat{P}}\hat{\sigma}^{\bullet} = (\hat{\hat{P}}\hat{\sigma})^{\bullet} = -\hat{\hat{\Gamma}}\hat{\hat{P}}\hat{\sigma} = -\hat{\hat{P}}\hat{\hat{\Gamma}}\hat{\sigma} \tag{27}$$

In the second case, the sign of the evolution is changed:

$$\hat{\hat{P}}\hat{\sigma}^{\bullet} = (\hat{\hat{P}}\hat{\sigma})^{\bullet} = -\hat{\hat{\Gamma}}\hat{\hat{P}}\hat{\sigma} = +\hat{\hat{P}}\hat{\hat{\Gamma}}\hat{\sigma} \tag{28}$$

In the second case, a π pulse applied in the middle of two equal delays leads to refocusing of the effect of the relaxation mechanism. Since relaxation superoperators are derived from double commutators, we present some simple rules that clearly show the effect of a π pulse on a given relaxation rate. We assume we have two operators $\hat{A}$, $\hat{B}$ whose cross correlation is a source of relaxation. The relaxation superoperator is then given by

$$\hat{\hat{\Gamma}} = [\hat{A}, [\hat{B},]] + [\hat{A}^{\dagger}, [\hat{B}^{\dagger},]] \tag{29}$$

A π pulse on any coherence σ can be written as

$$\hat{\hat{P}}\hat{\sigma} = \hat{P}\hat{\sigma}\hat{P}^{\dagger} \tag{30}$$

where $\hat{\hat{P}}$ is the superoperator affecting a π-rotation and $\hat{P}$ is the corresponding operator.

We assume that the two operators $\hat{A}$ and $\hat{B}$ are affected by the π pulse in the following way:

$$\hat{\hat{P}}\hat{A} = s_A\hat{A}, \quad \hat{\hat{P}}\hat{B} = s_B\hat{B}$$

or

$$\hat{\hat{P}}\hat{A} = s_A\hat{A}^\dagger, \quad \hat{\hat{P}}\hat{B} = s_B\hat{B}^\dagger \tag{31}$$

with $s_A, s_B = \pm 1$. These equations are true for all operators of Table III except for the case discussed below. Then it follows:

$$\hat{\hat{\Gamma}}\hat{\hat{P}}\hat{\sigma} = s_A s_B \hat{\hat{P}}\hat{\hat{\Gamma}}\hat{\sigma} \tag{32}$$

Thus, if $s_A s_B = 1$ we have $[\hat{\hat{\Gamma}}, \hat{\hat{P}}] = 0$; if $s_A s_B = -1$ we have $[\hat{\hat{\Gamma}}, \hat{\hat{P}}]_+ = 0$.

Thus, it is sufficient to know whether the π pulse inverts $\hat{A}(s_A = -1)$ or not ($s_A = 1$). The same holds for $\hat{B}$. We can pictorially represent the inversion of an interaction as depicted in Fig. 3 by drawing horizontal lines, the upper representing positive evolution of the operators $\hat{A}$ or $\hat{B}$ and the lower if it has been inverted. For a given cross-correlated relaxation rate, the sequence is segmented in the periods

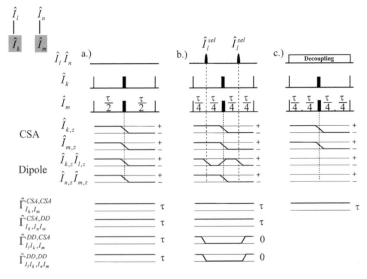

FIG. 3. Pictorial presentation of how application of π-pulses and decoupling sequences can affect the evolution of cross-correlated relaxation rates. The evolution of, e.g., $\Gamma_{I_l I_k, I_m}^{DD,CSA}$ cross-correlated relaxation is given by the product of the transformation properties of the individual tensors operators $\hat{I}_{k,z}\hat{I}_{l,z}$ and $\hat{I}_{m,z}$ under application of π-pulses. Similarly to scalar heteronuclear coupling, heteronuclear dipolar coupling does not evolve under decoupling of transitions of spins $\hat{I}_l$ and $\hat{I}_n$ in (c).

in which there is no sign change either for $\hat{A}$ or for $\hat{B}$. Then the weighted sum over the times is formed with the weights being $s_A s_B$ for each time segment.

In the autocorrelated case $s_A = s_B$ and therefore $s_A s_B = 1$. Therefore, autocorrelated relaxation is not affected by the application of a π-pulse and cannot be inverted by it. For the cross-correlated case, we consider here two examples:

(a) Dipole–dipole cross-correlated relaxation:

$$\hat{\hat{\Gamma}} = \left[H_z^N N_z, \left[H_z^C C_z, \right] \right] \tag{33}$$

Application of a $\pi(^{15}N)$-pulse or a $\pi(^{13}C)$-pulse leads to $s_A s_B = -1$ and therefore refocusing of the cross-correlated relaxation. A $\pi(^1H)$-pulse, however, does not affect the relaxation since $s_A s_B = +1$.

(b) CSA–dipole cross-correlated relaxation:

$$\hat{\hat{\Gamma}} = \left[\sqrt{\tfrac{3}{8}} N^+, \left[\sqrt{\tfrac{3}{8}} H_z^N N^-, \right] \right] + \left[\sqrt{\tfrac{3}{8}} N^-, \left[\sqrt{\tfrac{3}{8}} H_z^N N^+, \right] \right] + \left[N_z, \left[H_z^N N_z, \right] \right] \tag{34}$$

Application of a $\pi(^{15}N)$-pulse or a $\pi(^{13}C)$-pulse leads to $s_A s_B = +1$:

$$\hat{\hat{P}} N^- = N^{-\dagger} = N^+$$

whereas a $\pi(^1H)$-pulse leads to refocusing of the interaction ($s_A s_B = -1$).

Figure 3 depicts the evolution of each of the Hamiltonians of interest during the period τ.

4. Applications: Structural Information in Proteins

Cross-correlated relaxation has been used in several applications that we will discuss in the Sections 4 to 7. The applications are summarized in Table VI, in which the respective references and a short description of the experiments are given. In proteins, experiments that investigate the cross-correlated relaxation mechanism have been developed to determine structural information and to determine dynamic information. We will first concentrate on the structural aspects of cross-correlated relaxation. A number of experiments discussed in the following have been developed to determine the backbone angle ψ from both $\Gamma_{N_i H_i^N, C_{i-1}^\alpha H_{i-1}^\alpha}^{DD,DD}$ and $\Gamma_{C_{i-1}^\alpha H_{i-1}^\alpha, C_{i-1}'}^{DD,CSA}$ cross-correlated relaxation rates. Their dependence on ψ is shown in Figs. 4 and 5. We will discuss the original sequence proposed in Ref. 1 and improvements developed thereafter.

TABLE VI

REPORTS EXPLOITING CROSS-CORRELATED RELAXATION IN H–X–Y–H AND H–X–Y MOIETIES

Molecules, information	Type of experiment	Cross-correlated relaxation rate	Ref.

Proteins

DD, DD   **DD, CSA**

$$\mathrm{H}_{\alpha,i-1} \quad \mathrm{H}_i^N \quad \mathrm{H}_{\alpha,i-1} \qquad \mathrm{H}_{\alpha,i-1}$$

Conformation, angle ψ

$$\mathrm{C}_{\alpha,i-1}\!-\!\mathrm{N}_i \quad \mathrm{C}_{\alpha,i-1}\!-\!\mathrm{N}_i \quad \mathrm{C}_{\alpha,i-1}\!=\!\mathrm{C}_i$$

	Type of experiment	Cross-correlated relaxation rate	Ref.
a	CT-DQZQ-HN(CO)CA N,C_α multiquantum	$\Gamma^{DD,DD}_{NH^N,C_aH_a}$	1,16
b	CT-DQZQ-HN(CO)CA C', C_α multiquantum	$\Gamma^{DD,CSA}_{C_a,H_a,C'}$	35,36
c	CT-DQZQ HN(CO)CA N,C_α multiquantum	$\Gamma^{DD,DD}_{NH^N,C_aH_a}$ $\Gamma^{DD,CSA}_{C_aH_a,C'}$	34
d	Quantitative CTΓ-NH(CO)CA N, C_α multiquantum	$\Gamma^{DD,DD}_{NH^N,C_aH_a}$ $\Gamma^{DD,CSA}_{NH^N,C'}$	20,69
e	quant. Γ-HNCA N, C_α multiquantum	$\Gamma^{DD,DD}_{NH^N,C_aH_a}$	38
f	Experiment described in Refs. 35, 36 and (b) together with $^3\Delta\delta(C_\alpha(N^H, N^D))$	$\Gamma^{DD,CSA}_{C_aH_a,C'}$	27

Unfolded proteins

| g | Experiment described in Refs. 19 and (d) | $\Gamma^{DD,DD}_{NH^N,C_aH_a}$ | 70 |

DD, DD

Conformation, angle ϕ

$$\mathrm{H}_i^N \quad \mathrm{H}_{\alpha,i} \qquad \mathrm{H}_i^N\!-\!\mathrm{H}_{\alpha,i}$$
$$\mathrm{N}_i\!-\!\mathrm{C}_{\alpha,i} \qquad \mathrm{N}_i$$

| h | RT-SQ H^N | $\Gamma^{DD,DD}_{NH^N,H^NH_a}$ | 40 |
| i | Quantitative Γ-HNCA N, C_α multiquantum | $\Gamma^{DD,DD}_{NH^N,CH_\alpha}$ | 38 |

DD, DD

Conformation, angles ϕ and Ψ

$$\mathrm{H}_{\alpha,i-1} \qquad \mathrm{H}_{\alpha,i}$$
$$\mathrm{C}_{\alpha,i-1}\!-\!\mathrm{C}_{\alpha,i}$$

| j | Quantitative Γ-HNCA | $\hat{\Gamma}^{DD,DD}_{C^\alpha_{i-1}H^\alpha_{i-1},C^\alpha_i H^\alpha_i}$ | 39 |

TABLE VI (*continued*)

Molecules, information	Type of experiment	Cross-correlated relaxation rate	Ref.
Dynamics, peptide plane			
k	CT-DQZQ HNCO N,C′ multiquantum	$\Gamma^{DD,CSA}_{NH^N,C'}$	15
l	CT-DQZQ HNCO	$\Gamma^{DD,DD}_{C'C_a,NH^N} + \Gamma^{DD,DD}_{C'H^N,NC_a}$ $\Gamma^{CSA,DD}_{N,NC_a} + \Gamma^{CSA,DD}_{C',C'C_a}$ $\Gamma^{CSA,DD}_{C',NH^N} + \Gamma^{CSA,DD}_{N,C'H^N}$	58
m	CT-DQZQ HNCO	$\Gamma^{CSA,DD}_{C',C',C_\alpha}$ $\Gamma^{CSA,DD}_{C',C',N}$ $\Gamma^{CSA,DD}_{C',C'N^H}$	60
n	CT-HNCO C′ singlequantum	$\Gamma^{CSA,DD}_{C',H^NC'}$	71
o	CT-HNCO N,C′ multiquantum	$\Gamma^{CSA,CSA}_{C',N}$ $\Gamma^{DD,CSA}_{NH^N,C'}$	56
Dynamics, side chain angle χ_1 in unfolded proteins			
p	Quantitative Γ-HCCH COSY C,C double quantum coherence	$\Gamma^{DD,DD}_{C_iH_i,C_{l+1}H_{i+1}}$	41
Protein, long-range order			
q	H,N-HSQC H single quantum coherence	$\Gamma^{Curie,DD}_{e^-,NH^N}$	72

DD, DD

$$\underset{C'_n}{\overset{H'_n}{|}} \quad \underset{C'_{n+1}}{\overset{H'_{n+1}}{|}}$$

$$C'_n —— C'_{n+1}$$

RNA

Conformation, P			
r	Quantitative Γ-HCCH COSY C,C double quantum coherence	$\Gamma^{DD,DD}_{C_iH_i,C_{i+1}H_{i+1}}$	19
s	Forward directed quantitative Γ-HCCH-TOCSY C,C double quantum coherence	$\Gamma^{DD,DD}_{C_iH_i,C_{i+1}H_{i+1}}$	46
t	CT-HSQC	$\Gamma^{CSA,DD}_{C1',C1'H1'} + \Gamma^{CSA,DD}_{C3',C3'H3'}$	47

(*continued*)

TABLE VI (continued)

Molecules, information	Type of experiment	Cross-correlated relaxation rate	Ref.
	DD, CSA		
Conformation, ζ	$\begin{array}{ccc} \text{H3}'_i & \text{H4}'_i & \text{H2}'_i \\ \vert & \vert & \vert \\ \text{C3}'_i\text{—P}_{i+1} & \text{C4}'_i\text{—P}_{i+1} & \text{C2}'_i\text{—P}_{i+1} \end{array}$		
	DD, CSA		
Conformation, α	$\begin{array}{cc} \text{H5}'_i & \text{H4}'_i \\ \vert & \vert \\ \text{C5}'_i\text{—P}_i & \text{C4}'_i\text{—P}_i \end{array}$		
u	CT-DQZQ-HCP	$\Gamma^{DD,CSA}_{CH,P}$	48
Carbohydrate Conformation			
v		$\Gamma^{DD,DD}_{C_iH_i,C_{i+1}H_{i+1}}$ $\Gamma^{DD,DD}_{C_iH_i,C_{i+2}H_{i+2}}$	49
Transferred cross correlation Conformation of bound ligands			
w	Experiment described in (e) Determination of conformation of bound ligand in fast exchange	$\Gamma^{DD,DD}_{C_iH_i,C_{i+1}H_{i+1}}$	61
x	Experiment described in (c) Determination of conformation of bound ligand in fast exchange	$\Gamma^{DD,DD}_{NH^N,C_aH_a}$	62

4.1. Backbone Angle ψ

The protein backbone angle ψ is difficult to determine by liquid state NMR spectroscopy. The scalar coupling constants $^3J(H^N_{k+1}, C^\alpha_k)$[22] and $^3J(N_{k+1}, N_k)$ coupling[23] as well as distance measurements between the protons H^α_k and H^N_{k+1}[24]

[22] A. C. Wang and A. Bax, J. Am. Chem. Soc. **117,** 1810 (1995).
[23] F. Löhr and H. Rüterjans, J. Magn. Reson. **132,** 130 (1998); K. Theis, A. J. Dingley, A. Hoffmann, J. G. Omichinski, and S. Grzesiek, J. Biomol. NMR **10,** 403 (1997).
[24] K. Wüthrich, "NMR of Proteins and Nucleic Acids." Wiley, New York, 1986.

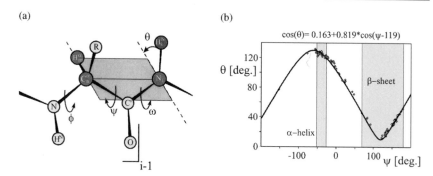

FIG. 4. (a) Schematic representation of the peptide backbone with backbone angles ϕ, ψ, ω, and θ of amino acids $i-1$ and i in a protein. Assuming planar geometry, the measured angle θ can be correlated with the backbone angle ψ. The plane, which is spanned by the atoms H^{α}_{i-1}, C^{α}_{k-1}, and N_i, is highlighted graphically. (b) Correlation of the angles θ and ψ in the protein backbone. Values of pairs of θ and ψ found in the protein rhodniin are depicted as diamonds. The following bond lengths and bond angles have been assumed to derive $\cos(\theta) = 0.163 + 0.819 \times \cos(\psi - 119°)$: $NH^N = 1.03$ Å, $C^{\alpha}H^{\alpha} = 1.09$ Å, $NC^{\alpha} = 1.47$ Å, $NC' = 1.33$ Å, and $C'C^{\alpha} = 1.52$ Å. Tetrahedral symmetry and planarity of the peptide backbone have also been assumed.

are often too inaccurate to restrain the angle accurately. Measurement of $^3J(C^{\alpha}_{k+1}$, $C^{\alpha}_{k})$ provides a semiquantitative measure of ψ.[25] In a different approach, the relative displacement of the ^{1}H and ^{15}N resonance frequencies in $1:1$ mixtures of D_2O and H_2O have been measured.[26,27] This displacement is caused by the three-bond NH/ND isotope effect on the C^{α} chemical shifts and is a function of the backbone torsion angle ψ.

From the measurement of dipole–dipole cross-correlated relaxation $\Gamma^{DD,DD}_{N_i H^N_i, C^{\alpha}_{i-1} H^{\alpha}_{i-1}}$ and several $\Gamma^{CSA,DD}$ of double $\hat{N}^{\pm}_i \hat{C}^{\pm}_{\alpha,i-1}$ (DQ) and zero $\hat{N}^{\pm}_i \hat{C}^{\mp}_{\alpha,i-1}$ quantum (ZQ) coherences between the nitrogen and the neighboring C^{α} atoms, the backbone angle ψ can be determined.

Correlation of the projection angle θ between the two dipole tensors and the torsion angle ψ, shown in Fig. 4b, reveals that regions of different secondary structure elements such as α helices and β sheets can be differentiated. For an isotropic overall reorientation of the protein, the cross-correlated relaxation rate $\Gamma^{DD,DD}_{N_i H^N_i, C^{\alpha}_{i-1} H^{\alpha}_{i-1}}$ shows a $3\cos^2\theta - 1$ dependence on the projection angle θ (Fig. 5). The dependence of the $\Gamma^{DD,CSA}_{C^{\alpha}_{i-1} H^{\alpha}_{i-1}, C'_{i-1}}$ rate on ψ is also given in Fig. 5.

[25] M. Hennig, W. Bermel, H. Schwalbe, and C. Griesinger, *J. Am. Chem. Soc.* **122**, 6268 (2000).
[26] M. Ottiger and A. Bax, *J. Am. Chem. Soc.* **119**, 8070 (1997).
[27] R. Sprangers, M. J. Bottomley, J. P. Linge, J. Schultz, M. Nilges, and M. Sattler, *J. Biomol. NMR* **16**, 47 (2000).

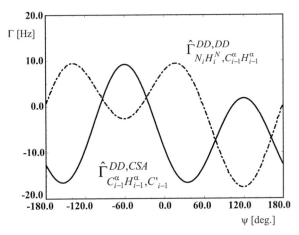

FIG. 5. Calculated cross-correlated relaxation rates $\Gamma^{DD,DD}_{N_iH_i^N,C_{i-1}^\alpha H_{i-1}^\alpha}$ and $\Gamma^{DD,CSA}_{C_{i-1}^\alpha H_{i-1}^\alpha,C'_{i-1}}$ as a function of the backbone angle ψ. The following formulae have been used:

$$\Gamma^{DD,DD}_{N_iH_i^N,C_{i-1}^\alpha H_{i-1}^\alpha} = \frac{1}{5}\gamma_H^2\gamma_N\gamma_C\left(\frac{\mu_0}{4\pi}\right)^2 \hbar^2 r_{CH}^{-3} r_{NH}^{-3}(3\cos^2\theta - 1)$$

with $\theta = 0.163 + 0.819\cos(\psi - 119°)$;

$$\Gamma^{DD,CSA}_{C_{i-1}^\alpha H_{i-1}^\alpha,C_{i-1}} = \frac{4}{15}\gamma_H\gamma_C\gamma_C B_0\left(\frac{\mu_0}{4\pi}\right)\hbar r_{CH}^{-3}\frac{1}{2}(\sigma_x(3\cos^2\theta_x - 1) + \sigma_y(3\cos^2\theta_y - 1)$$

$$+ \sigma_z(3\cos^2\theta_z - 1))\text{ with }\cos\theta_x = -0.3095 + 0.3531\cos(\psi - 120°);$$

$$\cos\theta_y = -0.1250 - 0.8740\cos(\psi - 120°); \cos\theta_z = -0.9426\sin(\psi - 120°).$$

In these equations; the following natural constant were assumed:

$$\gamma_H = 26.75 \times 10^7[sA/kg]; \gamma_N = 2.75 \times 10^7[sA/kg]; \gamma_C = 6.73 \times 10^7[sA/kg];$$

$$\hbar = 1.05 \times 10^{-34}[Js]; \mu_0 = 10^{-7}[NA^{-2}]; r_{CH} = 1.09 \times 10^{-10}[m];$$

$$r_{NH} = 1.02 \times 10^{-10}[m]; \omega_c = 2 \cdot \pi \times 150 \times 10^6[Hz];$$

τ_c was assumed to be 4.6 ns.

4.1.1. J-Resolved Constant Time Γ Experiment to Measure $\Gamma^{DD,DD}_{N_iH_i^N,C_{i-1}^\alpha H_{i-1}^\alpha}$. The

pulse sequence, which is shown in Fig. 6, is derived from an HN(CO)CA experiment.[28] DQ/ZQ coherences between the nitrogen and the neighboring C^α atom are excited by two simultaneous 90° pulses on ^{15}N and $^{13}C^\alpha$ for a constant time τ'' and evolve during t_2.

Application of the pictographical analysis introduced in Section 3.3 and in Fig. 3 shows that CSA–dipole and dipole–dipole cross-correlated relaxation rates evolve during τ''. Experimental data of the DQ/ZQ HN(CO)CA experiment is

[28] A. Bax and M. Ikura, *J. Biomol. NMR* **1**, 99 (1991).

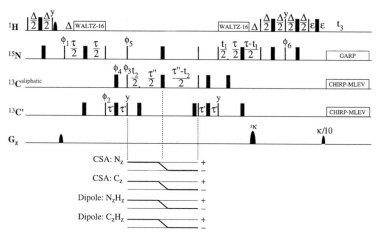

FIG. 6. Pulse sequence derived from an HN(CO)CA for the measurement of N–H^N, C$^\alpha$–H$^\alpha$ projection angles. DQ and ZQ coherences evolve chemical shift during t_1. Delays are as follows: $\Delta = 5\,$ms, $\tau = 35\,$ms, $2\tau' = 9\,$ms, $\tau'' = 26\,$ms, $\varepsilon = 1.2\,$ms. G3 and G4 pulses[63] have been used as selective 90° and 180° pulses on C$^\alpha$ and C' resonances. Default phase is x. $\phi_1 = x, -x; \phi_2 = 2(x), 2(-x); \phi_3 = 4(x), 4(-x); \phi_5 = 8(x), 8(-x); \phi_{rec} = \phi_1 + \phi_2 + \phi_3 + \phi_5$. Quadrature in t_2 is achieved by variation of phases ϕ_3 and ϕ_4 in the STATES-TPPI manner. Echo–antiecho coherences are selected during t_1 by inversion of phases $\phi_6 = -y$ together with the sign of the second gradient.[64] The phases ϕ_3 and ϕ_5 are shifted by 90° in subsequent FIDS and stored independently to separate DQ and ZQ coherences evolving during t_2. Aliphatic and carbonyl resonances are decoupled during acquisition using MLEV expanded CHIRP pulses.[65]

shown in Fig. 7 for the thrombin inhibitor protein rhodniin consisting of 103 amino acids. For K96, the outer two multiplet lines are weak and the inner are strong; for S90, all four multiplet lines have similar intensities; and for L40, the inner two lines are weak and the outer strong. In addition, each line of the multiplet has a different intensity, which stems from different CSA contributions of the C$^\alpha$ and N spins.

The cross-correlated relaxation rates $\Gamma^{DD,DD}_{N_i H^N_i, C^\alpha_{i-1} H^\alpha_{i-1}}$ can be extracted from the intensities of the lines $\alpha\alpha$, $\alpha\beta$, $\beta\alpha$, and $\beta\beta$ with the procedure described above from the line intensities in the multiplet. The line intensities of the four multiplet components of the N,C$^\alpha$ double quantum coherence are affected in the following way:

$$\Gamma^{DQ}_{\alpha\alpha} = +\Gamma^a + \Gamma^c_{NH,CH} + \Gamma^c_{N,NH} + \Gamma^c_{C,NH} + \Gamma^c_{N,CH} + \Gamma^c_{C,CH} + W_2 + \Gamma_1$$

$$\Gamma^{DQ}_{\alpha\beta} = +\Gamma^a - \Gamma^c_{NH,CH} - \Gamma^c_{N,NH} - \Gamma^c_{C,NH} + \Gamma^c_{N,CH} + \Gamma^c_{C,CH} + W_0 + \Gamma_1$$

$$\Gamma^{DQ}_{\beta\alpha} = +\Gamma^a - \Gamma^c_{NH,CH} + \Gamma^c_{N,NH} + \Gamma^c_{C,NH} - \Gamma^c_{N,CH} - \Gamma^c_{C,CH} + W_0 + \Gamma_1$$

$$\Gamma^{DQ}_{\beta\beta} = +\Gamma^a + \Gamma^c_{NH,CH} - \Gamma^c_{N,NH} - \Gamma^c_{C,NH} - \Gamma^c_{N,CH} - \Gamma^c_{C,CH} + W_2 + \Gamma_1$$

$$(35a)$$

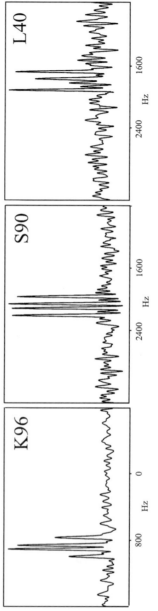

FIG. 7. 1D strips from the 3D DQ/ZQ HN(CO)CA for the residues K96, S90, and L40 in rhodniin.

and the ZQ spectrum

$$\Gamma_{\alpha\alpha}^{ZQ} = +\Gamma^a - \Gamma_{NH,CH}^c + \Gamma_{N,NH}^c - \Gamma_{C,NH}^c - \Gamma_{N,CH}^c + \Gamma_{C,CH}^c + W_0 + \Gamma_1$$
$$\Gamma_{\alpha\beta}^{ZQ} = +\Gamma^a + \Gamma_{NH,CH}^c - \Gamma_{N,NH}^c + \Gamma_{C,NH}^c - \Gamma_{N,CH}^c + \Gamma_{C,CH}^c + W_2 + \Gamma_1$$
$$\Gamma_{\beta\alpha}^{ZQ} = +\Gamma^a + \Gamma_{NH,CH}^c + \Gamma_{N,NH}^c - \Gamma_{C,NH}^c + \Gamma_{N,CH}^c - \Gamma_{C,CH}^c + W_2 + \Gamma_1$$
$$\Gamma_{\beta\beta}^{ZQ} = +\Gamma^a - \Gamma_{NH,CH}^c - \Gamma_{N,NH}^c + \Gamma_{C,NH}^c + \Gamma_{N,CH}^c - \Gamma_{C,CH}^c + W_0 + \Gamma_1$$

$$(35b)$$

Because of the nature of the J-resolved constant time Γ experiment, the relaxation rate of each multiplet component is directly reflected in the intensity of the signal by $I_{\mu\nu} \propto \exp(-\Gamma_{\mu\nu}T)$. Correspondingly, the cross-correlated relaxation rate can be extracted from the multiplet intensities and also the integrals according to

$$\Gamma_{NH,CH}^c = \frac{1}{4T} \ln\left(\frac{I^{DQ}(\alpha\beta)I^{DQ}(\beta\alpha)}{I^{DQ}(\alpha\alpha)I^{DQ}(\beta\beta)}\right) - \frac{1}{2}(W_2 - W_0)$$

$$\Gamma_{NH,CH}^c = \frac{1}{4T} \ln\left(\frac{I^{ZQ}(\alpha\alpha)I^{ZQ}(\beta\beta)}{I^{ZQ}(\alpha\beta)I^{ZQ}(\beta\alpha)}\right) - \frac{1}{2}(W_2 - W_0)$$

$$(36)$$

Variation of the constant time delay T allows for reliability check of the data.[29]

The proton–proton cross relaxation rate can be determined and is found to be small compared to the one-bond relaxation rates between proton and the heteronucleus.

The four cross-correlated relaxation rates $\Gamma_{N,NH}^c$, $\Gamma_{C,NH}^c$, $\Gamma_{N,CH}^c$, and $\Gamma_{C,CH}^c$ can be extracted from Eqs. (35a) and (35b) in a method similar to the way in which the dipole–dipole cross-correlated relaxation rates $\Gamma_{NH,CH}^c$ were obtained. The single dipole (NH)–CSA cross-correlation rates are given by

$$\Gamma_{N,NH}^c = \frac{1}{8T} \ln\left(\frac{I^{DQ}(\alpha\beta)I^{DQ}(\beta\beta)}{I^{DQ}(\alpha\alpha)I^{DQ}(\beta\alpha)} \frac{I^{ZQ}(\alpha\beta)I^{ZQ}(\beta\beta)}{I^{ZQ}(\alpha\alpha)I^{ZQ}(\beta\alpha)}\right)$$

$$\Gamma_{N,CH}^c = \frac{1}{8T} \ln\left(\frac{I^{DQ}(\alpha\beta)I^{DQ}(\beta\beta)}{I^{DQ}(\alpha\alpha)I^{DQ}(\beta\alpha)} \frac{I^{ZQ}(\alpha\alpha)I^{ZQ}(\beta a)}{I^{ZQ}(\alpha\beta)I^{ZQ}(\beta\beta)}\right)$$

$$(37)$$

and similar for the dipole (CH)–CSA cross-correlated relaxation rate,

$$\Gamma_{C,CH}^c = \frac{1}{8T} \ln\left(\frac{I^{DQ}(\beta\beta)I^{DQ}(\beta a)}{I^{DQ}(\alpha\alpha)I^{DQ}(\alpha\beta)} \frac{I^{ZQ}(\beta\beta)I^{ZQ}(\beta\alpha)}{I^{ZQ}(\alpha\alpha)I^{ZQ}(\alpha\beta)}\right)$$

$$\Gamma_{N,CH}^c = \frac{1}{8T} \ln\left(\frac{I^{DQ}(\beta\beta)I^{DQ}(\beta a)}{I^{DQ}(\alpha\alpha)I^{DQ}(\alpha\beta)} \frac{I^{ZQ}(\alpha\alpha)I^{ZQ}(\alpha\beta)}{I^{ZQ}(\beta\beta)I^{ZQ}(\beta\alpha)}\right)$$

$$(38)$$

[29] B. Brutscher, N. R. Skrynnikov, T. Bremi, R. Brüschweiler, and R. R. Ernst, *J. Magn. Reson.* **130**, 346 (1998).

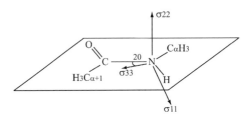

FIG. 8. Size and orientation of the ^{15}N CSA tensor as determined from solid state NMR data.

The orientation and size of CSA tensors are well known from solid-state NMR studies for amides[30] and for aliphatic carbons[31] and can be used for these studies in high-resolution NMR. A review of investigations of CSA tensors of all kind of heteronuclei that have been determined by means of solid-state NMR is given in Ref. 32.

The following average values for the main components of the ^{15}N CSA tensor for a peptide have been reported[32]: $\sigma_{11} = 223 \pm 7$ ppm, $\sigma_{22} = 79 \pm 8$ ppm, $\sigma_{33} = 55 \pm 9$ ppm, and $\Delta\sigma = \sigma_\| - \sigma_{\hat{A}} = 156$ ppm. The orientation of the ^{15}N CSA tensor is indicated in Fig. 8. The ^{13}C CSA tensor for aliphatic carbons shows small anisotropy values. One finds the following values for L-threonine[31]: $\sigma_{11} = 69.0 \pm 0.4$ ppm, $\sigma_{22} = 58.9 \pm 0.4$ ppm, $\sigma_{33} = 52.6 \pm 0.3$ ppm. Other amino acids have been investigated in Ref. 33, showing that the CSA of the ^{13}C$^\alpha$ varies quite strongly. Therefore, DQ and ZQ spectra should show different rates depending on the ψ angle.

Cross-correlated relaxation between ^{13}C$^\alpha$ CSA and ^{15}N CSA does not affect the extraction procedure provided the rates are extracted from DQ and ZQ spectra individually. This holds because this cross-correlated relaxation affects all the submultiplet lines in the same way.

Yang and Kay[34] have developed an improved pulse sequence for the measurement of $\Gamma^{DD,DD}_{N_i H_i^N, C_{i-1}^\alpha H_{i-1}^\alpha}$ in larger protein systems. The pulse sequence is shown in Fig. 9. In contrast to the original experiment,[1] only the C$^\alpha$ but not the N chemical shift of the N,C$^\alpha$ DQ/ZQ coherence is evolved. Using C$^\alpha$ selective pulses allows for relaxation optimized constant time periods for the evolution of N,C$^\alpha$ DQ/ZQ coherence. Lastly, the number of multiplet components recorded per experiment is separated into groups of two in a very elegant manner by recording two data sets, in which the ^1J(N,H) coupling constants is either in phase or in antiphase.

[30] M. D. Lumsden, R. W. Wasylishen, K. Eichele, M. Schindler, G. H. Penner, W. P. Power, and R. D. Curtis, *J. Am. Chem. Soc.* **116**, 1403 (1994); C. J. Hartzell, M. Whitfield, T. G. Oas, and G. P. Drobny, *J. Am. Chem. Soc.* **109**, 5966 (1987).

[31] N. Janes, S. Ganapathy, and E. Oldfield, *J. Magn. Reson.* **54**, 111 (1983).

[32] T. M. Duncan, "A Compilation of Chemical Shift Anisotropies." The Farragut Press, Chicago, 1990.

[33] C. Ye, R. Fu, J. Hu, Lei Hou, and S. Ding, *Magn. Reson. Chem.* **31**, 699 (1993).

[34] D. Yang and L. E. Kay, *J. Am. Chem. Soc.* **120**, 9880 (1998).

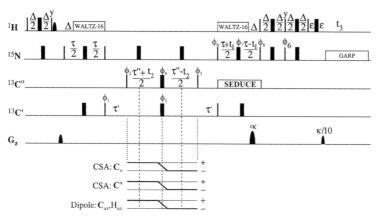

FIG. 9. 3D pulse sequence by Yang *et al.*[35] to measure $\Gamma^{CSA,DD}_{C'_i,C^\alpha_{i-1}H^\alpha_{i-1}}$ for the determination of backbone angle ψ. Delays are as follows: $\Delta = 4.6$ ms, $\tau = 24.8$ ms, $\tau' = 9$ ms, $\tau'' = 26$ ms, $\varepsilon = 0.25$ ms. Default phase is x. $\phi_1 = x, -x$; $\phi_2 = y, -y, -y, y$; $\phi_3 = 8(x), 8(-x)$; $\phi_4 = 4(x), 4(-x)$; $\phi_5 = 8(y)$, $8(-y)$; $\phi_6 = x$; $\phi_7 = 4(x), 4(-x)$; $\phi_8 = x$; $\phi_{rec} = 2(x), 2(-x), 2(x), 2(-x), 2(-x), 2(x), 2(-x), 2(x)$. Quadrature in t_2 is achieved by variation of phases ϕ_2 in the STATES-TPPI manner. Echo–antiecho coherences are selected during t_1 using gradient sensitivity enhancement. The phases ϕ_1 and ϕ_2 are shifted by 90° in subsequent FIDS and stored independently to separate DQ and ZQ coherences evolving during t_2. For further details, we refer to the original publication.

Addition/subtraction of alternate FIDs allows for editing the doublet of doublets. Since relatively long constant time periods are required to resolve the inner two lines of the doublet of doublets, which are separated by approximately 50 Hz, this editing allows for a rather short period of evolution of the N,C$^\alpha$ DQ/ZQ coherence.

4.1.2. J-Resolved Constant Time Γ Experiment to Measure $\Gamma^{CSA,DD}_{C',C^\alpha_{i-1}H^\alpha_{i-1}}$.
Yang *et al.* suggested measuring the backbone angle ψ in a protein based on cross-correlated relaxation between the H$^\alpha$–C$^\alpha$ dipolar and the C' chemical shift anisotropy interaction mechanism[35,36] The pulse sequence, which is shown in Fig. 10a, is derived from an HN(CO)CA experiment.[28] DQ/ZQ coherences between the C' and the neighboring C$^\alpha$ atom are excited by for a constant time $\tau'' = 1/^1J(C_\alpha, C_\beta)$ and evolve during t_2. In this experiment, the projection of the H$^\alpha$–C$^\alpha$ dipole tensor onto the three components of the C' CSA tensor is measured. Using solid-state CSA data on peptide carbonyl tensors by Teng *et al.*[37] the direction cosines of the dipole tensor are given by $\cos\theta_x = -0.3095 + 0.3531\cos(\psi - 120°)$; $\cos\theta_y = -0.1250 - 0.8740\cos(\psi - 120°)$; $\cos\theta_z = -0.9426\sin(\psi - 120°)$. The angular dependence of $\hat{\Gamma}^{CSA,DD}_{C',C^\alpha_{i-1}H^\alpha_{i-1}}$ on the backbone angle ψ is shifted (and inverted by

[35] D. Yang, R. Konrat, and L. E. Kay, *J. Am. Chem. Soc.* **119,** 11938 (1997).
[36] D. Yang, K. H. Gardner, and L. E. Kay, *J. Biomol. NMR* **11,** 213 (1998).
[37] Q. Teng, M. Iqbal, and T. A. Cross, *J. Am. Chem. Soc.* **114,** 5312 (1992).

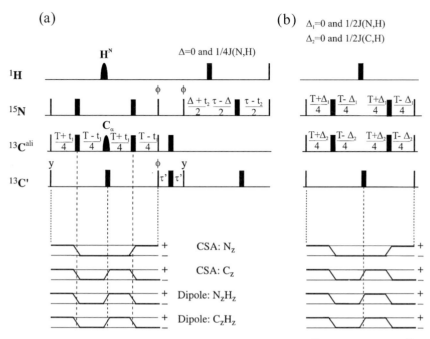

FIG. 10. DQ/ZQ evolution period in 3D pulse sequences by Yang et al.[34] and Pelupessy et al.[20] to measure cross-correlated relaxation rates.

$180°$) compared to $\hat{\Gamma}^{DD,DD}_{N_i H_i^N, C_{i-1}^\alpha H_{i-1}^\alpha}$, which provides information to resolve ambiguities between the measured cross-correlated relaxation rates and the angle ψ similar to information obtained from measurement of multiple J coupling constants around a given torsion angle.

Sprangers et al.[27] have proposed a modified pulse sequence to measure $\Gamma^{DD,DD}_{N_i H_i^N, C_{i-1}^\alpha H_{i-1}^\alpha}$ shown in Fig. 11. This pulse sequence is a quantitative Γ experiment discussed in Section 3.2. In addition, they propose to combine $\Gamma^{DD,DD}_{N_i H_i^N, C_{i-1}^\alpha H_{i-1}^\alpha}$ as a tool to restrain the backbone angle ψ with the analysis of the differences of the C^α chemical shift $^3\Delta\delta(C^\alpha(N^H, N^D))$ measured in H_2O and D_2O. This additional parameter allows differentiating between angle pairs giving rise to identical cross-correlated relaxation rates (shown in Fig. 12). The authors have also compared different approaches, namely J-resolved and quantitative Γ experiments, to measure the same cross-correlated relaxation. The experiments show that the two different approaches yield almost identical cross-correlated relaxation rates.

4.1.3. Quantitative Γ Experiment to Measure $\Gamma^{DD,DD}_{N_i H_i^N, C_{i-1}^\alpha H_{i-1}^\alpha}$. Pelupessy et al.[20] have developed a method that measures dipole–dipole cross-correlated relaxation in a nonevolution period of a pulse sequence (Fig. 9b). The underlying

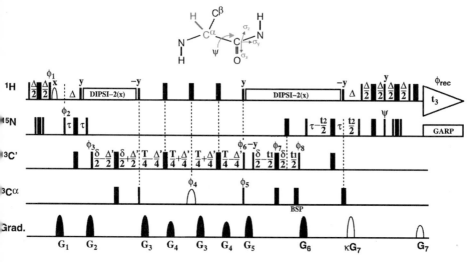

FIG. 11. 3D pulse sequence by Sprangers *et al.*[27] to measure $\Gamma^{CSA,DD}_{C'_i,C^\alpha_{i-1}H^\alpha_{i-1}}$ for the determination of backbone angle ψ. Delays are as follows: $\Delta = 5.2$ ms, $T = 28$ ms, $\zeta = 0$ ms (reference experiment) and 3.57 ms (cross experiment), $\tau = 15$ ms, $\delta = 9$ ms. Default phase is x. $\phi_1 = 4(y), 4(-y); \phi_2 = 8(x), 8(-x); \phi_3 = x + \text{BSP}; \phi_4 = 16(x), 16(y); \phi_5 = 2(x), 2(-x); \phi_6 = y$ (reference experiment), x (cross experiment); $\phi_7 = x, y, -x, -y; \phi_8 = x + \text{TPPI}; \phi_{rec} = x, 2(-x), x, -x, 2(x), -x, -x, 2(x), -x, x, 2(-x), x, -x, 2(x), -x, x, 2(-x), x, x, 2(-x), x, -x, 2(x), -x.$ Echo–antiecho coherences are selected during t_1 using gradient sensitivity enhancement. For further details, we refer to the original publication.

principle of transfer of coherence in a nonevolution period is identical to the experiment developed by Felli *et al.*[19] for the measurement of dipole–dipole cross-correlated relaxation in oligonucleotides. The quantitative experiments evolve cross-correlated relaxation for a nonevolution period. In this specific application, N,C^α DQ/ZQ coherence is excited using an HN(CO)CA experiment. Because of cross-correlated relaxation N,C^α DQ/ZQ coherence evolves as described in Eq. (39):

$$4N_x C_{\alpha x} C'_z \longrightarrow$$

$$4N_x C_{\alpha x} C'_z \left[\begin{array}{l} \cosh\left(\Gamma^{DD,DD}_{NH^N,C_\alpha H_\alpha} T\right) \cos(\pi^1 J(N, H^N)\Delta_1) \cos(\pi^1 J(C_\alpha, H_\alpha)\Delta_2) \\ - \sinh\left(\Gamma^{DD,DD}_{NH^N,C_\alpha H_\alpha} T\right) \sin(\pi^1 J(N, H^N)\Delta_1) \sin(\pi^1 J(C_\alpha, H_\alpha)\Delta_2) \end{array} \right]$$

$$+ 16N_y C_{\alpha y} C'_z H^N_z H_{\alpha z} \left[\begin{array}{l} \sinh\left(\Gamma^{DD,DD}_{NH^N,C_\alpha H_\alpha} T\right) \cos(\pi^1 J(N, H^N)\Delta_1) \cos(\pi^1 J(C_\alpha, H_\alpha)\Delta_2) \\ - \cosh\left(\Gamma^{DD,DD}_{NH^N,C_\alpha H_\alpha} T\right) \sin(\pi^1 J(N, H^N)\Delta_1) \sin(\pi^1 J(C_\alpha, H_\alpha)\Delta_2) \end{array} \right]$$

$$(39)$$

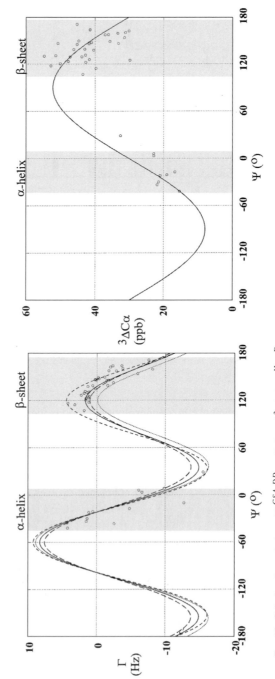

FIG. 12. Calculated correlation of $\Gamma^{CSA,DD}_{C'_i,C^\alpha_{i-1}H^\alpha_{i-1}}$ (*left*) and $^3\Delta\delta[C_\alpha(N^H, N^D)]$ (*right*) as a function of the backbone angle ψ and experimental data. Overall correlation time $\tau_c = 4.6$ ns.

Equation (39) can be derived from Eq. (17) using the relation $\hat{I}_k^\alpha + \hat{I}_k^\beta = 1$ and $\hat{I}_k^\alpha - \hat{I}_k^\beta = 2\hat{I}_{k,z}$. Setting in two experiments $\Delta_1 = 0$ or $1/(2\,^1J(N, H^N))$ and $\Delta_2 = 0$ or $1/(2\,^1J(C_\alpha, H_\alpha))$ allows one to select for the cosh- and sinh-modulated operator $4N_x C_{\alpha x} C_z'$ in two experiments from which the cross-correlated relaxation rate can be derived (see Section 5.1). Equation (39) then simplifies to

$$4N_x C_{\alpha x} C_z' \xrightarrow{\Delta_1 = \Delta_2 = 0} -4N_x C_{\alpha x} C_z' \cosh\left(\Gamma_{NH^N, C_\alpha H_\alpha}^{DD, DD} T\right)$$
$$+ 16N_y C_{\alpha y} C_z' H_z^N H_{\alpha z} \sinh\left(\Gamma_{NH^N, C_\alpha H_\alpha}^{DD, DD} T\right) \tag{40}$$

4.2. Backbone Angle ϕ and Measurement of Projection Restraints Depending on ϕ_i and ψ_i

Additional experiments have been proposed to measure the backbone angle ϕ. Pelupessy et al.[38] have measured $\Gamma_{N_i H_i^N, C_i^\alpha H_i^\alpha}^{DD, DD}$ and $\Gamma_{N_i H_i^N, C_{i-1}^\alpha H_{i-1}^\alpha}^{DD, DD}$ in a quantitative Γ-HNCA experiment (Fig. 13a). In order to differentiate between the intraresidual and interresidual N–C^α correlation, the C^α chemical shift evolution and the quantitative correlation evolution has been concatenated in the proposed experiments. In an HNCA-derived experiment, Chiarparin et al.[39] have measured cross-correlated relaxation rates of the adjacent C^α–H^α dipole tensors. The $\Gamma_{C_{i-1}^\alpha H_{i-1}^\alpha, C_i^\alpha H_i^\alpha}^{DD, DD}$ depend on the dihedral angle spanned by two planes defined by the adjacent $H_{\alpha, i-1} C_{\alpha, i-1} C_{\alpha, i}$, and $C_{\alpha, i} C_{\alpha, i} H_{\alpha, i}$ atoms and depend on the backbone angles ϕ and ψ. This pseudo-torsion angle is different for α helices and β sheets and can therefore be used in structure calculation protocols.

Crowley et al.[40] have proposed an experiment in which $\Gamma_{N_i H_i^N, H_i^N H_i^\alpha}^{DD, DD}$ is measured to distinguish between negative and positive ϕ angles that cannot be differentiated on the basis of $^3J(H^N, H_\alpha)$ coupling constants alone (Fig. 13b).

CH–CH dipole–dipole cross-correlated relaxation rates have been used to measure torsion angles in CH_2–CH_2 groups.[41] The experiment used is a slightly modified 3D version of the experiment of ref. 19 discussed in section 5.1. Nevertheless, because of the higher complexity of the spin system in a CH_2–CH_2 group with respect to a CH–CH moiety and the incomplete transfer achievable between carbon and protons via coupling constant in the reference experiment, the extraction of the cross-correlated relaxation rates from the NMR data can only be obtained through an iterative fitting procedure that takes into consideration the full relaxation matrix of the six spins in a CH_2–CH_2 group. This method has been applied to the conformational investigation of a 2'-aminoethoxy-modified oligonucleotide, and the

[38] P. Pelupessy, E. Chiarparin, R. Ghose, and G. Bodenhausen, J. Biomol. NMR 14, 227 (1999).
[39] E. Chiarparin, P. Pelupessy, R. Ghose, and G. Bodenhausen, J. Am. Chem. Soc. 122, 1758 (2000).
[40] P. Crowley, M. Ubbink, and G. Otting, J. Am. Chem. Soc. 122, 2968 (2000).
[41] T. Carlomagno, M. J. J. Blommers, J. Meiler, B. Cuenoud, and C. Griesinger, submitted.

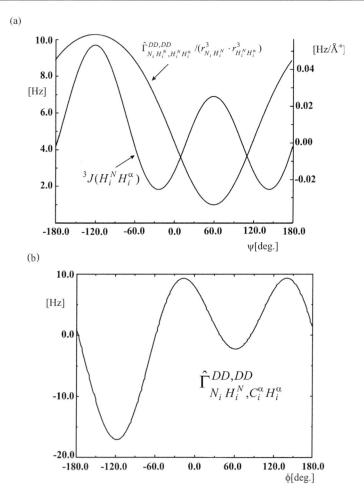

FIG. 13. (a) Cross-correlated relaxation rates $(3\cos^2\theta_{N_i H_i^N, H_i^N H_i^\alpha} - 1)/(r^3_{N_i H_i^N} \cdot r^3_{H_{iN} H_i^\alpha})$ as proposed by Crowley *et al.*[40] and $^3J(H_i^N H_i^\alpha)$ coupling constants and (b) cross-correlated relaxation rates $\hat{\Gamma}^{DD,DD}_{N_i H_i^N, C_i^\alpha H_i^\alpha}$ as proposed by Pelupessy *et al.*[38] as a function of protein backbone angle ϕ. (b) The following formula and a τ_c of 4.6 ns have been used:

$$\hat{\Gamma}^{DD,DD}_{N_i H_i^N, C_i^\alpha H_i^\alpha} = \frac{1}{5}\gamma_H^2\gamma_N\gamma_C\left(\frac{\mu_0}{4\pi}\right)^2 \hbar^2 r_{CH}^{-3} r_{NH}^{-3}(3\cos^2\theta - 1)$$

with $\theta = -0.164 + 0.807\cos(\phi - 62°)$.

experimental data have allowed the determination of the equilibrium population of the gauche+, gauche−, and trans conformations of the NH_3^+–CH_2–CH_2–O sub-unit. This approach can potentially be used to investigate side-chain conformation in proteins.

4.3. Cross-Correlated Relaxation Rates for Structure Refinement

Cross-correlated relaxation rates have been used as restraints in X-PLOR protocols to refine the backbone angle ψ. The cross-correlated relaxation rate is translated into a potential energy of the form:

$$V^c_{A^1 A^2, B^1 B^2} = k\left[\Gamma^{c,\text{theo}}(\theta_{A^1 A^2, B^1 B^2}) - \Gamma^{c,\exp}_{A^1 A^2, B^1 B^2}\right]$$

$$= k\left[\frac{\kappa}{r_A^2 r_B^2}\{3(\vec{A}\,\vec{B})^2 - (\vec{A})^2(\vec{B})^2\} - \Gamma^{c,\exp}_{A^1 A^2, B^1 B^2}\right] \tag{41}$$

with

$$\kappa = \frac{\gamma_{A^1}\gamma_{A^2}}{(r_A)^3}\frac{\gamma_{B^1}\gamma_{B^2}}{(r_B)^3}\left(\frac{\hbar\mu_0}{4\pi}\right)^2 J(0) \quad \text{and} \quad \vec{A} = \vec{A}^1 - \vec{A}^2, \vec{B} = \vec{B}^1 - \vec{B}^2$$

The potential is translated into a force on atom $X = A^1, B^1, A^2,$ or B^2 by differentiation with respect to the respective atom positions:

$$\vec{F}_X = -\vec{\nabla}_X V^c_{A^1 A^2, B^1 B^2} \tag{42}$$

Cross-correlated relaxation rates have been introduced into structure calculation for the protein rhodniin for the refinement of the angle ψ. The starting structures for the Γ-refinement have been generated using the simulated annealing protocol for X-PLOR.[42] It consists of a 32.5 ps high temperature phase at 2000 K, a first 25 ps cooling phase to 1000 K, and a second 10 ps cooling phase to a final temperature of 100 K. NOE restraints (1645 unambiguous and 65 ambiguous) have been taken into account, as well as 22 dihedral restraints for the ϕ angles and 28 dihedral restraints for the χ_1 angles. In addition, 34 hydrogen bonds have been identified by slow proton exchange, of which 16 hydrogen bonds are implemented as ambiguous. The 30 structures with the lowest total energy out of 200 structures are chosen for further refinement. The ψ angles for the refined structures are shown in Fig. 14 together with $\Gamma^c_{NH,CH}$ as a function of ψ.

Most of the angles adopt a unique value. Further analysis has been discussed in more detail in Ref. 16.

5. Applications: Structural Information in RNA and Carbohydrates

5.1. Quantitative Γ-HCCH for Determination of Pseudorotation Phase P

Dipole–dipole cross-correlated relaxation $\Gamma^{DD,DD}_{C_i H_i, C_j H_j}$ between adjacent CH pairs in the ribose ring can be used to determine the pseudorotation phase of the

[42] A. T. Brünger, "X-PLOR Version 3.851: A System for X-ray Crystallography and NMR" Yale University Press, New Haven, CT, 1992.

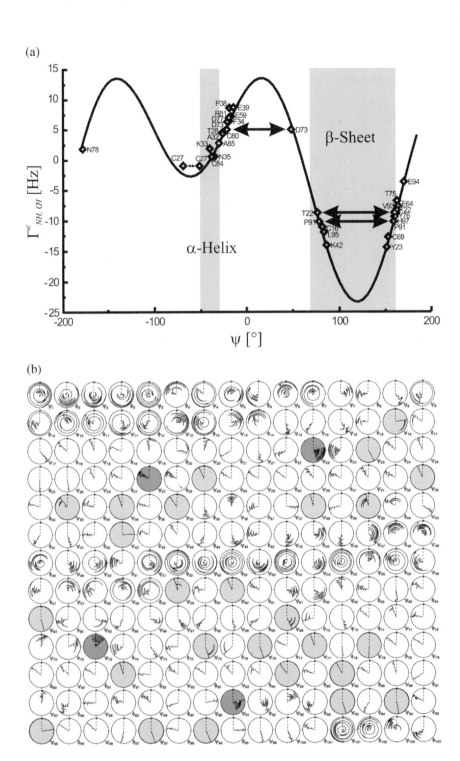

ribosyl moiety in RNA. The experiment proposed by Felli et $al.$[19] is a $quantitative$ Γ-HCCH NMR experiment (Fig. 15), which uses cross-correlated relaxation as a means for coherence transfer.

Cross-correlated relaxation $\Gamma_{C_i H_i, C_j H_j}^{DD,DD}$ of the double and zero quantum coherence (DQ/ZQ) $4H_{iz}C_{ix}C_{jy}$ generated at time point a creates the DQ/ZQ operator $4H_{jz}C_{jx}C_{iy}$. The evolution of chemical shift and of heteronuclear scalar coupling is refocused. In the second part of the experiment, the operator $4H_{jz}C_{jx}C_{iy}$ is transferred to give rise to a cross peak at $(\omega_{H_j}, \omega_{C_i})$. $\Gamma_{C_i, C_j H_j}^{CSA,DD}$ and $\Gamma_{H_i, C_j H_j}^{CSA,DD}$ cross-correlated relaxation are refocused by application of the $180°$ carbon and proton pulses during the mixing time τ_M. The following transfers are achieved in the sequence of Fig. 15:

$$
\begin{aligned}
4H_{iz}C_{ix}C_{jy} \longrightarrow\ &4H_{iz}C_{ix}C_{jy}\Big[\cosh\left(\Gamma_{C_i H_i, C_j H_j}^{c} \tau_M\right) \cos^2(\pi J_{CH}\Delta') \\
&- \sinh\left(\Gamma_{C_i H_i, C_j H_j}^{c} \tau_M\right) \sin^2(\pi J_{CH}\Delta')\Big] \\
&- 4H_{jz}C_{jx}C_{iy}\Big[\sinh\left(\Gamma_{C_i H_i, C_j H_j}^{c} \tau_M\right) \cos^2(\pi J_{CH}\Delta') \\
&+ \cosh\left(\Gamma_{C_i H_i, C_j H_j}^{c} \tau_M\right) \sin^2(\pi J_{CH}\Delta')\Big]
\end{aligned}
\tag{43}
$$

The last term produces a cross peak at $(\omega_{Ci}, \omega_{Hj})$ due to coherence transfer between $4H_{iz}C_{ix}C_{jy}$ and $4H_{jz}C_{jx}C_{jy}$. In the experiment with $\Delta' = 0$, the intensity of the cross peak (I^{cross}) is proportional to $\sinh(\Gamma_{C_i H_i, C_j H_j}^{c} \tau_M)$, whereas for $\Delta' = 1/(2J_{CH})$, the intensity of the cross peak (I^{ref}) is proportional to $\cosh(\Gamma_{C_i H_i, C_j H_j}^{c} \tau_M)$. By comparing the intensity of the cross peak measured in the two experiments, one can determine

$$
\frac{I^{cross}}{I^{ref}} = \tanh\left(\Gamma_{C_i H_i, C_j H_j}^{c} \tau_M\right)
\tag{44}
$$

The dependence of cross-correlated relaxation on the projection angle between the two dipole vectors can be exploited to determine local conformations in ribose rings.

Figure 16 shows the cross-correlation rates $\Gamma_{C1'H1',C2'H2'}^{DD,DD}$, $\Gamma_{C2'H2',C3'H3'}^{DD,DD}$ and $\Gamma_{C3'H3',C4'H4'}^{DD,DD}$ as a function of pseudo rotation phase (P) and amplitude (v^{max}).[43,44]

[43] C. Altona and M. Sundaralingam, $J.$ $Am.$ $Chem.$ $Soc.$ **94**, 8205 (1972).

[44] J. L. Markley, A. Bax, Y. Arata, C. W. Hilbers, R. Kaptein, B. D. Sykes, P. E. Wright, and K. Wüthrich, $J.$ $Biomol.$ NMR **12**, 1 (1998).

FIG. 14. (a) Determined cross-correlated relaxation rates $\hat{\Gamma}_{N_i H_i^N, C_{i-1}^\alpha H_{i-1}^\alpha}^{DD,DD}$ as a function of the backbone angle ψ_k found in structure calculations after refinement with $\hat{\Gamma}_{n_i H_i^N, C_{i-1}^\alpha H_{i-1}^\alpha}^{DD,DD}$ relaxation rates. (b) Circle diagrams of ϕ- and ψ-angles of all 30 structures are shown after refinement with $\hat{\Gamma}_{n_i H_i^N, C_{i-1}^\alpha H_{i-1}^\alpha}^{DD,DD}$ relaxation rates. Shaded with light gray are all restrained angles; shaded with darker gray are the four angles that assume two values after refinement.

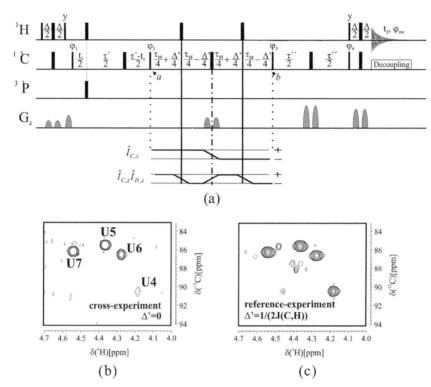

FIG. 15. (a) Pulse sequence of the 2D *quantitative* Γ-HCCH. Narrow and thick bars represent 90° and 180° pulses. The default phase for pulses is x. $\Delta' = 0$ ms for the *quantitative* Γ-HCCH-cross experiment and $\Delta' = 3.36$ ms for the *quantitative* Γ-HCCH-reference experiment. $\Delta = 3.2$ ms, $\tau' = 1/(4^1 J_{cc}) = 6.25$ ms, $\tau_M = 1/(^1 J_{cc}) = 25$ ms. ^{13}C-decoupling was applied during acquisition with $\gamma B_1(2\pi) = 2.5$ kHz. The relaxation delay was 1.5 s. Quadrature detection in ω_1 was achieved by States-TPPI phase incrementation of phase φ_1. 128 scans per t_1 (34 complex points, spectral width: 6024 Hz) increment were recorded with 2K points in t_2 (spectral width: 6010 kHz). The total time for one experiment was 3.5 hours. The phase cycle employed was $\varphi_1 = x, -x; \varphi_2 = 4(x), 4(-x); \varphi_3 = 8(x), 8(-x); \varphi_4 = x, x, -x, -x; \varphi_{rec} = x, -x, -x, x, -x, x, x, -x, -x, x, x, -x, x, -x, -x, x$ (b) *H2', C1'* fingerprint region of the *quantitative* Γ-HCCH cross experiment and of the *quantitative* Γ-HCCH reference experiment (c). The spectra are acquired on the 5'-CGCUUUUGCG-3' hairpin, ^{13}C labeled in the uridine residues by chemical synthesis.[60] The spectra are shown, for clarity, with different thresholds (1 : 4). The assignment of the cross peaks is also shown. The cross peak of U4 in (b) is negative.

Discrimination between sugar pucker modes C2'-endo and C3'-endo,[43] which are shaded in gray in Fig. 16, can be achieved from observation of opposite signs of the two rates $\Gamma^{DD,DD}_{C1'H1',C2'H2'}$ and $\Gamma^{DD,DD}_{C3'H3',C4'H4'}$. Analysis of the relative signs and magnitudes of the two rates $\Gamma^{DD,DD}_{C1'H1',C2'H2'}$ and $\Gamma^{DD,DD}_{C3'H3',C4'H4'}$ provides a method to

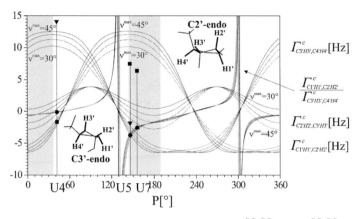

FIG. 16. The calculated cross-correlated relaxation rates $\Gamma^{DD,DD}_{C1'H1',C2'H2'}$, $\Gamma^{DD,DD}_{C2'H2',C3'H3'}$, and $\Gamma^{DD,DD}_{C3'H3',C4'H4'}$ and the ratio $\Gamma^{DD,DD}_{C1'H1',C2'H2'}/\Gamma^{DD,DD}_{C3'H3',C4'H4'}$ are shown in solid lines as a function of the pseudorotation phase P for $\tau_c = 1.5$ ns and for four values of the amplitude ν^{max} (30°, 35°, 40°, 45°). The experimental data for $\Gamma^{DD,DD}_{C1'H1',C2'H2'}$ (squares), $\Gamma^{DD,DD}_{C3'H3',C4'H4'}$ (triangles), and the ratio $\Gamma^{DD,DD}_{C1'H1',C2'H2'}/\Gamma^{DD,DD}_{C3'H3',C4'H4'}$ (circles) are overlaid on the graph.

distinguish between the two main sugar pucker conformations. However, as can be observed from the various plots with different ν^{max}, the absolute values of the rates are affected by variations of the sugar pucker amplitude. Figure 16 shows the ratio of cross-correlated relaxation rates:

$$\frac{\Gamma^{DD,DD}_{C1'H1',C2'H2'}}{\Gamma^{DD,DD}_{C3'H3',C4'H4'}} = \frac{\left(S^c_{1',2'}\right)(3\cos^2\theta_{1',2'} - 1)}{\left(S^c_{3',4'}\right)(3\cos^2\theta_{3',4'} - 1)} \tag{45}$$

which is less sensitive to variations of ν^{max}, does not depend on τ_c, and, if the fluctuations of the respective dipole tensors are comparable, it is also less sensitive to $S^c_{i,j}$. The ratio $\Gamma^{DD,DD}_{C1'H1',C2'H2'}/\Gamma^{DD,DD}_{C3'H3',C4'H4'}$ is a measure of the sugar puckering mode, while closer analysis of individual rates reveals information on fluctuations of the sugar pucker on various time scales.

By combining the quantitative approach[19] to extract cross-correlated relaxation with resolution enhancement methods using restricted coherence transfer in a so-called forward directed TOCSY,[45] Richter et al. could determine the ribose sugar conformation for all but two residues in a uniformly ^{13}C,^{15}N-labeled 25-mer

---

[45] H. Schwalbe, J. P. Marino, S. J. Glaser, and C. Griesinger, J. Am. Chem. Soc. **117**, 7251 (1995); S. J. Glaser, H. Schwalbe, J. P. Marino, and C. Griesinger, J. Magn. Reson., Ser. B **112**, 160 (1996); J. P. Marino, H. Schwalbe, S. J. Glaser, and C. Griesinger, J. Am. Chem. Soc. **118**, 7251 (1996).

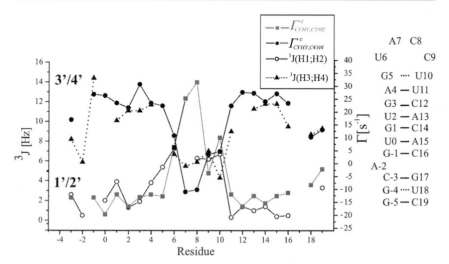

FIG. 17. ^3J(H,H) coupling constants as determined from the HCC-TOCSY-CCH-E.COSY[45] and the cross-correlated relaxation rates as determined from the 3D *forward directed quantitative* Γ-HCCH-TOCSY.[46] The secondary structure of the 25-mer RNA is shown on the right. 1'2' and 3'4' denote coupling constants and cross-correlated relaxation rates in the H–C–C–H moiety, respectively.

RNA.[46] For this system, a comparison between the determined cross-correlated relaxation rates and ^3J(H,H) determined using a *forward directed* HCC-TOCSY-CCH-E.COSY experiment[45] is shown in Fig. 17.

Oligonucleotides may adopt a global conformation for which the assumption of an isotropic overall reorientation is no longer valid. In this case, as discussed in Section 2.3.3., the formulation of the spectral density is more complicated and the observed cross-correlated relaxation rates depend on the orientation of the dipole tensors within a nucleotide relative to the axis of the diffusion tensor. Figure 18 shows a calculation of the dependence of the observed cross-correlated relaxation rates and the orientation within the diffusion tensor system assuming an axially symmetric diffusion tensor and a ratio of $D_\perp / D_\parallel = 5 : 1$. It can be shown that even for this highly anisotropic diffusion tensor, cross-correlated relaxation rates allow for differentiation between C2'-endo and C3'-endo sugar pucker conformation.

Boisbouvier et al.[47] have shown both experimentally and using DFT quantum chemical calculations that $\Gamma_{C1',C1'H1'}^{CSA,DD}$ and $\Gamma_{C3',C3'H3'}^{CSA,DD}$ depend on the sugar pucker mode.

[46] C. Richter, C. Griesinger, I. Felli, P. T. Cole, G. Varani, and H. Schwalbe, *J. Biomol. NMR* **15,** 241 (1999).

[47] J. Boisbouvier, B. Brutscher, A. Pardi, D. Marion, and J.-P. Simorre, *J. Am. Chem. Soc.* **122,** 6779 (2000).

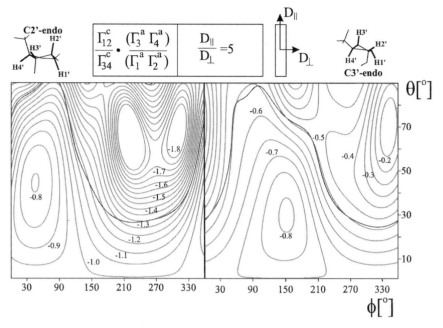

FIG. 18. Cross-correlated relaxation rates as a function of the relative orientation of the diffusion tensor described by angles θ and ϕ and the nucleotide assuming C2'-endo or C3'-endo conformation. Axially symmetric anisotropy of $D_\perp / D_\parallel = 5 : 1$ has been assumed. The dark lines mark the ratio of cross-correlated relaxation in the absence of any anisotropy.

5.2. J-Resolved Constant Time Γ Experiment for Determination of Phosphodiester Backbone Angles α and ζ

Richter et al.[48] have designed an experiment that uses cross-correlated relaxation of $^1H, ^{13}C$-dipolar coupling and ^{31}P-chemical shift anisotropy to determine the phosphodiester backbone angles α and ζ. Since ^{31}P is not bound to NMR-active nuclei, NOE information is sparse and vicinal scalar coupling constants cannot be exploited. Therefore, the experiments by Richter et al. provide a useful addition to the tools to determine the phosphodiester conformation in solution. The cross-correlated relaxation rates can be obtained from the modulation (shown in Fig. 20a) of the two submultiplets of an ^1H-coupled constant time spectrum of $^{13}C, ^{31}P$ double- and zero-quantum coherence (DQC and ZQC, respectively).

The orientation of the ^{31}P CSA is shown in Fig. 16. The experiment determines the projection angles of a dipole tensor, e.g., the C2'H2' dipole tensor, onto the components of the ^{31}P CSA tensors marked in Fig. 19.

[48] C. Richter, B. Reif, C. Griesinger, and H. Schwalbe, J. Am. Chem. Soc. 122, 12728 (2000).

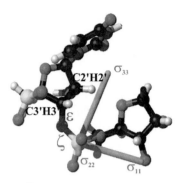

FIG. 19. The orientation of the ^{31}P CSA tensor as determined in diethyl phosphate[67] in the dinucleotide molecular frame.

The size and the orientation of the ^{31}P CSA tensor has been calibrated from single crystal solid state NMR data of barium diethylphosphate.[67] The projection angles $\theta_{CH,\sigma22}$ and $\theta_{CH,\sigma33}$ between the ^{1}H,^{13}C-dipolar coupling and the components of the ^{31}P-CSA tensor depend on a number of torsion angles. The dependence of $\Gamma^{DD,CSA}_{(C3'_i,H3'_i),(P_{i+1})}$ on the backbone angles ε and ζ is shown in blue in Fig. 20b and of $\Gamma^{DD,CSA}_{(C2'_i,H2'_i),(P_{i+1})}$ on ε and ζ (for the ribofuranoside ring in C3'-endo conformation) is shown in green. For $\tau_c = 2.5 \pm 0.2$ ns, the rates vary from -10 Hz to 20 Hz. With an experimental uncertainty of 1.5 Hz determined from measurement of the $\Gamma^{DD,CSA}_{(C',H'),(C')}$ in a nondecoupled ^{1}H,^{13}C-HSQC (see Table I), the circled areas in Fig. 20b can be defined for the angular pairs ε, ζ that are in agreement with the observed cross-correlated relaxation rates $\Gamma^{DD,CSA}_{(C3'_i,H3'_i),CSA(P_{i+1})} = 4.6$ Hz and $\Gamma^{DD,CSA}_{(C2'_i,H2'_i),(P_{i+1})} = 9.5$ Hz for U4. For U4, ε is found to be $-146°$.[68] The angle ζ must therefore be close to $-100°$. For the angle α the sum of $\Gamma^{DD}_{(C5'_i,H5'_i),(P_i)} + \Gamma^{DD}_{(C5'_i,H5''_i),(P_i)}$ representing the arithmetic average of the two dipole tensors can be measured. Together with scalar coupling constant information for β, the backbone angle α can be defined.

Ravindranathan et al.[49] provide a very interesting investigation of the conformation of glucose and methyl-α-D-glucose in the solid state by heteronuclear local field NMR and compared measured cross-correlated relaxation rates in solution. The torsion angle H1–C1–C2–H2 in the solid state for β-glucose is found to be $170° \pm 5°$ and for α-glucose to be $40° \pm 15°$, while in solution torsion angles of $159° \pm 10°$ and $57° \pm 7°$ are found, respectively.

[49] S. Ravindranathan, X. Feng, T. Karlsson, G. Widmalm, and M. H. Levitt, J. Am. Chem. Soc. 122, 1102 (2000).

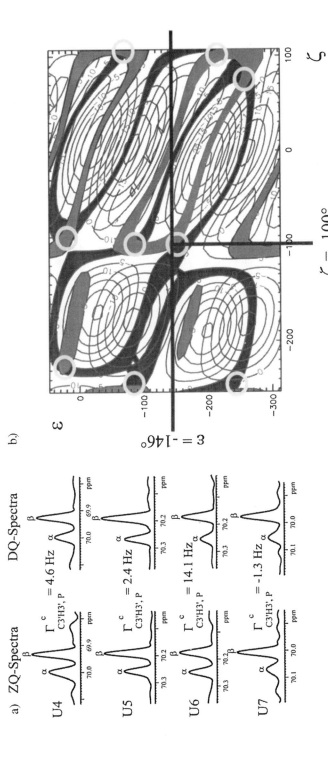

FIG. 20. (a) ZQ and DQ spectra taken at the positions $\Omega_2 = H3'$ for ^{13}C-U labeled[66] nucleotides U4, U5, U6, and U7 of the RNA oligonucleotide 5'-CGCUUUGCG-3'. (b) Dependence of torsion angles ϵ and ζ on $\Gamma^{DD,CSA}_{C2'_i,H2'_i;P_{i+1}}$ (in light gray) and $\Gamma^{DD,CSA}_{C3'_i,H3'_i;P_{i+1}}$ (in dark gray). From the interpretation of $^3J(H,H)$ coupling constants and the C,H-dipolar coupling cross-correlated relaxation rates the pseudo rotation phase $P = 44°$ and the amplitude $\nu^{max} = 44°$. The circles indicate the conformational regions, which fulfill the experimental cross-correlated relaxation rates. Analysis of scalar $^3J(C2'_i, P_{i+1})$, $^3J(H3'_i, P_{i+1})$, and $^3J(C4'_i, P_{i+1})$ coupling constants yield $\epsilon = -146°$. [68] The observed cross-correlated relaxation rates are in agreement with an angle $\zeta = -100°$.

6. Dynamics

Besides providing structural information, cross-correlated relaxation data represent a powerful probe of the local dynamics of a macromolecule, because of their dependence on the order parameter S^2. In case of anisotropic internal motions, the order parameters of the various cross-correlated relaxation rates assume different values. Cross-correlated relaxation rates can be measured between vectors with different orientations in the three-dimensional space; considerable data could be collected to define the local internal reorientations of molecular fragments. Prerequisite for such an analysis is the knowledge of the geometry of the molecular fragments and of the physical constants, such as CSA principal axis values and orientations. In the most favorable case a whole set of auto- and cross-correlated relaxation rates for a certain molecular fragment could be fitted to a model of anisotropic motions to get the amplitudes and the time scale of these motions. Several attempts have been made in this respect in the past few years, but up to now only a partial understanding of the mechanisms of molecular motions has been achieved. A comprehensive overview of dynamics of molecular fragments is beyond the scope of this review article. Important contributions however, are summarized in Table VII.

Investigations based on measurement of cross-correlated relaxation rates have focused on description of the peptide plane motions. This molecular fragment is particularly interesting since it has a fixed geometry and a considerable number of cross-correlated relaxation rates can be measured with good spectral resolution. Moreover, the dynamical behavior of every peptide plane in a protein gives a precise description of the internal motion of the backbone of the protein and potentially indicates secondary structure-related motions. In addition, the NMR data allow the testing of motional models derived from MD calculations.[50–52] More recently, the dynamics of the peptide plane and associated relaxation properties have been investigated. Interest in this area stems from the fact that by using different relaxation probes around a defined geometry such as the planar peptide plane in proteins, important information about anisotropic motion can be derived. In addition, these NMR data allow testing of motional models derived from MD calculations.[53–55]

[50] T. Bremi and R. Brüschweiler, *J. Am. Chem. Soc.* **119**, 6672 (1997).
[51] S. F. Lienin, T. Bremi, B. Brutscher, R. Brüschweiler, and R. R. Ernst, *J. Am. Chem. Soc.* **120**, 9870 (1998).
[52] M. Buck and M. Karplus, *J. Am. Chem. Soc.* **121**, 9645 (1999).
[53] T. Bremi and R. Brüschweiler, *J. Am. Chem. Soc.* **119**, 6672 (1997).
[54] S. F. Lienin, T. Bremi, B. Brutscher, R. Brüschweiler, and R. R. Ernst, *J. Am. Chem. Soc.* **120**, 9870 (1998).
[55] M. Buck and M. Karplus, *J. Am. Chem. Soc.* **121**, 9645 (1999).

TABLE VII
REPORTS EXPLOITING CROSS-CORRELATED RELAXATION IN C–H$_n$ AND N–H
MOIETIES

Molecules, Information	Ref.
Proteins	
(a) Measurement of variation of $^{13}C^\alpha$-CSA that depends on 2nd structure in proteins	73
(b) Measurement of variation of H^N-CSA that depends on 2nd structure in proteins	74
(c) Measurement of variation of the ^{15}N-CSA that depends on 2nd structure in proteins	75
(d) Measurement of variation of the ^{15}N-CSA from ^{15}N autocorrelated relaxation data	76
(e) Side-chain dynamics from dipole–dipole cross-correlated relaxation in methylene side chains	77
(f) Cross-correlated relaxation effects in CHD$_2$-labeled methyl groups.	78
(g)Cross-correlated relaxation effects in methyl-groups	79
(h) Measurement of variation of the ^{15}N-CSA from ^{15}N autocorrelated relaxation data	80
(i) Measurement of longitudinal and transverse cross-correlated relaxation to differentiate between chemical exchange and internal dynamics	81,82
(j) Cross-correlated cross relaxation ^{13}CH$_2$ and ^{13}CH$_3$ subsystems called SIIS	83
(k) Determination of ^{15}N CSA tensor from ^{15}N relaxation data	84

Several experiments have been proposed to measure $\Gamma^{DD,CSA}_{NH^N,C'} + \Gamma^{DD,CSA}_{C'H^N,N}$,[15] the $\Gamma^{DD,CSA}_{CH^N,C}$,[71] $\Gamma^{CSA,CSA}_{N,C'}$ and $\Gamma^{DD,DD}_{N,H^N,C',H^N}$,[49,56] the $\Gamma^{DD,CSA}_{CH^N,C'}$,[57] and the rates $\Gamma^{DD,DD}_{C',C_\alpha,NH^N} + \Gamma^{DD,DD}_{C',H^N,NH_\alpha}$, $\Gamma^{CSA,DD}_{N,NC_\alpha} + \Gamma^{CSA,DD}_{C',C'C_\alpha}$, and $\Gamma^{CSA,DD}_{C'NH^N} + \Gamma^{CSA,DD}_{N,C'H^N}$.[58] Individual cross-correlated relaxation rates have been interpreted in refs. 15, 50, and 51. Carlomagno *et al.*[58] have attempted to fit more than one cross-correlated relaxation rates at a time to the GAF model.[50] This model describes the internal motion of a peptide plane as reorientation around three perpendicular axes with Gaussian distribution of amplitudes. The axis around which the largest amplitude of motions was found in molecular simulation crosses the two $C_{\alpha,i}$ and $C_{\alpha,i-1}$ nuclei, while

[56] M. Pellecchia, Y. Pang, L. Wang, A. V. Kurochkin, A. Kumar, and E. R. P. Zuiderweg, *J. Am. Chem. Soc.* **121**, 9165 (1999).

[57] M. W. F. Fischer, L. Zeng, Y. Pang, W. Hu, A. Majumdar, and E. R. P. Zuiderweg, *J. Am. Chem. Soc.* **119**, 12629 (1997).

[58] T. Carlomagno, M. Maurer, M. Hennig, and C. Griesinger, *J. Am. Chem. Soc.* **122**, 5105 (2000).

TABLE VIII

AVERAGED ^{13}C′ CHEMICAL SHIFT TENSORS^a

	Solid^b	α Helix	β Sheet	Loop	All
$\sigma_{11} - \sigma_{iso}$	73.6 ± 6.0	68.5 ± 15.1	69.1 ± 9.9	72.2 ± 13.5	70.5 ± 13.1
$\sigma_{22} - \sigma_{iso}$	4.0 ± 6.3	-7.9 ± 15.1	-3.2 ± 19.2	-4.1 ± 14.3	-4.8 ± 15.7
$\sigma_{33} - \sigma_{iso}$	-77.6 ± 3.8	-60.5 ± 9.7	-65.9 ± 16.8	-68.2 ± 9.5	-65.8 ± 12.0
σ_{iso}	170 ± 2.0	178.1 ± 1.4	174.9 ± 1.4	175.4 ± 1.8	175.9 ± 2.1
α	35.7 ± 1.5	35.8 ± 6.9	41.4 ± 9.9	41.2 ± 9.8	40.0 ± 9.4

^{a} Taken from Ref. 60
^{b} From Refs. 85, 37

the amplitudes of motions around the two remaining axes were found to be equal. Although it is possible to fit only one cross-correlated relaxation rate at a time to the GAF model of motion, the attempt to interpret all three cross-correlated relaxation rates of Ref. 58 with this model failed for most residues of the protein ubiquitin and particularly for those situated in β sheets. These deviations could be attributed either to changes in size and orientation of the ^{15}N and ^{13}C CSA tensors with respect to the values used in the fitting, which were derived from solid state studies,[77] or to the presence of more complicated motions, which cannot be described by the GAF model. Even in the interpretation of one cross-correlated relaxation rate at a time, as it is the case for the protein binase,[49,71] large motions have to be assumed to account for the measured values.[71] This is again particularly true for β-sheet regions. Interestingly, all the three $\Gamma^{DD,DD}_{C'C_\alpha,NH^N} + \Gamma^{DD,DD}_{C'H^N,NH_\alpha}$, $\Gamma^{CSA,DD}_{N,NC_\alpha} + \Gamma^{CSA,DD}_{C',C'C_\alpha}$, and $\Gamma^{CSA,DD}_{C'NH^N} + \Gamma^{CSA,DD}_{N,C'H^N}$ cross-correlated relaxation rates belonging to the residues of α helix 23–34 in ubiquitin[50] could be fitted to a GAF model of motion assuming relative large reorientations around axes perpendicular to the helix axis. Indication of possible cooperative motions in α helices were found before by Fischer *et al.*[59] in a study of the internal dynamics of flavodoxin via autocorrelated rates.

Individual ^{13}C′ CSA tensors for amino acids residues have been derived by Pang and Zuiderweg[60] from measurement of $\Gamma^{CSA,DD}_{C',C'N}$, $\Gamma^{CSA,DD}_{C',C'C_\alpha}$, and $\Gamma^{CSA,DD}_{C',H^N}$. The data summarized in Table VIII have been compared with solid-state data. The authors conclude that different peptide planes have very different properties and that these differences are mostly uncorrelated with protein secondary structure. The origins of these variations have not been completely identified, and anisotropic local motion might provide some explanation for the observed variations along the peptide backbone.

[59] M. W. F. Fischer, L. Zeng, A. Majumdar, and E. R. P. Zuiderweg, *Proc. Natl. Acad. Sci. USA* **95**, 8016 (1999).
[60] Y. Pang and E. R. P. Zuiderweg, *J. Am. Chem. Soc.* **122**, 4841 (2000).

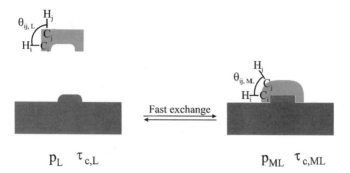

FIG. 21. Schematic representation of the binding of a ligand to a macromolecule. p_L, $\tau_{c,L}$, and $\theta_{ij,L}$ are molar fraction, correlation time, and projection angle between the CH vectors in the free ligand; ML represents the complex.

7. Transferred Cross Correlation

Transferred NOEs has proven to be a very valuable tool to study the conformation of small ligands when bound to macromolecules in fast exchange conditions. For K_d values greater than $10^{-6} M$, ligands are often in an equilibrium between a protein bound and a free form and all measured NMR parameters are averaged between the bound and the free state. The measurement of NMR parameters that directly depend on the correlation time gives structural information for the bound form, since the large molecular weight of the complex, and the correspondingly large correlation time, make the contribution of the bound form to the measured values predominant, in spite of the small population of the complex. Since cross-correlated relaxation rates can provide direct structural angular information and linearly depend on the correlation time, they compensate the lack of coupling constants data in ligand/macromolecule complexes in fast exchange regimes. Because of that they represent a unique and valuable tool in the context of SAR by NMR techniques and for the study of transient species in enzyme-catalyzed reactions.

This approach has been used in publications by Carlomagno et al.[61] and Blommers et al.[62] The averaged cross-correlated relaxation rate for dipole–dipole cross-correlated relaxation (for example, two CH dipoles) between two sites populated with population for the free ligand p_L and the bound ligand $p_{ML} = (1 - p_L)$ is

[61] T. Carlomagno, I. C. Felli, M. Czech, R. Fischer, M. Sprinzl, and C. Griesinger, *J. Am. Chem. Soc.* **121,** 1945 (1999).

[62] M. J. J. Blommers, W. Stark, C. E. Jones, D. Head, C. E. Owen, and W. Jahnke, *J. Am. Chem. Soc.* **121,** 1949 (1999).

[63] L. Emsley and G. Bodenhausen, *J. Magn. Reson.* **82,** 211 (1989); L. Emsley and G. Bodenhausen, *Chem. Phys. Lett.* **165,** 469 (1989).

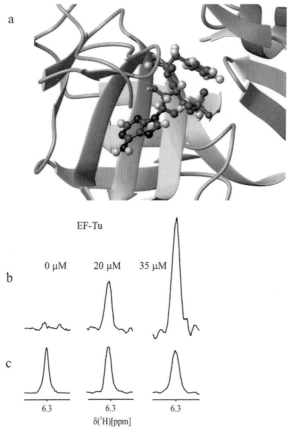

FIG. 22. Measurement of transferred cross-correlated relaxation in the complex of a tRNA mimetic Ant-Ado bound to EF-TU. (a) The structure of 3'-Ant-Ado in the bound state from the X-ray structure. The conformation of the ribose ring is C2' endo. (b) C2', H1' cross peak for 2'-Ant-Ado extracted from the quantitative Γ-HCCH cross experiment and multiplied by 4. (c) The same, but from the reference experiment divided by 2. The cross peak increases linearly with increasing concentration of the EF-TU ranging from 0, 20 to 35 μM. The reference peak stays relatively constant in this range of protein concentrations. The extraction of this and other cross-correlated relaxation rates supports that also the 2'Ant-Ado is bound in the C2'-endo conformation.

given in Eq. (46) and schematically shown in Fig. 21:

$$
\begin{aligned}
\left\langle \Gamma_{C_m H_m, C_n H_n}^{DD, DD} \right\rangle = {} & \frac{2}{5} \frac{\gamma_C^2 \gamma_H^2 \mu_0^2 \hbar^2}{(4\pi)^2 r_{C_m H_m}^3 r_{C_n H_n}^3} \left[S_{mn,L}^2 p_L \frac{3\cos^2 \theta_{mn,L} - 1}{2} \tau_{c,L} \right. \\
& \left. + S_{mn,ML}^2 p_{ML} \frac{3\cos^2 \theta_{mn,ML} - 1}{2} \tau_{c,ML} \right]
\end{aligned} \tag{46}
$$

For $\tau_{c,L} \ll \tau_{c,ML}$, the observed averaged cross-correlated relaxation rate $\langle \Gamma^{DD,DD}_{C_m H_m, C_n H_n} \rangle$ is dominated by the bound conformation, and therefore, precise information on the conformation of a bound ligand can be obtained. The dipole–dipole $\Gamma_{CH,CH}$ cross-correlated relaxation rate has been used to determine the sugar pucker in a tRNA analog bound to EF–Tu · GDP complex, while the dipole–dipole $\Gamma_{C_\alpha H_\alpha, N H^N}$ cross-correlated relaxation rate has been used in the conformational investigation of a phosphotyrosine peptide weakly bound to STAT-6. The experimental and structural data obtained for the complex Ant-Ado (tRNA mimetic)-EF-Tu · GDP are shown in Fig. 22.

[64] A. G. Palmer, J. Cavanagh, P. E. Wright, and M. Rance, *J. Magn. Reson.* **93**, 151 (1991); L. E. Kay, O. Keifer, and T. Saarinen, *J. Am. Chem. Soc.* **114**, 10663 (1992); J. Schleucher, M. Sattler, and C. Griesinger, *Angew. Chem., Int. Ed. Engl.* **32**, 1489 (1993); M. Sattler, M. G. Schwendinger, J. Schleucher, and C. Griesinger, *J. Biomol. NMR* **5**, 11 (1995).

[65] R. Fu and G. Bodenhausen, *Chem. Phys. Lett.* **245**, 415 (1995).

[66] S. Quant, R. W. Wechselberger, M. A. Wolter, K. Wörner, P. Schell, J. W. Engels, C. Griesinger, and H. Schwalbe, *Tetrahed. Lett.* **35**, 6649 (1994).

[67] H. Shindo, *Biopolymers* **19**, 509 (1980); J. W. Keepers, and T. L. James, *J. Am. Chem. Soc.* **104**, 929 (1982); J. Herzfeld, R. G. Griffin, and R. A. Haberkorn, *Biochemistry* **17**, 2711 (1978).

[68] C. Richter, B. Reif, K. Wörner, S. Quant, J. W. Engels, C. Griesinger, and H. Schwalbe, *J. Biomol. NMR* **12**, 223 (1998).

[69] E. Chiarparin, P. Pelupessy, R. Ghose, and G. Bodenhausen, *J. Am. Chem. Soc.* **121**, 6876 (1999).

[70] W. Peti, M. Hennig, L. Smith, and H. Schwalbe, *J. Am. Chem. Soc.* **122**, 12017 (2000).

[71] Y. Pang, L. Wang, M. Pellecchia, A. V. Kurochkin, and E. R. P. Zuiderweg, *J. Biomol. NMR* **14**, 297 (1999).

[72] J. Boisbouvier, P. Gans, M. Blackledge, B. Brutscher, and D. Marion, *J. Am. Chem. Soc.* **121**, 7700 (1999).

[73] N. Tjandra and A. Bax, *J. Am. Chem. Soc.* **119**, 9576 (1997).

[74] N. Tjandra and A. Bax, *J. Am. Chem. Soc.* **119**, 8076 (1997).

[75] N. Tjandra, A. Szabo, and A. Bax, *J. Am. Chem. Soc.* **118**, 6986 (1996).

[76] D. Fushman, N. Tjandra, and D. Cowburn, *J. Am. Chem. Soc.* **120**, 10947 (1998); D. Fushman, N. Tjandra, and D. Cowburn, *J. Am. Chem. Soc.* **120**, 7109 (1998).

[77] D. Yang, A. Mittermaier, Y.-K. Mok, and L. E. Kay, *J. Mol. Biol.* **276**, 939 (1998); D. Yang, Y.-K. Mok, D. R. Muhandiram, J. D. Forman-Kay, and L. E. Kay, *J. Am. Chem. Soc.* **121**, 3555 (1999).

[78] D. Yang and L. E. Kay, *J. Magn. Reson. Ser. B* **110**, 213 (1996).

[79] L. E. Kay and T. E. Bull, *J. Magn. Reson.* **99**, 615 (1992); L. E. Kay, and D. A. Torchia, *J. Magn. Reson.* **95**, 536 (1991); L. E. Kay, T. E. Bull, L. K. Nicholson, C. Griesinger, H. Schwalbe, A. Bax, and D. A. Torchia, *J. Magn. Reson.* **100**, 538 (1992).

[80] C. D. Kroenke, M. Rance, and A. G. Palmer III, *J. Am. Chem. Soc.* **121**, 10119 (1999).

[81] C. D. Kroenke, J. P. Loria, L. K. Lee, M. Rance, and A. G. Palmer III, *J. Am. Chem. Soc.* **120**, 7905 (1998).

[82] I. C. Felli, H. Desvaux, and G. Bodenhausen, *J. Biomol. NMR* **12**, 509 (1998).

[83] M. Ernst and R. R. Ernst, *J. Magn. Reson., Ser. A* **110**, 202 (1994); J. Engelke and H. Rüterjans, *J. Biomol. NMR* **11**, 165 (1998).

[84] J. Boyd and C. Redfield, *J. Am. Chem. Soc.* **120**, 9692 (1998).

[85] T. G. Oas, C. J. Hartzell, T. J. McMahon, G. P. Drobny, and F. W. Dahlquist, *J. Am. Chem. Soc.* **109**, 5956 (1987).

[3] Applications of Adiabatic Pulses in Biomolecular Nuclear Magnetic Resonance

By ĒRIKS KUPČE

1. Introduction

A great deal of development in nuclear magnetic resonance (NMR) is aimed at improving the sensitivity and quality of spectra by means of selective excitation or selective decoupling and by optimizing waveforms used in such experiments.[1,2] There are two types of radio frequency (RF) pulses commonly used in NMR. The phase-alternating RF pulses, which can be regarded as linearly polarized waveforms, are the simplest and also the most extensively used type of RF pulses. Most of the early developed composite pulse schemes, such as GARP,[3] WALTZ,[4] or DIPSI,[5] and selective excitation pulses, for example, BURP pulses,[6] SNOB pulses,[7] or Gaussian cascade pulses,[8] belong to the first class. On the other hand, waveforms based on frequency sweeps can be described in terms of circular polarization. Adiabatic frequency sweep is one of the most efficient methods for inversion of spin magnetization.[9–12] There are many important techniques in NMR, such as wide-band decoupling or isotropic mixing, which are based on recursive inversion of spin magnetization. In order to achieve the required efficiency of spin inversion, the frequency swept pulses must obey the rules of adiabatic following.

2. Principle of Adiabatic Following

According to the adiabatic theorem,[9] if the variation of the effective field with time is slow, the angle of the magnetization with the instantaneous direction of the

[1] R. Freeman, "Spin Choreography." Spektrum, Oxford, 1997.
[2] J. Cavanagh, W. J. Fairbrother, A. G. Palmer III, and N. J. Skelton, "Protein NMR Spectroscopy." Academic Press, San Diego, 1996.
[3] A. J. Shaka, P. B. Barker, and R. Freeman, *J. Magn. Reson.* **64**, 547 (1985).
[4] A. J. Shaka, J. Keeler, T. Frenkiel, and R. Freeman, *J. Magn. Reson.* **52**, 335 (1983).
[5] A. J. Shaka, C. J. Lee, and A. Pines, *J. Magn. Reson.* **77**, 274 (1988).
[6] H. Geen and R. Freeman, *J. Magn. Reson.* **93**, 93 (1991).
[7] Ē. Kupče, J. Boyd, and I. D. Campbell, *J. Magn. Reson. Ser. B* **106**, 300 (1995).
[8] L. Emsley and G. Bodenhausen, *Chem. Phys. Lett.* **165**, 469 (1990).
[9] A. Abragam, "The Principles of Nuclear Magnetism." Oxford University Press, Oxford, 1961.
[10] J. Baum, R. Tycko, and A. Pines, *Phys. Rev. A* **32**, 3435 (1985).
[11] M. Garwood and K. Uğurbil, *in* "NMR Basic Principles and Progress" (P. Diehl, E. Fluck, H. Günther, R. Kosfeld, and J. Seeling, eds.), Vol. 26, p. 109. Springer-Verlag, Berlin and New York, 1992.
[12] R. A. de Graaf and K. Nicolay, *Concepts Magn. Reson.* **9**, 247 (1997).

effective field is a constant of the motion. Mathematically the adiabatic condition is represented by Eq. (1).

$$\frac{\partial \theta}{\partial t} \ll \omega_{eff} \qquad (1)$$

In practice this means that nutation of magnetization around the effective field ω_{eff} must be much faster than the change in the direction of the RF field. The adiabatic condition can be quantified by introducing the adiabaticity factor, Q[10]:

$$Q = \frac{\omega_{eff}}{\dot{\theta}} \qquad (2)$$

Expressed in terms of RF field strength ω_1 and frequency offset $\Delta\omega$, Eq. (2) becomes

$$Q = \left(\frac{1}{2\pi}\right) \frac{\left(\omega_1^2 + \Delta\omega^2\right)^{3/2}}{\omega_1 \Delta\dot{\omega} + \Delta\omega\dot{\omega}_1} \qquad (3)$$

Now the adiabatic condition becomes simply $Q \gg 1$. According to Eq. (2) Q is dimensionless. However, in practice the $1/2\pi$ factor is usually omitted.[10] With this definition a good quality of magnetization inversion can be achieved with $Q \geq 4$.

The principle of adiabatic following is demonstrated in Fig. 1. If the RF field is switched on suddenly, the magnetization merely nutates around the effective field (see Fig. 1a). However, if the RF field is turned on smoothly, the magnetization follows the effective field (see Fig. 1b). This kind of adiabatic ω_1 sweep is often used to lock magnetization along a tilted axis in off-resonance ROESY experiments.[13] As shown in Fig. 1b, the magnetization stays locked along the effective field for most of the time and is smoothly returned to the Z axis at the end of the spin lock. Finally, if the frequency of the RF field is swept through the resonance condition, spin magnetization becomes inverted at the end of the sweep (see Fig. 1c). Note that adiabaticity Q is at minimum when the RF field is on-resonance ($\Delta\omega = 0$):

$$Q_0 = \omega_1^2/k \qquad (4)$$

where k is the sweep rate. This is the most critical point in the whole adiabatic sweep.

There are many ways in which RF amplitude can be turned on and off smoothly.[13-16] For instance, the RF amplitude of a waveform that employs a linear

[13] H. Desvaux, P. Berthault, N. Birliakis, M. Goldman, and M. Piotto, *J. Magn. Reson. Ser. A* **113,** 47 (1995).

[14] Ē. Kupče and R. Freeman, *J. Magn. Reson. Ser. A* **115,** 273 (1995).

[15] Ē. Kupče and R. Freeman, *J. Magn. Reson. Ser. A* **117,** 246 (1995).

[16] J. M. Bohlen and G. Bodenhausen, *J. Magn. Reson. Ser. A* **102,** 293 (1993).

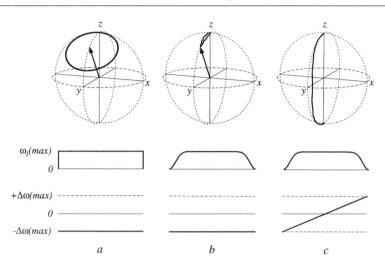

FIG. 1. The principle of adiabatic following as viewed in the RF reference frame. (a) The RF field is switched on suddenly. The magnetization nutates around the effective field. (b) The RF field is turned on gradually. The magnetization follows the effective field, stays locked for some time, and then is returned back to the equilibrium. (c) The RF field is swept through the resonance frequency adiabatically. The magnetization follows the effective field and is inverted at the end of the sweep.

frequency sweep (chirp) can be apodized using the so-called WURST (*w*ide-band *u*niform *r*ate *s*mooth *t*runcation) modulation[14]:

$$\omega_1(t) = \omega_1(\max)[1 - |\sin(\beta t)|^n] \tag{5}$$

where $-\pi/2 < \beta t < \pi/2$. Alternatively, the RF amplitude can be kept constant while the frequency sweep function is adjusted to achieve efficient inversion.[10,15] Finally, both the RF amplitude and frequency sweep modes can be chosen in a consistent way.[17–19] For instance, one can choose to keep the minimum adiabaticity Q_0 constant for all frequencies within the sweep range. This is the so-called constant adiabaticity (CA) sweep,[18] sometimes also called the offset independent adiabaticity sweep.[19] From Eq. (4) it follows that for any RF amplitude modulation, the optimum frequency sweep function is just an integral of RF power:

$$\Delta\omega(t) = \frac{1}{Q_0} \int \omega_1^2(t)dt \tag{6}$$

One of the most remarkable properties of adiabatic pulses is their exceptional tolerance to inhomogeneity of the RF field (see Fig. 2). It is because of this property

[17] Ē. Kupče and G. Wagner, *J. Magn. Reson. Ser. B* **109,** 329 (1995).
[18] Ē. Kupče and R. Freeman, *J. Magn. Reson. Ser. A* **118,** 299 (1996).
[19] A. Tannus and M. Garwood, *J. Magn. Reson. Ser. A* **120,** 133 (1996).

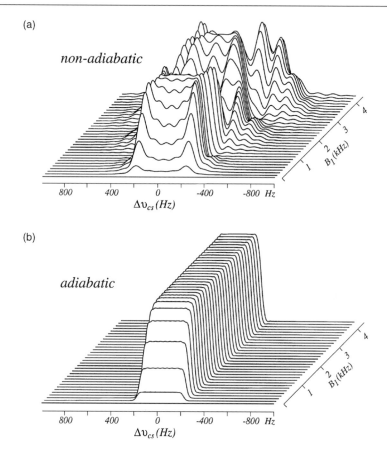

FIG. 2. Comparison of the tolerance to mis-setting of the RF field strength for (a) phase alternating pulses (I-BURP-2) and (b) adiabatic pulses (hyperbolic secant).

that adiabatic pulses are very common in NMR imaging,[11,12,20] where the RF homogeneity is typically very poor. This property of adiabatic pulses allows one to design experiments that do not require any calibration of RF pulses.[21] Unfortunately, such an approach is quite costly in terms of the total experiment time. As a result it has not found widespread applications in biomolecular NMR, where relaxation times are relatively short and the total number of pulses in typical (3D) experiments can be very large.

[20] R. A. de Graaf, "In Vivo NMR Spectroscopy." Wiley, New York, 1998.
[21] P. C. M. van Zijl, T.-L. Hwang, M. O'Neil Johnson, and M. Garwood, *J. Am. Chem. Soc.* **118,** 5510 (1996).

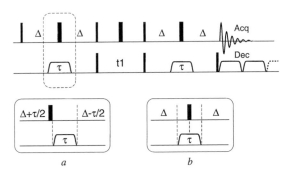

FIG. 3. Pulse sequence for adiabatic HSQC experiment. The two methods of implementation of adiabatic pulses (dotted boxes) are (a) sequential and (b) centered.

An equally useful property of adiabatic pulses is that their active bandwidth is essentially unlimited. In fact, the only limitation here is pulse duration. There are many examples in ^{19}F, ^{31}P, and heavy metal NMR, where the chemical shift range of substances of interest is very large. Even simple ^{13}C experiments such as HSQC[2] or INADEQUATE[1] often suffer badly from inadequate performance of the conventional hard pulses.

3. Adiabatic Inversion

Let us consider the conventional ^{13}C HSQC experiment at high magnetic field strength. For the given, limited RF power level the active bandwidth of a rectangular inversion pulse is four times lower than that of an ordinary excitation pulse. For instance, assuming that a typical chemical shift range for ^{13}C is 200 ppm, the frequency band that needs to be covered at 900 MHz is in excess of 45 kHz. A relatively short 180° inversion pulse of 25 μs covers a band only 20 kHz wide. Furthermore, because the errors introduced by poor inversion pulses are cumulative, the performance of multiple pulse experiments degrades very quickly. For the same reason miscalibration of inversion pulses has much greater impact on the experiment than that of excitation pulses.

A simple and also a very efficient solution to these problems is replacing the most critical rectangular inversion pulses by adiabatic pulses (see Fig. 3).[22–24] Note that as long as the adiabatic pulses are relatively short, it does not matter whether they are applied simultaneously,[22,23] with the proton inversion pulses (centered)

[22] K. Ogura, H. Terasawa, and F. Inagaki, *J. Magn. Reson. Ser. B* **112**, 63 (1996).
[23] C. Zwahlen, P. Legault, S. J. F. Vincent, J. Greenblatt, R. Konrat, and L. Kay, *J. Am. Chem. Soc.* **119**, 6711 (1997).
[24] Ē. Kupče and R. Freeman, *J. Magn. Reson.* **127**, 36 (1997).

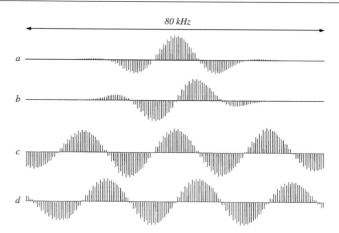

FIG. 4. The offset dependence for the first increment of ^{13}C HSQC spectra of ^{13}CH$_3$I in CDCl$_3$ recorded on INOVA 400; (a) and (b) are the real and imaginary parts of the first complex data point in t_1 acquired using rectangular inversion pulses; (c) and (d) the same as in (a) and (b), but using WURST-20 inversion pulses. No effort has been made to correct the phase errors introduced by evolution of magnetization during the 14 μs long 180° H-1 pulse in the middle of the t_1 period. The ^{13}C carrier frequency was incremented in 81 steps of 1 kHz each. The length of the rectangular 90° ^{13}C pulse was 12.0 μs.

or sequentially.[24] The effect of this modification on performance of the ^{13}C–^{1}H HSQC experiment is shown in Fig. 4.

In order to ensure that adiabatic pulses are as short as possible, the maximum allowed RF power must be used. Often it is also very convenient to use the same RF power level for adiabatic sweep as for the rectangular pulses. This allows one to simplify the experiment setup. For instance, assuming a linear frequency sweep with bandwidth of π/τ_{180} and $Q = 4$, from Eq. (4) we can estimate the required length of the adiabatic pulse to be 8 τ_{180}, where τ_{180} is the length of rectangular 180° pulse at the given power level. For $\tau_{180} = 25$ μs this gives a 200 μs long inversion pulse and more than adequate inversion bandwidth. Such waveforms can be created automatically ("on-the-fly") from within the pulse program. As an example, a ^{13}C HSQC spectrum of an alkaloid saponin acquired at 800 MHz using the procedure described above is shown in Fig. 5. Although the ^{13}C chemical shift range in this spectrum exceeds 40 kHz, all signals appear with fair intensities.

Similarly adiabatic inversion pulses can be implemented in more sophisticated experiments.

4. Chemical Shift Refocusing with Adiabatic Pulses

Employing the adiabatic pulses in experiments that require refocusing of magnetization is somewhat more complicated. The problem is that adiabatic pulses in

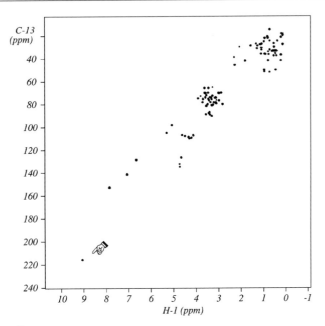

FIG. 5. The ^{13}C HSQC spectrum of saponin recorded at 800 MHz on INOVA spectrometer. Note the peak at ca 216 ppm that appears undistorted and with full intensity. The sample is courtesy of Dr. C. Roumestand.

general are not good refocusing pulses. The reason for this can be demonstrated using the so-called instant flip approximation, which assumes that the spin magnetization is inverted at the moment when RF frequency reaches the larmor frequency of the given spin. As shown in Fig. 6a, the refocusing for spins with different larmor frequencies is achieved at different times. The problem can be "fixed" by applying the adiabatic pulse twice, as shown in Fig. 6b.[24,25] The second pulse now acts as a "trim" pulse, which corrects the phase errors introduced by the first pulse. It can be shown that refocusing by a pair of adiabatic pulses is exact.[25] Alternatively, a composite adiabatic pulse can be used in the conventional way (see Fig. 6c).[26]

5. J-Refocusing with Adiabatic Pulses

As already discussed in the case of the HSQC experiment, adiabatic pulses can be used in isotope filtering procedures to refocus the heteronuclear J couplings. If the pulse duration T_p is considerably shorter than $1/J$, its length can be

[25] K. Stott, J. Stonehouse, J. Keeler, T.-L. Hwang, and A. J. Shaka, J. Am. Chem. Soc. **117**, 4199 (1995).
[26] T.-L. Hwang, P. C. M. van Zijl, and M. Garwood, J. Magn. Reson. **124**, 250 (1997).

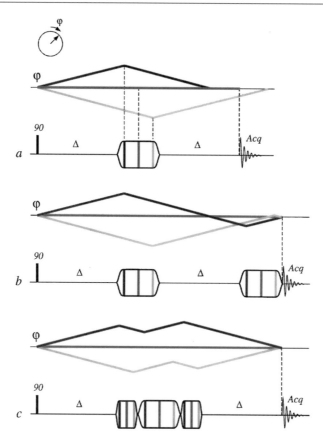

FIG. 6. Explanation of chemical shift refocusing using phase evolution diagrams and the instant flip approximation. Three representative cases are considered, shown as bold vertical bars of different shades for spins flipped at different times during the sweep. (a) A single adiabatic pulse does not achieve refocusing; (b) two pulses refocus magnetization exactly; (c) refocusing using a composite adiabatic pulse. The leading and trailing adiabatic pulses are twice as short as the central pulse. As a result the experiment (c) requires higher RF power than that in (b).

neglected. However, if T_p becomes comparable with $1/J$, the refocusing is no longer accurate.[24] As in the case of chemical shift refocusing, two pulses are required to achieve exact J-refocusing over the full active bandwidth, except in this case one of the adiabatic pulses needs to be sweep reversed. The reason for this is the 180° H-1 pulse that reverses the phase evolution of spin isochromats. Again, the mechanism of J-refocusing can be represented in terms of the instant flip approximation, as shown in Fig. 7.

In practice, the ^{13}C–^{1}H spin–spin couplings typically are far from uniform. Namely, the one-bond ^{13}C–^{1}H couplings in aromatic systems are typically

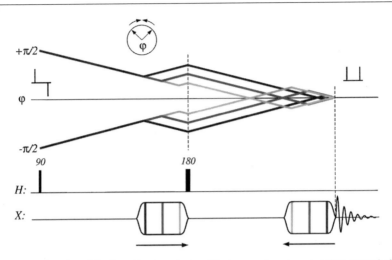

FIG. 7. Refocusing of the J coupling in an isotope filtering experiment requires time reversal of the second adiabatic pulse. The arrows below adiabatic pulses indicate the frequency sweep direction.

160–180 Hz, whereas in the aliphatic systems they are in the range between 120 and 150 Hz. Often one can observe a relatively good correlation between the ^{13}C chemical shifts and the ^{13}C–^{1}H coupling constants. Indeed, such correlation has been reported for proteins [Eq. (7)] and RNA [Eq. (8)]:[23]

$$^{1}J_{CH} = 0.365\delta_C + 120.0 \tag{7}$$

$$^{1}J_{CH} = 0.710\delta_C + 101.0 \tag{8}$$

As a result the $^{1}J_{CH}$ couplings can be refocused more accurately simply by matching the sweep rate of the adiabatic pulse with the natural distribution of the J couplings, as shown schematically in Fig. 8.[23,24]

For example, let us estimate the optimum length of an adiabatic inversion pulse for the ^{13}C–^{1}H HSQC experiment shown in Fig. 3 with sequential implementation of the adiabatic sweep. Assuming that the ^{13}C carrier frequency is centered at 70 ppm and the sweep range is set to 200 ppm, the required length of the adiabatic pulse T_p for a linear frequency sweep is calculated as

$$T_P = 0.25 \left(\frac{J_{max} - J_{min}}{J_{max} J_{min}} \right) \tag{9}$$

From Eq. 7 we find $T_p = 0.919$ ms, which is independent of the spectrometer frequency. Note that the sweep direction in this experiment is from aliphatics (smaller $^{1}J_{CH}$) to aromatics (larger $^{1}J_{CH}$).

Using correlations such as Eqs. (7) and (8) provides a better approximation than using an average coupling constant, which assumes uniform distribution of

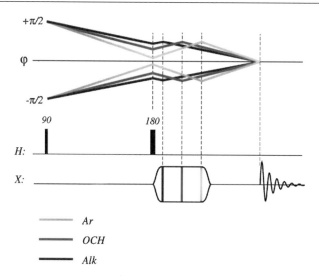

FIG. 8. Matching the direction and rate of the adiabatic frequency sweep with the natural distribution of the $^1J_{CH}$ couplings. The alkyl groups with smaller $^1J_{CH}$ couplings need to be refocused first, while the carbon nuclei in the aromatic region typically have larger couplings and need to be refocused last. The frequency sweep direction is from low (Alk) to high (Ar).

J couplings. Even more accurate results can be obtained by considering evolution of the $^1J_{CH}$ couplings in the presence of the RF field during the adiabatic sweep. This is important in experiments where quantitative measurements are required, for instance measurements of intermolecular nuclear Overhauser effects (NOEs) in protein–RNA complexes.[23]

6. Heteronuclear Decoupling

6.1. Techniques of Adiabatic Decoupling

The adiabatic HSQC experiment would not be complete without equally efficient decoupling scheme. Already on 600 MHz instruments the conventional decoupling schemes have reached their limits in applications where the full chemical shift range of ^{13}C needs to be covered. On the other hand, adiabatic decoupling allows covering essentially unlimited bandwidth. Indeed, ^{13}C decoupling of a frequency range in excess of 1 MHz has been demonstrated using the WURST scheme.[14,27] This provides an important solution for NMR of biomolecules at very high magnetic fields. As shown in Fig. 9, the entire ^{13}C chemical shift range at

[27] Ē. Kupče, R. Freeman, G. Wider, and K. Wüthrich, *J. Magn. Reson. Ser. A* **120,** 264 (1996).

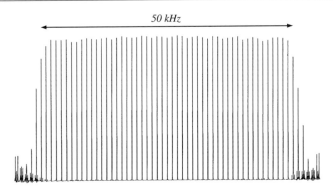

FIG. 9. Adiabatic (WURST-40) decoupling at 900 MHz (H-1) frequency. The test sample is 1% $^{13}CH_3I$ in $CDCl_3$. The average (RMS) amplitude of the decoupling field is 2.75 kHz. The decoupler frequency was incremented over a 60-kHz-wide band in 1 kHz steps. The spectra were recorded on a Varian INOVA spectrometer.

900 MHz can easily be decoupled using a moderate RF RMS amplitude of only 2.75 kHz. The same technique has been used for decoupling in Fig. 5.

The secret of the success of adiabatic decoupling is in its flexibility. Unlike the conventional decoupling schemes,[28] adiabatic decoupling can be adjusted to the magnitude of spin–spin couplings, which allows avoiding use of excessive RF power. A more detailed discussion of recent developments in spin-decoupling techniques has been published by Freeman and Kupče.[29]

6.2. Sideband Suppression in Heteronuclear Decoupling

One of the problems with any decoupling technique is the appearance of decoupling sidebands.[28] Asynchronous decoupling, which is used routinely with the conventional decoupling schemes, only partially addresses the problem and does not apply to adiabatic decoupling. Fortunately, it is relatively easy to manipulate with sidebands that arise from adiabatic decoupling. For example, changing the sweep direction in alternate scans makes the phase of decoupling sidebands absorptive and reduces their appearance.[24] On the other hand, the intensity of sidebands can be made more uniform by increasing the peak amplitude of the waveform. This is achieved by inserting delays between pulses[30] or by choosing a suitable amplitude function.[31-33] Complete suppression of decoupling sidebands requires somewhat more sophisticated techniques.[24,30,34]

[28] A. J. Shaka and J. Keeler, *Prog. NMR Spectrosc.* **19,** 47 (1987).
[29] R. Freeman and Ē. Kupče, *NMR in Biomedicine* **10,** 372 (1997).
[30] Ē. Kupče, R. Freeman, G. Wider, and K. Wüthrich, *J. Magn. Reson. Ser. A* **122,** 81 (1996).
[31] Z. Starčuk, Jr., K. Bartušek, and Z. Starčuk, *J. Magn. Reson. Ser. A* **107,** 24 (1994).
[32] M. R. Bendal, *J. Magn. Reson. Ser. A* **112,** 126 (1995).
[33] Ē. Kupče and R. Freeman, *Chem. Phys. Lett.* **250,** 523 (1996).

The intensity and position of decoupling sidebands directly depends on the length of adiabatic (or composite) inversion pulse T_d used as the basic building block in any decoupling scheme. In general, decoupling sidebands appear at multiples of $1/T_d$ and their distance from the main line increases with shorter T_d.[15] More importantly, decreasing pulse length T_d reduces the intensity of sidebands, unfortunately at the expense of increased RF power level. Although in many situations this may be acceptable, in general, the tolerable sideband level very much depends on the spectral dynamic range. Clearly, more efficient sideband suppression techniques are necessary in situations when high temperature stability is required, e.g., the DOSY experiment,[35] or in situations where small signals need to be measured quantitatively in the vicinity of very strong signals, for example in nuclear Overhauser effect spectroscopy (NOESY)-type experiments.[27,30]

It is often argued that decoupling schemes that use high RF peak amplitude, e.g., STUD,[31,32] produce less intense and more uniform decoupling sidebands. Although this is true for the traditional test samples employing two-spin systems such as formic acid, in practice, the distribution of the J couplings is often nonuniform. As a result the decoupling schemes based on linear frequency sweeps, e.g., WURST[14,15] or chirp,[36,37] are, in fact, more efficient. Not only do they require less RF peak power, but also the appearance of decoupling sidebands is reduced. This is achieved by matching the frequency sweep direction and rate with the distribution of $^1J_{CH}$ couplings in the sample, as already discussed in the case of J refocusing (see Section 5).

The matched adiabatic decoupling (MAD)[34] attempts to keep the defocusing of spin isochromats during decoupling at minimum for all spins within the actual decoupling bandwidth. In order to achieve this, the spins with larger couplings need to be flipped first. Therefore, in a typical $^{13}C-^1H$ spin system the correct sweep direction for MAD is from aromatics to aliphatics. The optimum decoupling pulse length, T_d is calculated as[24]:

$$T_d = 0.5/J_{min} - 0.5/J_{max} \qquad (10)$$

where J_{max} and J_{min} are estimated from Eq. (7) or (8). For instance, assuming that ^{13}C carrier frequency is centered at 80 ppm, from Eq. (7) we can estimate that for a 160 ppm wide sweep the optimum duration of the decoupling pulse $T_d = 1.364$ ms. An example of sideband suppression using MAD is shown in Fig. 10.

In cases where there is no reasonable correlation between coupling constants and chemical shifts, the required dispersion of J-refocusing times can be introduced artificially (see Fig. 11a), using a time-ordering pulse (TOP). The TOP pulse ensures that the phase excursion of individual spin isochromats is independent

[34] È. Kupče, *J. Magn. Reson.* **129**, 219 (1997).

[35] H. Barjat, G. A. Morris, and A. G. Swanson, *J. Magn. Reson.* **131**, 131 (1998).

[36] V. J. Basus, P. D. Ellis, H. D. W. Hill, and J. S. Waugh, *J. Magn. Reson.* **35**, 19 (1979).

[37] R. Fu and G. Bodenhausen, *Chem. Phys. Lett.* **245**, 415 (1995).

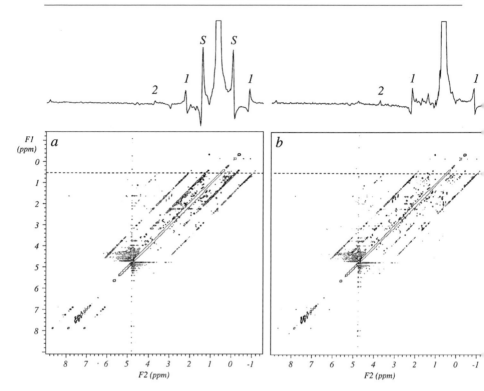

FIG. 10. Reduction of decoupling sidebands by matching the speed and direction of adiabatic sweep with the distribution of $^1J_{CH}$ couplings in the sample (see also Fig. 8); The two ^{13}C–^1H NOESY–HSQC spectra of $^{13}C,^{15}N$-labeled human ubiquitin were recorded at 600 MHz ^1H frequency with a short mixing time of 10 ms in order to enhance the appearance of decoupling sidebands. Both spectra were plotted close to the noise level. (a) The sweep direction is wrong and strong subharmonic sidebands (S) are clearly visible; (b) the sweep direction is matched with the distribution of the $^1J_{CH}$ couplings. Only the first (1) and second (2) order sidebands appear. The traces on top of the spectra were taken along the dotted line.

of the time when the spin is flipped during the frequency sweep. The effect of such reordering of J evolution is that the subharmonic sidebands are effectively eliminated. It can be shown that the required length of TOP is half the length of the decoupling pulse, $T_p = 0.5T_d$.[34]

Elimination of the principal and the higher harmonics requires systematic shifting of the modulation pattern introduced by decoupling.[30,34] This, in turn, can be achieved by inserting delays τ on both sides of the TOP pulse (see Fig. 11a):

$$\tau = nT_d/N \tag{11}$$

where N is the total number of increments.

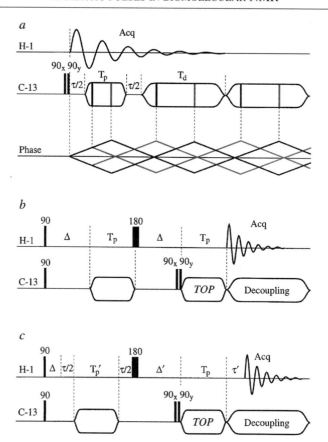

FIG. 11. Techniques for elimination of decoupling sidebands. (a) A time ordering pulse (TOP) is inserted before beginning of adiabatic decoupling. The length of the adiabatic time-ordering (TOP) pulse T_p is set to half the length of the decoupling pulse, $T_p = 0.5T_d$. This has an effect of eliminating the subharmonic sidebands over the active decoupling bandwidth. Incrementing the delay τ according to Eq. (11) eliminates the principal and higher order harmonics. The two 90° pulses remove any residual antiphase magnetization. (b) The same technique is implemented into a HSQC type experiment. Only the last coherence transfer step is shown. The TOP pulse has an effect of eliminating the subharmonic sidebands in a single step. (c) The same as in (b), except extra delays are inserted in order to suppress the principal and higher harmonics. The total length of the delays on both sides of the 180° ^{1}H pulse must be equal and remains constant throughout the experiment. Although not necessarily required, it is convenient to set $T'_p = T_p$. The τ' delay is incremented according to Eq. (11). At the same time the Δ' delay is decremented by the same amount. The τ delay is initially set to $(N - 1)T_d/N$ and is decremented N steps of T_d/N, while the Δ delay is incremented by T_d/N, so that $\Delta' + \Delta = 0.5/J_{av}$, where J_{av} is the average $^1J_{CH}$ coupling.

There are several ways in which the necessary time shifting of decoupling with respect to acquisition can be implemented. The τ delay can be incremented either sequentially or randomly. However, the most efficient sequence of time shifts can be derived using the bit-reversal reordering algorithm known from the fast Fourier transform.[39,40] This determines the order in which the delays are incremented. For example, in the case of sequential increments $n = 0, 1, 2, \ldots, N - 1$. With the bit-reversal reordering algorithm and $N = 4$ we find $n = 0, 2, 1, 3$. For $N = 8$ a more complex sequence of delays is obtained with $n = 0, 4, 2, 6, 1, 5, 3, 7$. Note that sequential implementation requires a full cycle of increments to be completed. Using the bit-reversal reordering algorithm ensures that the sideband suppression is always at its best, provided an even number of scans is acquired.

In practice it may not be straightforward to implement variable delays during the acquisition. The problem can be resolved by starting the decoupling well before acquisition. One possible scheme named SEAD (sideband elimination by adiabatic defocusing) is shown in Fig. 11(b) and (c), where the sideband suppression technique has been incorporated into the last step of an HSQC experiment. The main task here is to keep the total delay on both sides of the proton 180° pulse constant. In order to shift the decoupling with respect to acquisition, the τ' delay is incremented according to Eq. (11), while the τ delay is decremented at the same time. The two 90° ^{13}C pulses just prior to the TOP pulse serve to remove any possible remaining components of antiphase magnetization.[30,38]

The quality of sideband suppression that can be achieved using the SEAD technique is shown in Fig. 12. The mixing time of the NOESY-HSQC experiment has been set to 10 ms in order to enhance the sideband appearance and the spectrum is plotted close to the noise level. A four-step cycle proved to be sufficient for complete elimination of the decoupling sidebands. Some very weak residual responses may come from interference effects due to homonuclear $^1J_{CC}$ couplings.

6.3. Interference Effects in Adiabatic Decoupling

Usually the presence of homonuclear scalar couplings is ignored in both the theoretical treatment of spin decoupling and in practical implementations. However, most of the modern biomolecular NMR experiments use samples highly enriched in ^{13}C. Furthermore, the $^{13}C-^{13}C$ couplings in proteins are typically in the range between 35 and 55 Hz, which are not negligible compared to the typical one-bond $^{13}C-^1H$ couplings (120–180 Hz). The homonuclear couplings are known to interfere with the decoupling process and can seriously compromise the quality of the spectra.[28] The reason for such interference is very simple. Although

[38] M. R. Bendal and T. E. Skinner, *J. Magn. Reson.* **129**, 30 (1997).
[39] Ē. Kupče, Abstracts of the 41st Experimental NMR Conference, Asilomar, p. 124 (2000).
[40] R. N. Bracewell, "The Fourier Transform and Its Applications." McGraw-Hill, New York, 1986.

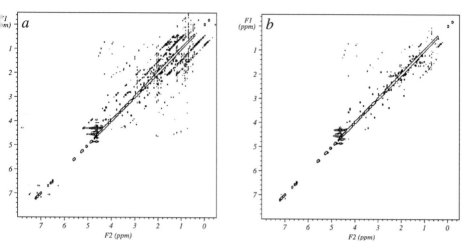

FIG. 12. Elimination of decoupling sidebands in a $^{13}C-^{1}H$ NOESY-HSQC experiment using the scheme shown in Fig. 11b. The two spectra of $^{13}C,^{15}N$-labeled protein (courtesy of Jens Duus, Carlsberg Laboratories) were recorded at 800 MHz ^1H frequency on Varian INOVA spectrometer using a short mixing time of 10 ms in order to enhance the appearance of decoupling sidebands. (a) Reference spectrum; (b) spectrum acquired using the SEAD technique (see Fig. 11) with $N = 4$. Slight reduction of cross peak intensities is due to the extra delays in the pulse sequence.

the evolution of the heteronuclear couplings is repeatedly refocused by the train of inversion pulses, the same pulses are, in general, transparent to homonuclear couplings, which continue to evolve. This gives rise to increased residual splittings and increased level of decoupling sidebands. Such interference is characteristic of composite pulse decoupling schemes[28] and also of adiabatic decoupling schemes that use pulses with high RF peak amplitudes, for instance, hyperbolic secant pulses.[31,32]

Linear frequency sweep provides an efficient mechanism for decoupling of homonuclear couplings. Because the spins are flipped sequentially, the homonuclear coupling is essentially refocused in the course of adiabatic sweep (see Fig. 13). The optimum sweep bandwidth in this case is twice the chemical shift difference between the two coupled spins, $\Delta\omega_{AX}$. The effect of homonuclear J refocusing on the quality of heteronuclear decoupling is demonstrated in Fig. 14. Unfortunately, the problem remains unsolved for strongly coupled spin systems.

If the decoupling pulses become too short (<0.6 ms), another interference mechanism comes into play. The coherence transfer by cross-polarization, or TOCSY, can generate similar artifacts.[5,41] The mechanism of TOCSY transfer will be discussed later in this chapter.

[41] P. B. Barker, A. J. Shaka, and R. Freeman, *J. Magn. Reson.* **65,** 535 (1985).

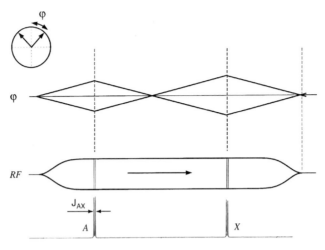

FIG. 13. Homonuclear decoupling during the adiabatic frequency sweep. As shown schematically by the phase evolution diagram, the evolution of the homonuclear J coupling is refocused by flipping the coupled spins sequentially. Complete homonuclear decoupling is achieved only if the sweep bandwidth is twice the chemical shift difference between spins A and X.

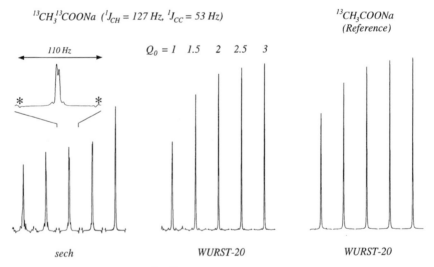

FIG. 14. Interference effects in ^1H–^{13}C heteronuclear decoupling caused by evolution of homonuclear $^1J_{CC}$ coupling in doubly ^{13}C-labeled sodium acetate ($^1J_{CC} = 53$ Hz, $^1J_{CH} = 127$ Hz). Heteronuclear decoupling using hyperbolic secant pulses (on the left) enhances residual splittings and sidebands, especially at low power settings. Linear frequency sweep used in WURST-20 scheme (in the middle) achieves simultaneous decouplings of both homonuclear $^1J_{CC}$ coupling and heteronuclear $^1J_{CH}$ coupling. The decoupling is essentially undisturbed, as seen from comparison with the reference spectrum (on the right) recorded using singly ^{13}C-labeled sodium acetate. The adiabaticity factor, Q_0 was incremented in all three experiments as $Q_0 = 1$, 1.5, 2, 2.5, and 3.

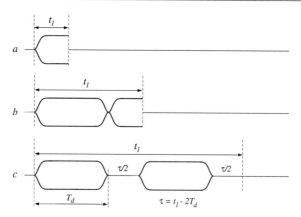

FIG. 15. Heteronuclear decoupling in the indirectly detected dimension using a pair of adiabatic pulses; (a) and (b) while the t_1 evolution time is shorter than $2T_d$ the adiabatic pulses are applied as a continuous decoupling sequence which is simply truncated after the t_1 period; (c) when $t_1 > 2T_d$ a delay $\tau/2 = t_1/2 - T_d$ is inserted on both sides of the second adiabatic pulse.

6.4. Adiabatic Decoupling in the Indirect Dimension

Clean 2D spectra can only be obtained if the decoupling is efficient in both dimensions. Although it is quite a common practice to use the same decoupling techniques in the indirectly detected dimension, F_1, a more efficient way of refocusing the heteronuclear J couplings is to use a single refocusing pulse in the middle of the evolution period, t_1. As we already discussed in the case of adiabatic HSQC experiment, rectangular $180°$ pulses are very inefficient if the bandwidth is large. Besides, the intensity of the sidebands in F_1 directly depends on how accurate the magnetization inversion is over the required bandwidth. The F_1-decoupling scheme employed to record the spectra shown in Figs. 10 and 12 is based on use of a pair of adiabatic pulses as explained in Fig. 15.[24,30] For short t_1 periods the adiabatic pulses are applied as if they were a decoupling waveform, which is truncated as soon as the t_1 period is over. When the t_1 period becomes longer than $2T_d$, the pair of adiabatic pulses is applied as in the case of adiabatic refocusing (see Fig. 6b). A relatively high adiabaticity of $Q_0 = 4$ was used to ensure a high accuracy of refocusing.

It has been demonstrated that a single adiabatic inversion pulse is sufficient for indirect decoupling in constant time experiments.[42]

---

[42] R. Konrat, D. R. Muhandiram, N. A. Farrow, and L. E. Kay, *J. Biomol. NMR* **9**, 409 (1997).

7. Homonuclear Decoupling

7.1. Homonuclear Decoupling in Indirect Dimension

As already mentioned in the previous section, adiabatic decoupling schemes are J-sensitive and therefore typically require substantially less power as compared to the conventional decoupling methods. This is what makes the homonuclear band selective decoupling so efficient.[17,43] The homonuclear couplings are typically relatively small as compared to the chemical shift differences between the groups of spins. This allows use of very low power levels to achieve efficient homonuclear decoupling. In turn, use of low RF power ensures minimum disturbances, such as Bloch–Siegert shifts,[43–45] sidebands, and other artifacts in the decoupled spectra.

It is very common to use ^{13}C–^{13}C homonuclear decoupling in the indirectly detected dimension to decouple CO[46,47] and/or C_β[48] form C_α in protein NMR experiments, such as HNCA,[49] HN(CO)CA,[49] HCCH-TOCSY,[50] and related experiments.[51] Use of low RF power also ensures minimum interference between adiabatic pulses in multiple band decoupling experiments. Often as many as four bands are decoupled simultaneously in protein NMR experiments (see Fig. 16).[49,50] Note that the bandwidth and pulse length for adiabatic decoupling are completely independent. In contrast to conventional decoupling schemes, this allows one to use the same pulse length for all decoupling bands even if they are of different width.

7.2. Homonuclear Decoupling during Acquisition

The ^1H–^1H homonuclear decoupling is typically used in the directly detected dimension. Therefore it must be applied during acquisition and requires time sharing.[52] This prevents use of the digital filters because of to the high factor of oversampling that is typically used on commercial systems. However, homonuclear decoupling in the time-shared mode implies that sidebands will inevitably be generated at the frequencies of $\Delta\nu_{SW}/2$, where $\Delta\nu_{SW}$ is the spectral width.[43] For this reason it is useful to apply the homonuclear decoupling close to the middle of the spectrum and to use slight oversampling. The decoupler duty cycle, d needs to

[43] Ē. Kupče, H. Matsuo, and G. Wagner, in "Biological Magnetic Resonance" (N. R. Krishna and L. J. Berliner, eds.), p. 149. Kluwer Academic/Plenum Publishers, New York, 1998.

[44] F. Bloch and A. Siegert, Phys. Rev. **57**, 522 (1940).

[45] D. Suter, G. C. Chingas, R. A. Harris, and A. Pines, Mol. Phys. **61**, 1327 (1987).

[46] M. A. McCoy and L. Mueller, J. Am. Chem. Soc. **114**, 2108 (1992).

[47] M. A. McCoy and L. Mueller, J. Magn. Reson. **98**, 674 (1992).

[48] Ē. Kupče and G. Wagner, J. Magn. Reson. Ser. B **110**, 309 (1996).

[49] H. Matsuo, Ē. Kupče, H. Li, and G. Wagner, J. Magn. Reson. Ser. B **113**, 91 (1996).

[50] H. Matsuo, Ē. Kupče, and G. Wagner, J. Magn. Reson. Ser. B **113**, 190 (1996).

[51] Y. Lin and G. Wagner, J. Biomol. NMR **15**, 227 (1999).

[52] B. L. Tomlinson and H. D. W. Hill, J. Chem. Phys. **59**, 1775 (1973).

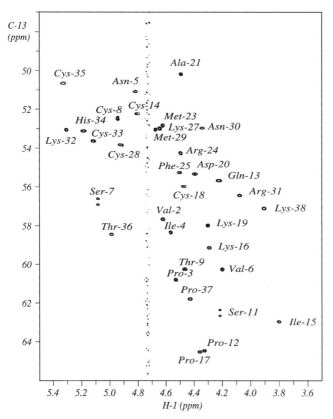

FIG. 16. $^1H–^1H$ and $^{13}C–^{13}C$ homonuclear decoupled $^1H_\alpha–^{13}C_\alpha$ HSQC spectrum of 0.3 mM solution of 39 residue protein agitoxin (courtesy of Prof. G. Wagner, Harvard Medical School, Boston, MA) in H_2O/D_2O (9 : 1) recorded at 600 MHz ^1H frequency. The $^1H–^1H$ homonuclear decoupling was implemented using 20 ms long CA-WURST-10 pulses covering 3 ppm wide bands applied at NH and $C_\beta H$ proton frequencies. Opposite sense frequency sweeps centered at 7.75 and 1.75 ppm were used to minimize modulation sidebands and the water response, as discussed in the text. The C–C homonuclear decoupling was implemented in a similar fashion using 5 ms long pulses covering bands from 160 to 180 ppm (CO), 63 to 67 (Thr), 20 to 40 (C_β), and 12 to 16 ppm (Ala).

be as low as possible, because the sensitivity reduction, r due to restricted sampling time directly depends on d:

$$r = 1 - \sqrt{1 - d} \qquad (12)$$

In practice, we typically use a 5% duty cycle, which makes the corresponding sensitivity loss negligible (ca. 2.2%). The gain in resolution and sensitivity due to the homonuclear decoupling typically outweighs the loss, especially in relatively small molecules. This is shown in Figs. 16 and 17.

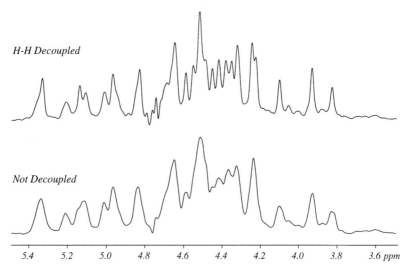

FIG. 17. The effect of H–H homonuclear decoupling on the linewidth of $C_\alpha H$ protons in the H-1 spectrum of a 39 residue protein, agitoxin. The experimental conditions are equivalent to those in Fig. 16.

7.3. Homonuclear Decoupling in Partially Oriented Molecules

In larger proteins the benefits of ^1H–^1H homonuclear decoupling can be masked by the increased linewidth. However, it becomes beneficial again in oriented systems, where proton linewidth increases dramatically due to multiple dipolar (through space) ^1H–^1H couplings.[53] As demonstrated in Fig. 18, not only do the spectra benefit from increased resolution and sensitivity, but more accurate measurement of heteronuclear dipolar couplings becomes possible. Note that the measurement accuracy is not affected by the Bloch–Siegert frequency shifts.[44,45]

7.4. Suppression of Modulation Sidebands in Homonuclear Decoupling

One of the problems in homonuclear decoupling is modulation sidebands.[17,43] They surround every peak in the decoupled spectrum, irrespective of the coupling. The modulation sidebands appear at the same frequencies as the decoupling sidebands, i.e., at multiples of $1/T_d$. However, the phase of modulation sidebands is opposite on both sides of the main peak. As a result the intensities of sidebands in homonuclear decoupled spectra can be asymmetric.[43] Because of the proximity of the decoupling field to the detection region, the modulation sidebands in homonuclear decoupling often are considerably stronger than the decoupling sidebands.

[53] C. W. Vander Kooi, Ē. Kupče, E. R. P. Zuiderweg, and M. Pellecchia, *J. Biomol. NMR* **15**, 335 (1999).

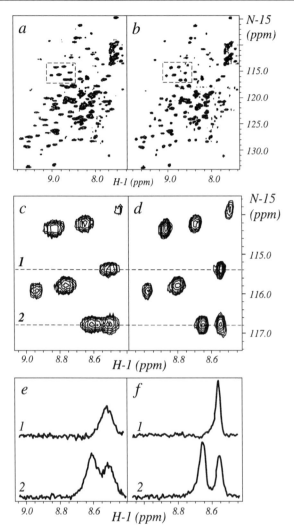

FIG. 18. ^{15}N–^{1}H correlation spectra of 0.7 mM DnaK substrate binding unit (residues 386–561) dissolved in magnetically aligned liquid crystalline bicelles. Only the slowly relaxing ^{15}N doublet component is observed. The spectra were measured at 800 MHz H-1 frequency on Varian INOVA spectrometer (a) without and (b) with homonuclear ^{1}H decoupling during acquisition; (c) and (d) represent zoomed regions marked by dotted rectangles in spectra (a) and (b), respectively; (e) and (f) show traces taken along the dotted lines in (c) and (d), correspondingly. The homonuclear ^{1}H–^{1}H decoupling was achieved using 1.6 ms long WURST-4 pulses applied from 0 to 5 ppm during acquisition. The decoupling pulse length T_d is determined from the observed proton linewidth, LW as $T_d = 0.2/LW$.

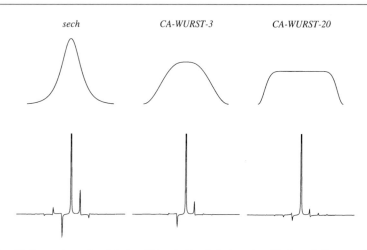

FIG. 19. Comparison of modulation sidebands (simulation) produced by hyperbolic secant (on the left), CA-WURST-3 (in the middle), and WURST-20 (on the right) homonuclear decoupling and relative amplitudes of corresponding adiabatic inversion pulses (top traces). The main lines are truncated at the 50% level.

Fortunately, it is also somewhat easier to suppress them. There are several possible strategies.

First, it is always advantageous to use a decoupling scheme that gives the lowest sidebands to start with. The intensity of modulation sidebands directly depends on the peak amplitude of the decoupling pulses (see Fig. 19). Therefore, once again, using pulses with high RF peak amplitude, such as hyperbolic secant or CA-Gaussian, is not recommended.

One possible way of suppressing the modulation sidebands is to use a compensating decoupling field symmetrically to the observation band.[54,55] Although in the center of the detection band the modulation sidebands seem to have vanished (see Fig. 20c), the intensity of the central line is not restored to 100%. The modulation sidebands have simply been relocated to a different place. They now appear far from the region of interest—at the frequencies $\pm\Delta\omega_d$ where the decoupling fields are applied. This method has also two other drawbacks. First, use of a compensating decoupling field requires doubling of the RF amplitude. Second, the compensating field may overlap with the spectral regions of interest.

In some cases, use of compensating decoupling may be dictated by the spectral properties of the compounds of interest. For example, ^1H–^1H homonuclear decoupling of NH and $C_\beta H$ in proteins can be centered around water signal at 4.75 ppm

[54] M. A. McCoy and L. Mueller, *J. Magn. Reson.* **99**, 18 (1992).
[55] S. Zhang and D. G. Gorenstein, *J. Magn. Reson.* **132**, 81 (1998).

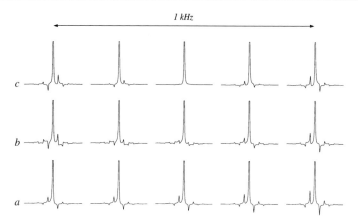

FIG. 20. Offset dependence of modulation sidebands recorded at 500 MHz (^{1}H) using a doped H$_2$O/D$_2$O (9:1) sample; (a) single decoupling field, (b) double decoupling field with the same sweep direction, and (c) double decoupling field with frequency swept in opposite directions. The transmitter offset was incremented in five 250 Hz steps. The decoupling bandwidth in all spectra was 1.5 kHz wide and centered at ±1.5 kHz; the pulse length of T_d = 20 ms (as for a 10 Hz ^{1}H–^{1}H coupling) and duty cycle of 5% were used. The RF peak amplitude was 5.82 kHz (RMS field strength of 0.58 kHz) in (a) and 11.61 kHz (RMS field strength of 0.82 kHz) in (b) and (c). The modulation sidebands appear at multiples of 50 Hz on both sides of the main line. Hyperbolic secant decoupling was used in order to enhance the appearance of sidebands.

with symmetric decoupling bands covering 3 ppm wide spectral regions from 0 to 3 ppm and from 6.5 to 9.5 ppm (see Fig. 17). This helps to reduce the artifacts otherwise introduced by the strong residual water peak.

Unlike decoupling sidebands, the modulation sidebands can be reduced by means of asynchronous decoupling.[28,56] The latter requires that decoupling always be started at a different point. Use of the bit-reversal algorithm discussed in Section 6 can also be adopted here. The number of steps that is necessary to eliminate all modulation sidebands can potentially be reduced to two, provided a suitable decoupling waveform can be found. Indeed, the harmonics other than the principal sidebands are very weak in the case of CA-WURST-3 decoupling (see Fig. 19).

One of the problems with asynchronous decoupling is that it actually enhances the subharmonic decoupling sidebands (see Fig. 21). Fortunately, the decoupling sidebands typically are negligible as compared to the modulation sidebands and in most applications can safely be ignored.

Finally, postacquisition data processing is another alternative to eliminate the modulation sidebands.[56] Like most of the other methods, it does not restore the intensity of the main signal.

[56] J. Weigelt, A. Hammarström, W. Bermel, and G. Otting, *J. Magn. Reson. Ser. B* **110**, 219 (1996).

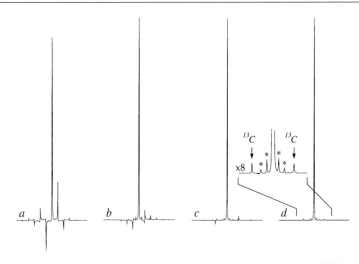

FIG. 21. Asynchronous homonuclear decoupling using bit-reversal reordering (BRR) algorithm; experimental conditions as described in Fig. 20. The decoupling field was applied to the low frequency doublet of uracil ($^3J_{HH} = 7.6$ Hz), while the high-frequency signal was observed. (a) Synchronous decoupling using hyperbolic secant pulses produces very strong modulation sidebands; (b) synchronous CA-WURST-20 decoupling; (c) asynchronous 2-step BRR decoupling; (d) asynchronous 4-step BRR decoupling. While asynchronous decoupling reduces the strong modulation sidebands, the weaker decoupling sidebands marked by (*) are not suppressed (see the inset). The ^{13}C satellites are marked by arrows.

8. Isotropic Mixing

8.1. *Adiabatic Mixing in Static Samples*

Not surprisingly, adiabatic pulses can also be used for isotropic mixing.[57] Unlike in the case of decoupling, the bandwidth that can be achieved in an adiabatic TOCSY experiment is quite limited and there is no real advantage over the conventional mixing sequences as far as the bandwidth is concerned. The main advantage of adiabatic mixing is its low sensitivity to RF inhomogeneity and miscalibration.

An important parameter, which needs to be properly adjusted in adiabatic TOCSY experiments, is the modulation depth (μ). As shown in Fig. 22, as the modulation depth increases the mixing bandwidth easily is propagated along the diagonal, but not along the antidiagonal. In practice it is reasonable to use intermediate values of μ. The performance of the conventional (DIPSI-2) and adiabatic (CA-WURST-8) mixing schemes is compared in Fig. 23. It should be noted that

[57] Ē. Kupče, P. Schmidt, M. Rance, and G. Wagner, *J. Magn. Reson.* **135**, 361 (1998).

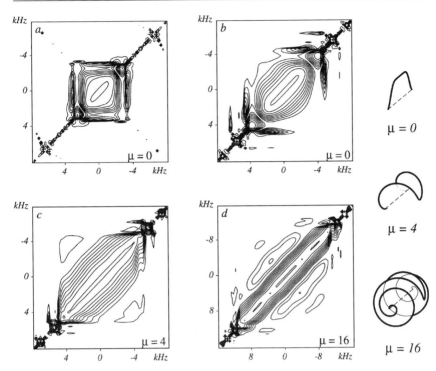

FIG. 22. The TOCSY coherence transfer efficiency calculated for WURST-8 waveforms with different modulation depth (μ) assuming $J = 10$ Hz; (a) $\mu = 0$, MLEV-4 phase cycle, $T_p = 0.1$ ms, $B_1(\text{max}) = 6.95$ kHz, B_1 (RMS) $= 5.57$ kHz; (b) the same as in (a), but with a 20-step phase cycle; (c) $\mu = 4$, sweep bandwidth 16 kHz, $T_p = 0.25$ ms, $B_1(\text{max}) = 5.6$ kHz, B_1 (RMS) $= 4.52$ kHz, $Q_0 = 2$; (d) $\mu = 8$, sweep bandwidth 32 kHz, $T_p = 0.25$ ms, $B_1(\text{max}) = 9.7$ kHz, B_1 (RMS) $= 7.84$ kHz, $Q_0 = 3$. On the right hand side the effect of modulation depth on the appearance of the mixing pulse shape is shown for different magnitudes of μ.

in order to suppress cross-relaxation artifacts a somewhat lower WURST index must be used. The "clean" mixing has been demonstrated for the CA-WURST-2 sequence.[57] Unlike the conventional approach, which requires inserting delays into a mixing sequence, changing the WURST index does not increase the average (RMS) power requirements for the given mixing bandwidth.

The resistance of adiabatic mixing sequences to RF miscalibration increases with μ and deeper modulation might be justified in situations when RF homogeneity is low (see Fig. 24). Obviously, when it comes to comparison of the effective bandwidth between various adiabatic mixing schemes, the same modulation depth should be used in order to avoid unnecessary confusion.

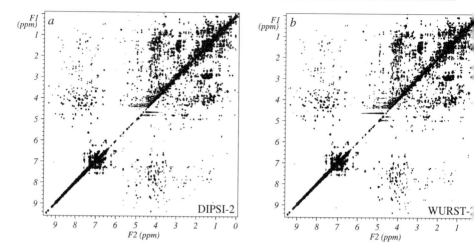

FIG. 23. Comparison of 2D TOCSY spectra of 1mM solution of hen lysozyme in H_2O/D_2O (9 : 1), recorded at 900 MHz using (a) DIPSI-2 mixing scheme and (b) WURST-2 mixing, 54 ms mixing time.

8.2. TOCSY with Magic Angle Spinning (MAS) in Liquids

Typically the NMR spectra of liquids are recorded using static samples. However, in certain situations the spectral resolution can be dramatically increased if the sample is spun at the magic angle. This allows one to remove line broadening due to nonuniform distribution of magnetic susceptibility, both within the sample and in the proximity of the detection coils. Such broadening is usually observed in heterogeneous samples and also in microprobes designed to work with very small amounts of liquids. For example, as little as 40 μl of sample is required with Varian's nanoprobe.[58,59]

Although the vast majority of liquid NMR experiments show very little or no degradation in performance, experiments employing spin lock periods, such as isotropic mixing and spin decoupling, can be affected dramatically (see Fig. 25). The reason for this degradation is interference between the sample spinning and the RF field.[60] Similar effects are well known from NMR in solids.[61,62]

[58] T. M. Barabara, J. Magn. Reson. Ser. A 109, 265 (1994).
[59] S. K. Sarkar, R. S. Garigipati, J. L. Adams, and P. Keifer, J. Am. Chem. Soc. 118, 2305 (1996).
[60] Ē. Kupče, P. Keifer, T. Barabara, K. Mehr, D. Rice, and M. Delepierre, Abstracts of the 41st Experimental NMR Conference, Asilomar, p. 124 (2000).
[61] D. Marks and S. Vega, J. Magn. Reson. Ser. A 118, 157 (1996).
[62] X. Wu and K. W. Zilm, J. Magn. Reson. Ser. A 104, 154 (1993).

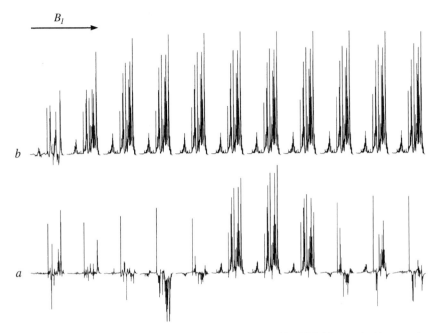

FIG. 24. Tolerance of 1D TOCSY spectra to miscalibration of the RF field strength; (a) composite pulse mixing (DIPSI-2) with optimum B1 field strength of 3.75 kHz; the RF power level was incremented in 1 dB steps from 2.16 to 6.84 kHz; (b) WURST-8 mixing with $T_p = 0.25$ ms and sweep bandwidth of 16 kHz, the RF power was increased from 3.96 kHz (3.2 kHz RMS) to 12.52 kHz (10.12 kHz RMS). The spectra were recorded at 400 MHz with a 5mM solution of the 14 kDa protein villin 14T in D_2O; 60 ms mixing time.

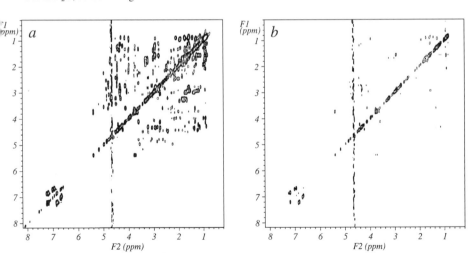

FIG. 25. Comparison of 2D TOCSY spectra of 1 mM solution of a protein from scorpion venom dissolved in 40 μl of D_2O and recorded in gXH nanoprobe at 500 MHz and 2.5 kHz MAS speed using 60 ms mixing time; (a) adiabatic (WURST-2) mixing scheme and (b) composite pulse (DIPSI-2) mixing.

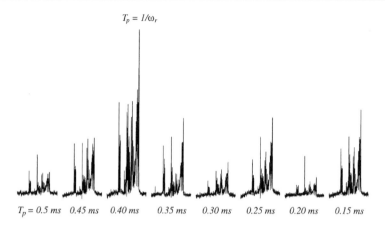

$T_p = 1/\omega_r$

$T_p = 0.5\ ms$ $0.45\ ms$ $0.40\ ms$ $0.35\ ms$ $0.30\ ms$ $0.25\ ms$ $0.20\ ms$ $0.15\ ms$

FIG. 26. The effect of WURST-2 mixing pulse length T_p on the signal intensity of the first increment of a 2D ^{1}H–^{1}H TOCSY experiment shown in Fig. 25a. The maximum signal is observed at the rotational resonance condition, when the pulse repetition rate is equal to the rotation frequency ω_r.

For a homonuclear spin system the isotropic mixing Hamiltonian H_{mix} in the presence of coherent (CW) irradiation for a static sample is written as[63,64]:

$$H_{\mathrm{mix}} = -\Delta\omega_I I_x - \Delta\omega_s S_x + J(I_x S_x + I_y S_y + I_z S_z) - \omega_1 I_x - \omega_1 S_x \qquad (13)$$

When the sample is spun at the magic angle, the ω_1 and $\Delta\omega$ terms become time dependent because of inhomogeneity of both the B_0 and B_1 fields:

$$\omega(t) = \sum_k \omega_k e^{ik\omega_r t} \qquad (14)$$

When the sample spinning speed matches the repetition rate of inversion pulses in the spin lock sequence, the sample can be regarded as if it is static in the RF reference frame. This is known as the rotational resonance condition.[64,65] The time-dependent terms of Eq. (14) can be regarded as a distortion of the amplitude and frequency profile of the mixing waveform. Provided such distortions are not very large, the isotropic mixing proceeds essentially undisturbed. Adiabatic pulses are particularly tolerant of such distortions. More importantly, the supercycles, which are an essential part of any mixing sequence, are efficient only if all the inversion pulses are equal, and this is only the case when the rotational resonance condition is met.[60]

The effect of sample spinning rate on the efficiency of the TOCSY experiment is demonstrated in Fig. 26. The maximum signal is observed when the rotational resonance is brought about.

[63] L. Müller and R. R. Ernst, *Mol. Phys.* **38**, 963 (1979).
[64] M. Levitt, *J. Chem. Phys.* **94**, 30 (1991).
[65] T. G. Oas, R. G. Griffin, and M. H. Levitt, *J. Chem. Phys.* **89**, 692 (1988).

9. Conclusions

Adiabatic pulses provide a very efficient means for improving the performance of many experiments used in biomolecular NMR. Not only are these pulses virtually insensitive to calibration errors and to poor RF homogeneity, but also essentially unlimited bandwidth can be inverted with a very high degree of accuracy. This has many useful applications such as wide-band spin decoupling, homonuclear decoupling, adiabatic mixing, line narrowing in partially oriented systems, J-filtering experiments, and wide-band refocusing. The implementation of adiabatic pulses into NMR experiments is straightforward and automatic experiment setup is feasible.

[4] Scalar Couplings Across Hydrogen Bonds

By STEPHAN GRZESIEK, FLORENCE CORDIER, and ANDREW J. DINGLEY

Introduction

The concept of a hydrogen bond (H-bond) has long been recognized[1-3] as a very useful model to explain the weak attractive forces between an electronegative acceptor atom and a hydrogen atom attached to a second electronegative atom. Compared to covalent bonds, the activation energies for the formation of an H-bond and the bond energies themselves are small, such that H-bonds are readily formed and broken between various partners under conditions where donor and acceptor groups undergo diffusive processes under ambient temperatures in many common solvents. This property explains the key role that H-bonds play in the formation and stabilization of biomacromolecular structures and as participants in many chemical reactions.[4,5]

In biological systems, H-bond donors and acceptors are predominantly nitrogen and oxygen atoms. However, sulfur, metallic cofactors, and sometimes π-electron-rich aromatic groups are also involved in hydrogen bonding. In most cases, evidence for individual hydrogen bonds in biomacromolecular structures is derived from the spatial proximity of donor and acceptor groups once the structure of a molecule is determined by diffractive or nuclear magnetic resonance (NMR) techniques. More detailed information about H-bonds from X-ray diffraction is

[1] M. Huggins, Thesis, University of California (1919).

[2] W. Latimer and W. Rodebush, *J. Am. Chem. Soc.* **42,** 1419 (1920).

[3] L. Pauling, "The Nature of the Chemical Bond." Cornell University Press, 1960.

[4] A. Fersht, "Enzyme Structure and Mechanism." 2nd Ed. W. H. Freeman, New York, 1985.

[5] G. A. Jeffrey and W. Saenger, "Hydrogen Bonding in Biological Structures." Springer, New York, 1991.

particularly hard to obtain since the weak scattering density of the hydrogen atom can only be detected in structures solved at highest resolutions, i.e., better than ~1.0 Å. Such structures currently comprise about 0.6% of all crystallographic structures entered in the Brookhaven protein data bank. Only in these cases is it possible to ascribe individual spatial positions to the hydrogen atoms that are independent of the use of standard covalent geometries. Neutron diffraction has yielded high resolution information on the position of either proton or deuterium nuclei in H-bonds for a small number of biomacromolecules for which large crystals can be grown. Recent developments in this field[6] have reduced the required size of crystals to about 1 mm^3.

Numerous NMR observables for individual nuclei in hydrogen-bonded moieties have been shown to correlate with the stability and relative geometry of individual H-bonds.[7] Such parameters include the reduced hydrogen exchange rates with the solvent,[8,9] fractionation factors,[8,10–12] isotope shifts for the substitution of the hydrogen bonded proton by ^{2}H and ^{3}H,[13,14] the isotropic and anisotropic chemical shifts of the hydrogen-bonded proton and of other nuclei within the donor and acceptor groups,[15–22] the size of the electric field gradient at the position of the proton within the H-bond as observed by the deuterium quadrupolar coupling constant,[23,24] and the sequential one-bond $^1J_{C'N}$ coupling constants in proteins.[25,26]

[6] N. Niimura, Curr. Opin. Struct. Biol. **9**, 602 (1999).

[7] E. D. Becker, in "Encyclopedia of Nuclear Magnetic Resonance" (D. M. Grant and R. K. Harris, eds.), p. 2409. 1996.

[8] A. Hvidt and S. Nielsen, Adv. Prot. Chem. **21**, 287 (1966).

[9] G. Wagner, Q. Rev. Biophys. **16**, 1 (1983).

[10] S. Loh and J. Markley, Biochemistry **33**, 1029 (1994).

[11] A. LiWang and A. Bax, J. Am. Chem. Soc. **118**, 12864 (1996).

[12] P. Bowers and R. Klevitt, J. Am. Chem. Soc. **122**, 1030 (2000).

[13] G. Gunnarsson, H. Wennerström, W. Ega, and S. Forsen, Chem. Phys. Lett. **38**, 96 (1976).

[14] L. J. Altman, D. Laungani, G. Gunnarsson, H. Wennerström, and S. Forsen, J. Am. Chem. Soc. **100**, 8264 (1978).

[15] R. R. Shoup, H. T. Miles, and E. D. Becker, Biochem. Biophys. Res. Commun. **23**, 194 (1966).

[16] V. Markowski, G. R. Sullivan, and J. D. Roberts, J. Am. Chem. Soc. **99**, 714 (1977).

[17] G. Wagner, A. Pardi, and K. Wüthrich, J. Am. Chem. Soc. **105**, 5948 (1983).

[18] N. Asakawa, S. Kuroki, H. Kuroso, I. Ando, A. Shoji, and T. Ozaki, J. Am. Chem. Soc. **114**, 3261 (1992).

[19] A. McDermott and C. I. Ridenour, in "Encyclopedia of Nuclear Magnetic Resonance" (D. M. Grant and R. K. Harris, eds.), p. 3820. 1996.

[20] N. Tjandra and A. Bax, J. Am. Chem. Soc. **119**, 8076 (1997).

[21] M. Tessari, H. Vis, R. Boelens, R. Kaptein, and G. W. Vuister, J. Am. Chem. Soc. **119**, 8985 (1997).

[22] A. Takahashi, S. Kuroki, I. Ando, T. Ozaki, and A. Shoji, J. Mol. Struct. **442**, 195 (1998).

[23] J. Boyd, T. K. Mal, N. Soffe, and I. D. Campbell, J. Magn. Reson. **124**, 61 (1997).

[24] A. C. LiWang and A. Bax, J. Magn. Reson. **127**, 54 (1997).

[25] N. Juranic, P. K. Ilich, and S. Macura, J. Am. Chem. Soc. **117**, 405 (1995).

[26] N. Juranio, V. A. Likic, F. G. Prendergast, and S. Macura, J. Am. Chem. Soc. **118**, 7859 (1996).

TABLE I

OBSERVED TRANS HYDROGEN BOND COUPLINGS IN NUCLEIC ACIDS

Structure	Size (Hz)	N_d	N_a	Type	Ref.
(A) $^{h2}J_{NN}$	6–7	G-N1	C-N3	Watson–Crick	31,32,35,63
	7–8	U-N3	A-N1	Watson–Crick	31,73
	7–8	T-N3	A-N1	Watson–Crick	32,35,63
	6–8	T-N3	A-N7	Hoogsteen	32
	10–11	C-N3$^+$	G-N7	Hoogsteen	32
	5.5	U-N3	A-N7	Reverse Hoogsteen	73
	5.0	G-N1	A-N1	GA mismatch	73
(B) $^{h1}J_{HN}$	3–4	G-N1	C-N3	Watson–Crick	32,35,63
	2–3	T-N3	A-N1	Watson–Crick	32,35,63
	2.5–3	T-N3	A-N7	Hoogsteen	32
	<2	C-N3$^+$	G-N7	Hoogsteen	32
(C) $^{h2}J_{NN}$	2.5	A-N6	A-N7	A-A mismatch	40
	6–8	G-N2	G-N7	G$_4$-tetrad	41
	nd^a	G-N2	A-N7	A$_2$G$_4$-hexad	42
	nd	G-N2	G-N7	A$_2$G$_4$-hexad	42
	nd	C-N4	A-N7	GCGC-tetrad	42
(D) $^{h3}J_{NC}$	~0.2^b	G–N1	G-C6	G$_4$-tetrad	41
(E) $^{h4}J_{NN}$	~0.14^b	G-N1	G-N1	G$_4$-tetrad	44

^{a} nd, Not determined.
^{b} Sign not determined experimentally.

Electron-mediated, scalar couplings have been discovered in biological macro-molecules that are active between magnetic nuclei on both sides of the hydrogen bridge (Table I and II). These couplings are closely related to similar inter- and intramolecular couplings across hydrogen bridges in small model compounds.[27–30]

[27] R. Crabtree, P. Siegbahn, O. Eisenstein, A. Rheingold, and T. Koetzle, *Acc. Chem. Res.* **29**, 348 (1996).

[28] I. G. Shenderovich, S. N. Smirnov, G. S. Denisov, V. A. Gindin, N. S. Golubev, A. Dunger, R. Reibke, S. Kirpekar, O. L. Malkina, and H.-H. Limbach, *Ber. Bunsenges. Phys. Chem.* **102**, 422 (1998).

[29] O. Kwon and S. Danishefsky, *J. Am. Chem. Soc.* **120**, 1588 (1998).

[30] N. S. Golubev, I. G. Shenderovich, S. N. Smirnov, G. S. Denisov, and H.-H. Limbach, *Chem. Eur. J.* **5**, 492 (1999).

TABLE II

OBSERVED TRANS HYDROGEN BOND COUPLINGS IN PROTEINS

Structure	Size (Hz)	Donor	Acceptor	References
(A) $C^{\alpha}-H^{\alpha}$ / N—H ···· O=C / O=C $^{h3}J_{NC}$	−0.2 to −0.9	Backbone amide	Backbone carbonyl or side-chain carboxylate	45–48
(B) N—H ···· O=C $^{h2}J_{HC}$	−0.6 to 1.3	Backbone amide	Backbone carbonyl	49,50
(C) N—H ···· O=C C^{α} $^{h3}J_{HC\alpha}$	0–1.4	Backbone amide	Backbone carbonyl	51
(D) N—H ···· N $^{h2}J_{NN}$	8–11	Histidine-Nε2	Histidine-Nε2	52
(E) N—H ···· O=P $^{h3}J_{NP}$	4.6[a]	Backbone amide	GDP-Phosphate	53
(F) N—H ···· O=P $^{h2}J_{HP}$	3.4[a]	Backbone amide	GDP-Phosphate	53
(G) N—H ···· S ···· Me $^{h2}J_{HMe}$	0.3–4[a]	Backbone amide	Cysteine-S coordinating ^{113}Cd or ^{119}Hg	54,74

[a] Sign not determined experimentally.

The scalar coupling effect can be used to "see" individual H-bonds within a biological macromolecule. This means that the frequencies of all three H-bond partners, i.e., of the donor, the acceptor, and the proton itself, can be correlated by a single two- or three-dimensional NMR experiment. Thus, in favorable cases, the hydrogen bond connectivity pattern of nucleic acids and proteins (and therefore the secondary structure) can be established directly via a COSY experiment. In addition to this direct structural use, the size of the scalar coupling has been shown to be influenced by the relative geometry of the H-bond partners. Therefore valuable information about the "strength" of individual H-bonds can be derived from an analysis of the size of the coupling constants.

Because scalar couplings had been thought to be associated with normal covalent bonds, the initial observations of H-bond couplings came as a surprise. However, it seems now well established[28,31–34] that the H-bond couplings are the result of the same nucleus → electron → nucleus polarization mechanism as any other covalent scalar coupling. Therefore, the H-bond couplings behave experimentally in exactly the same way as their more usual covalent counterparts.

It is the purpose of this article (1) to give an overview over the range of groups for which such couplings have been observed so far and to summarize their sizes, (2) to discuss experiments for their detection and to point at experimental problems, and (3) to suggest possible applications.

H-Bond Scalar Couplings in Nucleic Acids

Imino–N–H·· ·N-Aromatic H-Bonds

Scalar couplings across H-bonds are expected to be strongest when the number of bonds between the two interacting nuclei is small. It is therefore no surprise that the strongest H-bond couplings in biomolecules are observed within nucleic acid–base pairs when both donor and acceptor are ^{15}N-labeled nitrogen atoms. Figure 1A shows as an example the Watson–Crick base pairing scheme for U-A (T-A) and G-C pairs in RNA (DNA) where the central hydrogen bond connects the imino nitrogen N3 of U (T) to the aromatic acceptor nitrogen N1 of A and N1 of G is hydrogen bonded to the acceptor N3 of C. In such a situation, trans H-bond couplings only need to bridge one ($^{h1}J_{HN}$) or two ($^{h2}J_{NN}$) bonds (including the H-bond) in order to be observable (see also Table I). The symbol $^{hn}J_{AB}$ was introduced by Wüthrich and co-workers as a notation[35] for trans H-bond scalar couplings between nuclei A and B in order to emphasize that one of the n bonds connecting the two nuclei in the chemical structure is actually an H-bond.

Figure 2 shows as an example an E.COSY type detection of both $^{h1}J_{NN}$ and $^{h2}J_{NN}$ couplings in the U-A base pairs of the 69 nucleotide left terminal domain of the potato spindle tuber viroid. The experiment used is a normal, two-dimensional

[31] A. Dingley and S. Grzesiek, *J. Am. Chem. Soc.* **120,** 8293 (1998).

[32] A. J. Dingley, J. E. Masse, R. D. Peterson, M. Barfield, J. Feigon, and S. Grzesiek, *J. Am. Chem. Soc.* **121,** 6019 (1999).

[33] C. Scheurer and R. Brüschweiler, *J. Am. Chem. Soc.* **121,** 8661 (1999).

[34] S. Perera and R. Bartlett, *J. Am. Chem. Soc.* **122,** 1231 (2000).

[35] K. Pervushin, A. Ono, C. Fernandez, T. Szyperski, M. Kainosho, and K. Wüthrich, *Proc. Natl. Acad. Sci. U.S.A.* **95,** 14147 (1998).

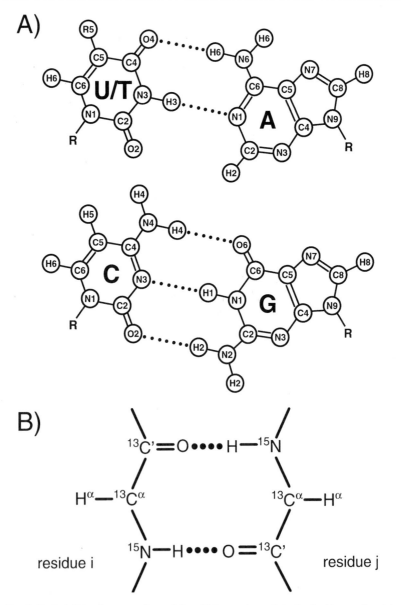

FIG. 1. Typical H-bond patterns in nucleic acid base pairs and in the backbone of proteins. (A) Watson–Crick U-A (T-A) and G-C nucleotide base pairs in RNA (DNA). R5 = H (RNA) or CH₃ (DNA). (B) Antiparallel β-sheet conformation of two amino acids in a protein.

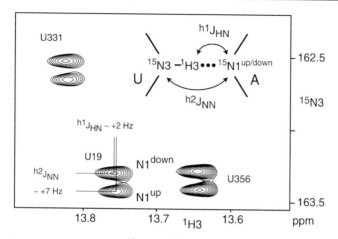

FIG. 2. Simultaneous observation of $^{h2}J_{NN}$ and $^{h1}J_{HN}$ couplings in the U-A base pairs of the left terminal domain of the potato spindle tuber viroid.[31,39] A ^{1}H–^{15}N E.COSY TROSY was recorded on a sample of the 1.6 mM uniformly ^{13}C/^{15}N-enriched 69-nucleotide RNA domain. Excitation of the ^{15}N resonances was restricted to the uridine ^{15}N3 region by means of selective sinc pulses centered at 163 ppm. The data matrix consisted of 150*(t_1) × 1024*(t_2) data points with acquisition times of 150 ms (t_1) and 71 ms (t_2). Data were acquired at 25° on a 600-MHz Bruker DMX instrument, total experimental time 18 hr.

^{1}H–^{15}N TROSY scheme[36] detecting the narrow components of the uridine imino ^{1}H3–^{15}N3 doublets in both the ^{1}H and ^{15}N dimension. Besides additions for a water flip-back WATERGATE,[37,38] the only further modification of the TROSY scheme is that all the ^{15}N pulses are applied as band selective pulses that only affect the chemical shift region of the uridine ^{15}N3 donor nuclei (~160 ppm). Therefore the spin states of the adenosine ^{15}N1 nuclei resonating around 230 ppm are not disturbed by radio frequency pulses during the entire experiment. Since the U-^{15}N3 and the A-^{15}N1 chemical shift regions are well separated, this can easily be achieved with most proposed selective pulses; in this case the 90° and 180° ^{15}N pulses were applied as sinc pulses with a duration of 300 and 600 μs, respectively. As a result, the uridine imino ^{15}N3 and ^{1}H3 spins evolve during the t_1 and t_2 evolution periods of the experiment under the influence of the scalar coupling to the unperturbed adenosine ^{15}N1 nucleus. Clearly visible in Fig. 2 are E.COSY type splittings of the uridine ^{15}N3–^{1}H3 correlations that are the result of the $^{h2}J_{NN}$ and the $^{h1}J_{HN}$ couplings to the adenosine ^{15}N1 nucleus, which is either in the up or in the down state during the evolution periods of the ^{15}N3

[36] K. Pervushin, R. Riek, G. Wider, and K. Wüthrich, *Proc. Natl. Acad. Sci. U.S.A.* **94**, 12366 (1997).
[37] M. Piotto, V. Saudek, and V. Sklenar, *J. Biomol. NMR* **2**, 661 (1992).
[38] S. Grzesiek and A. Bax, *J. Am. Chem. Soc.* **115**, 12593 (1993).

and [1]H3 nuclei. A quantification of these splittings yields values of approximately 7 Hz ($^{h2}J_{NN}$) and 2–3 Hz ($^{h1}J_{HN}$), respectively. In contrast to the HNN-COSY detection scheme[31] described below, the E.COSY TROSY used for this illustration does not provide the resonance frequencies of the acceptor nuclei. However, E.COSY provides the relative signs of the $^{h2}J_{NN}$ and the $^{h1}J_{HN}$ couplings. The negative slope of the E.COSY pattern in Fig. 2 indicates that $^{h2}J_{NN}$ and $^{h1}J_{HN}$ have the same sign since the gyromagnetic moments of ^{1}H and ^{15}N have opposite signs. As the $^{h1}J_{HN}$ couplings have been determined to be positive,[35] this indicates that in accordance with theoretical predictions[32,33] the $^{h2}J_{NN}$ coupling is also positive.

In general, the size of $^{h2}J_{NN}$ couplings between imino ^{15}N nuclei and aromatic ^{15}N nuclei in nucleic acid base pairs ranges between 5 and 11 Hz, whereas $^{h1}J_{HN}$ couplings between the imino proton and aromatic ^{15}N acceptor nuclei have been observed in the range of 2 to 4 Hz. Table I(A, B) lists ^{15}N-imino donor/aromatic ^{15}N-acceptor combinations in various base pair types for which trans H-bond couplings have been observed so far. The results of several density functional theory simulations[32,39] show that with the exception of charged Hoogsteen $C^+ \cdot G$ base pairs, the size of the $^{h1}J_{HN}$ and $^{h2}J_{NN}$ couplings does not depend strongly on the base pair type. The differences in J values listed in Table I(A,B) are rather the result of different donor–acceptor geometries in the different base pair types (see below).

Amino–N–H· · ·N-Aromatic H-Bonds

Because of intermediate-fast rotations around the C–N bond, the two amino protons of the bases in DNA and RNA are often only observed as a single broad resonance. In cases where this rotation is sufficiently slowed down, observations of $^{h2}J_{NN}$ and $^{h1}J_{HN}$ couplings between the amino ^{15}N or ^{1}H nuclei and an aromatic ^{15}N acceptor are possible by detection via the amino protons [Table I(C)]. Such observations of $^{h2}J_{NN}$ couplings have been made for amino groups in A-A mismatch pairs[40] and in guanosine tetrads.[41] Whereas $^{h2}J_{NN}$ couplings of ~7 Hz have been measured for the guanosine tetrads, the A–A mismatch pairs showed $^{h2}J_{NN}$ couplings of only 2.5 Hz. Very likely, these differences in the coupling constant values reflect differences in the H-bond geometries of the different base pair types. In cases where the amino proton is too broad to be detected or exchanged against deuterium, an alternative detection of $^{h2}J_{NN}$ couplings has been proposed.[42,43]

[39] M. Barfield, A. Dingley, J. Feigon, and S. Grzesiek, in preparation.
[40] A. Majumdar, A. Kettani, and E. Skripkin, *J. Biomol. NMR* **14,** 67 (1999).
[41] A. Dingley, J. Masse, F. Feigon, and S. Grzesiek, *J. Biomol. NMR* **16,** 279 (2000).
[42] A. Majumdar, A. Kettani, E. Skripkin, and D. J. Patel, *J. Biomol. NMR* **15,** 207 (1999).
[43] M. Hennig and J. Williamson, *Nucleic Acids Res.* **28,** 1585 (2000).

$^{h1}J_{HN}$ couplings between the hydrogen-bonded amino proton and its nitrogen acceptor have also been observed in the G_4 quartets (A. Dingley, unpublished results). With values of 2–3 Hz for these couplings, there is no indication of any fundamental difference in the mechanism of magnetization transfer across H-bonds from either imino or amino groups to aromatic nitrogen acceptors.

Imino–N–H· · ·Carbonyl H-Bonds

Besides the aromatic nitrogen acceptor atoms, many nucleic acid base pairs involve also the oxygen atoms of carbonyl groups as H-bond acceptors (e.g., Fig. 1A). Although the size of scalar couplings across the H-bridge to the magnetic oxygen isotope ^{17}O is likely to be on the same order as the couplings to a nitrogen ^{15}N acceptor, the fast relaxation of this quadrupolar nucleus would prevent such an observation in solution. Instead, H-bond couplings to the next possible nucleus of the carbonyl acceptor group, the carbonyl ^{13}C carbon have been observed in proteins (see below) and more recently also in the nucleic acid base pairs of guanosine quartets [Table I(D)]. The observed three-bond couplings ($^{h3}J_{NC}$) from the imino ^{15}N donor to the carbonyl ^{13}C acceptor nucleus are only 0.2 Hz in size and thus smaller than the average of the analogous couplings in proteins [Table II(A), Fig. 4]. Nevertheless, because of the favorable relaxation properties of the G-quartet system, all the expected $^{h3}J_{NC}$ correlations could be detected across H-bridges. The information obtained from this experiment is sufficient to establish the G-quartet structure and the number of nonequivalent G quartets within the entire molecule as well as to derive all the H-bond partners for the single guanine bases within each quartet.

Very recently four-bond couplings $^{h4}J_{NN}$ of about 0.14 Hz have also been observed in a similar G-quartet system[44] where the scalar couplings connect the imino $^{15}N1$ nuclei in the guanosine N1–H1· · ·O6=C6–N1 hydrogen bonds [Table I(E)].

H-Bond Scalar Couplings in Proteins

Amide–N–H· · ·Carbonyl H-Bonds

The predominant H-bond in proteins is the bridge between the backbone amide proton of one amino acid and the backbone carbonyl oxygen atom of a second amino acid. The canonical patterns of hydrogen bonds in protein secondary structures are given by $H_i· · ·O_k$, $O_i· · ·H_k$, $H_{i-2}· · ·O_{k+2}$, $O_{i-2}· · ·H_{k+2}$ for antiparallel β-sheets, by $H_i· · ·O_k$, $O_i· · ·H_{k+2}$, $H_{i+2}· · ·O_{k+2}$, $O_{i+2}· · ·H_{k+4}$ for parallel β sheets, by $O_i· · ·H_{i+4}$ for α helices, and by $O_i· · ·H_{i+3}$ for 3_{10} helices, where H_i and O_k

[44] A. Liu, A. Majumdar, W. Hu, A. Kettani, E. Skripkin, and D. Patel, *J. Am. Chem. Soc.* **122**, 3206 (2000).

stand for the backbone amide proton and oxygen atoms of the hydrogen bonded residues i and k, respectively. Figure 1C shows as an example the two H-bonds between two amino acids that are part of an antiparallel β sheet. As in the case of the nucleic acid H-bonds to carbonyl, on the acceptor side the oxygen nucleus is not accessible for the detection of trans H-bond couplings. Interactions that have been detected so far are $^{h3}J_{NiC'k}$, $^{h2}J_{HiC'k}$, and $^{h3}J_{HiC\alpha k}$ couplings that are active between the amide ^{15}N or ^1H nucleus of residue i and the carbonyl or alpha ^{13}C nucleus of residue k [Table II(A–C)]. The values determined for the $^{h3}J_{NiC'k}$ coupling constants cover a range of -0.2 to -0.9 Hz [Table II(A)]. Although the absolute size of these couplings is quite small and their use is therefore limited to small to midsize proteins, meanwhile a number of studies[45–49] have been carried out showing that complete backbone hydrogen bond networks can be detected and that the size of the couplings correlates with donor–acceptor distances.[45,47] Ranges of -0.6 to 1.3 Hz have been observed for $^{h2}J_{HiC'k}$ couplings.[49,50] Thus, these values are similar to the size of $^{h3}J_{NiC'k}$ couplings. Surprisingly, even three-bond couplings of the $^{h3}J_{HiC\alpha k}$ type have now been observed[51] with values of up to 1.4 Hz [Table II(C)].

Amide–N–H· · ·Carboxylate H-Bonds

H-bond $^{h3}J_{NC'}$ couplings of -0.1 to -0.4 Hz have also been detected for backbone amides H-bonded to the carboxylate side chains of glutamic and aspartic acids[47] [Table II(A)]. Apparently, the magnetization transfer across these H-bonds is very similar to the backbone amide to carbonyl H-bonds.

Histidine–Imidazole–N–H· · ·N H-Bonds

For a pair of H-bonded histidine side chains, $^{h2}J_{NN}$ couplings in the range of 8 to 11 Hz have been observed[52] between a protonated $^{15}N\varepsilon 2$-nucleus and an unprotonated $^{15}N\varepsilon 2$ [Table II(D)]. Thus, these values are close to the values for $^{h2}J_{NN}$ couplings found for the imino and amino N–H· · ·N$_{aromatic}$ H-bonds of nucleic acid base pairs.

[45] F. Cordier and S. Grzesiek, *J. Am. Chem. Soc.* **121**, 1601 (1999).

[46] G. Cornilescu, J.-S. Hu, and A. Bax, *J. Am. Chem. Soc.* **121**, 2949 (1999).

[47] G. Cornilescu, B. E. Ramirez, M. K. Frank, G. M. Clore, A. M. Gronenborn, and A. Bax, *J. Am. Chem. Soc.* **121**, 6275 (1999).

[48] Y. X. Wang, J. Jacob, F. Cordier, P. Wingfield, S. J. Stahl, S. Lee-Huang, D. Torchia, S. Grzesiek, and A. Bax, *J. Biomol. NMR* **14**, 181 (1999).

[49] A. Meissner and O. Sørensen, *J. Magn. Reson.* **143**, 387 (2000).

[50] F. Cordier, M. Rogowski, S. Grzesiek, and A. Bax, *J. Magn. Reson.* **140**, 510 (1999).

[51] A. Meissner and O. Sørensen, *J. Magn. Reson.* **143**, 431 (2000).

[52] M. Hennig and B. Geierstanger, *J. Am. Chem. Soc.* **121**, 5123 (1999).

Amide–N–H· · ·Phosphate H-Bonds

Very recently,[53] H-bond couplings involving protein amide groups as donors and phosphate groups as acceptors have been described in a molecular complex of Ras p21 and GDP [Table II(E,F)]. It was found that depending on the geometry both $^{h2}J_{HP}$ and $^{h3}J_{NP}$ can be as large as 3–5 Hz. Given the larger extent of phosphorus d orbitals participating in the phosphate oxygen bonds, as compared to the more restricted carbon p orbitals participating in the carbonylic oxygen bonds, it is perhaps not surprising that the H-bond couplings to ^{31}P in phosphates are larger than the analogous couplings to ^{13}C in the carbonyl groups. The detection of such couplings is particularly important because long-range information on the position of phosphorus in biomolecular complexes is very hard to obtain by traditional high-resolution NMR methods.

Amide–N–H· · ·S· · ·Metal ion H-Bonds

Probably the earliest observations of H-bond couplings in proteins were made by Summers and coworkers[54] in ^{113}Cd- or ^{199}Hg-substituted rubredoxin [Table II(G)]. In this protein, J interactions of up to 4 Hz connected backbone amide protons and the cysteine-coordinated metal atom. Apparently these J couplings were mediated via the hydrogen bonds from the backbone amides to the S atoms of the cysteine residues.

H-Bond Scalar Couplings in Other Systems

H-bond couplings have also been detected as intra- and intermolecular interactions in small model systems. A number of observations indicate that intramolecular H-bonds involving hydroxyl groups can yield couplings between the OH protons and protons on the other side of the H-bridge on the order of a few Hertz.[27,29] For solutions of hydrogen bonded clusters of F and $(HF)_n$, $^{h2}J_{FF}$ couplings of up to 146 Hz and $^{h1}J_{FH}$ couplings of up to 41 Hz have been reported[28] when the lifetime of these H-bonds is increased by the use of cryogenic solvents. Similarly, $^{h2}J_{FN}$ couplings of 96 Hz and one-bond couplings of 20–80 Hz from the proton to donor and acceptor atoms have been observed in almost symmetric H-bridges of the type F· · ·H· · ·N in a binary system consisting of ^{15}N-collodine and HF.[30]

[53] M. Mishima, M. Hatanaka, S. Yokoyama, T. Ikegami, M. Wälchli, Y. Ito, and M. Shirakawa, *J. Am. Chem. Soc.* **122**, 5884 (2000).

[54] P. R. Blake, J.-B. Park, M. W. Adams, and M. F. Summers, *J. Am. Chem. Soc.* **114**, 4931 (1992).

"Through-Space" Couplings

The observation of the H-bond scalar couplings closely ties in with a large number of earlier reports of scalar couplings that were attributed to a "through-space" electronic interaction because the number of intervening covalent bonds would have been too large to explain the effect.[55,56] It is particularly instructive that some of these "through-space" couplings were observed in cases[54,56] where the forces between the interacting atoms are repulsive. Clearly repulsive forces can also lead to electronic correlations that mediate scalar magnetization transfer. Therefore, the observation of the H-bond couplings alone does not implicate that the electronic interactions across the H-bridge are attractive and covalent in nature.[57]

Experiments

Since the H-bond scalar couplings are based on the same forces as their more usual covalent counterparts, their quantum-mechanical description and in particular the spin–spin coupling Hamiltonian used for the description of NMR experiments is identical. Therefore all the methods of detection, quantification, and transfer of magnetization described for other scalar couplings[58,59] can be applied in analogous ways to the H-bond scalar couplings.

$^{h2}J_{NN}$ Couplings

The conceptually simplest scheme for the detection of coupling partners and the quantification of $^{h2}J_{NN}$ correlations [Table I(A), II(D)] is the nonselective HNN-COSY experiment.[31,35] In this scheme, the ^{15}N-evolution period of a ^1H–^{15}N TROSY has been replaced by a homonuclear ^{15}N–^{15}N COSY scheme. As described in detail in the original publications, this scheme transfers magnetization from the proton (H) within the hydrogen bridge via the covalent $^1J_{HN}$ coupling to the ^{15}N nucleus of the H-bond donor (N_d) and then via the $^{h2}J_{NN}$ coupling to the ^{15}N acceptor nucleus (N_a). After a ^{15}N frequency labeling period (t_1), the magnetization is transferred back to the original proton and its oscillations are detected in the receiver during the t_2 period. As in similar quantitative J-correlation schemes, this leads to resonances in the two-dimensional spectrum at frequency

[55] L. Petrakis and C. Sederholm, *J. Chem. Phys.* **35**, 1243 (1961).

[56] M. Kainosho and A. Nakamura, *Tetrahedron* **25**, 4071 (1969).

[57] T. Ghanty, V. Staroverov, P. Koren, and E. Davidson, *J. Am. Chem. Soc.* **122**, 1210 (2000).

[58] A. Bax, G. W. Vuister, S. Grzesiek, F. Delaglio, A. C. Wang, R. Tschudin, and G. Zhu, *Methods Enzymol.* **239**, 79 (1994).

[59] G. Vuister, M. Tessari, Y. Karimi-Nejad, and B. Whitehead, *in* "Modern Techniques in Protein NMR" (N. Krishna and L. Berliner, eds.), Vol. 16, p. 195. Kluwer Academic, 1998.

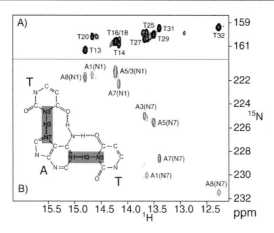

FIG. 3. Part of a quantitative-J_{NN} HNN-COSY spectrum of an uniformly $^{13}C/^{15}N$-labeled intramolecular DNA triplex.[32] This triplex consists of 5 Hoogsteen–Watson–Crick T·A-T and 3 Hoogsteen C$^+$· G-C base triplets. The spectral region corresponds to the 10 imino resonances of the Hoogsteen–Watson–Crick T·A-T triplets. Positive contours depict diagonal resonances (A); negative contours (B) correspond to cross peaks resulting from H-bond scalar ^{15}N–^{15}N magnetization transfer. Resonances are labeled with assignment information. *Inset* in (B) shows the chemical structure of the T·A-T triplet.

positions ω_{Na}–ω_H (cross peaks) and ω_{Nd}–ω_H (diagonal peaks). The value of the $^{h2}J_{NN}$ coupling constant is then derived from the intensity ratio of the cross and diagonal peaks by the formula $|^{h2}J_{NN}| = \text{atan}[(-I_{Na}/I_{Nd})^{1/2}]/(\pi T)$ where T is the COSY transfer time.

Figure 3 shows as an example the region of the Hoogsten–Watson–Crick T·A-T triplets of an intramolecular DNA-triplex of 32 nucleotides.[32] Panel A shows the diagonal resonances corresponding to the thymidine $^1H3$–$^{15}N3$ correlations, whereas panel B depicts the cross resonances corresponding to the Watson–Crick $^1H3(T)$–$^{15}N1(A)$ and the Hoogsteen $^1H3(T)$·$^{15}N7(A)$ correlations. Because of the considerable size of the $^{h2}J_{NN}$ couplings, the sensitivity of the HNN-COSY experiment is sufficient that all H-N···N type hydrogen bonds could be observed for this intramolecular triplex[32] and even for an RNA domain of 69 nucleotides[31] in overnight experiments at ~1.5 mM concentrations. Note that for large molecules with short transverse relaxation times of the ^{15}N donor ($\pi T_2^{h2} J_{NN} \ll 1$), the sensitivity for the observation of cross peaks is highest when the total COSY transfer time is set to a value of approximately T_2.

Two experimental difficulties should be pointed out for the HNN-COSY. The first is that the quantification of the coupling constants from the intensity ratio of cross and diagonal peaks is made difficult by the different line widths of the

^{15}N donor and acceptor resonances (Fig. 3A,B). The different line widths are the result of the different relaxation mechanisms for both nuclei. Whereas the width of donor ^{15}N resonances is narrowed by the partial cancellation of the ^{1}H–^{15}N$_d$ dipole and ^{15}N$_d$ CSA relaxation in the TROSY scheme, the very large (300–400 ppm) CSA[60] of the ^{15}N acceptor nuclei in the nucleic bases dominates the relaxation for the acceptor resonances. As a practical consequence, it is not sufficient to approximate the intensity ratios of the cross and diagonal peaks by their amplitude ratios. Instead, the intensities of both donor and acceptor resonances need to be determined by an appropriate peak integration scheme in order to derive accurate values for the coupling constants. Since the amplitude of the oscillations in the ^{15}N time domain is proportional to the magnetic excitation and to the peak integral, time domain fitting routines are a practical alternative to the peak integration in the frequency domain.[31,61]

The second, more severe problem stems from the frequency separation of ^{15}N donor and acceptor resonances and the limited strength of the available ^{15}N radio frequency pulses. In Watson–Crick and Hoogsteen imino–aromatic–N hydrogen bridges, frequency separations of the ^{15}N donor and acceptor resonances are in the range of 50 to 70 ppm. For typical ^{15}N radio frequency pulse strengths of about 6 kHz and a magnetic field strength of 14 tesla, errors in $^{h2}J_{NN}$ values resulting from the finite excitation bandwidth have been estimated to be on the order of 10%.[31] Smaller errors can be achieved by the use of stronger radio frequency pulses, lower magnetic field strengths, or by ^{15}N excitation schemes that cover only selected regions of the donor and acceptor resonances in the different base pair types. Frequency separations between amino and aromatic ^{15}N resonances are typically in the range of 100 to 130 ppm. As has been pointed out by Majumdar and co-workers,[40] such ^{15}N frequency separations are too large to be covered effectively by the nonselective ^{15}N pulses of the homonuclear HNN-COSY. They therefore designed a pseudo-heteronuclear H(N)N-COSY experiment where selective ^{15}N pulses excite the amino and aromatic ^{15}N resonances separately and that yields excellent sensitivity.[40] An inconvenience of the latter experiment is that the resonances corresponding to the amino ^{15}N nuclei are not detected and a separate spin–echo difference experiment was used to quantify the $^{h2}J_{NN}$ values. A slightly improved version of this pseudo-heteronuclear H(N)N-COSY[41] remedies this problem by the use of phase-coherent ^{15}N pulses such that both amino and aromatic ^{15}N resonances can be detected in a single experiment.

A further modification of the HNN-COSY scheme has been proposed for the observation of $^{h2}J_{NN}$ correlations where the hydrogen nucleus in the H-bond is

[60] J. Hu, J. Facelli, D. Alderman, R. Pugmire, and D. Grant, *J. Am. Chem Soc.* **120,** 9863 (1998).
[61] F. Delaglio, S. Grzesiek, G. W. Vuister, G. Zhu, J. Pfeifer, and A. Bax, *J. Biomol. NMR* **6,** 277 (1995).

unobservable.[42,43] This situation is often found due to intermediate conformational exchange of hydrogen bonding amino groups or due to exchange of the proton with the solvent. In some cases, the HNN-COSY can then be started and detected on a carbon-bound proton in the vicinity of the acceptor. This is possible for adenosine and guanosine $^{15}N7$ acceptors as well as for adenosine $^{15}N1$ acceptors that can be connected to the H8 or H2 proton by means of the $^2J_{H8N7}$ or $^2J_{H2N1}$ couplings of approximately 11–15 Hz,[62] respectively.

Another ingenious modification of the HNN-COSY scheme involves the replacement of the homonuclear ^{15}N COSY transfer by a ^{15}N-TOCSY transfer.[44] As the homonuclear TOCSY transfer is twice as fast as the COSY transfer, a significant sensitivity increase is achieved. The application is, however, limited to cases where the ^{15}N donor and acceptor resonances are at similar frequencies such that the power of the ^{15}N-TOCSY radio frequency pulses need not to be too strong. A very interesting application for this scheme was presented for the sensitive detection of very small [0.14 Hz, Table I(E)] $^{h4}J_{N1N1}$ couplings in imino-carbonyl hydrogen bridges of guanosine tetrads.[44]

$^{h1}J_{HN}$ Couplings

Because of their smaller size as compared to $^{h2}J_{NN}$ and due to the faster relaxation times of the proton resonances, the measurement of $^{h1}J_{HN}$-couplings is more challenging. At present, three methods have been described for the quantification of $^{h1}J_{HN}$-couplings: an E.COSY experiment,[35] a quantitative J-correlation TROSY scheme based on the more favorable relaxation properties of ^1H–^{15}N zero-quantum coherences,[63] and a quantitative COSY scheme that also yields the frequencies of the nitrogen acceptor nuclei.[32]

$^{h3}J_{NC}$ Couplings

The detection of these weak ($|^{h3}J_{NC}| < 0.9$ Hz) correlations in proteins can be achieved with a conventional HNCO experiment where the nitrogen to carbonyl dephasing and rephasing delays are set to longer values than for the detection of the sequential one-bond $J_{NiC'i-1}$ correlations. In a standard HNCO experiment, the time for the two INEPT transfers between in-phase N_i^y and antiphase $2N_i^x C_{i-1}^{'z}$ magnetization is usually set to values of about 25–30 ms. In an HNCO experiment suitable for the transfer by $^{h3}J_{NC'}$, this time is set to 133 ms $\approx 2/^1J_{NC'}$ such that the one-bond transfer from N_i^y to $2N_i^x C_{i-1}^{'z}$ is approximately refocused.[45,46] In such a situation, the resulting HNCO spectrum contains mostly correlations which

[62] J. H. Ippel, S. S. Wijmenga, R. de Jong, H. A. Heus, C. W. Hilbers, E. de Vroom, G. A. van der Marcel, and J. H. van Boom, *Magn. Reson. Chem.* **34**, S156–S176 (1996).

[63] K. Pervushin, C. Fernandez, R. Riek, A. Ono, M. Kainosho, and K. Wüthrich, *J. Biomol. NMR* **16**, 39 (2000).

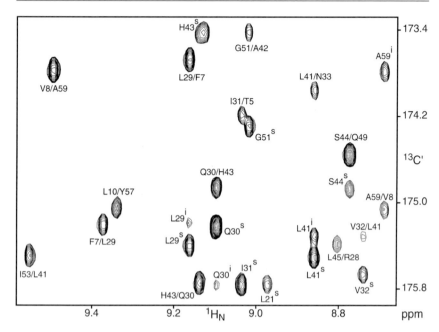

FIG. 4. Selected region of the two-dimensional, long-range, quantitative-$J_{NC'}$ H(N)CO-TROSY spectrum recorded on a 3.5 mM sample of ^{2}H/^{15}N/^{13}C-labeled c-Src SH3 in complex with the polyproline peptide RLP2 at 95% H_2O/5% D_2O, 25°C.[64] The data matrix consisted of 65* (t_1) × 512* (t_2) data points (where n^* refers to complex points) with acquisition times of 39 ms (t_1) and 53 ms (t_2). Total measuring time was 17 hr on a Bruker DMX600 instrument. Cross peaks marked as Res$_i$/Res$_j$ are due to $^{3h}J_{NiC'j}$ trans-hydrogen bond scalar couplings between the ^{15}N nucleus of residue i and ^{13}C' nucleus of residue j. Residue names marked by the superscript "s" denote not completely suppressed, sequential, one-bond correlations between the ^{15}N nucleus of residue i and ^{13}C' nucleus of residue $i - 1$. The superscript 'i' demarks intraresidue two-bond ^{15}N$_i$–^{13}C'$_i$ correlations.

result from smaller, long-range $J_{NC'}$ couplings. Figure 4 shows the results of this experiment on the c-Src SH3 domain[64] carried out as a two-dimensional H(N)CO where only amide proton and carbonyl ^{13}C frequencies were detected. Clearly visible, for example, are the correlations from the amide ^{1}H-(^{15}N) of residue F7 to the carbonyl ^{13}C nucleus of residue L29 and from the amide ^{1}H–(^{15}N) of residue L29 to the carbonyl ^{13}C of residue F7 that correspond to the antiparallel β sheet arrangement of these two amino acids (Fig. 1B). Quantification of the long-range $J_{NC'}$ coupling constants can be achieved by comparison of the cross-peak intensities with the intensities of sequential amide-carbonyl correlations measured in a second reference experiment.[45,46] In the latter experiment transfer is tuned to the

[64] F. Cordier, C. Wang, S. Grzesiek, and L. Nicholson, *J. Mol. Biol.* **304,** 497 (2000).

sequential $^1J_{NC'}$ coupling while the relaxation losses are kept identical to the long-range experiment. A slightly modified version of this long-range HNCO using highly selective [^{13}C] carbonyl pulses can also be used for the detection of $^{h3}J_{NC'}$ correlations in nucleic acids.[41]

The quantitative long-range HNCO experiment only yields the absolute value of $^{h3}J_{NC'}$. In agreement with density functional theory simulations,[33] the sign of the $^{h3}J_{NC'}$ coupling in proteins was determined as negative by a zero-quantum/double-quantum technique.[47]

Because of the small size of the $^{h3}J_{NC'}$ couplings, the sensitivity of the long-range HNCO experiment is rather low. Whereas about 83% of all expected backbone to backbone hydrogen bonds could be observed in a 12-hr experiment on a 1.6 mM sample of uniformly ^{13}C/^{15}N labeled human ubiquitin (8.6 kDa), for larger systems, the shorter ^{15}N transverse relaxation times together with the long ^{15}N–^{13}C dephasing and rephasing times of 266 ms clearly limit the detection of $^{h3}J_{NC'}$ correlations. Several sensitivity improvements have been proposed.

An obvious enhancement in sensitivity can be obtained by the use of an HNCO-TROSY[65] at higher magnetic field strengths (>14 T), which makes use of the cancellation of dipolar and CSA relaxation mechanisms for the down-field component of the amide ^{15}N doublet. Such an approach was successful to detect $^{h3}J_{NC'}$ correlations in the 30 kDa ribosome inactivating protein MAP30[48] using uniform ^{2}H, ^{13}C, ^{15}N enrichment and a protein concentration of only 0.7 mM. Experience from this experiment shows that the number of detected $^{h3}J_{NC'}$ correlations (65) was not limited by the transverse ^{15}N relaxation time, but by the incomplete back-exchange of amide deuterons against protons and by the long amide proton T_1 relaxation times. Deuteration is essential for sensitivity gains by the HNCO-TROSY approach, because in protonated samples proton–proton spin flips strongly reduce the line narrowing gain of TROSY.[66] For example, for protonated ubiquitin at 25°C and 14 T magnetic field strength, the "TROSY T_2" and the T_2 of the amide ^{15}N singlet (using ^{1}H-decoupling) are almost identical (170–180 ms). In contrast, in perdeuterated (>85%, nonexchangeable hydrogen positions), but amide protonated ^{13}C,^{15}N-labeled ubiquitin, the "TROSY-^{15}N-T_2" increases to 340–400 ms. Clearly, larger enhancements can be achieved at higher magnetic field strengths than 14 T.[36,66]

A second sensitivity improvement of the long-range HNCO has been proposed for nondeuterated protein samples.[67] In such samples, the transverse relaxation rate of the amide ^{15}N nucleus is increased to some extent by the scalar coupling to the

[65] M. Salzmann, K. Pervushin, G. Wider, H. Senn, and K. Wüthrich, *Proc. Natl. Acad. Sci. U.S.A.* **95**, 13585 (1998).
[66] G. Kontaxis, G. Clore, and A. Bax, *J. Magn. Reson.* **143**, 184 (2000).
[67] A. Liu, W. Hu, S. Qamar, and A. Majumdar, *J. Biomol. NMR*, **17**, 55 (2000).

$^{13}C_\alpha$ nucleus and its relatively short longitudinal relaxation time. This relaxation mechanism is commonly referred to as "scalar relaxation of the second kind."[68] A similar increase in imino ^{15}N transverse relaxation rates due to the scalar relaxation effect has been observed in nucleic acids.[41] The effect is particularly pronounced for smaller macromolecules where ^{13}C T_1 relaxation times are short. In the case of proteins, the scalar contribution to the ^{15}N relaxation can be removed with relative ease by continuous decoupling of the $^{13}C_\alpha$ nucleus. Using this approach for the 12 kDa FK506 binding protein, a sensitivity increase by about 50% was observed as compared to the conventional long-range HNCO experiment.[67]

$^{h2}J_{HC'}$ and $^{h3}J_{HC\alpha}$ Couplings

At present, two schemes have been proposed to detect $^{h2}J_{HC'}$ correlations in proteins. The first uses very selective carbonyl pulses to separate the $^{h2}J_{HC'}$ correlations from interfering covalent two- and three-bond H_N–^{13}C-carbonyl scalar couplings.[50] The second scheme utilizes a nonselective E.COSY–HNCO–TROSY approach[49] that also yields the sign of the coupling constants. A very similar experiment was used to detect $^{h3}J_{HC\alpha}$ couplings.[51]

$^{h3}J_{NP}$ and $^{h2}J_{HP}$ Couplings

Direct observations of $^{h3}J_{NP}$ correlations can be obtained by an HNPO or HNPO-TROSY experiment[53] that is analogous to the HNCO or HNCO-TROSY experiment, the only difference being that the ^{13}C carbonyl pulses are replaced by ^{31}P pulses. For the quantification of the $^{h3}J_{NP}$ and $^{h2}J_{HP}$ couplings, appropriate spin–echo difference HSQC schemes were proposed.[53]

Determination of Upper Limits for Couplings

In all the listed quantitative J-correlation schemes, the coupling constants are derived from the ratio of the cross peak (I_c) to reference peak (I_r) intensities that have been normalized to the respective number of scans (NS) in the different experiments. For small couplings ($|\pi J T| \ll 1$), where T is the transfer time, the relation is always approximately given by:

$$|J| \approx |(I_c^* NS_r/(I_r^* NS_c)|^{1/2}/(\pi T)$$

In cases where no cross peaks are detected, an upper limit of the coupling constant can be derived if a reference peak is observed. In this situation, the unobserved cross peak intensity must be smaller than the threshold level (I_t) at which the experimental data are plotted. For a convenient representation of the

[68] A. Abragam, "The Principles of Nuclear Magnetism." Clarendon Press, Oxford, 1961.

data, usually the threshold level is chosen at values of about 2.5 times the rmsd of the spectral noise. It is then clear that the coupling constant for the unobserved correlation must be smaller in absolute size then the threshold coupling constant J_t:

$$|J| < J_t = |(I_t^* NS_r / (I_r^* NS_c)|^{1/2} / (\pi T)$$

This relation has proven useful in many cases where one wants to decide whether the unobserved coupling is significantly different in value from another properly observed correlation. It should be kept in mind that the noise in both cross and reference experiments is proportional to the square root of the respective number of scans. This implies that the threshold coupling constant is inversely proportional to the fourth root of the number of scans in the cross experiment:

$$|J_t| = 2.5^{1/2} (NS_r / NS_c)^{1/4} SN_r^{-1/2} / (\pi T)$$

where $SN_r = I_r / \text{Noise}_r$ is the signal-to-noise (rmsd) ratio of the reference experiment. Thus, the chance to observe a cross peak increases only very slowly as the number of scans is increased, and it is often a better idea to try to increase the sensitivity by other means, e.g., by using larger concentrations of the macromolecule, or by deuteration if this is feasible. On the other hand, this estimate for the threshold is a convenient tool to judge what experimental time will be needed to observe a certain coupling once a short reference experiment has been recorded.

Relation to Chemical Shift

Clear correlations of the chemical shift of the protons within the H-bond and the trans-H-bond coupling constants have been observed both for nucleic acids[32] and for proteins.[45] In nucleic acids N-H···N H-bonds, $^{h1}J_{HN}$ is found to decrease whereas both $^{h2}J_{NN}$ and $^1J_{HN}$ increase linearly with increasing imino proton chemical shifts (Fig. 5A). Similarly, in protein N-H···O=C H-bonds, values of $|^{h2}J_{HC'}|$ and $|^{h3}J_{NC'}|$ increase linearly with increasing amide proton chemical shifts (Fig. 5B). For proteins containing a large number of aromatic side chains, these correlations are considerably improved, when the proton chemical shifts are corrected for ring current effects (F. Cordier, unpublished results). Therefore, these correlations could make it possible to separate the proton chemical shift into different contributions that can be understood in a quantitative way.

The isotropic[9,17,69–71] and also the anisotropic chemical shift[20,21] of amide protons have been used as an indicator for the strength of the hydrogen bond in

[69] D. S. Wishart, B. D. Sykes, and F. M. Richards, *J. Mol. Biol.* **222**, 311 (1991).
[70] I. D. Kuntz, P. A. Kosen, and E. C. Craig, *J. Am. Chem. Soc.* **113**, 1406 (1991).
[71] N. E. Zhou, B.-Y. Zhu, B. D. Sykes, and R. S. Hodges, *J. Am. Chem. Soc.* **114**, 4320 (1992).

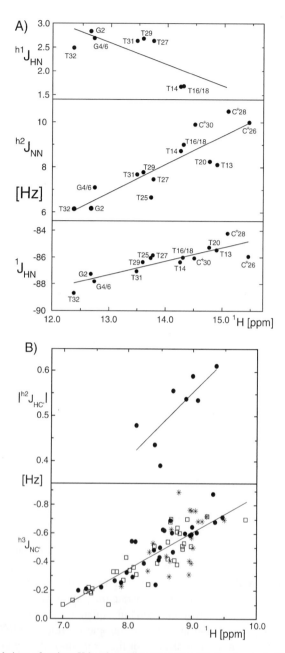

FIG. 5. Correlations of various H-bond coupling constants with the chemical shift of the proton in the hydrogen bridge. (A) Correlations between the imino proton chemical shift and $^{h1}J_{HN}$, $^{h2}J_{NN'}$ and $^{1}J_{HN}$ couplings in the intramolecular DNA triplex.[32] (B) Correlations between the amide proton chemical shift and $^{h2}J_{HC'}$ and $^{h3}J_{NC'}$ couplings in proteins. Filled circles, open asterisks, and rectangles correspond to data from human ubiquitin, from the c-Src SH3 domain, and from protein G, respectively. Proton chemical shifts are corrected for ring current effects.

proteins. In particular, there is a strong correlation between the hydrogen bond length and the isotropic shift of the amide proton.[9,17,69–71] Short hydrogen bond lengths correspond to larger values for the isotropic chemical shift. Thus, the observed correlations between H-bond coupling constants and the isotropic proton chemical shifts indicate that both parameters are mainly determined by the relative geometry of the partners in the H-bond.

Dependence on Geometry

H-Bond Lengths

Comparisons of the measured coupling constants to the geometry of the H-bond are hampered by the limited availability of very high resolution diffraction data. Especially, no crystallographic data are available for most of the nucleic acids for which H-bond couplings have been measured. In the available high resolution structures of nucleic acids, the variation of N1–N3 distances in Watson–Crick base pairs is clearly very limited. A survey of a number of such structures showed[32] that typical N1–N3 distances are 2.92 ± 0.05 Å and 2.82 ± 0.05 Å for G-C and A-T (A-U) base pairs in DNA and RNA. The shorter crystallographic donor acceptor distances in A-T (A-U) base pairs as compared to G-C base pairs coincide with an increase in the value of the $^{h2}J_{NN}$ coupling constants from about 6–7 Hz (G-C) to about 7–8 Hz (A-T, A-U) in the NMR investigations.[31,32,35] Density functional theory simulations[39] indicate that the different chemical nature of the donor and acceptor groups in Watson–Crick G-C and A-T (A-U) and in Hoogsteen A · T base pairs have a very limited influence on the $^{h2}J_{NN}$ coupling constants (≤ 0.2 Hz, for N–N distances from 2.7 to 4.0 Å). Therefore, the observed differences in $^{h2}J_{NN}$ values for G-C, A-T, A-U, and A · T base pairs should be largely due to the differences in donor–acceptor distances. Neglecting angular variations, it follows that at N–N distances of 2.8 to 2.9 Å, a change in the value of $^{h2}J_{NN}$ by 1 Hz corresponds to a change in donor–acceptor distance of 0.07 ± 0.01 Å. Decreases of $^{h2}J_{NN}$ coupling constants by about 1 Hz have been observed at the ends of helical stems and have been interpreted as a corresponding increase in the ensemble average of the donor–acceptor distance.[32]

In contrast to N–H· · ·N H-bonds in Watson–Crick base pairs, the geometry of N–H· · ·O=C H-bonds in proteins is more varied. A survey of a number of crystallographic structures[72] showed that typical values for N· · ·O and H· · ·O distances and for N–H· · ·O and H· · ·O=C angles are 2.99 ± 0.14 Å

[72] E. N. Baker and R. E. Hubbard, *Prog. Biophys. Molec. Biol.* **44**, 97 (1984).

[73] J. Wöhnert, A. J. Dingley, M. Stoldt, M. Görlach, S. Grzesiek, and L. R. Brown, *Nucleic Acids Res.* **27**, 3104 (1999).

[74] P. R. Blake, B. Lee, M. F. Summers, M. W. Adams, J. B. Park, Z. H. Zhou, and A. Bax, *J. Biomol. NMR* **2**, 527 (1992).

2.06 ± 0.16 Å, $157 \pm 11°$, $147 \pm 9°$ in the case of α helices and 2.91 ± 0.14 Å, 1.96 ± 0.16 Å, $160 \pm 10°$, $151 \pm 12°$ in the case of β-sheet conformations, respectively. Thus, on average the N· · ·O and H· · ·O distances are 0.08–0.10 Å shorter in β sheets as compared to α helices. This coincides with an increase in the average strength of the observed $^{h3}J_{NC'}$ coupling from -0.38 ± 0.12 Hz for α-helical conformations to -0.65 ± 0.14 Hz for β sheets in ubiquitin.[45] An exponential correlation between the coupling constant and the N· · ·O distance was described for protein G by the Bax group[47] where $^{h3}J_{NC'}$ is given as -59×10^3 Hz $\times \exp[-4 \times R_{NO}/\text{Å}]$. Again neglecting angular dependencies, it follows from this relation that for a typical α-helical or β-sheet conformation, a variation in the value of $^{h3}J_{NC'}$ by 0.1 Hz corresponds to a change in the N· · ·O distance of 0.05 or 0.07 Å, respectively.

H-Bond Angles

At present, the knowledge about the angular dependencies of the H-bond coupling constants is limited. Density functional theory simulations[33] indicate that the $|^{h3}J_{NC'}|$ value has a maximum for an N–H· · ·O angle of 180° and it drops by about 20 to 30% for a decrease in this angle to 140°. The dependence of $^{h3}J_{NC'}$ on the H· · ·O=C angle is less well understood. Experimental evidence shows that the $^{h3}J_{NC'}$ couplings in the nucleic acid G-tetrad are weaker than the average of $^{h3}J_{NC'}$ couplings observed in proteins.[41] The principal difference in the geometry of the two systems is that the H· · ·O=C angle in the G-tetrad is only $125 \pm 5°$, whereas this angle has typical values around 150° in proteins (see above). This might hint at a similar angular dependency as for the N–H· · ·O angle, namely that a straight H· · ·O=C conformation yields stronger $^{h3}J_{NC'}$ couplings than a bent conformation. Support for this interpretation comes from observations of $^{h3}J_{NP}$ couplings in N–H· · ·O–P H-bonds.[53] In this study a $^{h3}J_{NP}$ coupling of 4.6 Hz was observed for an H· · ·O–P angle of 173°, but the values of $^{h3}J_{NP}$ dropped below 0.35 Hz for conformations where the H· · ·O–P angle was smaller than 126°.

Conclusion

Many of the possible applications of the H-bond couplings are clearly directed to the establishment of chemical shift assignment and secondary or tertiary structure information. Although information about H-bonds in the backbone of proteins can often be derived from nuclear overhauser effect (NOE) data, the observation of an H-bond coupling yields unambiguous evidence about the existence of the H-bond. The exponential dependence of the coupling constant values on the H-bond distance should yield very strong distance constraints in structure calculations. In contrast to the backbone H-bonds, the establishment of backbone to side-chain carboxylate H-bonds[47] or H-bonds between histidine side chains[52] by means of long-range NOEs is considerably more difficult. In all cases, the good spectral dispersion of ^1H_N, ^{15}N, [^{13}C]carbonyl, and [^{13}C]carboxylate resonances makes the identification of the donor and acceptor groups by H-bond J-couplings rather easy. Important applications may be the unambiguous establishment of H-bond networks in the active site of enzymes or ribozymes, during the folding process, as well as the detection of intermolecular H-bonds in macromolecular interactions.

Compared to proteins, nucleic acids have a lower abundance of protons. As a consequence, structural information from NOEs is often more limited. Because the few long-range interactions in folded nucleic acids generally involve hydrogen bonds, the detection of the H-bond couplings not only for Watson–Crick base pairs but also for a wide variety of nonstandard base pairs and for phosphates yields tertiary structure information that might not be obtainable by any other NMR parameter. In this context, it is remarkable that the $^{h2}J_{NN}$, $^{h2}J_{HP}$, and $^{h3}J_{NP}$ couplings are large and that sensitivity issues are only a minor problem.

A number of applications can be imagined from a quantitative analysis of the H-bond couplings: (1) Very strong couplings should be detected in systems that have been proposed as low barrier hydrogen bonds. (2) The good correlation with the proton chemical shift is particularly striking. A quantitative understanding of this phenomenon might lead to an understanding of the proton shift itself. (3) At present, it is unclear how the size of the couplings relates to the energy of the hydrogen bond. If a simple correlation exists, the J values could be used as a direct indicator of the H-bond stability. (4) The correlation between the size of the H-bond couplings and the donor–acceptor distances is well established. A good understanding of the angular dependencies is missing at present. Changes in the size of the coupling constants can be detected when the macromolecules are subjected to different physicochemical conditions such as changes in temperature, pH, ligand binding, or protein folding. First results on peptide binding of the c-Src SH3 domain[64] indicate that these changes in coupling constants can be rationalized as an induced fit of the protein hydrogen bond network during ligand binding.

Acknowledgments

We thank Profs. Juli Feigon, Michael Barfield, Linda K. Nicholson, and Masatsune Kainosho for many stimulating discussions, as well as Chunyu Wang, James Masse and Robert D. Peterson for the preparation of the c-Src SH3 and DNA triplex samples. A.J.D. acknowledges funding by the Australian National Health and Medical Research Council C.J. Martin Fellowship (Regkey 987074). F.C. is a recipient of an A. v. Humboldt fellowship. This work was supported by DFG grant GR1683/1-1 and SNF grant 31-61'757.00 to S.G.

[5] On-Line Cell High-Pressure Nuclear Magnetic Resonance Technique: Application to Protein Studies

By KAZUYUKI AKASAKA and HIROAKI YAMADA

High-Pressure NMR Techniques

Two Basic Methods of High-Pressure NMR

There are two general methods for carrying out the high-pressure, high-resolution nuclear magnetic resonance (NMR) experiments. One uses a pressure vessel or autoclave made of nonmagnetic materials (the high-pressure probe method)[1] in which a detection coil containing the sample tube is housed. The other uses a pressure-resisting sample cell made of thick-walled glass,[2,3] Wespel,[4] sapphire,[5] or synthetic quartz[6] (the *high-pressure cell method*). The *high-pressure probe method* originated in 1954 when Benedeck and Purcell measured spin relaxation and self-diffusion coefficients up to 10 kbar,[7] but it was only in 1972[1] that Jonas made it useful for high-resolution spectroscopic work, which was followed by Oldenziel and Trappeniers,[8] Merbach and co-workers,[4] Shimokawa and Yamada,[9] Funahashi and co-workers,[10] Nakahara and Emi,[11] van Eldik and co-workers,[12] Markley and co-workers[13] and Ernst and Schick and co-workers.[14] When applied to proteins, the high-pressure probe method has an advantage in the relative ease in attaining very high pressure ($\sim$9 kbar).[15] On the other hand, the most severe disadvantage of the high-pressure probe method for protein studies is in the difficulty

[1] J. Jonas, *Rev. Sci. Instrum.* **43**, 643 (1972).

[2] H. Yamada, *Chem. Lett.* 747 (1972).

[3] H. Yamada, *Rev. Sci. Instrum.* **45**, 640 (1974).

[4] H. Vanni, W. L. Earl, and A. E. Merbach, *J. Magn. Reson.* **29**, 11 (1978).

[5] J. L. Urbauer, M. R. Ehrhard, R. J. Bieber, P. F. Flynn, and A. J. Wand, *J. Am. Chem. Soc.* **118**, 11329 (1996).

[6] H. Yamada, M. Umeda, T. Maeda, and A. Sera, *Chem. Lett.* 437 (1996).

[7] G. B. Benedek and E. M. Purcell, *J. Chem. Phys.* **22**, 2003 (1954).

[8] J. G. Oldenziel and N. J. Trappeniers, *Physica* **82A**, 565 (1976).

[9] S. Shimokawa and E. Yamada, *J. Magn. Reson.* **51**, 103 (1983).

[10] M. Ishii, S. Funahashi, K. Ishihara, and M. Tanaka, *Bull. Chem. Soc. Jpn.* **62**, 1852 (1986).

[11] M. Nakahara and K. Emi, *J. Chem. Phys.* **99**, 5418 (1993).

[12] A. Zhal, A. Neubrand, S. Aygen, and R. van Eldik, *Rev. Sci. Instrum.* **65**, 882 (1994).

[13] E. Prehoda, S. Mooberry, and J. L. Markley, *in* "Protein Dynamics, Function and Design" (J.-F. Lefevre and O. Jardetzky, eds.), Plenum Press, New York, 1998.

[14] R. Wiedenbruch, M. Schick, A. Pampei, B. H. Meier, R. Meyer, and R. E. Ernst, *J. Phys. Chem.* **99**, 13088 (1995); M. Schick, Ph.D. Thesis, Eidegenoessische Technische Hochschule (1998).

[15] E. L. Ballard, C. Reiner, and J. Jonas, *J. Magn. Reson.* **123A**, 81 (1996); L. Ballard, Y. Yu, C. Reiner, and J. Jonas, *J. Magn. Reson.* **133**, 190 (1998).

in attaining a sufficiently high magnetic field homogeneity for performing experiments in an ^1H_2O-rich environment and the limitation in the versatility in pulse sequences, both of which are crucial for modern NMR applications to proteins.

On-Line Variable-Pressure Cell Method

Historical Background. In 1972, another high-pressure, high-resolution NMR method suitable for liquid samples was introduced independently by the laboratories of Jouanne and Heidberg[16] (up to about 50 MPa using quartz) and of Yamada[2] (up to 167 MPa using a borosilicate glass). In the first version of the Yamada cell, a thick-walled glass cell was made as a sealed pressure-resisting glass tube.[2] The technique utilizes thermal expansion of an organic solvent on solid-to-liquid transition to generate pressure in a sealed glass tube. A mercury layer separated the pressurizing liquid from the sample solution. In the early 1980s, Morishima used this technique extensively in conjunction with a commercial spectrometer to study pressure effects on the heme environment in heme proteins.[17] Later, Akasaka utilized the same technique to study pressure denaturation of a protein.[18] However, the sealed glass tube method has limitations. First, pressure can be adjusted only approximately, because it depends on both the amount of the organic solvent confined in the cell and the temperature of measurement. Second, the method requires a new sample for each new pressure, which makes it difficult to apply the method to a limited amount of a protein sample.

In 1974 Yamada[4] introduced a more versatile on-line variable pressure high-pressure cell technique with a glass capillary cell, which was followed by Land and Luedemann[19] and by Williams *et al.*[20] Land and Luedemann used it extensively for studying dynamics of simple liquids.[21] Yamada and Luedemann have published several reviews on the technique.[22-24] Later the same technique was employed by Wagner for studying pressure effect on ring-flip motions in basic pancreatic trypsin inhibitor (BPTI).[25] More recently, a pressure cell made of sapphire was used by

[16] J. Jouanne and J. Heidberg, *J. Magn. Reson.* **7,** 1 (1972).
[17] I. Morishima, *in* "Current Perspectives in High Pressure Biology." Academic Press, London, 1987.
[18] T. Yamaguchi, H. Yamada, and K. Akasaka, *J. Mol. Biol.* **250,** 689 (1995).
[19] E. W. Land and H.-D. Luedemann, *Prog. NMR Spectrosc.* **25,** 507 (1993).
[20] R. K. Williams, C. A. Fyfe, R. M. Epand, and D. Bruck, *Biochemistry* **17,** 1506 (1978); R. K. Williams, C. A. Fyfe, D. Bruck, and L. van Veen, *Biochemistry* **18,** 757 (1978).
[21] H.-D. Luedemann, *Polish J. Chem.* **70,** 387 (1996).
[22] H. Yamada, *in* "NMR Basic Principles and Progress 24" (J. Jonas, ed.). Springer-Verlag, Berlin Heidelberg, 1990.
[23] H. Yamada, K. Kubo, I. Kakihara, and A. Sera, *in* "High Pressure Liquids and Solutions" (Y. Taniguchi, M. Senoo, and K. Hara, eds.), p. 49. Elsevier Science B. V., 1994.
[24] W. E. Price and H.-D. Luedemann, *in* "High-Pressure Techniques in Chemistry and Physics: A Practical Approach" (W. B. Holzapfel and N. S. Isaacs, eds.), p. 225. Oxford University Press, Oxford, 1997.
[25] G. Wagner, *FEBS Lett.* **112,** 180 (1980).

Wand and co-workers for a multidimensional NMR study of a small protein up to $\sim$1 kbar.[5] The pressure range for the glass tube method has been limited to $\sim$2 kbar. The pressure range has been extended to $\sim$4 kbar by using synthetic quartz instead of glass for the cell material.[26,27]

Advantage of High-Pressure Cell Method. The advantage common to all the high-pressure cell methods is that no modification of NMR hardware is required, so that any commercially available NMR probe and spectrometer can be employed for high pressure experiments *with all their advanced capabilities.* After our introduction of the on-line cell high-pressure technique to high-resolution NMR studies of proteins,[28] a number of applications have been made successfully to various protein systems, opening a new field of research on protein structure, dynamics, and folding as a function of pressure. Despite the small sample volume, the combination of the high resolution and high sensitivity inherent in a high field spectrometer (e.g., 750 MHz) with advanced pulse sequence capabilities has effectively overcome the sensitivity problem. Although sensitivity is still a limiting factor for many challenging problems, future technical developments of this technique seem to be in order.

Apart from constructing a proper high-pressure line, the success of this method depends crucially on making cells that can endure pressures up to at least a few kilobar and also on sample handling. All these procedures require skills and techniques.

Choice of Cell Materials. Tensile strength determines the maximum pressure that a cell can endure. The tensile strengths of most ordinary glasses as received from a factory are about $10-100$ kg cm^{-2}, but a strengthened Pyrex glass has a tensile strength of $3000-4000$ kg cm^{-2}.[29] The reason for the discrepancy between the theoretical strength and the actual strength is considered to be minute surface flows or cracks acting as stress concentrators. Thus, good surface condition is of prime importance for making a good cell, and actually it is possible to make Duran or Pyrex cells that endure up to 2 kbar by careful pre-etching of the surface as seen below. Good synthetic quartz is known to have a tensile strength $>21,000$ kg cm^{-2}.[22] This means that theoretically it is possible to make quartz cells that can endure pressures above 10 kbar.

Both borosilicate glass (Duran or Pyrex) and synthetic quartz are commercially available.

Design Principle of High-Pressure Cell. The maximum pressure that a cylindrical cell can endure actually depends on the ratio of the outer diameter to the

[26] H. Yamada, K. Nishikawa, M. Sugiura, and K. Akasaka, *International Conference on High-Pressure Science and Technology, Abstracts,* p. 413. The Japan Society of High Pressure Science and Technology, Kyoto, Japan, 1997.

[27] M. W. Lassalle, H. Yamada, and K. Akasaka, *J. Mol. Biol.* **298,** 293 (2000).

[28] K. Akasaka, T. Tezuka, and H. Yamada, *J. Mol. Biol.* **271,** 671 (1997).

[29] N. S. Isaacs, "Liquid Phase High Pressure Chemistry," Chapter 1. John Wiley, Chichester, UK, 1981.

inner diameter of the cell, and can be roughly predicted for an elastic cylinder model by

$$P_{max} = \sigma_T \left[(d_0/d_i)^2 - 1 \right] \left[(d_0/d_i)^2 + 1 \right]^{-1}$$

where σ_T represents the tensile strength of the cylinder material and d_0, d_i are the outside and inside diameters of the cylinder, respectively.[29] To compromise between the attainable pressure and the filling factor of the sample solution, a cylinder with $d_0/d_i \sim 3$ to 3.5 is considered appropriate. A cylindrical cell with $d_0/d_i = 3.0 - 3.5$ mm/0.8–1.0 mm is good for use in a 5 mm probe and a cell with $d_0/d_i = 6.0$ mm/1.5 mm is good for use in an 8 mm probe. We have so far used a sample cell with the former geometry for most of our experiments in conjunction with a 5 mm inverse-detected triple-resonance probe with x,y,z gradient on a spectrometer operating at 750 MHz for ^{1}H (Bruker DMX-750).

Design of High-Pressure Cell System. Figure 1 shows schematically the design details of the entire cell system. A total of about 100 μl of the sample solution is transferred into the cell assembly, of which some 20 μl fills the body of the cell ($d_0/d_i = 3.0$–3.5 mm/0.8–1.0 mm). The cell part is protected with a polytetrafluoroethylene (PTFE) tube of 5.0 mm o.d. The upper part of the cell capillary is glued to a SUS-316 nozzle with an epoxy adhesive. The top of the capillary is capped with a PTFE connector tube, which is connected to a larger PTFE tube filled with the sample solution. The sample solution is separated from kerosene by a layer of two freely movable PTFE pistons. The space between the two pistons is filled with an inert solvent perfluoropolyether to ensure separation of the sample solution and kerosene. All are housed in a separator cylinder made of Cu–Be. Pressure is mediated through the movable pistons to the sample solution and is maintained at a constant value equal to that of the kerosene during measurements with a help of the valves.

Setting Up the Line. Figure 2 shows a schematic of the entire setup of the high-pressure line. Before the cell is set into the NMR probe, the cell is connected to the separator. Then the entire connection (Fig. 1) is inserted from the top of the superconducting magnet into the probe inside. A ^{2}H lock signal is monitored to secure the correct positioning. Then the top end of the assembly is connected with flexible SUS-316 tubing ($d_0/d_i = 3.1$ mm/0.6 mm, length $\sim$6 m) on top of the 17.6 tesla magnet, which is connected to a remotely located hand pump (Fig. 2). To avoid any effect from bubbles remaining in the sample cell, $\sim$30 bar is usually applied for the lowest pressure rather than 1 bar. The spectra at 30 bar and at 1 bar are indistinguishable.

A suitable guide system is necessary for exact positioning of the sample cell in the probe. This may be done with the help of a commercially obtained sample spinner attached to the PTFE protector. No sample spinning is needed.

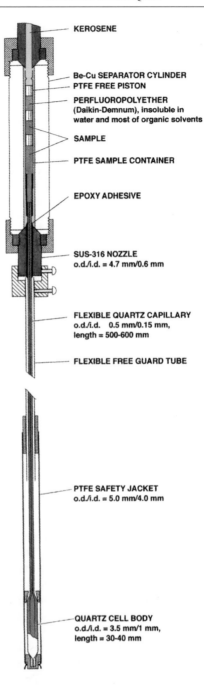

KEROSENE

Be-Cu SEPARATOR CYLINDER
PTFE FREE PISTON

PERFLUOROPOLYETHER
(Daikin-Demnum), insoluble in
water and most of organic solvents

SAMPLE

PTFE SAMPLE CONTAINER

EPOXY ADHESIVE

SUS-316 NOZZLE
o.d./i.d. = 4.7 mm/0.6 mm

FLEXIBLE QUARTZ CAPILLARY
o.d./i.d.   0.5 mm/0.15 mm,
length = 500-600 mm

FLEXIBLE FREE GUARD TUBE

PTFE SAFETY JACKET
o.d./i.d. = 5.0 mm/4.0 mm

QUARTZ CELL BODY
o.d./i.d. = 3.5 mm/1 mm,
length = 30-40 mm

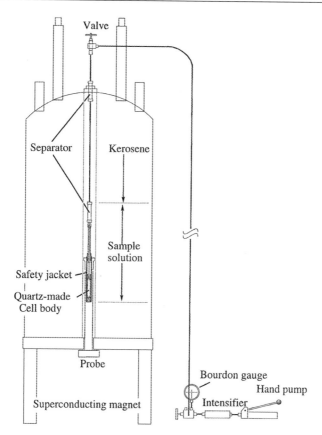

FIG. 2. A schematic view of the on-line cell high pressure NMR system developed at Kobe University (the magnet and the high pressure system). The essential feature is contained in the original design.[3] The present setup uses a Bruker DMX-750 NMR spectrometer with a superconducting magnet (17. 6 tesla, Japan Magnet Technology). The protein solution is contained in a cell body made of quartz tube (inner diameter about 1 mm, outer diameter about 3 mm) and is connected to the pressure mediator (kerosene) through pistons in a Be–Cu cylinder. The pressure is controlled manually with a hand pump placed remotely from the magnet.

FIG. 1. Design details of the high pressure cell. The high pressure cell has a body part of ∼30 mm length with $d_0/d_i = 3$–3.5 mm/0.8–1.0 mm connected to a long tail ($d_0/d_i = 0.5$ mm/0.15 mm, length 500–600 mm) with an open end. The open end of the long tail is inserted into a SUS-316 nozzle and fixed there with an epoxy adhesive. The nozzle is coupled with a separator–safety cylinder assembly. An inner PTFE tube and double pistons inside the separator functions to separate the sample solution and the pressure transmitting fluid (kerosene) and to prevent the adhesive in the nozzle from being damaged by the sample solvent. The entire length is settled in the superconducting magnet (17. 6 tesla).

Maximum Attainable Pressure. The pressure of the sample solution is set at a desired value on a Bourdon tube gauge with a hand pump. The pressure can be maintained at a constant value at least for a few days.

The maximum endurable pressure for glass tubes is usually set at 2 kbar. The endurable pressure is increased greatly by using synthetic quartz for the cell material. A quartz cell enduring 4.5 kbar has been made, for which at present the cell dimension becomes a little smaller than for 2 kbar with the same d_0/d_i ratio. Consequently, the signal-to-noise ratio for a 4 kbar cell becomes a little less than that for a 2 kbar cell. Judging from the tensile strength, a quartz cell might be made in the future that can endure higher pressure.

Temperature Control. The temperature of the sample solution is maintained by the airflow through the probe with a temperature controller provided commercially for the NMR spectrometer. Calibration of the actual temperature of the sample solution is necessary, and this can be made using an independent measurement of a water or methanol sample in an equivalently shaped sample cell with a thermistor directly immersed into the sample cell.

Resolution Adjustment. The resolution can be adjusted by maximizing the ^2H lock signal of 5–10% deuterons in the solvent. The field/frequency lock is used during measurements. The resolution attainable for a quartz cell enduring up to 4 kbar is excellent with the best expected resolution of the spectrometer, usually limited by intrinsic line widths of the protein signals. As an example, one-dimensional proton spectra of hen lysozyme at 1 bar and 2000 bar are shown in Fig. 3.

Active Sample Volume and Signal-to-Noise Ratio. For a sample cell with $d_0/d_i = 3-3.5$ mm/0.8–1.0 mm, the NMR active volume would be less than 20 μl, which is less than one-tenth of the volume in a standard NMR tube. This means that the signal-to-noise ratio should be at least an order of magnitude smaller than in a regular measurement. Therefore, the on-line cell high-pressure NMR for proteins can be used most successfully with the highest frequency spectrometer available or with a spectrometer with a cryoprobe using a superconducting material for a radio frequency coil.

Caution in Sample Preparation. For the preparation of a protein solution for high-pressure studies, a general procedure for measurements under normal conditions at 1 bar should be followed. In addition, a few extra cautions must be taken. (1) Buffers with the least dissociation volume (e.g., maleic acid, MES, Tris) should be chosen. (2) For shift reference, the methyl proton signal of 2,2′-dimethyl-2-silapentane-5-sulfonate (DSS) or the methyl proton signal of trimethylsily propionate-d_4 (TSP) can be used as internal reference, but it is always recommended to have one or more additional shift references. As the second reference, dioxane is recommended for its relative inertness with biopolymers and the simple nature of the signal (singlet at $\sim$3.75 ppm). An additional reference could be a signal from the buffer material used, e.g., Tris. The independence of mutual shifts of these

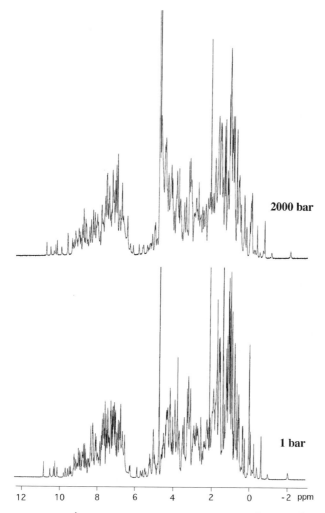

FIG. 3. One-dimensional ^{1}H NMR spectra of hen lysozyme in 90%^{1}H$_2$O/10%^{2}H$_2$O, pH 3.8 at 1 bar (measured in a 5 mm Shigemi tube) and at 2000 bar (measured in a high pressure cell), measured at 750 MHz at 25°. Chemical shift is referenced to TSP.

reference signals on pressure is a good indication that they work properly as shift references. For chemical shift references for other nuclei, the indirect referencing procedure[30] is recommended.

[30] D. S. Wishart, C. G. Bigam, J. Yao, F. Abildgaad, H. J. Dyson, E. Oldfield, J. L. Markley, and B. D. Sykes, *J. Biomol. NMR* **6**, 135 (1995).

Pulse Sequences. Practically any pulse sequences that the hardware of the spectrometer permits can be used for high-pressure NMR experiments, including ^{1}H one-dimensional, ^{1}H–^{1}H two-dimensional, ^{1}H-heteronuclear two-dimensional, or ^{1}H-heteronuclear three-dimensional measurements and diffusion measurements with field-gradient pulses. There is no particular loss of quality factor compared to the regular measurement. The signal-to-noise ratio due to the limited filling factor limits the pulse sequence. All measurements can be performed in proton-rich aqueous environments (90–95% ^{1}H$_2$O/10–5% ^{2}H$_2$O).

Cell Burst. The cell can be used repeatedly (more than 20 times) in a favorable case, but an accidental cell burst could occur as noted by the (complete) loss of signals. However, as the PTFE safety jacket contains the broken quartz and the sample solution, no damage occurs to the probe. This is because the experiment is performed below the boiling temperature, and the total amount of the solution in the probe is quite small ($\sim$100 μl).

Future Development of Technique. Currently, there is a need to attain higher sensitivity in the signal detection. Two lines of progress can be anticipated. One obviously involves increasing the sample volume, which can be done by employing a larger probe size (e.g., 8 mm) than hitherto used (5 mm). A 3-fold increase in the signal-to-noise ratio is expected. Another is the use of a cryoprobe for the high-pressure measurement. Because the high-pressure cell method can be combined with any commercial spectrometer, it can also be used with a cryoprobe setup. This will increase the signal-to-noise ratio about threefold. If combined with the probe size mentioned above, a total increase of the signal-to-noise ratio by some tenfold would be expected. This will extend the range of applicability of high-pressure NMR enormously, making on-line cell high-pressure NMR an indispensable tool for investigating the fundamental nature of proteins in the 21st century.

Thermodynamic Background

Protein in solution can be a dynamic entity spanning a wide range of conformational space. In general, it can exist in equilibrium among a variety of conformers from the native conformation (N) to the denatured conformation (D), although NMR usually detects only one conformer N (native), which is usually a major species under physiological conditions. The equilibrium can be manipulated by external variables, namely temperature (T) and pressure (P) along with solvent conditions (C), such as hydrogen ion concentration, ionic strength, and denaturant. At fixed C, Gibbs free energy difference (ΔG) between any two states, e.g., A and B, is given as a function of T and P by

$$\Delta G = \Delta G_0 - \Delta S_0(T - T_0) - \Delta C_p \left[T \left(\ln \frac{T}{T_0} - 1 \right) + T_0 \right]$$
$$+ \Delta V_0(p - p_0) - \frac{\Delta \kappa}{2}(p - p_0)^2 + \Delta \alpha(p - p_0)(T - T_0) \qquad (1)$$

where $\Delta G = G_B - G_A$ and $\Delta G_0 = \Delta G_0(C, T_0, P_0)$, the Gibbs energy difference under a reference condition (C, T_0, P_0).[31] Equation (1) indicates that pressure is primarily coupled with the volumetric properties of a protein, namely partial molar volume change under a reference condition $\Delta V_0 = V_B - V_A = \Delta V(C, T_0, P_0)$, the compressibility change $\Delta \kappa = -(\partial \Delta V / \partial P)_T$ and the thermal expansivity change $\Delta \alpha = (\partial \Delta V / \partial T)_P = -(\partial \Delta S / \partial P)_T$, in contrast to temperature which is coupled to the entropy change $\Delta S_0 = \Delta S(C, T_0, P_0)$ and the heat capacity change $\Delta C_p = T(\partial \Delta S / \partial T)_P$ along with the thermal expansivity change $\Delta \alpha$.

At constant temperature, we have a simplified version of Eq. (1) by neglecting the $\Delta \alpha$ term,

$$\Delta G = -RT \ln K = \Delta G_0 + \Delta V_0(P - P_0) - (\Delta \kappa / 2)(P - P_0)^2 \qquad (2)$$

In a relatively low pressure range, the last term of Eq. (2) is often neglected. Experimentally, we can evaluate ΔG_0 and ΔV_0 (and $\Delta \kappa$, if possible) from the pressure dependence of the equilibrium constant K according to Eq. (2).

There is general agreement that a change in partial molar volume of a protein (ΔV) on unfolding is contributed by two terms, i.e., loss of the cavities (ΔV_{cav}) that are usually formed within a folded structure because of imperfect packing of atoms, and change in the volume of the hydrated layer of the protein (ΔV_{hyd}),[32] namely,

$$\Delta V = V_B - V_A = \Delta V_{cav} + \Delta V_{hyd} \qquad (3)$$

ΔV_{cav} is always negative, while ΔV_{hyd} can be positive or negative depending on temperature and pressure. At room temperature, most reported ΔV values for unfolding lie in the range of -30 to -100 ml/mol, making proteins unfold at high pressure [Eq. (2)].

The volume of a folded protein fluctuates about an average value, and high pressure favors a smaller volume. Here, the isothermal compressibility coefficient is defined by

$$\beta_T = -(1/V_0)(\partial V / \partial P)_T \qquad (4)$$

(The sign convention makes the volume decrease yield a positive value of β_T.) If we neglect the hydration contribution and consider only the contribution from the protein, compression or positive β_T corresponds to a volume decrease of the polypeptide moiety, causing average decrease in internuclear distances within the protein architecture. Changes in internuclear distances may be followed site-specifically, in principle, as changes in chemical shift, nuclear overhauser effect (NOE), spin relaxation, and spin coupling constant in high-resolution NMR.

[31] A. S. Hawley, *Biochemistry* **10**, 2436 (1963).
[32] K. Gekko, *in* "Encyclopedia of Molecular Biology" (T. E. Creighton, ed.), p. 553. John Wiley & Sons, Inc., New York, 1999.

Macroscopically, the isothermal compressibility coefficient (β_T) of a protein is related to the amplitude of the volume fluctuation (δV) through the relation

$$\langle (\delta V)^2 \rangle = kTV\beta_T \tag{5}$$

where $\langle (\delta V)^2 \rangle$ is the ensemble average of squared volume fluctuation, k the Boltzmann constant, T the absolute temperature, and V the volume of the protein.[33] This relation gives a basis for discussing amplitude of structural fluctuation based on compressibility measurement.[34] Microscopically, the same relation may be extended to discuss amplitude of structural fluctuation of a protein based on pressure-induced structural changes detected by NMR.[35]

Equations (1)–(5) give thermodynamic relations that are macroscopic, and most studies on pressure effects on proteins have been treated macroscopically, even though they may monitor spectroscopic properties that primarily reflect site-specific structural changes. NMR can give not only thermodynamic but also structural and dynamic information that is potentially site-specific. Morishima[17] and Wagner,[25] using the high-pressure glass cell method, and Jonas[36] and Markley,[37] using the high-pressure probe method, have exploited considerably the potential utility of NMR spectroscopy for proteins from site-specific structural and dynamic viewpoints as well as from thermodynamic viewpoints.

Pressure Effect on Protein Structure

NMR Parameters

In this section, we discuss application of the on-line high-pressure NMR method to proteins in the native fold. Because compressibility coefficients of native proteins are positive [$\Delta\beta > 0$, Eq. (4)],[34] we expect that the average packing density of atoms somehow increases within the native fold at high pressure. Since covalent bond lengths are considered invariant under pressure at least within kilobar ranges, weak chemical bonds (hydrogen bonds, ionic bonds, and van der Waals interactions) along with "cavities" with or without water molecules are considered as primary sources of compressibility. The compression would generally be accompanied by changes in average torsion angles. These structural changes may be reflected in the NMR spectrum, as usual, as changes in chemical shift, nuclear Overhauser effect, spin–spin coupling constants, line shape, and spin relaxation,

[33] A. Cooper, *Proc. Natl. Acad. Sci. U.S.A.* **73**, 2740 (1976).
[34] K. Gekko and Y. Hasegawa, *Biochemistry* **25**, 6563 (1986).
[35] K. Akasaka, H. Li, H. Yamada, R. Li, T. Thoresen, and C. K. Woodward, *Protein Sci.* **8**, 1946 (1999).
[36] J. Jonas and A. Jonas, *Annu. Rev. Biophys. Biomol. Struct.* **23**, 287 (1994).
[37] C. A. Royer, A. P. Hinck, S. N. Loh, K. E. Prehoda, X. Peng, J. Jonas, and J. L. Markley, *Biochemistry* **32**, 5222 (1993).

all of which can be measured, in principle site-specifically, by utilizing the on-line high-pressure cell NMR method.

Pressure Effects on Secondary Structure: Hydrogen Bonds and Backbone Torsion Angles

Pressure-Induced Chemical Shifts of Amide Protons and Nitrogens. Chemical shift is among the most sensitive parameters to pressure, particularly when measurements are made on a spectrometer operating at high frequency (e.g., 750 MHz for ^{1}H). Backbone amide groups maintain the secondary structure of a protein by forming hydrogen bonds with carbonyls. Pressure-induced changes in chemical shifts of amide protons can be measured site-specifically either on ^{1}H/^{1}H two-dimensional spectra (e.g., nuclear Overhauser effect spectroscopy (NOESY) and total correlation spectroscopy (TOCSY)) or on ^{1}H/^{15}N two-dimensional spectra heteronuclear single-quantum correlation (HSQC) (if the sample is uniformly ^{15}N-labeled) in an ^{1}H$_2$O-rich environment (e.g., 90% ^{1}H$_2$O/10% ^{2}H$_2$O) at neutral to low pH. As a typical example, the pressure dependence of an ^{1}H/^{15}N HSQC spectrum is shown in Fig. 4A (see color insert), in which spectra of uniformly ^{15}N-labeled HPr[38] measured at 500 bar intervals are superimposed. Practically all the cross peaks shift their positions with pressure both in ^{1}H and ^{15}N frequency domains. The spectral changes are *fully reversible* with pressure up to 2 kbar, and the continuity of shifts with pressure makes the extension of cross peak assignments at 1 bar to those at 2000 bar straightforward. The result demonstrates that the protein, initially in the native conformation at 1 bar, changes its conformation *continuously* with pressure, while the protein remains *folded*. The size of the shift varies from one amide group to another, suggesting that the conformational change takes place nonuniformly over the entire protein structure.

Figure 4B shows pressure-induced changes of individual H^N and N^H chemical shifts, expressed as differential shifts between 1 bar and 2000 bar, i.e., $\Delta\delta = \delta$ (2000 bar) $- \delta$ (1 bar), for the case of BPTI.[35]

Structural Changes Inferred from Proton Shifts. The shifts of H^N occur preferentially to the lower field (positive $\Delta\delta$ values) with a rms shift of 0.101 ppm/ 2 kbar.[39] Pressure will generally decrease the N–H$\cdots$O=C hydrogen bond distance and will increase the polarization of the N–H bond. Thus, the low field shift trend of H^N may be understood as arising from a decreased shielding of the magnetic field on the H^N nucleus due to an increased polarization of the N–H bond and also from the increased susceptibility effect from the increased proximity of the C=O group.[40,41] Therefore, the pressure-induced shift of H^N gives at least a

[38] H.-R. Kalbitzer, A. Goerler, P. Dubovskii, H. Li, W. Hengstenberg, C. Kowolik, H. Yamada, and K. Akasaka, *Protein Sci.* **9**, 693 (2000).

[39] H. Li, H. Yamada, and K. Akasaka, *Biochemistry* **37**, 1167 (1998).

[40] T. Asakura, K. Taoka, M. Demura, and M. P. Williamson, *J. Biomol. NMR* **6**, 227 (1995).

TABLE I
AVERAGE VALUES OF PRESSURE-INDUCED CHEMICAL SHIFTS $\Delta\delta_p{}^a$ OF AMIDE PROTONS IN PROTEINS

	$\Delta\delta_p$ (ppm)		
Proteins	For all amide protons[b]	For hydrogen–bonded protons[c]	For non-hydrogen-bonded protons[d]
Melittin (in water)[e]	0.108 (24 : 26)	—	0.108 (24 : 26)
BPTI[f]	0.075 (52 : 58)	0.059 (28 : 28)	0.098 (20 : 20)
Protein G[g]	0.050 (55 : 56)	0.039 (34 : 34)	0.069 (21 : 22)
HPr[h]	0.051 (83 : 88)	0.033 (26 : 26)	0.060 (57 : 61)
Cytochrome c[i]	0.056 (61 : 104)	0.049 (29 : 44)	0.062 (32 : 60)
Lysozyme[j]	0.079 (61 : 129)	0.053 (27 : 64)	0.099 (34 : 65)

[a] $\Delta\delta_p$ is the pressure-induced chemical shift of an amide proton, *i.e.*, the chemical shift difference between 2000 bar and 30 bar.

[b] The ratio of the number of amide protons observed to the total number of amide protons in the protein is given in parenthesis.

[c] The ratio of the number of internally hydrogen bonded amide protons observed to the total number of internally hydrogen bonded amide protons in the protein is given in parenthesis.

[d] The ratio of the number of internally non-hydrogen-bonded amide protons observed to the total number of internally non-hydrogen-bonded amide protons in the protein is given in parenthesis.

[e] K. Akasaka, H. Li, P. Dubovskii, H.-R. Kalbitzer, and H. Yamada, *in* "Structure, Dynamics and Function of Biological Macromolecules" (O. Jardetzky, ed.). IOS Press, Amsterdam, in press (2001).

[f] From H. Li, H. Yamada, and K. Akasaka, *Biochemistry* **37**, 1167 (1998).

[g] From H. Li, H. Yamada, K. Akasaka, and A. M. Gronenborn, *J. Biol. NMR* **18**, 207 (2000).

[h] From H. R. Kalbitzer, A. Gorler, H. Li, P. V. Dubovskii, W. Hengstenberg, C. Kowolik, H. Yamada, and K. Akasaka, *Protein Sci.* **9**, 693 (2000).

[i] S. Ohji *et al.*, unpublished results.

[j] T. Tezuka *et al.*, unpublished results.

qualitative measure of the pressure-induced change in the N–H$\cdots$O=C hydrogen bond distance by pressure.

One of the rules found in Fig. 4B is that the shifts of the non-hydrogen-bonded H^N (probably hydrogen bonded to water; dotted columns) are larger (0.098 ± 0.079 ppm/2 kbar) than the H^N hydrogen bonded to carbonyls (filled columns; 0.058 ± 0.056 ppm/2 kbar).[39] The trend is generally observed in several other proteins[38,42] (Table I). The data for melittin in water (random-coiled) support the notion that the shifts are large for the amide groups exposed to water. Since exposed

[41] D. Sitkoff and D. Case, *Prog. Nucl. Magn. Reson. Spectrosc.* **32**, 165 (1998).

[42] K. Akasaka, H. Li, P. Dubovskii, H.-R. Kalbitzer, and H. Yamada, *in* "Structure, Dynamics and Function of Biological Macromolecules" (O. Jardetzky, ed.). IOS Press, Amsterdam, in press (2001).

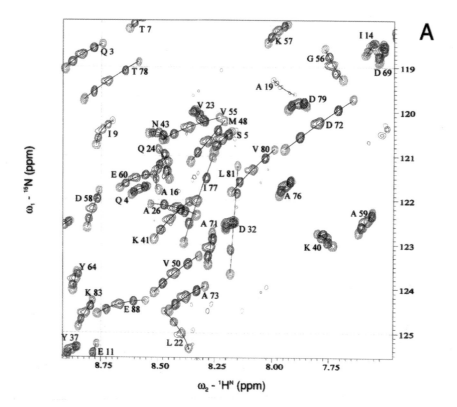

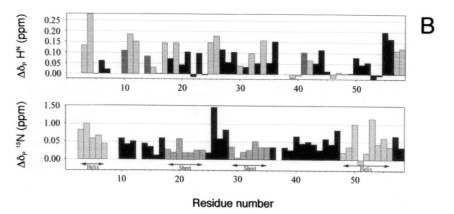

Fig. 4. (**A**) Superposition of ^{15}N/^{1}H HSQC spectra of uniformly ^{15}N-enriched histidine containing protein (HPr) from *S. carnosus* measured at 40, 500, 1000, 1500, and 2000 bar at 25°. The protein is dissolved to 14.3 mg/ml in 8.5% ^{1}H$_2$O/12.5% ^{2}H$_2$O, 10 m*M* Tris-HC1, pH 7.14 (after Ref. 38). Chemical shifts are referenced to TSP. (**B**) Histograms of pressure-induced changes in chemical shifts for individual amide protons H^N (upper) and amide nitrogens ^{15}N (lower) of BPTI, shown as differential shifts between 1 bar and 2000 bar, $\Delta\delta = \delta$ (2000 bar) $- \delta$ (1 bar), in 90% ^{1}H$_2$O/10% ^{2}H$_2$O, pH 4.6, 36°.

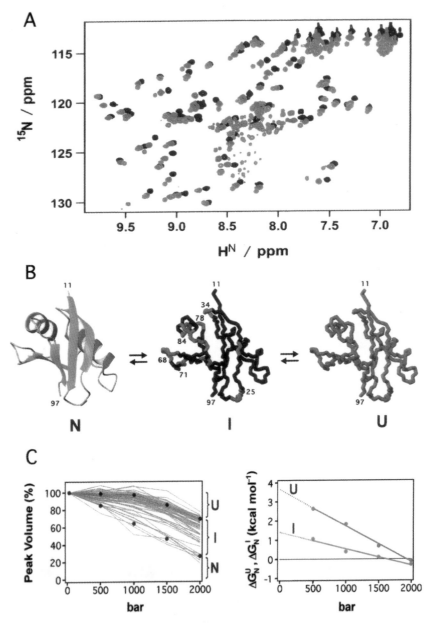

FIG. 10. Conformational equilibrium in the Ras-binding domain of RalGEF as revealed from high-pressure NMR study. (A) ^{15}N/^{1}H HSQC spectra of uniformly ^{15}N-labeled RalGEF-RBD (1.2 mM, pH 7.3) measured at 30 bar (blue), 1000 bar (green), and 2000 bar (red). Note that new peaks appear above 1000 bar in the region around 7.9–8.6 ppm for ^{1}H and 120–128 ppm for ^{15}N, typical for a "random-coiled" and hydrated polypeptide chain. (B) Equilibrium conformations of RalGEF-RBD; N (native), I (intermediate), and U (unfolded).The red color in I and U represents "melted" regions of the polypeptide chain. (C) (*Left*) Plot of individual cross-peak volumes in the ^{15}N/^{1}H HSQC spectra against pressure, showing changes of the N, I, U fractions with pressure. (*Right*) Plot of free energy differences (ΔG) between I and N and between U and N against pressure according to Eq. (2). (After ref. 56.)

amide groups are necessarily hydrogen bonded with water, the result indicates that pressure strengthens the hydrogen bonding interaction of an amide group with water.

For the NH groups forming internal hydrogen bonds with carbonyls, i.e., N–H· · ·O=C in BPTI, the "folding shift" $[\Delta\delta_f(^1H) = \delta_{observed} - \delta_{random}]$ of H^N is fairly well correlated with the H· · ·O distance (d_N). For BPTI, we have[39]

$$\Delta\delta_f(^1H) = 19.9d_N^{-3} - 2.3 \text{ (ppm) (correlation coefficient} = 0.77) \qquad (6)$$

where d_N is expressed in angstroms. Based on Eq. (6) the H· · ·O distance shortened by pressure varies considerably from site to site in the range from ∼0 to 0.11 Å/ 2 kbar, showing that the compaction takes place *nonuniformly* over the secondary structure. The average distance shortened at 2 kbar is 0.020 Å or about 1% of a typical H· · ·O distance (∼2 Å). Since similar shift values are obtained for other proteins (see Table I), the estimated average shortening of the peptide hydrogen bond would be fairly general. The results indicate that the N–H· · ·O=C hydrogen bond is one of the sources of protein compressibility.

Structural Changes Inferred from ^{15}N *Shifts.* ^{15}N chemical shift is another sensitive probe for the protein backbone structure.[43] The low field shift of ^{15}N resonances by pressure is also generally expected with the decrease of N· · ·O distances in N–H· · ·O=C hydrogen bonds.[44] However, the variation in ^{15}N shift among different amide groups in Fig. 4B is not correlated well with the variation in hydrogen bond distance changes inferred from H^N shifts. On the other hand, for the β sheet part of BPTI, the ^{15}N shift at 1 bar $(\Delta\delta_f(^{15}N_i)) = \delta_{observed} - \delta_{random}$ is well correlated with the backbone torsion angle ψ_{i-1}, giving an empirical relation[35]

$$\Delta\delta_f(^{15}N_i) = 36.04 - 0.2325\psi_{i-1}\text{(ppm)(correlation coefficient} = -0.91) \qquad (7)$$

where i refers to the ith residue. According to Eq. (7), the average pressure shift for the β sheet region of BPTI (0.295 ppm/2 kbar) corresponds to a decrease in ψ_{i-1} by an average of 1.2 degrees at 2 kbar, giving a measure of a structural change of the β sheet part of the polypeptide backbone by pressure.[35] In many proteins, the average ^{15}N pressure-induced shifts are often smaller for the β sheet part than for the helical and loop parts (see Table II).

^{13}C *Chemical Shifts.* $^{13}C^\alpha$ and $^{13}C^\beta$ chemical shifts are known to be sensitive measures of secondary structure. $^{13}C^\alpha$ chemical shifts can be measured as a function of pressure by performing two-dimensional HNCA and HNCO, and $^1H/^{13}C$ HSQC experiments. From pressure-induced $^{13}C^\alpha$ chemical shifts, pressure-induced changes in ϕ and ψ angles in α helices of BPTI are estimated to be, on average, about 2 degrees at 2 kbar and are negatively correlated.[45]

[43] H. Li and E. Oldfield, *J. Biomol. NMR* **4**, 341 (1994).
[44] N. Asakawa, T. Kameda, S. Kuroki, H. Kurosu, S. Ando, and A. Shoji, *Ann. Rep. NMR Spectroscopy* **35**, 55 (1998).
[45] H. Li, H. Yamada, K. Akasaka, and G. Montelione, unpublished results.

TABLE II

Average Values of Pressure-Induced Chemical Shifts $\Delta\delta_P{}^a$
of Amide Nitrogens of Proteins

		$\Delta\delta_P$ (ppm)		
Proteins	For all amide $^{15}N^b$	For amide ^{15}N in sheet^c	For amide^{15}N in helix^d	For amide ^{15}N in loop^e
BPTI^f	0.468 (52 : 58)	0.295 (14 : 14)	0.553 (14 : 14)	0.519 (24 : 30)
Protein G^g	0.427 (55 : 56)	0.405 (30 : 31)	0.422 (16 : 16)	0.507 (9 : 9)
HPr^h	0.472 (83 : 88)	0.195 (22 : 22)	0.598 (35 : 38)	0.536 (26 : 28)
DHFR^i	0.465 (121 : 159)	0.396 (42 : 49)	0.580 (26 : 36)	0.464 (33 : 43)

^{a} $\Delta\delta_P$ is the pressure-induced chemical shift of an amide nitrogen, *i. e.*, the chemical shift difference between 2000 bar and 30 bar.
^{b} The ratio of the number of amide nitrogens observed to the total number of amide nitrogens in the protein is given in parenthesis.
^{c} The ratio of the number of amide nitrogens observed in the sheet to the total number of amide nitrogens in the sheet is given in parenthesis.
^{d} The ratio of the number of amide nitrogens observed in the helix to the total number of amide nitrogens in the helix is given in parenthesis.
^{e} The ratio of the number of amide nitrogens observed in the loop to the total number of amide nitrogens in the loop is given in parenthesis.
^{f} From K. Akasaka, H. Li, H. Yamada, R. Li, T. Thoresen, and C. K. Woodward, *Protein Sci.* **8**, 1946 (1999).
^{g} From H. Li, H. Yamada, K. Akasaka, and A. M. Gronenborn, *J. Biol. NMR* **18**, 207 (2000).
^{h} From H. R. Kalbitzer, A. Gorler, H. Li, P. V. Dubovskii, W. Hengstenberg, C. Kowolik, H. Yamada, and K. Akasaka, *Protein Sci.* **9**, 693 (2000).
^{i} R. K. Kitahara, S. Sareth, H. Yamada, E. Ohmae, K. Gekko, and K. Akasaka, *Biochemistry* **39**, 12789 (2000).

$^{3h}J_{NC'}$ *Coupling Constant.* The hydrogen bond has a small covalent character that gives rise to a small $^{3h}J_{NC'}$ coupling constant (<1 Hz) between the ^{15}N and ^{13}C nuclei of the hydrogen-bonded pair ($^{15}N-H\cdots O=^{13}C)^{46}$. The $^{3h}J_{NC'}$ value is shown to be an excellent measure of the peptide hydrogen bond distance.[47] If a change in hydrogen bond distance occurs at high pressure, it should cause a change in the $^{3h}J_{NC'}$ coupling constant. The first measurement of $^{3h}J_{NC'}$ coupling under pressure was performed on a $\{^{13}C, {}^{15}N\}$-double labeled protein, the immunoglobulin binding domain of protein G.[48] Although the change is quite small (on the order of 0.1–0.2 Hz) at 2 kbar, careful measurement shows detectable changes (both increased and decreased) in the $^{3h}J_{NC'}$ coupling constant at 2 kbar. The result

[46] F. Cordier and S. Grzesiek, *J. Am. Chem. Soc.* **121**, 1601 (1999).
[47] G. Cornilescu, B. E. Ramirez, M. K. Frank, G. M. Clore, A. M. Gronenborn, and A. Bax, *J. Am. Chem. Soc.* **121**, 6275 (1999).
[48] H. Li, H. Yamada, K. Akasaka, and A. Gronenborn, unpublished results.

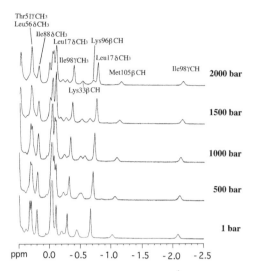

FIG. 5. Pressure dependence of the high-field region of the ^{1}H NMR spectrum of hen lysozyme in 90%^{1}H$_2$O/10%^{2}H$_2$O, pH 4.6, 36° (after Ref. 28).

gives the first direct evidence that electronic overlap through hydrogen bond is altered by pressure.

Changes in $^1J_{HN}$ by pressure have also been reported,[38] but the effect is too small to allow discussion of changes in hydrogen bonding state.

Pressure Effects on Tertiary Structure

Chemical Shifts of Side-Chain Protons. ^{1}H chemical shift is a sensitive measure of tertiary structure. Figure 5 shows the methyl proton region of the one-dimensional ^{1}H NMR spectrum of hen lysozyme in 99.8% ^{2}H$_2$O at varying pressure from 1 bar to 2000 bar.[28] Most signals shift their positions toward high field with increasing pressure, while the overall dispersion of signals characteristic of a folded structure is maintained. The spectral changes demonstrate that the average tertiary structure is altered by pressure continuously within the folded manifold: namely, it reflects the ensemble nature of the folded lysozyme. In this particular example, the high-field shifts of the methyl proton signals are a result of increasing ring current shifts with pressure from nearby Trp rings, meaning a compaction of the hydrophobic core.[29] A more thorough analysis can be made on smaller proteins such as gurmarin[49] and BPTI[50] based on ^{1}H/^{1}H two-dimensional NMR measurements (TOCSY and NOESY), in which signals are detected from practically all the side-chain protons.

[49] K. Inoue, H. Yamada, T. Imoto, and K. Akasaka, *J. Biomol. NMR* **12,** 35 (1998).
[50] H. Li, H. Yamada, and K. Akasaka, *Biophys. J.* **77,** 2801 (1999).

Nuclear Overhauser Effect. Since the intensity of NOE has a reciprocal 6th power dependence on interproton distance, changes in NOE between side-chain protons would give clear evidence that average distances between side chains are altered by pressure. Figure 6 shows a part of the NOESY spectrum for BPTI where considerable changes in NOE intensities are observed between side-chain protons at high pressure.[50] Not only increased NOE, but also decreased NOE has been observed in BPTI with increasing pressure, suggesting that not only the compaction, but also local expansion can take place within the native fold under pressure. It would, in principle, be possible to quantify pressure-induced average distance changes between side chains based on these NOE measurements.

Pressure Effect on Protein Dynamics

Microscopic Compressibility and Conformational Fluctuation

Macroscopically, compressibility of a protein measured under equilibrium is related to volume fluctuation through Eq. (5). Microscopically, site-specific changes in internuclear distance and torsion angle as depicted from static NOE and ^{15}N or ^{13}C chemical shift measurements with pressure are similarly considered to be related to fluctuation in internuclear distance and torsion angles.[35] The fluctuation inferred therein should contain all frequencies of motions from ~psec to ~msec or even longer.[35] The method can be used to identify regions of large-amplitude fluctuation in globular proteins.[35] In HPr, such regions are found surrounding the active site.

Effect of Pressure on psec–nsec Dynamics

Rapid (psec–nsec) backbone dynamics of a protein can be most conveniently studied by ^{15}N spin relaxation measurements of uniformly ^{15}N-labeled proteins. Because pressure leads to a more limited interatomic space, one might expect that its dynamics are sensitively affected by pressure. To test this expectation, spin relaxation (T_1, T_2, NOE) measurements were carried out at 30 bar and 2000 bar on BPTI, which remains fully folded at 2000 bar.[51] Subsequent analysis based on the Lipari–Szabo model showed that the NH vectors fluctuate at 2000 bar much as they do at 30 bar in the time range of psec–nsec with an average order parameter of ~0.85.[51] Practical invariance of psec–nsec dynamics with pressure is also true in a simpler system of an isolated α helix.[52] The observation is actually consistent with the remarkable linearity of pressure-induced chemical shifts of BPTI (Fig. 7A,B) that shows the invariance of the amplitude of volume fluctuation

[51] S. Sareth, H. Li, H. Yamada, C. K. Woodward, and K. Akasaka, *FEBS Lett.* **470**, 11 (2000).
[52] V. Y. Orekhov, P. V. Dubovskii, A. S. Arseniev, H. Yamada, and K. Akasaka, *J. Biomol. NMR* **17**, 257 (2000).

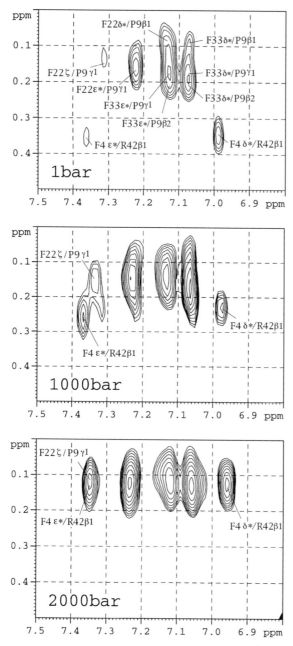

FIG. 6. Comparison of ^{1}H/^{1}H NOESY at 1 bar, 1000 bar, and 2000 bar for BPTI. The aromatic–aliphatic region at 36° with a mixing time of 150 ms at 750 MHz. The increase of NOE intensities between Phe-22 and Pro-9, between Phe-4 and Lys-42, and between Phe-4 and Lys-42 with pressure, gives direct evidence for core compaction. (After Ref. 50.)

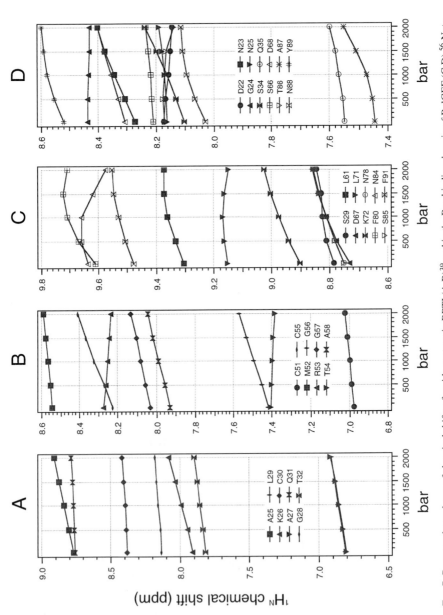

FIG. 7. Pressure dependence of chemical shifts of amide protons in BPTI (A,B)[39] and in the Ras-binding domain of RalGEF (C,D).[56] Note the linearity in A, B and the nonlinearity in C, D.

$\{\langle(\delta V)^2\rangle\}^{-1/2}$ with pressure.[35] In fact, pressure-induced changes in average torsion angles (ϕ, ψ) are within a couple of degrees at 2000 bar,[5] whereas the fluctuation of the NH vectors span much larger angles (± 20 degrees) as calculated from order parameters. Therefore, the effect of pressure (2 kbar) on BPTI is to shift the population slightly within the native ensemble, giving in both a linear pressure shift and pressure invariance in psec–nsec dynamics.

Pressure Effect on Slow (~msec) Protein Dynamics

Although pressure has little effect on psec–nsec dynamics, it has a profound effect on motions in slower time range (~msec). Earlier, Wagner found retardation of flip-flop motions of certain Phe and Tyr rings of BPTI about their C^β–C^γ axes.[25] The phenomenon was reexamined more closely by performing two-dimensional ^{1}H/^{1}H TOCSY NMR measurements under varying pressure[50]; the cross-sectional spectra give well-resolved signals of the (3,5) ring protons of Tyr-35 as a function of pressure (Fig. 8A). Spectral simulation determines the flip rate, which decreases dramatically from 620 sec^{-1} at 2000 bar to 12 sec^{-1} at 30 bar (Fig. 8A).

A plot of the logarithms of the flip rates against pressure gives activation volumes $\Delta V^\ddagger$ of 85 ± 20Å^3 (or 51.2 ml/mol) and 46 ± 9Å^3 (or 27.7 ml/mol), for Tyr-35 and Phe-45, respectively, at 57° (Fig. 8E). This means that a volume fluctuation occurs within the protein core, transiently creating an internal void of 50–80 Å^3 around these rings. It is interesting to note that these volumes are outside the range of the statistically expected volume fluctuation (~30 Å^3) of BPTI,[50] illustrating the case that the ~msec fluctuation of the protein causing the ring flip is well outside the range of fluctuation represented by an average dynamic behavior of the protein. Such a slow and rare fluctuation could be more important for protein function than the average fluctuation in the psec–nsec range.

Pressure Effect on Conformational Equilibrium

Low-Lying Excited States Inferred from Nonlinear Pressure Shifts

The ^{15}N and ^{1}H chemical shifts of amide groups between 1 and 2000 bar is surprisingly linear with pressure for BPTI[35] (Fig. 7C,D) as well as for other proteins such as lysozyme[28] and gurmarin.[49] The constant slope would imply that for these proteins the compressibility is independent of pressure. However, for many other proteins, pressure-induced shifts are slightly (HPr, Fig. 4A) or markedly (RalGEF-RBD, Fig. 7C,D) nonlinear. The nonlinearity would mean that for these proteins local compressibility (e.g., change of internuclear distance per unit pressure) is not constant over the range of pressure, but varies with pressure. This would mean that, as pressure is increased, conformation(s) having a compressibility different from that of the basic N become populated. However, the new conformer N' is

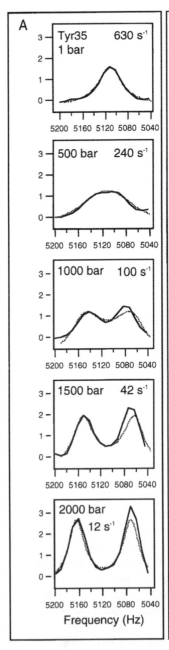

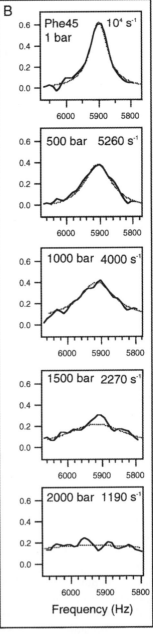

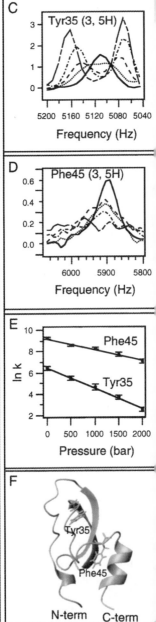

in rapid exchange with N so that it does not appear as a separate signal from that of N, giving only *nonlinear shifts*. Since obviously the population of N' is much less than that of N at 1 bar, it is considered to be a low-lying excited state of N. The reason why it becomes populated with increasing pressure is because it has a smaller partial volume than N. The low-lying excited states could, in general, be important for proteins to perform their biological function.

Two-State Approximation to Conformational Transition

When the rate of exchange (k) between two conformations, such as between native (N) and denatured (D) conformations, is slow compared to the frequency separation (v) of their signals ($k \ll v$), the two signals can be observed as separate peaks. In this case, their integrated signal intensities may be used to evaluate fractions of the two conformers in equilibrium, which are then subjected to thermodynamic analysis according to Eq. (1) or (2). In contrast to most other spectroscopic methods, NMR has an advantage that we do not need to record baselines for 0% N or 0% D, because the base lines are *a priori* known to be null, as long as the signals are not overlapped. This allows us to evaluate equilibrium constants in a relatively low range of pressure or temperature without going into extreme conditions that realize 0% N.

Utilizing this property of NMR spectroscopy, variable-pressure experiments can yield reliable values of ΔV and ΔG_0 of the transition, provided the signal-to-noise ratio is sufficiently good. In general, ΔV is not constant over the pressure range studied (1 to several kilobar), and one must evaluate $\Delta\beta$ and ΔV_0 separately[53] according to Eq. (2) or more completely according to Eq. (1). By careful measurement of integral intensities of His resonances in the one-dimensional ^{1}H NMR spectra, Lassalle *et al.* preformed a complete thermodynamic analysis based on Eq. (1) on staphylococcal nuclease.[27] The result gives a free energy landscape on the P-T plane, as shown in Fig. 9, with all relevant parameters (ΔG_0, ΔV_0, ΔC_p, ΔS_0, $\Delta\beta$, $\Delta\alpha$) in Eq. (1).[27]

[53] K. E. Prehoda and J. L. Markley, *in* "High-Pressure Effects in Molecular Biophysics and Enzymology" (J. L. Markley, D. B. Northrop, and C. A. Royer, eds.), p. 33. Oxford University Press, Oxford, 1996.

FIG. 8. Pressure effects on flip rates of Tyr-35 and Phe-45 rings of BPTI about their C^β–C^γ axes. Measurements are made in 90% ^{1}H$_2$O/10% ^{2}H$_2$O, 200 mM acetate d_3 buffer, pH 4.6 at 57°. Experimental (—) and simulated (· · · ·) 3H and 5H spectra at various pressures of (A) Tyr-35 and (B) Phe-45. Superposition of 3H and 5H signals at various pressures, obtained as TOCSY slice spectra at the 2H signal position. 1 bar (—), 500 bar (· · · ·), 1000 bar (— — — —), 1500 bar (-.-.-.-), 2000 bar (——), of (C) Tyr-35 and (D) Phe-45. (E) Plot of the logarithm of the flip rate, k (sec^{-1}), of the Tyr-35 and Phe-45 rings against pressure. (F) Ribbon model showing locations of Tyr-35 and Phe-45 in the tertiary structure of BPTI. (After Ref. 50.)

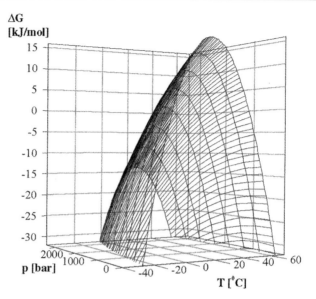

FIG. 9. Free energy landscape of staphylococcal nuclease (in 99. 8% ^2H_2O, pH 5. 3) on the P–T plane [Eq. (1)] determined from ^1H NMR signals of His-8, with the following thermodynamic parameters: $\Delta G_0 = 13.18 \pm 2$ kJ mol^{-1} (at 24°); $\Delta V_0 = -41.9 \pm 6.3$ ml mol^{-1} (at 24°); $\Delta\kappa = 0.02 \pm 0.003$ ml mol^{-1} bar^{-1}; $\Delta\alpha = 1.33 \pm 0.2$ ml mol^{-1} K^{-1}; $\Delta C_p = 13.12 \pm 2$ kJ mol^{-1} K^{-1}; $\Delta S_0 = 0.32 \pm 0.05$ kJ mol^{-1} K^{-1} (at 24°). (After Ref. 27.)

Multistate Approach to Conformational Transition

Use of nonoverlapping signals in one-dimensional spectra or cross peaks in heteronuclear two-dimensional spectra gives multiple probes for conformational population distributed over an entire protein structure. This may lead, in general, to site-dependent multiple equilibrium constants with multiple thermodynamic parameters, as first suggested by using several side-chain signals in one-dimensional ^1H NMR spectra.[54] Hitchens and Bryant[55] evaluated ΔV values for individual amide groups by measuring off-line pressure dependence of hydrogen exchange rates in T4 lysozyme by NMR. The multiple nature of transitions by pressure can be examined more thoroughly by using heteronuclear two-dimensional NMR spectroscopy, such as $^{15}N/^1H$ HSQC. Measurement of cross-peak intensities for individual amino acid residues as a function of pressure gives, in principle, a number of transition curves equal to the number of amino acid residues making up the protein molecule. If all their transition curves do not coincide with each other, it

[54] J. Zhang, X. Peng, A. Jonas, and J. Jonas, *Biochemistry* **34,** 8631 (1995).
[55] T. K. Hitchens and R. G. Bryant, *Biochemistry* **37,** 5878 (1998).

would mean that the transition is not a two-state transition. Careful analysis of integrated cross-peak intensities would tell us detailed features of the conformational transition, which is not possible by other spectroscopic methods. Although BPTI and lysozyme would need more than several kilobars for pressure denaturation, many other proteins show local melting behavior even at less than 1 kbar, followed by a melting of the rest of the molecule at higher pressure, meaning a multistate transition.

A good example is found in the HSQC spectra of RalGEF-RBD, the Ras-binding domain of the Ral guanine-nucleotide exchange factor (Fig. 10A, see color insert).[56] Integral intensities of some cross peaks in the $^{15}N/^{1}H$ HSQC spectrum begin to decrease selectively at relatively low pressures ($\sim$500 bar), with a concomitant increase of new peaks in the narrow areas of 7.9–8.6 ppm for ^{1}H and 120–128 ppm for ^{15}N, typical for "random-coiled" and hydrated polypeptide chain. The resultant species is locally melted selectively in a few loop regions (Fig. 10B, the red portion in I). Further increase to 2 kbar makes the entire molecule to start melting (Fig. 10B, U). Since the entire process is reversible with pressure, N, I, and U species must be in equilibrium at all pressures between 30 bar and 2000 bar, as depicted in Fig. 10B in which the melted part is colored red, while the intact (native) part is colored blue.

In Fig. 10 C, left, cross-peak volumes of all the signals of the "native" ensemble are plotted as a function of pressure, after a correction for the signal intensity due to the compression of the solvent (water).[57] The data represent conformational transitions with individual amide groups as probes. Clearly the data indicate that the protein undergoes a multistate transition. Actually, the transition curves are grouped into two major groups with rapid-decaying and slow-decaying properties with respect to pressure, the rapid-decaying groups showing the N to I transition and the slow-decaying groups showing the I to U transition. The three species N, I, and U are in equilibrium at all pressures (Fig. 10B), and, according to Eq. (2), Gibbs energy differences between I and N and between U and N are plotted as a function of pressure in Fig. 10C, right, from which may be found that $\Delta G_0 = 1.5$ and 3.7 kcal/mol and $\Delta V_0 = -36$ ml/mol and -78 ml/mol for I and U relative to N, respectively. Namely, I is an excited state of RalGEF-RBD lying 1.5 kcal/mol above N at 1 bar, and U lies 3.7 kcal/mol above N at 1 bar.

The populations of I and U at 1 bar estimated from the ΔG_0 values are $<10\%$ and $<1\%$, respectively, which are too small for NMR detection. However, these rare conformational species are rendered detectable by NMR at high pressure, simply because the transition from N to I involves a volume change of -36 ml/mol

[56] K. Inoue, H. Yamada, K. Akasaka, C. Herrmann, W. Kremer, T. Maurer, R. Doerker, and H.-R. Kalbitzer, *Nature Struct. Biol.* **7**, 547 (2000).
[57] IAPWS Release on the Skelton Tables 1985 for the Thermodynamic Properties of Ordinary Water Substance, p. 210 (1994).

and the transition from I to U involves a further decrease by -42 ml/mol. These volume changes are accounted for, at least in part, by the collapse of cavities existing in N.[56] As this example illustrates, pressure provides a unique means for increasing the population of rare conformational species that are barely detectable at 1 bar by taking advantage of differences in partial molar volume among different conformational states. Combined with the on-line cell high-pressure NMR method, it offers a general means for analyzing structures and thermodynamics of these rare conformational species in proteins.

Comparison of High-Pressure NMR to Hydrogen Exchange Method

Until recently, hydrogen exchange combined with NMR had been the method by which some excited states are monitored site-specifically, giving protection factors for individual amide groups.[58,59] Now high-resolution high-pressure NMR can monitor such states directly in the NMR spectrum as shown above by taking advantage of the difference in partial molar volume. A direct relationship between the excited state responsible for hydrogen exchange and the excited state stabilized by pressure was inferred earlier by Fuentes and Wand in apocytochrome b_{562}.[60] Using the on-line high-pressure cell NMR method, Kuwata *et al.* have found in β-lactoglobulin that the "open" conformers responsible for hydrogen exchange and the locally unfolded conformers identified by high-pressure NMR have nearly identical ΔG_0 values, indicating that they are likely to be the same.[61]

Although both hydrogen exchange and high-pressure NMR offer methods to detect the presence of excited states or rare conformational species in proteins in solution, the drawback of the hydrogen exchange method is that the method cannot directly observe the excited state structure. The on-line high-pressure cell NMR method offers a method to observe those species spectroscopically and analyze their structures in atomic detail. The method would have enormous potential utility in examining novel structures of proteins that are rare under physiological conditions, but are important in biological function.

[58] C. K. Woodward, I. Simon, and E. Tuechsen, *Mol. Cell. Biochem.* **48,** 135 (1982).
[59] S. W. Englander and L. Mayo, *Annu. Rev. Biophys. Biomol. Struct.* **21,** 243 (1992).
[60] E. J. Fuentes and A. J. Wand, *Biochemistry* **37,** 9877 (1998).
[61] K. Kuwata, H. Li, H. Yamada, C. A. Batt, Y. Goto, and K. Akasaka, *J. Mol. Biol.* **305,** 1073 (2001).

[6] Maximum Entropy Reconstruction, Spectrum Analysis and Deconvolution in Multidimensional Nuclear Magnetic Resonance

By JEFFREY C. HOCH and ALAN S. STERN

Introduction

Although the discrete Fourier transform (DFT) played an enabling role in the development of modern nuclear magnetic resonance (NMR) spectroscopy, it has well-known limitations, most notably difficulties providing high-resolution spectra from short data records. A variety of modern non-DFT methods for spectrum analysis have been developed over the past two decades that can produce better results than the DFT. Maximum entropy reconstruction (MaxEnt) is one particularly versatile method. In contrast to other modern methods that assume Lorentzian line shapes, such as those based on linear prediction, MaxEnt is suitable for spectra containing arbitrary line shapes. Far from being simply a substitute for the DFT, MaxEnt affords unique opportunities for experiment design to improve spectral quality or reduce data acquisition times. Given the cost of state-of-the-art NMR instruments, even modest reductions in data acquisition time can be significant. Other capabilities of MaxEnt, such as the ability to perform stable deconvolution, enable novel experiments that would otherwise be impractical.

Spectrum analysis of time-series data is integral to every modern application of NMR, including high-resolution studies of molecules in solution, solid-state studies, and imaging. The story of the development of modern pulsed NMR and the central role played by the discrete Fourier transform is widely known.[1,2] Less widely known is that computationally efficient algorithms for computing the DFT (i.e., the fast Fourier transform), although already published, were not known to Ernst and Anderson at the time.[3,4] The proposal to use the DFT to improve the efficiency of NMR experiments (over frequency- or field-swept techniques) was thus particularly audacious, since the barriers to computing the Fourier transform of a free induction decay were high.[2]

Today modern spectrometers positively bristle with integrated computers, some with the power to compute spectra for acquired data in near real time. Nevertheless,

[1] R. R. Ernst, *in* "Computational Aspects of the Study of Biological Macromolecules by Nuclear Magnetic Resonance Spectroscopy" (J. C. Hoch, F. M. Poulsen, and C. Redfield, eds.), pp. 1–25. Plenum Press, New York, 1991.

[2] W. T. Anderson, *in* "Encyclopedia of Nuclear Magnetic Resonance" (D. M. Grant and R. K. Harris, eds.), pp. 168–176. John Wiley & Sons, Chichester, 1996.

[3] R. R. Ernst, personal communication (1988).

[4] J. W. Cooley and J. W. Tukey, *Math. Comp.* **19,** 297 (1965).

METHODS IN ENZYMOLOGY, VOL. 338

the DFT remains the most widely used method for computing spectra. This is despite the fact that the DFT is not very good at computing high-resolution spectra from short data records, a shortcoming that has significant implications in multidimensional NMR, since the amount of instrument time required grows in direct proportion to the number of indirect time samples. Techniques such as apodization and zero-filling are only of modest help.

Extrapolation by linear prediction (LP) is a far more intelligent way to extend typical free induction decays (FIDs) beyond the measured data than appending zeros, and it is rapidly supplanting the practice of zero-filling.[5] LP extrapolation belongs to a class of modern methods that model the data as a sum of exponentially decaying sinusoids (giving rise to Lorentzian lines in the frequency domain) plus random noise. Also belonging to this class are linear prediction singular value decomposition (LPSVD),[6] maximum likelihood,[7,8] Bayesian,[9] and the filter diagonalization methods.[10,11] Although all of these are capable of providing high-resolution spectra from short samples of data with the appropriate characteristics, they are less suitable for spectra containing highly non-Lorentzian peak shapes or for imaging applications. They try to fit a Lorentzian peak to every signal in the data, even if it arises from nonexponential decay of a signal, nonrandom noise, or some other artifact, resulting in a spectrum that appears normal but may contain biased peak frequencies or other defects. These problems can be particularly pronounced for noisy data.[12]

Maximum entropy reconstruction belongs to a different class of methods that make very few assumptions about the characteristics of the signal. In practice, it works well applied to spectra that are "nearly black," i.e., that contain a sparse collection of peaks with a baseline equal to zero.[13] Because the peaks are not assumed to have any particular characteristics, MaxEnt can be used for spectral analysis in solid-state NMR and other applications that have non-Lorentzian peak shapes.

MaxEnt computes spectral estimates by solving an inverse problem: Instead of computing a spectrum directly from the data, MaxEnt constructs trial spectra, computes the hypothetical time domain signal that would give rise to that spectrum, and then compares this with the actual data for consistency. The hypothetical or mock data are said to be consistent with the experimental data when the level

[5] Y. Zeng, J. Tang, and C. A. Bush, *J. Magn. Reson.* **83**, 473 (1989).

[6] R. De Beer and D. van Ormondt, *NMR Basic Prin. Prog.* **26**, 201 (1992).

[7] S. Sibisi, *in* "Maximum Entropy and Bayesian Methods" (P. F. Fougere, ed.), pp. 351–358. Kluwer Academic, Dordrecht, 1990.

[8] R. A. Chylla and J. L. Markley, *J. Biomol. NMR* **3**, 515 (1993).

[9] G. L. Bretthorst, *J. Magn. Reson.* **88**, 533, 552, and 571 (1990).

[10] V. A. Mandelshtam and H. S. Taylor, *J. Chem. Phys.* **107**, 6756 (1997).

[11] H. Hu, Q. N. Van, and A. J. Shaka, *J. Magn. Reson.* **134**, 76 (1998).

[12] P. Koehl, C. Ling, and J. F. Lefèvre, *J. Magn. Reson., Ser. A* **109**, 32 (1994).

[13] D. L. Donoho, I. M. Johnstone, J. C. Hoch, and A. S. Stern, *J. R. Stat. Soc. B* **54**, 41 (1992).

of disagreement [computed using a chi-squared (χ^2) statistic, for example] is approximately equal to the noise level. From among all the consistent trial spectra, MaxEnt selects the one with the highest entropy. Advantages of this inverted approach to spectrum analysis are that the experimental data can be sampled at arbitrary times, small amounts of missing or corrupted data can be tolerated, and arbitrary functions can be stably deconvolved.

Like virtually all modern spectral methods—including LP—MaxEnt is non-linear. This necessitates special care when quantifying peak intensities and in applications such as difference spectroscopy. Nevertheless, the advantages, broad applicability, and versatility of MaxEnt are compelling reasons to add it to the repertoire of spectral analysis tools in biomolecular NMR.

Introduction to MaxEnt

The MaxEnt reconstruction of the spectrum of a complex-valued time series $\mathbf{d}$ is the spectrum $\mathbf{f}$ which maximizes the entropy $S(\mathbf{f})$, subject to the constraint that the mock data $\mathbf{m}$, given by the inverse DFT of the spectrum, is consistent with the time series $\mathbf{d}$. Consistency is defined by the condition

$$C(\mathbf{f}, \mathbf{d}) \leq C_0 \tag{1}$$

where $C(\mathbf{f}, \mathbf{d})$ is the unweighted chi-squared statistic,

$$C(\mathbf{f}, \mathbf{d}) = \sum_{i=0}^{M-1} |m_i - d_i|^2 = \sum_{i=0}^{M-1} |k_i \cdot \mathrm{IDFT}(\mathbf{f})_i - d_i|^2 \tag{2}$$

and C_0 is an estimate of the noise level. For generality we include the convolution kernel $\mathbf{k}$ in the transformation from the trial spectrum to mock data. When $\mathbf{k}$ is everywhere equal to one, the mock data are simply the inverse DFT (IDFT) of the trial spectrum. Otherwise, the spectrum of the mock data is given by $\mathbf{f}$ convolved with the spectrum of $\mathbf{k}$, so $\mathbf{f}$ is the deconvolved spectrum. A definition of the entropy $S(\mathbf{f})$ applicable to complex-valued spectra can be derived from consideration of the entropy of an ensemble of spin-$\frac{1}{2}$ particles,[14] or by extending the Shannon information formula[15] to the complex domain along with suitable differentiability conditions.[16] The two approaches yield equivalent formulations for the entropy of a complex spectrum

$$S(\mathbf{f}) = -\sum_{n=0}^{N-1} \frac{|f_n|}{def} \log\left(\frac{|f_n|/def + \sqrt{4 + |f_n|^2/def^2}}{2}\right) - \sqrt{4 + |f_n|^2/def^2} \tag{3}$$

[14] G. J. Daniell and P. J. Hore, *J. Magn. Reson.* **84**, 515 (1989).
[15] C. E. Shannon, *Bell Syst. Tech. J.* **27**, 379 (1948).
[16] J. C. Hoch, A. S. Stern, D. L. Donoho, and I. M. Johnstone, *J. Magn. Reson.* **86**, 236 (1990).

where *def* is a scale factor. In principle, the quantum-mechanical derivation prescribes the value of *def* (it depends on the sensitivity of the spectrometer and the number of spins in the sample), but it is more convenient to treat *def* as an adjustable parameter. Essentially, it determines the scale at which the nonlinearity of MaxEnt becomes pronounced.

Computing MaxEnt

A standard method for solving the constrained optimization problem is to convert it into an equivalent unconstrained optimization problem, through the introduction of a Lagrange multiplier λ to create the objective function $Q(\mathbf{f}) = S(\mathbf{f}) - \lambda C(\mathbf{f}, \mathbf{d})$. The MaxEnt reconstruction corresponds to a critical point of Q, that is, a value of $\mathbf{f}$ for which ∇Q is zero. Since the entropy and chi-squared statistic are both everywhere convex, there is a unique, global solution to the constrained optimization problem. Furthermore, the solution satisfies $C = C_0$, provided that the trivial solution $\mathbf{f} = \mathbf{0}$ does not satisfy the constraint.

Quite often problems of this sort can be solved efficiently by following the gradient of the objective function (steepest ascent or conjugate gradients). Unfortunately, the objective function Q is difficult to optimize, and the step size using these methods converges to near zero long before the critical point of Q is reached. Practical computation of MaxEnt reconstructions thus requires a fairly complex optimizer. Methods employing multiple search directions and a variable metric to determine the step size and direction have proven effective and are described in detail elsewhere.[17,18] Software implementing one variant is available from the authors.

The parameters *def* and C_0 affect the results of MaxEnt reconstructions. In principle, C_0 should be equal to the noise level in the data, which is not always easy to ascertain beforehand. Set C_0 too large, and the result is a flat, featureless spectrum. Too small, and the result is a spectrum that resembles the zero-filled DFT of the data. A simple method for estimating C_0 that can be used with linearly sampled data is to compute the DFT of the data, without apodization or zero-filling. The noise level can be estimated by computing the root-mean-square (RMS) value for a blank region of the spectrum (it may be necessary to scale the result by a normalization factor, depending on the normalization constants used in the DFT).

Very large values of *def* can render the MaxEnt reconstruction similar to the zero-filled DFT. In the large-*def* limit, the entropy given by Eq. (3) becomes proportional to the power divided by def^2 (plus a constant), so MaxEnt reconstruction becomes minimum power reconstruction. Since the power is the same in the time

[17] J. Skilling and R. Bryan, *Mon. Not. R. Astron. Soc.* **211**, 111 (1984).
[18] J. C. Hoch and A. S. Stern, "NMR Data Processing," Chapter 5. Wiley-Liss, New York, 1996.

and frequency domains, the result is equivalent to zero-filling (which extends the data but adds no power). In general, *def* determines the level at which the nonlinearity of MaxEnt becomes pronounced. If *def* is set too small, the onset of noticeable nonlinearity occurs in the range spanned by the noise, leading to reconstructions with characteristically spiky baselines. The best results are obtained when *def* has a value smaller than the smallest significant feature in the spectrum, but larger than the noise.

Using MaxEnt to reconstruct multidimensional spectra raises several issues. What is the proper way to apply MaxEnt along more than one dimension at a time? How does one handle the other dimensions that are not being processed by MaxEnt? It is easiest to answer these questions in the context of a specific example, say a three-dimensional experiment. The first thing to note is that the number of points in t_3, the acquisition dimension, is generally not limited by experiment time. Consequently, there is little incentive to use MaxEnt for processing this dimension. Once the data have been transformed to the frequency domain in the acquisition dimension, by applying a DFT possibly augmented with LP extrapolation, further processing can be simplified by handling the separate $t_1–t_2$ planes independently. One may want to use MaxEnt to process either or both of the indirect dimensions. We have found that the nonlinear phase distortions introduced by MaxEnt, although not apparent when viewing a spectrum, render the reconstruction unsuitable for further processing in any other dimension. In practice, this means that MaxEnt must be the *last* time-to-frequency transformation applied. So if MaxEnt is to be used in t_1, then the t_2 dimension must be transformed to the frequency domain first. And if MaxEnt is to be used for both t_1 and t_2, then it must be applied to both dimensions simultaneously, rather than one at a time.

There is no conceptual difficulty in using MaxEnt to process more than one dimension at a time. The IDFT operation appearing in Eq. (2), and the corresponding sums in Eqs. (2) and (3), can be applied along as many dimensions as one likes. There is a practical limitation, however. Computing a MaxEnt reconstruction can be very demanding, requiring intermediate storage equivalent to several times the size of the final output spectrum. Typically the t_1 and t_2 dimensions of a three-dimensional experiment contain sufficiently few points that the reconstruction can be carried out on a modern workstation, but for larger data sets the storage requirements may be an obstacle.

Applying MaxEnt to reconstruct separately each t_1 row, or each $t_1–t_2$ plane, raises another problem. The extent of the nonlinearity of the reconstructions depends on the parameter λ, which will vary from row to row (or from plane to plane), resulting in distorted peak shapes. One way to avoid this is to process the entire data set as a single unit, extending the sums in Eqs. (2) and (3) to cover the entire spectrum, rather than working on a single row or plane at a time. Of course, such an approach will entail drastically increased data storage requirements. Fortunately the problem has a simpler solution.[19] All that is needed is to use a fixed value of λ rather than iterating to attain $C = C_0$. The correct value for λ can be determined by choosing a representative row, then computing the normal MaxEnt reconstruction with an appropriate value for C_0. In quantitative applications, it is important to use one of these methods—full data set MaxEnt reconstruction or constant-λ reconstruction—to ensure that the nonlinearity is uniform across the spectrum. Only then can the nonlinearity be calibrated in a reliable way.

Properties of MaxEnt

In general the MaxEnt reconstruction must be computed numerically, but there is a special case that has an analytic solution. The solution can give some insights into how MaxEnt works. When N (the number of points in the reconstructed spectrum) is equal to M (the number of experimental data points) and the mock FID is equal to the inverse DFT of the trial spectrum (i.e., there is no convolution kernel), the constraint statistic $C(\mathbf{f}, \mathbf{d})$ can be computed in the frequency domain using Parseval's theorem.[20] The Lagrange condition for a critical point of Q then becomes

$$\mathbf{0} = \nabla Q = \nabla S - 2\lambda(\mathbf{f} - \mathbf{F}) \tag{4}$$

where $\mathbf{F}$ is the DFT of the data $\mathbf{d}$. The solution is

$$|f_n| = \delta_\lambda^{-1}(|F_n|), \quad \text{phase}(f_n) = \text{phase}(F_n) \tag{5}$$

where δ_λ is the function

$$\delta_\lambda(x) = x - s'(x)/2\lambda \tag{6}$$

and $s(x)$ is the contribution of a spectral component with magnitude x to the overall entropy S [Eq. (3)]. The function δ_λ^{-1} is plotted in Fig. 1 for various values of λ. For this special case, MaxEnt reconstruction is equivalent to applying the nonlinear transformation δ_λ^{-1} point by point to the DFT of the experimental data. The transformation depends on the value of λ and has the effect of scaling every point in the spectrum down. Smaller values are scaled down more than larger values. The difference in scale factor for small and large values becomes more significant when λ is small; with large values of λ, more relative weight in Q is placed on the constraint, and the transformation becomes more nearly linear.

Although this result strictly applies only to the special case, it helps to explain some characteristic features of MaxEnt reconstructions in general. The nonlinear transformations in Fig. 1 will naturally suppress noise close to the baseline

[19] P. Schmieder, A. S. Stern, G. Wagner, and J. C. Hoch, *J. Magn. Reson.* **125**, 332 (1997).
[20] D. L. Donoho, I. M. Johnstone, A. S. Stern, and J. C. Hoch, *Proc. Natl. Acad. Sci. U.S.A.* **87**, 5066 (1990).

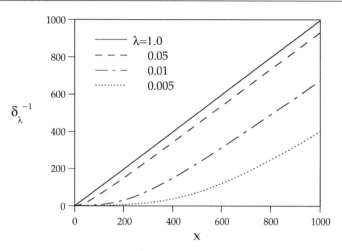

FIG. 1. The nonlinear transformation $\delta_\lambda^{-1}(x)$ relating the intensities in the analytic ($N = M$) MaxEnt reconstruction to intensities in the DFT spectrum (arbitrary units). Reprinted by permission of John Wiley & Sons, Inc., from "NMR Data Processing," J. C. Hoch and A. S. Stern, copyright © 1996, Wiley-Liss Inc.

more than noise superimposed on top of a broad peak. Figure 2 shows a MaxEnt reconstruction demonstrating this effect. Another important consequence is that with MaxEnt reconstructions, an apparent increase in signal-to-noise ratio does not necessarily correspond to increased sensitivity.[20] Since the nonlinear transformation applies to both signal and noise, peaks that are comparable in height to the noise will be reduced by the same amount as the noise. Thus, although the ratio between the highest signal peaks and the noise increases, small peaks are just as difficult to distinguish from noise as before. It should be emphasized that in the more general case, for example when an approximately known decay is deconvolved from the reconstructed spectrum, MaxEnt may deliver real gains in sensitivity. However, when utilizing nonlinear methods of spectrum analysis, a prudent investigator will not automatically assume that increases in signal-to-noise ratio reflect true gains in sensitivity. In a later section we will discuss a method for assessing sensitivity that can be applied to MaxEnt or any form of spectrum analysis.

Applications of MaxEnt

Resolution Enhancement

The most notorious shortcoming of the DFT is its inability to produce high-resolution spectra from short data records. Zero-filling improves digital resolution

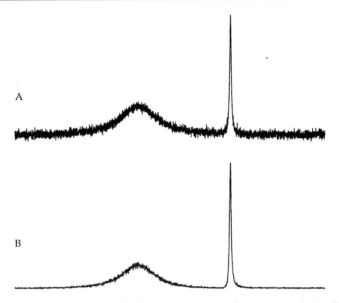

FIG. 2. A typical characteristic of MaxEnt reconstructions is that noise near the baseline is suppressed more than noise superimposed on a broad peak, which can be explained by the nonlinear transformation in Fig. 1. (A) is the DFT spectrum and (B) is the MaxEnt reconstruction; the value of λ was chosen to exaggerate the effect. Reprinted by permission of John Wiley & Sons, Inc., from "NMR Data Processing," J. C. Hoch and A. S. Stern, copyright © 1996, Wiley-Liss Inc.

(that is, the interval between consecutive points in the spectrum, which is not the same as the resolution of the spectrum), often at the expense of introducing truncation artifacts, and apodization forces trade-offs between sensitivity and resolution. (For a discussion of the DFT and its application in NMR, see ref. 21.) MaxEnt is quite adept at reconstructing high-resolution spectra from short data records, as illustrated in Fig. 3. In this example 8192-point spectra are computed from 1024 (complex) data points, using (A) zero-filling and DFT, (B) zero-filling, apodization, and DFT, (C) LP extrapolation and DFT, and (D) MaxEnt. The truncation artifacts near the base of the methyl resonances are diminished by apodization and LP extrapolation, but they are completely eliminated by MaxEnt. Comparison of the region around 2 ppm reveals an accompanying loss of resolution using apodization that is not encountered with MaxEnt. Similar trends are obtained using shorter data sets more typical of the indirect dimensions of multidimensional NMR experiments.

[21] J. C. Hoch and A. S. Stern, "NMR Data Processing," Chapters 2 and 3. Wiley-Liss, New York, 1996.

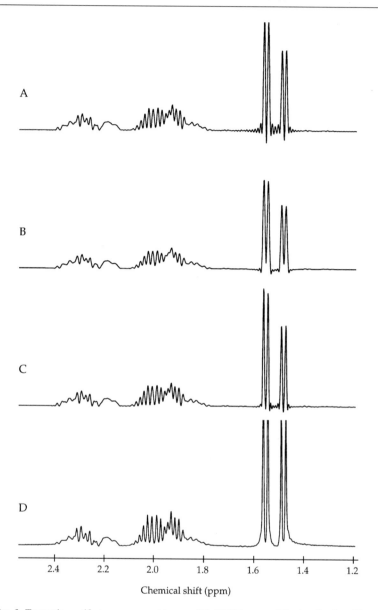

FIG. 3. Truncation artifacts are apparent in zero-filled DFT spectra (A). Apodization (B) and LP extrapolation (C) reduce the artifacts. MaxEnt reconstruction (D) eliminates them and at the same time improves resolution. Reprinted by permission of John Wiley & Sons, Inc., from "NMR Data Processing," J. C. Hoch and A. S. Stern, copyright © 1996, Wiley-Liss Inc.

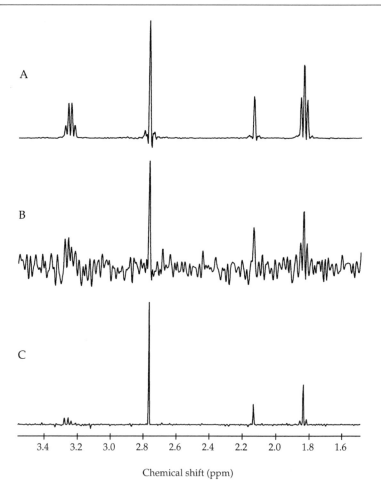

A

B

C

3.4   3.2   3.0   2.8   2.6   2.4   2.2   2.0   1.8   1.6

Chemical shift (ppm)

FIG. 4. DFT spectra computed from data obtained using (A) 90° and (B) 7° observe pulses. The MaxEnt reconstruction (C), computed using the same data as (B), has higher signal-to-noise ratio than (B), but the *sensitivity*—the ability to distinguish signal from noise—is not greatly affected. Reprinted by permission of John Wiley & Sons, Inc., from "NMR Data Processing," J. C. Hoch and A. S. Stern, copyright © 1996, Wiley-Liss Inc.

Improving S/N

MaxEnt reconstruction is capable of remarkable improvements in signal-to-noise ratio, but as the analytic solution makes clear, the improvements can be purely cosmetic. Figure 4 demonstrates the dramatic improvements that are possible. Shown in panel A is a spectrum computed from data obtained using a 90° observe pulse. In Fig. 4B and C the data was obtained using a 7° observe pulse to reduce the signal-to-noise ratio. The spectrum in Fig. 4B was computed using sine-bell apodization and DFT, the one in Fig. 4C using MaxEnt reconstruction. Although the signal-to-noise ratio for the largest peak is substantially improved, the weak quartet near 3.2 ppm is no longer clearly distinguishable. Sensitivity—the ability to distinguish signal from noise—must be assessed by means other than signal-to-noise ratio. In a later section we discuss the use of *in situ* analysis for quantifying sensitivity and calibrating nonlinearities.

Handling Corrupted Data

In part because it makes fewer assumptions about the characteristics of the signal, MaxEnt reconstruction is less susceptible than LP and related methods to bias or artifacts when the signal has unusual characteristics, such as nonrandom noise, nonexponential decay, or missing or corrupted values. In fact, because individual points can be left out of the computation of $C(\mathbf{f}, \mathbf{d})$, MaxEnt reconstruction can tolerate modest dropouts or corrupted data with virtually no ill effects, provided that the corrupted data points can be identified. Figure 5 illustrates the impact of missing data on spectral estimates. Here a perfectly good FID was artificially mangled by setting two disjoint stretches of 10 points to zero. The DFT of the data exhibits truncation artifacts throughout the spectrum (Fig. 5A). The artifacts are reduced in the MaxEnt reconstruction (Fig. 5B), and if the affected points in the FID are left out of the computation of the reconstruction—they represent a small fraction of the data—the spectrum that results is nearly flawless (Fig. 5C).

Nonlinear Sampling

Ernst introduced the concept of the so-called matched filter to NMR.[22] The matched filter is a function that follows the decay envelope of the FID; multiplying the FID by the matched filter results in optimal signal-to-noise ratio in the DFT spectrum. In essence, the matched filter places more emphasis on the portion of the data where the signal-to-noise ratio (in the time domain) is highest. The same idea can be applied to sampling in the time domain: Instead of sampling at uniform intervals, the signal can be sampled more frequently when it is strong, and less frequently when it is weak.[23] This prevents the use of LP or the DFT, since they require data sampled at uniform intervals, but MaxEnt handles such data with aplomb. Nonlinear sampling provides additional flexibility in balancing the trade-offs among acquisition time, resolution, and sensitivity. The potential for time savings accrues mainly in the indirect time dimensions of multidimensional experiments, since the relaxation delay between transients renders the total

[22] R. R. Ernst, *Adv. Magn Reson.* **2,** 1 (1966).
[23] J. C. J. Barna, E. D. Laue, M. R. Mayger, J. Skilling, and S. J. P. Worrall, *J. Magn. Reson.* **73,** 69 (1987).

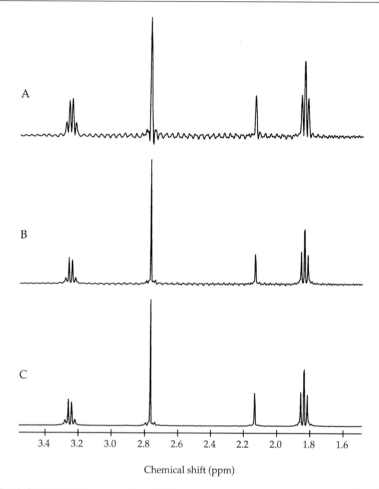

FIG. 5. Setting two disjoint stretches of 10 data values to zero results in characteristic artifacts in the DFT spectrum (A), which are diminished in the MaxEnt reconstruction (B). Leaving the 20 corrupted values out of the computation of the MaxEnt reconstruction yields a pristine spectrum (C). Reprinted by permission of John Wiley & Sons, Inc., from "NMR Data Processing," J. C. Hoch and A. S. Stern, copyright © 1996, Wiley-Liss Inc.

acquisition time insensitive to the number of samples acquired in the direct time dimension. In contrast, the total experiment time increases in direct proportion to the number of indirect time samples.

Examples of nonlinear sampling schedules are depicted in Fig. 6. A uniform grid is used to signify evenly spaced points in the various time dimensions. A large dot indicates a sampled time interval, small dots indicate times not sampled.

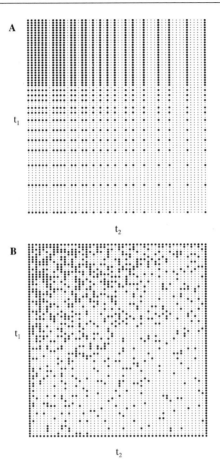

FIG. 6. Exponential sampling in two time dimensions can be performed using the same schedule for each row or column (A), or an element of randomness can be introduced while adhering to an exponential probability distribution (B). Here sampled time values are indicated by large dots, and times not sampled by small dots.

Figure 6A illustrates a fixed exponential sampling schedule in two dimensions, and panel B illustrates randomly selected points chosen from a two-dimensional exponential distribution. Reduced sampling of constant-time dimensions, where there is no decay, can be achieved using random sampling. For COSY-type spectra, an optimal sampling schedule can be constructed from a sine-modulated exponential distribution.[24]

[24] P. Schmieder, A. S. Stern, G. Wagner, and J. C. Hoch, *J. Biomol. NMR* **3,** 569 (1993).

Comparisons of spectra obtained using conventional linear sampling and non-linear sampling are shown in Figs. 7 and 8. Figure 7 shows the methyl region of the $^{13}C-^1H$ HMQC spectrum of a 16-residue peptide. Figure 7A is the spectrum computed using LP extrapolation, windowing, and DFT, from a 128-point FID. Figure 7B was computed using MaxEnt and 128 samples, this time using exponential sampling in t_1. The increase in resolution can be seen from the appearance of the 7 Hz proton–carbon splittings, which are much more apparent. Panel C is the MaxEnt reconstruction using only 64 points in t_1, sampled exponentially. While the resolution is still better than the DFT spectrum, data acquisition required only half the time.

Figure 8 illustrates the benefits of nonlinear sampling in more than one dimension, and the flexibility MaxEnt offers for allocating the available instrument time between sampling the time dimensions and signal averaging. Shown are cross sections computed for a three-dimensional HNCO experiment for the protein villin 14T. Figure 8A is from the linearly sampled DFT spectrum, using 42 points in t_1 and 48 points in t_2, requiring 26 hr to collect, and zero-filled to 128 and 256, respectively. Figure 8B was computed using MaxEnt reconstruction from data containing 16 points nonlinearly sampled in t_1 and 90 points linearly sampled in t_2, requiring 19 hr. The spectrum in panel C was reconstructed using data nonlinearly sampled both in t_1 and t_2, 16 points in t_1 and 45 in t_2, requiring 9.5 hr. Figure 8B demonstrates that MaxEnt reconstruction can improve resolution for comparable acquisition time, and Fig. 8C that it can reduce the time required to obtain useful spectra.

Deconvolution

The convolution theorem, which states that multiplication of two functions in the time domain corresponds to their convolution in the frequency domain, is the basis for the use of apodization to enhance signal-to-noise ratio or resolution. Multiplication (or division) of the time domain data results in changes in the line shape in the frequency domain, which can be designed to enhance sensitivity or resolution. Using apodization to enhance resolution of noisy data unavoidably produces degradation of the signal-to-noise ratio: more weight is applied to the data at long times, where the signal-to-noise ratio (in the time domain) is lower, resulting in noise amplification.

The problem is even more difficult if the FID is divided by a function having values close or equal to zero. The result can be extreme noise amplification or numerical overflow. Thus, the conceptually simple task of removing the modulation due to spin–spin coupling by dividing the signal by a cosine or sine modulation of the appropriate frequency is, at best, extremely delicate. By solving the spectrum as an inverse problem, MaxEnt reconstruction avoids the difficulties of dividing by zero. The function to be deconvolved multiplies the mock data before the constraint

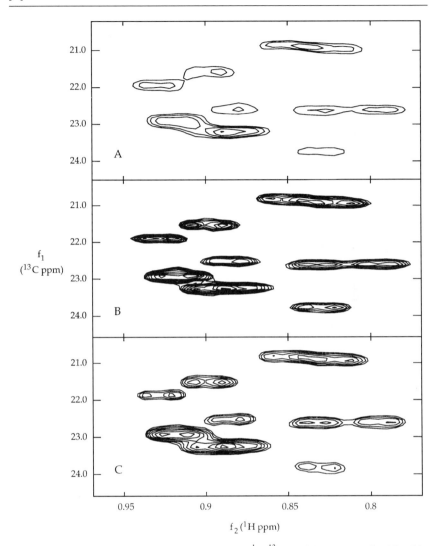

FIG. 7. The methyl region of the natural abundance ^1H–^{13}C HMQC spectrum of a 16-residue peptide, containing 12 methyl groups, obtained using linear and nonlinear sampling in t_1. In (A), 128 points were collected using normal linear sampling, and the DFT spectrum was computed following LP extrapolation and windowing. In (B), 128 points were sampled using an exponential schedule, followed by MaxEnt reconstruction. For (C), 64 points were exponentially sampled, followed by MaxEnt reconstruction. The data used to compute (C) require half the acquisition time of (A) and (B), yet the sensitivity and resolution compare favorably. The number of points sampled and the processing (LP extrapolation, windowing, DFT) in t_2 were the same for all three spectra. Reprinted by permission of John Wiley & Sons, Inc., from "NMR Data Processing," J. C. Hoch and A. S. Stern, copyright © 1996, Wiley-Liss Inc.

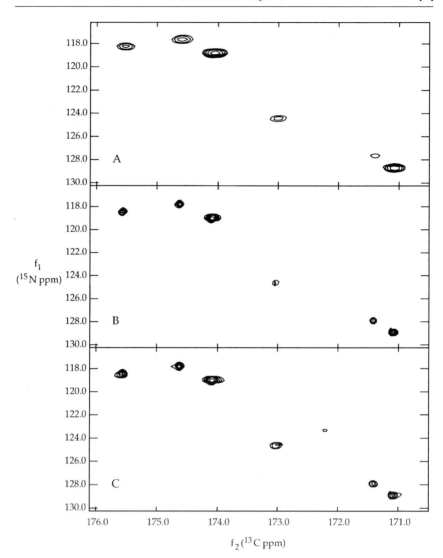

FIG. 8. Two-dimensional cross sections of HNCO spectra for villin 14T computed from data collected using linear (A) and nonlinear (B, C) sampling. Spectrum (A) was computed using 42 points in t_1 and 48 points in t_2, with LP extrapolation, windowing, and DFT processing in both dimensions. Spectrum (B) was obtained using 16 points in t_1 from an exponential schedule and processed using MaxEnt; 90 points were collected and DFT processing as in (A) was applied in t_2. The spectrum in (C) was computed using MaxEnt reconstruction from exponentially sampled data in both dimensions (16 points in t_1 and 45 in t_2). Data collection times are 26, 19, and 9.5 hr for (A), (B), and (C), respectively. Reprinted by permission of John Wiley & Sons, Inc., from "NMR Data Processing," J. C. Hoch and A. S. Stern, copyright © 1996, Wiley-Liss Inc.

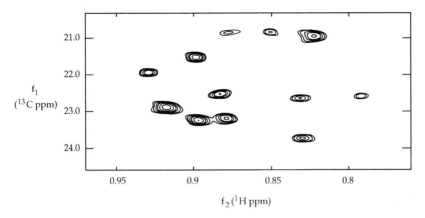

FIG. 9. MaxEnt reconstruction permits stable deconvolution of functions that have small or zero values, which would be impossible to divide into a data set. Here the 3.5-Hz cosine modulation, corresponding to a 7 Hz coupling, is deconvolved from the HMQC data used in Fig. 7. The simplification and improved resolution of the spectrum permits the 12 methyl cross peaks to be counted.

statistic is computed, and thus the reconstruction is the deconvolved spectrum. Figures 9 and 10 show examples of "virtual decoupling" and demonstrate that the ability to compute stable deconvolution can enable novel experiments. Figure 9 shows the same spectrum as Fig. 7B, but the 7 Hz modulation due to the ^{1}H$-^{13}$C coupling of methyl protons was deconvolved using MaxEnt. In addition to simplifying the spectrum and improving resolution, collapsing the doublets increases the signal-to-noise ratio. Individual peaks are easily recognized for each of the 12 methyl groups in the peptide, which contains six leucines. Figure 10 shows spectra obtained from a direct carbon-detected HCACO experiment.[25] Practical considerations make it difficult to decouple the carbon–carbon splitting during acquisition, but since the carbon–carbon coupling between α and carbonyl carbons in proteins is fairly uniform at approximately 55 Hz, MaxEnt performs the "virtual decoupling" quite admirably (Fig. 10B).

Quantification: *In Situ* Error Analysis

The nonlinear nature of modern non-DFT methods of spectrum analysis is responsible for much of their power, but it necessitates special care if they are to be used for quantitative applications. There is a simple method for calibrating the nonlinearities, called *in situ* error analysis,[20] which can be used in conjunction with any method for spectrum analysis. In addition to providing calibration curves

[25] Z. Serber, C. Richter, D. Moskau, J.-M. Böhlen, T. Gerfin, D. Marek, M. Häberli, L. Baselgia, F. Laukien, A. S. Stern, J. C. Hoch, and V. Dötsch, *J. Am. Chem. Soc.* **122**, 3554 (2000).

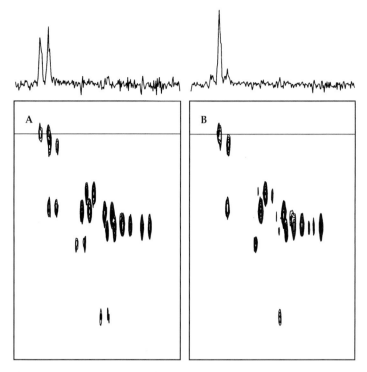

FIG. 10. "Virtual decoupling" of J modulation is particularly valuable when RF decoupling is not practical. In this carbon-detected HCACO spectrum, MaxEnt deconvolution of the J modulation caused by the 55 Hz carbon–carbon coupling both simplifies the spectrum and improves the signal-to-noise ratio. The one-dimensional traces at the top are cross sections at the position indicated by lines in the contour plots.

that can yield improved estimates of relative intensities, it can be used to quantify other spectral properties, such as resolution and the uncertainty in peak frequencies. The method involves the addition of synthetic peaks to the experimental data, with amplitudes that span the range of the real peaks and frequencies corresponding to otherwise empty regions of the spectrum (Fig. 11). For example, calibration curves for MaxEnt reconstructions of the spectrum in Fig. 11 computed using different parameter values for the reconstruction are shown in Fig. 12. Signal peak volumes are corrected for nonlinearity by interpolation on a smooth polynomial fit to the synthetic peak volumes as a function of their known relative amplitudes. Uncertainties can be estimated using the deviations from the fitting polynomial. An advantage of *in situ* analysis is that it automatically includes the effects of nonrandom noise or other artifacts that may be present in the data.

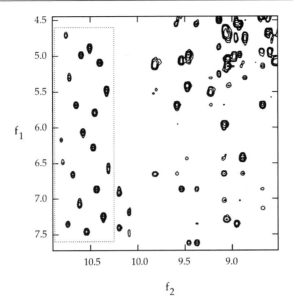

FIG. 11. Synthetic peaks added in the time domain to an experimental data set permit *in situ* calibration of nonlinearity and other effects of signal processing algorithms used in computation of the spectrum, as well as assessment of uncertainties in measured parameters such as peak frequencies and volumes. The frequencies of the synthetic peaks are chosen to fall in an otherwise empty region of the spectrum (designated by the dotted line). Reprinted by permission of John Wiley & Sons, Inc., from "NMR Data Processing," J. C. Hoch and A. S. Stern, copyright © 1996, Wiley-Liss Inc.

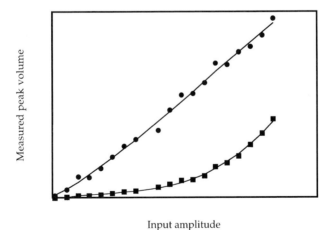

FIG. 12. *In situ* calibration curves fitted to the measured volumes of the synthetic peaks in Fig. 11, for two values of λ. (Note the resemblance to the theoretical curves plotted in Fig. 1.) Interpolation using these curves permits peak volumes obtained from MaxEnt reconstructions to be used in quantitative applications such as the determination of internuclear distances or relaxation rates.

Conclusions

MaxEnt reconstruction is a particularly versatile method of spectrum analysis, capable of providing high-resolution spectral estimates from short data records, improved signal-to-noise ratio, and stable deconvolution. Its ability to determine spectra from data sampled at arbitrary times enables the use of nonlinear sampling to improve resolution or reduce data acquisition time and permits computation of useful spectra when data are missing or corrupted. Model-based methods, such as Bayesian analysis or filter diagonalization, offer many of the same capabilities but are not suited to the computation of spectra containing non-Lorentzian lines. The availability of efficient algorithms and powerful computers has removed what were once perceived as significant barriers to the widespread application of MaxEnt reconstruction in NMR. The staggering cost of state-of-the art NMR spectrometers and the large number of important biomolecules remaining to be studied provide ample incentive for exploiting the capabilities of MaxEnt reconstruction to improve spectral quality and reduce instrument time.

Acknowledgments

We are grateful for the support and assistance of colleagues and collaborators who have contributed to various aspects of this work: Peter Connolly, David Donoho, Volker Dötsch, Kuo-Bin Li, John Osterhout, Peter Schmieder, Zach Serber, and Gerhard Wagner. Financial support was provided by the National Institutes of Health (GM47467), the National Science Foundation (MCB 9316938), and The Rowland Institute for Science.

[7] Magnetic Relaxation Dispersion Studies of Biomolecular Solutions

By BERTIL HALLE and VLADIMIR P. DENISOV

Introduction

In connection with nuclear spin relaxation, the term dispersion refers to the frequency dependence of relaxation rates arising from thermal molecular motions. The magnetic relaxation dispersion (MRD) resolves the total fluctuation amplitude of the spin–lattice coupling into contributions from motions on different time scales. This distribution of motional frequencies, known as the spectral density, is the essential information provided by spin relaxation experiments in the motional-narrowing regime. In most MRD experiments, the probing frequency is controlled by the strength of the static magnetic field applied during the evolution period. The spectral density associated with the secular part of the spin–lattice coupling can also be probed in a restricted low-frequency range by pulse-train and spin-lock experiments, which will not be discussed here.

The MRD method provides direct access to the spectral density over a wide frequency range: 2–3 decades can be covered with multiple-field measurements (MF-MRD), while 6–7 decades can be accessed with the field-cycling technique (FC-MRD). This allows molecular motions on a wide range of time scales to be studied. Furthermore, since the actual shape of the spectral density can be determined, dynamic models can be tested rather than assumed. The price for obtaining such detailed dynamic information is loss of sensitivity and resolution. At the low (detection) fields used in most MRD studies, only relatively abundant low-molecular-weight species can be studied, and the spectral resolution is rarely sufficient to monitor several resonances in the same experiment. In the biomolecular field, the MRD method complements high-resolution NMR spectroscopy by allowing direct studies of the exchange, local dynamics, and ordering of solvent species (water, cosolvents, ions) interacting with proteins and other biomolecules in solution as well as in more complex biological systems.

Although the first MRD studies of biomolecular solutions were performed three decades ago,[1–3] recent methodological developments have enabled novel applications and provided important new information. Whereas ^{1}H FC-MRD dominated the early applications, much of the recent work has employed MF-MRD to study quadrupolar nuclei such as [17]O and [23]Na. These rapidly relaxing nuclei cannot at present be studied over the wide frequency range available with ^{1}H FC-MRD, but, even within the more limited frequency window accessible with MF-MRD, they provide direct and unambiguous access to molecular dynamics. For example, whereas water [17]O MRD directly monitors water dynamics, the interpretation of water [1]H (and to a lesser extent ^{2}H) MRD data from aqueous biomolecular systems is severely compromised by fast proton exchange. In the following, we concentrate on the methodology of MRD studies using quadrupolar nuclei. Reviews with a more system-oriented focus have discussed MRD studies of protein hydration[4] and DNA hydration and ion binding.[5]

[1] S. H. Koenig and W. E. Schillinger, *J. Biol. Chem.* **244**, 3283 (1969).

[2] B. Blicharska, Z. Florkowski, J. W. Hennel, G. Held, and F. Noack, *Biochim. Biophys. Acta* **207**, 381 (1970).

[3] R. Kimmich and F. Noack, *Z. Naturforsch.* **25a**, 299 (1970).

[4] B. Halle, V. P. Denisov, and K. Venu, *in* "Biological Magnetic Resonance," Vol. 17 (N. R. Krishna and L. J. Berliner, eds.), Chapter 10. Kluwer Academic/Plenum, New York, 1999. We note two misprints: in Eq. (72), the power of the first term in the denominator should be 3/2; in Eq. (75), a factor β should be inserted on the left side.

[5] B. Halle and V. P. Denisov, *Nucleic Acid Sci. (Biopolymers)*, **48**, 210 (1998).

Experimental Methodology

Multiple-Field MRD

In low molecular weight liquids at room temperature, the relaxation dispersion occurs above 10 GHz and is therefore inaccessible with current NMR technology. In aqueous solutions of compact biomolecules with molecular weights in the range 5,000–50,000, however, long-lived (>1 ns) interactions of small molecules with biomolecules give rise to a dispersion in the range 1–100 MHz, which is accessible with conventional NMR spectrometers.

Conceptually, the simplest way to record a relaxation dispersion is to carry out relaxation measurements on a series of NMR spectrometers operating at different fixed magnetic fields. In a well-equipped NMR laboratory, relaxation data recorded on half a dozen spectrometers with fixed-field cryomagnets in the field range 2.35–18.8 T might be gathered in this way. However, even for low-γ nuclei such as ^{2}H and ^{17}O, this barely covers the upper half of the 1–100 MHz frequency range. Access to the lower part of this range is most conveniently provided by a field-variable electromagnet interfaced to a 90 or 100 MHz console.[6,7] MRD studies using such tunable, or multichannel, NMR spectrometers have been carried out since the 1960s. The early ^{1}H MRD measurements actually extended as far down as 50 kHz, but required large samples (50 ml) even for aqueous solutions of paramagnetic ions, where the proton density is high and the dispersion is strong.[8]

For MRD measurements in the frequency range 2–85 MHz, we use a modified Bruker MSL 100 spectrometer with three tuned preamplifiers and eight interchangeable probeheads. The console is interfaced to a 2 T iron-core electromagnet (Drusch EAR-35N) with a field-variable external field-frequency lock. The construction of an inexpensive field-variable external lock system for use in MRD studies has been described.[9] The lock works in parallel with a flux stabilizer, which eliminates high-frequency field variations. A peak-to-peak field stability of better than $\pm 0.2~\mu$T is achieved in this way.[6] For cryomagnets, a lock is not essential for MRD work. Besides the intrinsic field instabilities, external transients may cause field jumps on the order of 10 μT. The influence of such transients can be minimized by extensive grounding and shielding.

The low-frequency limit of MF-MRD is mainly determined by the frequency response of the spectrometer components. The range 0.1–1 MHz is technically difficult since it falls between the regimes of audio-frequency and radio-frequency (RF) electronics. However, modifications of a standard 90 MHz console to allow MRD measurements down to 0.5 MHz have been described.[7] The signal-to-noise

[6] I. Furó and B. Halle, *Phys. Rev. E* **51**, 466 (1995).
[7] R. Sitnikov, I. Furó, U. Henriksson, and F. Tóth, *Rev. Sci. Instrum.* **71**, 450 (2000).
[8] R. Hausser and F. Noack, *Zeitschr. f. Physik* **182**, 93 (1964).
[9] R. Sitnikov, I. Furó, and U. Henriksson, *J. Magn. Reson. A* **122**, 76 (1996).

ratio, which scales with the magnetic field strength roughly as $B_0^{3/2}$, may also impose a practical low-frequency limit.

To accurately characterize a Lorentzian dispersion step, relaxation data from approximately two decades of frequency are needed. With the MF-MRD technique, relaxation data rarely extend beyond two decades and usually comprise about 10 fields. To accurately determine the three parameters defining a Lorentzian dispersion step—the low- and high-frequency levels and the dispersion frequency—and to detect any deviation from Lorentzian shape, each relaxation rate must be determined accurately. By maintaining a high (>100) signal-to-noise ratio at all fields, the longitudinal relaxation rate can usually be determined to an accuracy of 1% or better using the $180-\tau-90$ inversion recovery pulse sequence with standard phase cycling carefully and calibrated pulse length. To reduce B_1 inhomogeneity, the sample volume should not exceed the length of the coil. The resulting increase of B_0 inhomogeneity presents no problem provided that it does not vary during the experiment. For ^{1}H measurements at high fields, radiation damping produces an apparent (initial) relaxation enhancement if inversion is imperfect. Such artifacts can be avoided by reducing the filling factor, reducing B_0 homogeneity, detuning the probe, inserting a homospoil pulse, or using the saturation recovery pulse sequence with low-power irradiation or an aperiodic excitation sequence.

Accurate temperature control is perhaps the most important factor in reducing systematic errors in MF-MRD experiments. Since the measured relaxation rate is essentially proportional to solvent viscosity, it follows that a temperature deviation of merely 0.4 K can produce a systematic error of 1% (at room temperature). For measurements on different spectrometers with different temperature controllers and different probe designs, it is therefore essential to use an accurate and robust procedure for temperature calibration, e.g., a thermocouple immersed in a dummy sample inserted at the same height in the probe as the actual sample. A measurement on a reference sample with known temperature-dependent relaxation rate is a reliable temperature check. With most standard temperature controllers and with a properly insulated probe, fluctuations in sample temperature can be kept below 0.1 K and temperature gradients below 0.2 K cm^{-1}.

Field-Cycling MRD

For large biomolecules or complexes, very high concentration or solvent viscosity, and for semisolid samples, the relaxation dispersion may extend into the kilohertz range and below. At such low fields, MRD studies can only be performed with the field-cycling (FC) technique. FC-MRD differs fundamentally from MF-MRD in that the static field is varied during the course of an individual relaxation experiment. The principal limitations of FC-MRD are the relatively low maximum attainable field and the time required to switch the field, factors which have so far precluded FC-MRD studies of fast-relaxing nuclei such as ^{17}O and ^{23}Na. Because

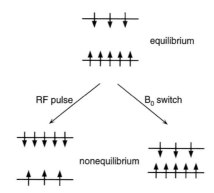

FIG. 1. A nonequilibrium longitudinal magnetization can be created by a coherent RF pulse that inverts the spin level populations, or by rapidly reducing the strength of the polarizing magnetic field so as to change the level spacing without altering the populations.

of their different strengths and weaknesses, the MF-MRD and FC-MRD techniques complement each other. Several excellent reviews of the FC-MRD technique and its diverse applications are available.[10-13]

FC-MRD experiments nearly always measure the longitudinal relaxation rate R_1. The initial nonequilibrium polarization required for a longitudinal relaxation experiment can be created either by manipulating the spin state populations with a coherent RF magnetic field without altering the Zeeman levels, as in an MF-MRD experiment at a given B_0 field, or by changing the level spacing without altering the populations, as in the cyclic variation of the B_0 field performed in FC-MRD (Fig. 1). In either case, the nonequilibrium state should preferably be established (by the RF pulse or the field switch) in a time short compared to the longitudinal relaxation time. Field switching can be accomplished mechanically or electronically. The mechanical approach, where the sample is pneumatically shuttled between two magnetic fields, has the advantage that the detection field can be provided by a high-field cryomagnet, with consequent improvements in sensitivity and resolution.[14] However, the long shuttling time (typically on the order of 100 ms) limits the mechanical approach to samples with relatively slow relaxation. With electronic switching of the magnet current, the maximum field is usually in range 0.5–2 T, but much shorter relaxation times can be measured. Our FC instrument (produced by Stelar s.n.c., Mede, Italy) has a maximum field

[10] R. Kimmich, *Bull. Magn. Reson.* **1,** 195 (1980).

[11] F. Noack, *Prog. NMR Spectrosc.* **18,** 171 (1986).

[12] F. Noack, *in* "Encyclopedia of Nuclear Magnetic Resonance" (D. M. Grant and R. K. Harris, eds.), p. 1980. Wiley, New York, 1995.

[13] F. Noack, S. Becker, and J. Struppe, *Annu. Rep. NMR Spectrosc.* **33,** 1 (1997).

[14] S. Wagner, T. R. J. Dinesen, T. Rayner, and R. G. Bryant, *J. Magn. Reson.* **140,** 172 (1999).

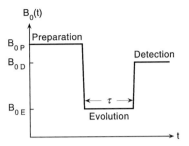

FIG. 2. Cyclic variation of the magnetic field used in FC-MRD to measure the longitudinal relaxation rate at a low evolution field B_{0E}.

of 0.5 T (21 MHz ^{1}H frequency) and a maximum switching rate of 100 T s^{-1}, allowing accurate measurements of longitudinal relaxation times down to a few milliseconds. Using the energy storage principle, switching rates as high as 10^3 T s^{-1} can be achieved.[11,15,16]

A field-cycling relaxation experiment generally consists of three periods (Fig. 2). The sample, initially equilibrated in a field B_{0P}, is brought into a nonequilibrium state by rapidly changing the field to a lower (or higher) value B_{0E}. After an evolution period of variable length τ, during which relaxation takes place in the field B_{0E}, the field is switched to a relatively high value B_{0D} for signal detection. The sensitivity of FC-MRD is comparable to MF-MRD at a field equal to the detection field B_{0D}.[11] Since the detection field is fixed, there is no need for retuning as the evolution field is changed. The accessible frequency range is limited from above by the maximum field of the FC magnet, usually a low-inductance air coil, and from below by interference from ambient fields and leakage currents. The low-frequency limit can be suppressed by ambient-field compensation.

As an illustration of the accuracy achievable with our FC instrument, Fig. 3 shows the water ^{1}H R_1 dispersion from a protein solution. The low-frequency FC-MRD data are seen to merge smoothly with the high-frequency MF-MRD data. Primarily because of the lower quality of the detection field, the accuracy of the FC-MRD data (2% on average) is inferior to that of the MF-MRD data (1%). FC-MRD data are also more prone to systematic error, since the restricted space available for thermal insulation of the probe in the FC magnet does not allow temperature control as precise as that in conventional probes.

Sensitivity and Relaxation Time Limitations

MRD studies have been performed with several nuclides, both spin-1/2 and quadrupolar. Apart from sample-related considerations, the main factors

[15] A. G. Redfield, W. Fite, and H. E. Bleich, *Rev. Sci. Instrum.* **39,** 710 (1968).
[16] C. Job, J. Zajicek, and M. F. Brown, *Rev. Sci. Instrum.* **67,** 2113 (1996).

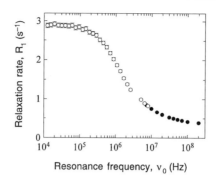

FIG. 3. Water ^{1}H MRD profile from a protein solution obtained with FC-MRD (○) and MF-MRD (●).

determining whether a particular nuclide is suitable for MRD work are sensitivity and relaxation time. The sensitivity at constant frequency, σ_ν, which is the relevant measure for MRD work, scales as $C\gamma I(I + 1)$, where C is the concentration of the observed nucleus, γ its magnetogyric ratio, and I its spin quantum number. Table I lists σ_ν and approximate T_1 values for several nuclides of interest for MRD. The sensitivity limitation is most severe for MF-MRD, where the signal-to-noise ratio must be high even at the lowest evolution field. In FC-MRD, the sensitivity can in principle be improved by using a higher detection field. On the other hand, FC-MRD has more limited applicability with regard to the relaxation time, which must not be shorter than the current switching time (a few milliseconds) or the sample shuttling time (about 100 ms). Very short relaxation times, on the order of 0.1 ms, are difficult to measure also with MF-MRD because of signal loss during the acquisition delay. On the other hand, long T_1 values are unfavorable when extensive signal averaging is required, since a repetition delay of about 5 T_1 must be inserted to allow the spin system to return to equilibrium between successive inversion recovery transients.

In Table I, σ_ν is given at a concentration of 1 M, except for ^{1}H, ^{2}H, and ^{17}O, for which σ_ν refers to the concentration in water, and for ^{23}Na and ^{111}Cd, where the concentrations refer to two recent studies that approach the practical limit. In the ^{23}Na study, which investigated sequence-specific sodium binding in the minor groove of DNA, a single relaxation measurement at a Na$^+$ concentration of 0.2 M required about a week at the lowest frequency (2 MHz).[17] In the ^{111}Cd study, which investigated ion pairing between Cd^{2+} and Mn^{2+}, a dispersion profile extending over 4 decades was recorded in 5 days.[14] Of course, whether a MRD study with a particular nuclide is feasible depends on several other factors besides σ_ν and T_1, such as probe characteristics, sample volume, and required accuracy.

[17] V. P. Denisov and B. Halle, *Proc. Natl. Acad. Sci. U.S.A.* **97**, 629 (2000).

TABLE I
CHARACTERISTICS OF NUCLIDES OF INTEREST FOR MRD STUDIES

Nuclide	Spin, I	Isotopic purity (%)	Sample concentration, C (M)	Relative sensitivity, σ_v	Approximate T_1 (s)[b]
^{1}H	1/2	100	110	100	4 [H_2O]
^{2}H	1	100[a]	110	41	0.4 [D_2O]
^{7}Li	3/2	92.5	1	1.6	7 [(Li^+)$_{aq}$]
^{13}C	1/2	100[a]	1	0.23	6 [DMSO]
^{14}N	1	99.6	1	0.17	1 [(NH_4^+)$_{aq}$]
^{15}N	1/2	100[a]	1	0.092	30 [(NH_4^+)$_{aq}$]
^{17}O	5/2	20[a]	55	16	0.006 [H_2O]
^{19}F	1/2	100	1	0.86	3 [(CF_3COOH)$_{aq}$]
^{23}Na	3/2	100	0.13	0.16	0.05 [(Na^+)$_{aq}$]
^{31}P	1/2	100	1	0.37	6 [($H_2PO_4^-$)$_{aq}$]
^{35}Cl	3/2	75.8	1	0.34	0.04 [(Cl^-)$_{aq}$]
^{81}Br	3/2	49.3	1	0.61	0.001 [(Br^-)$_{aq}$]
^{87}Rb	3/2	27.8	1	0.42	0.003 [(Rb^+)$_{aq}$]
^{111}Cd	1/2	12.8	3	0.074	1 [(Cd^{2+})$_{aq}$]
^{133}Cs	7/2	100	1	3.1	10 [(Cs^+)$_{aq}$]

[a] Enriched isotope.
[b] At room temperature for the indicated species.

Nuclides for Hydration Studies

For studies of biomolecular hydration, three stable nuclides are available: ^{1}H, ^{2}H, and ^{17}O (disregarding the radioactive ^{3}H isotope). Until recently, the vast majority of MRD studies of biomolecular solutions employed the ^{1}H isotope, which has the highest sensitivity and the widest accessible frequency range. These advantages, however, are offset by several drawbacks. First and foremost, ^{1}H relaxation generally contains a potentially confounding contribution from labile biomolecule protons. Because of proton exchange catalysis, this contribution is more pronounced away from neutral pH and in the presence of buffers.[18,19] In the past, exchange-averaged labile proton contributions to ^{1}H MRD data have frequently been unjustifiably ignored. Water protons are engaged in intramolecular as well as intermolecular dipole couplings. The relative contribution to the dipolar relaxation rate from intermolecular couplings is considerably smaller for water molecules buried inside proteins than in bulk water.[20] The dynamic coupling between the water and biomolecule ^{1}H spin systems brought about by cross-relaxation

[18] M. Eigen, *Angew. Chemie, Intl. Ed. Engl.* **3**, 1 (1964).
[19] E. Liepinsh and G. Otting, *Magn. Reson. Med.* **35**, 30 (1996).
[20] K. Venu, V. P. Denisov, and B. Halle, *J. Am. Chem. Soc.* **119**, 3122 (1997).

(*via* intermolecular dipole couplings) was long thought to be essential for analyzing ^{1}H MRD data,[21,22] but more recent work has shown that this effect is negligible for biomolecular solutions.[20]

Despite having two orders of magnitude lower receptivity and one order of magnitude faster relaxation than ^{1}H, the ^{2}H isotope can be used for water MRD studies (both MF and FC). Because of the shorter intrinsic relaxation time, ^{2}H is less susceptible than ^{1}H to labile hydrogen exchange averaging. Nevertheless, rapidly exchanging biopolymer deuterons can dominate water ^{2}H relaxation at low and high pH values[23–26] and can never be neglected *a priori*.

Unlike the hydrogen isotopes, water ^{17}O does not exchange with biomolecular atoms on the relaxation time scale and therefore reports exclusively on water molecules. An observed water ^{17}O dispersion in the 1–100 MHz range thus provides indisputable evidence for water molecules in long-lived association with the biomolecule.[27] Because of the low natural abundance, ca 20 mM in water, ^{17}O-enriched water is essential for accurate MRD work. ^{17}O-enriched H$_2$O and D$_2$O is commercially available at enrichment levels up to at least 60%. We generally use 20% enrichment, a compromise between cost and convenience. The fast quadrupolar relaxation of ^{17}O allows for efficient signal averaging and, in combination with the small magnetic moment, makes ^{17}O much less susceptible than ^{1}H to paramagnetic impurities. In the neutral pH range, the transverse water ^{17}O relaxation is not purely quadrupolar, but includes a contribution due to hydrogen-exchange modulation of the scalar coupling between ^{17}O and ^{1}H or ^{2}H in water.[28,29]

Analysis of Quadrupolar MRD Data

Relaxation Rates and Spectral Densities

The molecular information provided by MRD experiments in the motional-narrowing (or fast-exchange) regime is contained in the spectral density $J(\omega)$, which may be expressed as the cosine transform of the time autocorrelation function $C(\tau)$ of the appropriate lattice variable,

$$J(\omega) = \int_0^\infty d\tau \, \cos(\omega\tau) \, C(\tau) \tag{1}$$

[21] S. H. Koenig, R. G. Bryant, K. Hallenga, and G. S. Jacob, *Biochemistry* **17**, 4348 (1978).
[22] S. H. Koenig and R. D. Brown, *Progr. NMR Spectrosc.* **22**, 487 (1991).
[23] D. E. Woessner and B. S. Snowden, *J. Colloid Interface Sci.* **34**, 290 (1970).
[24] J. J. van der Klink, J. Schriever, and J. Leyte, *Ber. Bunsenges. Phys. Chem.* **78**, 369 (1974).
[25] L. Piculell and B. Halle, *J. Chem. Soc., Faraday Trans. 1* **82**, 401 (1986).
[26] V. P. Denisov and B. Halle, *J. Mol. Biol.* **245**, 698 (1995).
[27] B. Halle, T. Andersson, S. Forsén, and B. Lindman, *J. Am. Chem. Soc.* **103**, 500 (1981).
[28] S. Meiboom, *J. Chem. Phys.* **34**, 375 (1961).
[29] B. Halle and G. Karlström, *J. Chem. Soc., Faraday Trans. 2* **79**, 1031 (1983).

The various spin relaxation rates can usually be expressed as linear combinations of spectral densities.[30] For quadrupolar relaxation, all relaxation rates are determined by the three spectral density values $J(0)$, $J(\omega_0)$, and $J(2\omega_0)$. For example, the longitudinal (R_1) and transverse (R_2) relaxation rates for a spin-1 nucleus, usually obtained from inversion recovery and spin echo experiments, respectively, are given by

$$R_1 = 0.2\,J(\omega_0) + 0.8\,J(2\omega_0) \tag{2a}$$

$$R_2 = 0.3\,J(0) + 0.5\,J(\omega_0) + 0.2\,J(2\omega_0) \tag{2b}$$

For quadrupolar nuclei of half-integral spin ($I = 3/2, 5/2, \ldots$), both longitudinal and transverse relaxation involves $(I + 1/2)$ exponentials.[30] Under most conditions of interest in MRD work, the longitudinal relaxation rates, which only involve the nonsecular spectral densities $J(\omega_0)$ and $J(2\omega_0)$, do not differ sufficiently to cause a significant deviation from exponential inversion recovery. The data can then be analyzed in terms of an effective, average relaxation rate, which has the same form as Eq. (2a).[31] For spin-3/2 nuclei, such as ^{23}Na, it is often possible to separately determine the two transverse rates $R_{2f} = 0.5J(0) + 0.5J(\omega_0)$ and $R_{2s} = 0.5\,J(\omega_0) + 0.5\,J(2\omega_0)$. For nuclei of higher spin, such as ^{17}O with $I = 5/2$, also the transverse relaxation is usually effectively exponential and then takes the same form as Eq. (2b).[31]

The frequency-dependent relaxation data obtained from an MRD study are usually analyzed by fitting the parameters in a model-dependent spectral density function. The simplest physically motivated model is a Lorentzian plus constant,

$$J(\omega) = \alpha + \beta\,\tau_C/[1 + (\omega\tau_C)^2] \tag{3}$$

where the constant term α represents the high-frequency motions responsible for the decay of $J(\omega)$ to zero at frequencies above the investigated frequency window (Fig. 4).

Although the R_1 dispersion contains the same information as the spectral density function $J(\omega)$, it is sometimes desirable to extract the latter from the relaxation data without imposing restrictions on its functional form. Depending on the amount of data at hand, this can be done in different ways. If three linearly independent relaxation rates are measured, one can solve for the three spectral densities at each field. This is can often be done for spin-3/2 nuclei.[17] Because $J(0)$ is the same for all fields, one actually needs only two independent rates provided that the low-frequency plateau, where $R_1 = R_2 = J(0)$, has been established. More commonly, MRD studies only measure the (effective) longitudinal relaxation rate R_1. (In FC-MRD, transverse relaxation is almost never measured.) Even then,

[30] A. Abragam, "The Principles of Nuclear Magnetism." Clarendon Press, Oxford, 1961.
[31] B. Halle and H. Wennerström, *J. Magn. Reson.* **44,** 89 (1981).

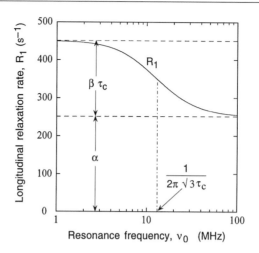

FIG. 4. Typical MF-MRD profile from a biomolecular solution. Long-lived interactions between the small spin-bearing molecule and the biomolecule give rise to a dispersion, centered at $\nu_0 = 1/(2\pi\sqrt{3}\tau_C)$. The parameters α and β determine the high-frequency and low-frequency plateaus, respectively, of the dispersion profile within the investigated frequency window.

$J(\omega)$ can be extracted provided that the magnetic field is uniformly incremented by a factor 2. The $n\,R_1$ values measured at n fields are then determined by $n + 1$ spectral density values, which can be solved for if the data extend into the low- or high-frequency plateau, where $J(\omega_0) \approx J(2\omega_0)$. Alternatively, if the dispersion step does not deviate significantly from Lorentzian shape, one can use the approximation $0.2\,J(\omega_0) + 0.8\,J(2\omega_0) \approx J(\sqrt{3}\omega_0)$,[32] which is accurate to better than 2% of $J(0)$ throughout the dispersion for the Lorentzian spectral density in Eq. (3). The correlation time τ_C can then be extracted directly from the frequency where the dispersion step has decayed to half its amplitude, $\omega_{1/2} = 1/(2\pi\sqrt{3}\tau_C)$ (Fig. 4). If nothing else, this is a useful criterion for a crude initial analysis of the R_1 dispersion profile.

In isotropic solutions, where the quadrupole coupling is averaged to zero, relaxation of an initial nonequilibrium magnetization outside the extreme-narrowing limit induces a dynamic coupling with higher odd-rank components of the spin density matrix. The interplay of these higher-rank components gives rise to multiexponential relaxation and also makes it possible to generate multiple-quantum coherences (MQCs).[33] At least two RF pulses are required to excite MQCs, the evolution of which can be indirectly detected after conversion to observable single-quantum coherence by a third pulse. In principle, the use of MQC filters offers

[32] K. Hallenga and S. H. Koenig, *Biochemistry* **15**, 4255 (1976).
[33] G. Jaccard, S. Wimperis, and G. Bodenhausen, *J. Chem. Phys.* **85**, 6282 (1986).

certain advantages for MRD studies. In practice, however, these advantages are offset by poor sensitivity.[34]

Exchange Averaging

In MRD studies of biomolecular solutions, the investigated nuclide usually resides in a small mobile molecule or ion that can diffuse between the bulk solution region and one or more sites on (or in) the biomolecule. The relaxation is then governed by molecular motions at two levels. At the level of spin dynamics, translational diffusion transfers magnetization between microenvironments with different local relaxation rates. If sufficiently fast, such diffusional exchange leads to spatial averaging of the local relaxation rates. At the level of orientational time correlation functions, molecular rotation (and translation, in the case of intermolecular couplings) averages out the anisotropic spin–lattice coupling and thus determines the local spin relaxation rates.

The theoretical framework for analyzing relaxation data from nuclei exchanging between discrete states is well established. It is not obvious, however, that a *discrete*-state exchange model provides a valid description of *continuous* diffusion in a spatially heterogeneous system such as a biomolecular solution.[35] There are two aspects to this issue. First, in the fast-exchange regime, the actual exchange mechanism is irrelevant, and it is only necessary that the dynamic perturbation induced by the biomolecule be relatively short-ranged. Second, the observed spin relaxation rate depends on the exchange mechanism only in the intermediate-exchange regime where the residence times are in the μs–ms range. Such long residence times only occur for specifically site-bound species, for which a discrete-state exchange (or jump) model is indeed appropriate.

The simplest description of exchange averaging of the longitudinal relaxation rate in a biomolecular solution is of the form

$$R_1(\omega_0) = (1 - f)R_0 + \sum_k f_k/[\tau_k + 1/R_{1k}(\omega_0)] \tag{4}$$

Provided that chemical shift differences can be ignored, an analogous result holds for R_2. The first term in Eq. (4) refers to nuclei residing in molecules that either are unperturbed by the biomolecule or have such short-lived encounters with it that the local relaxation rate remains in the extreme-narrowing regime throughout the MRD frequency window. The quantity R_0 is the average local relaxation rate for the fraction $1 - f$ of nuclei that belong to these categories. For uncharged molecules (but not for ions), it is helpful to subdivide the first term in Eq. (4) by writing $R_0 = [1 - f_s/(1 - f)]R_{bulk} + [f_s/(1 - f)]R_s$, where R_{bulk} is the relaxation rate

[34] A. M. Torres, S. M. Grieve, B. E. Chapman, and P. W. Kuchel, *Biophys. Chem.* **67**, 187 (1997).
[35] B. Halle and P.-O. Westlund, *Mol. Phys.* **63**, 97 (1988).

in the absence of biomolecules and R_s is the average relaxation rate for molecules in short-lived association with the biomolecule.

The second term in Eq. (4) refers to nuclei in molecules that have sufficiently long-lived encounters with the biomolecule that the local relaxation rate is frequency-dependent within the MRD window. Such encounters normally involve well-defined sites, each characterized by a mean residence time τ_k and a local longitudinal relaxation rate $R_{1k}(\omega_0)$. At any time, a fraction $f = \sum_k f_k$ of the nuclei reside in such sites. The simple form of the second term in Eq. (4) is valid provided that $f_k \ll 1$ and $R_0 \ll R_{1k}$.[36] This is nearly always the case in MRD studies of biomolecular solutions.

Difference MRD

Because the magnetization probed by MRD is exchange averaged, contributions to the relaxation dispersion from different sites cannot be separated or identified *a priori*. However, this can be done by means of a difference-MRD experiment, where the dispersion profiles from two structurally related biomolecules are compared. In the most informative type of difference-MRD experiment, the dispersion profile is recorded before and after a site-directed structural perturbation that eliminates one or more interaction sites without significantly affecting other parts of the biomolecule. Equation (4) then yields for the difference dispersion

$$R_1(\omega_0) = \sum_{\Delta k} f_k/[\tau_k + 1/R_{1k}(\omega_0)] \tag{5}$$

where the sum now includes only the eliminated sites.

A local structural perturbation can be induced in several ways. Site-directed mutagenesis is the method of choice for engineering internal cavities in proteins that may be occupied by water or other small molecules. For example, in the single-residue BPTI mutant G36S, a buried water molecule is replaced by the hydroxyl group in the side chain of serine-36 (Fig. 5).[37,38] Local covalent modifications can of course also be introduced by conventional chemical methods, e.g., selective reduction of disulfide bonds in proteins or introduction of modified bases in oligonucleotide duplexes. More accessible sites can be eliminated (or modified) by removing an intrinsic metal ion or cofactor or by adding a high-affinity substrate or inhibitor. If complete removal of an intrinsic ligand cannot be achieved, MRD profiles can be recorded at a series of ligand/biomolecule ratios and the results extrapolated to zero ligand concentration. This approach can be used, for example, for intrinsic multivalent metal ions that coordinate long-lived water molecules.[39,40] The

[36] Z. Luz and S. Meiboom, *J. Chem. Phys.* **40**, 2686 (1964).
[37] V. P. Denisov, B. Halle, J. Peters, and H. D. Hörlein, *Biochemistry* **34**, 9046 (1995).
[38] V. P. Denisov, J. Peters, H. D. Hörlein, and B. Halle, *Nature Struct. Biol.* **3**, 505 (1996).
[39] V. P. Denisov and B. Halle, *J. Am. Chem. Soc.* **117**, 8456 (1995).
[40] V. P. Denisov, B.-H. Jonsson, and B. Halle, *J. Am. Chem. Soc.* **121**, 2327 (1999).

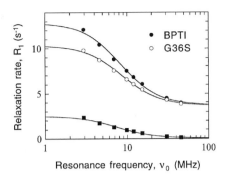

FIG. 5. Water ^{2}H MF-MRD profiles from aqueous solutions of wild-type BPTI and the G36S mutant. The difference dispersion at the bottom is due to the single buried water molecule in wild-type BPTI that is displaced in the mutant.[37]

strategy of water elimination by ligand binding is illustrated in Figs. 6 and 7 for a B-DNA dodecamer, where five primary hydration sites[41,42] and at least one sodium ion site[17] in the minor groove are displaced by the polyaromatic drug netropsin.

Dispersion Shape

The dispersion is produced by the frequency dependence of the local relaxation rates $R_{1k}(\omega_0)$ in sites where the nucleus experiences a relatively long correlation time. Within the motional-narrowing regime, these rates are usually expressible as linear combinations of the spectral density values $J_k(\omega_0)$ and $J_k(2\omega_0)$, as in Eq. (2a). For a small molecule rigidly bound to a biomolecule undergoing spherical-top rotational diffusion, the spectral density has the Lorentzian form

$$J_k(\omega) = \omega_{Qk}^2 \tau_{Ck}/[1 + (\omega\tau_{Ck})^2] \tag{6}$$

The effective correlation time τ_{Ck} is determined by the residence time τ_k in site k and the rotational correlation time τ_R of the biomolecule according to

$$1/\tau_{Ck} = 1/\tau_R + 1/\tau_k \tag{7}$$

The rigid-lattice quadrupole frequency ω_Q is defined through

$$\omega_{Qk}^2 = (3\pi^2/10)\{(2I+3)/[I^2(2I-1)]\}\chi_k^2(1+\eta_k^2/3) \tag{8}$$

with I the spin quantum number, χ the quadrupole coupling constant, and η the asymmetry parameter of the electric field gradient tensor.[30] For ^{2}H and ^{17}O in water, the recommended values for the quadrupole frequency are 8.7×10^5 s^{-1} and

[41] V. P. Denisov, G. Carlström, K. Venu, and B. Halle, *J. Mol. Biol.* **268**, 118 (1997).
[42] H. Jóhannesson and B. Halle, *J. Am. Chem. Soc.* **120**, 6859 (1998).

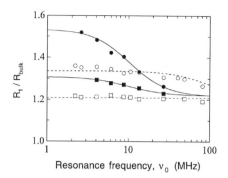

Fig. 6. Water ^{17}O MF-MRD profiles from solutions of the [d(CGCGAATTCGCG)]$_2$ duplex at 4° (open symbols)[41] and at −20° (filled symbols).[42] The profiles were recorded before (circles) and after (squares) addition of one equivalent of netropsin. The R_1 data have been normalized by the (frequency-independent) bulk water relaxation rate R_{bulk} at the respective temperatures.

7.6×10^6 s^{-1}, respectively.[4] When the electric field gradient is of intermolecular origin, as for monatomic ions, the quadrupole frequency may vary significantly among different sites. For ^{23}Na in a sodium ion at a primary solvation site in the narrow (AT-tract) minor groove of a B-DNA duplex, $\omega_{Qk} = 2.6 \times 10^6$ s^{-1}.[17] Equation (6) is readily generalized to nonspherical biomolecules with symmetric-top rather than spherical-top rotational diffusion. However, for globular proteins, the effect of anisotropic rotational diffusion on the shape of the relaxation dispersion is often negligible.

In general, small molecules do not tumble rigidly with the biomolecule, but undergo restricted rotational motions on time scales short compared to τ_R. If the

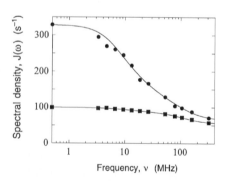

Fig. 7. Spectral density for the sodium counterions in a solution (0.34 M NaCl, pH 7.0, 4°) of the [d(CGCGAATTCGCG)]$_2$ duplex before (circles) and after (squares) addition of one equivalent of netropsin. The spectral density was derived from ^{23}Na MF-MRD longitudinal and transverse relaxation rates.[17]

local motions are much faster than the global motion (with correlation time τ_{Ck}) and remain in the extreme narrowing regime at the highest MRD frequency, then the generalization of Eq. (6) is[43,44]

$$J_k(\omega) = \omega_{Qk}^2 \{ (1 - S_k^2)\tau_{Ck}^{loc} + S_k^2 \tau_{Ck}/[1 + (\omega\tau_{Ck})^2] \} \tag{9}$$

where τ_{Ck}^{loc} is an effective correlation time for the local restricted rotation. Furthermore, S_k is the generalized second-rank orientational order parameter for site k, which has a maximum value of unity for a rigidly attached molecule.[4] When this spectral density is substituted into Eq. (2a), we may write

$$R_{1k}(\omega_0) = R_k^{loc} + b_k \, [0.2 \, j_k(\omega_0) + 0.8 \, j_k(2\omega_0)] \tag{10}$$

where $R_k^{loc} = \omega_{Qk}^2 (1 - S_k^2) \, \tau_{Ck}^{loc}$, $b_k = (\omega_{Qk} S_k)^2$, and $j_k(\omega) = \tau_{Ck}/[1 + (\omega\tau_{Ck})^2]$. Using this expression for $R_{1k}(\omega_0)$, we can transform Eq. (4) into the usual fast-exchange form, but with renormalized parameters:

$$R_1(\omega_0) = \alpha + \sum_k f_k b_k'[0.2 j_k'(\omega_0) + 0.8 j_k'(2\omega_0)] \tag{11}$$

where

$$\alpha = R_0 + \sum_k f_k[R_k^{loc}/(1 + R_k^{loc}\tau_k) - R_0] \tag{12}$$

Equation (11) is an exact result provided that the linear combination $0.2 j_k(\omega_0) + 0.8 j_k(2\omega_0)$ can be replaced by $j_k(c\omega_0)$, where c is a numerical constant. This is a highly accurate approximation for a Lorentzian spectral density ($c \approx \sqrt{3}$). The renormalized parameter b_k' is defined by

$$b_k' = b_k/\{(1 + R_k^{loc}\tau_k)^{3/2}[1 + R_{1k}(0)\tau_k]^{1/2}\} \tag{13}$$

The renormalized reduced spectral density is $j_k'(\omega) = \tau_{Ck}'/[1 + (\omega\tau_{Ck}')^2]$, with the renormalized effective correlation time τ_{Ck}' defined by

$$\tau_{Ck}' = \tau_{Ck}(1 + R_k^{loc}\tau_k)^{1/2}/[1 + R_{1k}(0)\tau_k]^{1/2} \tag{14}$$

Equation (11) shows that a given site produces a Lorentzian dispersion profile even outside the fast-exchange regime. With increasing residence time τ_k, the dispersion shifts to higher frequency ($\tau_{Ck}' < \tau_{Ck}$) and the dispersion step is reduced in magnitude ($b_k' < b_k$), but the shape is not affected. Of course, if several sites with different effective correlation times τ_{Ck} contribute, then the dispersion profile will be stretched out over a wider frequency range.

If all sites have the same effective correlation time τ_{Ck}, which would be the case if $\tau_k \gg \tau_R$ for all sites, and if also all sites are in the fast-exchange regime,

[43] B. Halle and H. Wennerström, *J. Chem. Phys.* **75,** 1928 (1981).
[44] G. Lipari and A. Szabo, *J. Am. Chem. Soc.* **104,** 4546 (1982).

meaning that $R_{1k}(0)\tau_k \ll 1$, then Eq. (11) reduces to

$$R_1(\omega_0) = \alpha + \beta \, [0.2 j(\omega_0) + 0.8 j(2\omega_0)] \qquad (15)$$

where $\beta = \sum_k f_k b_k$. This is the result obtained by substituting the spectral density (3) into Eq. (2a). The dispersion profile is then fully characterized by the three parameters α, β, and τ_C, as illustrated in Fig. 4. In the fast-exchange regime, sites of a particular type k contribute to the β parameter a quantity $f_k b_k = \theta_k N_k (\omega_{Qk} S_k)^2 / N_T$, where N_k is the number of k-sites per biomolecule, θ_k their mean occupancy, and N_T the overall ratio of small molecules to biomolecules in the sample. Usually, ω_{Qk} can be taken from solid-state data, while N_T is obtained from the sample composition (which requires an accurate determination of the biomolecule concentration). From a difference-MRD experiment that probes a single type of sites, we can thus obtain the product $\theta_k N_k S_k^2$. Since $S_k \leq 1$, this furnishes a lower bound on N_k for strong binding ($\theta_k = 1$) or a lower bound on θ_k for weak binding with known stoichiometry N_k.

The Lorentzian function is the fastest decaying spectral density that can result from incoherent molecular motions. On the other hand, experimental dispersion profiles are sometimes found to be stretched out over a wider frequency range than predicted by Eq. (15). Several factors can contribute to such dispersion stretching: (*i*) anisotropic biomolecule rotation, (*ii*) biomolecular self-association, and (*iii*) a distribution of sites with different residence times τ_k, leading to a distribution of effective correlation times τ'_{Ck}; see Eqs. (7) and (14). Depending on the circumstances, these effects can shift the dispersion to higher or lower frequency and/or stretch it over a wider frequency range.

Traditionally, stretched dielectric and magnetic relaxation dispersions have been accounted for in terms of empirical correlation time distributions. In connection with ^{1}H MRD studies of aqueous biological systems, a so-called Cole–Cole dispersion has often been used to fit stretched MRD profiles.[32] The Cole–Cole dispersion function was originally used to describe dielectric dispersion data[45] and can be inverted to yield a particular correlation time distribution.[46] When this dispersion function was modified so as to be dimensionally commensurate with the NMR spectral density, its physical meaning was lost. In fact, it can be shown that the modified Cole–Cole dispersion does not correspond to any correlation time distribution.[47] The significance of the effective correlation time extracted from a fit of the modified Cole–Cole dispersion to a stretched MRD profile is therefore obscure.

A physically consistent, model-free approach for analyzing stretched MRD profiles has been proposed.[47] Because the time correlation function $C(\tau)$ is even,

[45] K. S. Cole and R. H. Cole, *J. Chem. Phys.* **9**, 341 (1941).

[46] R. M. Fuoss and J. G. Kirkwood, *J. Am. Chem. Soc.* **63**, 385 (1941).

[47] B. Halle, H. Jóhannesson, and K. Venu, *J. Magn. Reson.* **135**, 1 (1998).

the cosine transform in Eq. (1) can be inverted to give

$$\int_0^\infty d\omega\, J(\omega) = (\pi/2)C(0) \tag{16}$$

This general property of the spectral density allows us to define two model-independent quantities in terms of the integral and amplitude of the (stretched) dispersion profile:[47]

$$\beta \equiv \sum_k f_k b_k = (\pi/2) \int_0^\infty d\omega_0 [R_1(\omega_0) - \alpha] \tag{17}$$

$$\langle \tau_C \rangle \equiv \sum_k f_k b_k \tau_{Ck} \Big/ \sum_k f_k b_k = [R_1(0) - \alpha]/\beta \tag{18}$$

where we have used Eq. (11) in the fast-exchange regime (without primes). This truly model-independent approach, which should not be confused with the so-called model-free approach used to derive the spectral density in Eq. (9), has the virtue of separating the static and dynamic information content of the MRD profile. The quantity β is a true equilibrium property, entirely independent of the rates of molecular motions, whereas the average correlation time $\langle \tau_C \rangle$ is a well-defined measure of the complex dynamics. Although it may be desirable to invoke a specific dynamic model at some stage of the analysis, the model-independent parameters β and $\langle \tau_C \rangle$ provide valuable guidance for any subsequent modeling. Figure 8 illustrates the use of the model-free approach.[47] The apparent correlation time τ_{Cole} obtained from a fit using the Cole–Cole dispersion function is essentially the inverse of the angular frequency where the spectral density has decayed to half of the zero-frequency value. As shown by the example in Fig. 8, τ_{Cole} may be considerably longer than the well-defined average correlation time $\langle \tau_C \rangle$. This has been shown to be the case for the ^{1}H dispersion from solutions of a large number of globular proteins.[48]

Residence Times

Even without a difference-MRD experiment, the dispersion profile separates contributions from interaction sites with long and short residence times. The observation of a dispersion step demonstrates unambiguously that a fraction of the nuclei reside in sites with correlation times on the order of the inverse dispersion frequency (Fig. 4). To contribute maximally to the relaxation dispersion, a site k must have a residence time τ_k much longer than the rotational correlation time τ_R of the biomolecule, but much shorter than the zero-frequency local relaxation time, $1/R_{1k}(0)$. If $R_{1k}(0) \gg R_k^{loc}$, as is usually the case, these conditions are expressed

[48] I. Bertini, M. Fragai, C. Luchinat, and G. Parigi, *Magn. Reson. Chem.* **38**, 543 (2000).

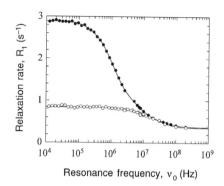

FIG. 8. Water ^{1}H MRD profiles from a BPTI solution (10 wt%, pH 4.5, 300 K) without added salt (○) and after addition of 0.7 M NaCl (●). The high salt concentration induces protein–protein interactions, leading to pronounced dispersion stretching. The model-free fit shown yields $\langle \tau_C \rangle = 15$ ns, whereas a fit using the Cole–Cole function yields an apparent correlation time $\tau_{Cole} = 65$ ns.[47]

by the inequalities

$$\tau_R \ll \tau_k \ll 1/(b_k \tau_R) \tag{19}$$

which may be said to define the "MRD window" on residence times. Of course, sites that do not obey these inequalities may still contribute to the dispersion, but do so with less than the maximum contribution $f_k b_k \tau_R$. If R_k^{loc} is neglected, violation of the right-hand (fast-exchange) inequality in Eq. (19) reduces the effective dispersion amplitude parameter by a relative amount $b'_k/b_k = [(1 + \tau_R/\tau_k)/(1 + \tau_R/\tau_k + b_k \tau_R \tau_k)]^{1/2}$. Similarly, the relative reduction of the effective correlation time is $\tau'_{Ck}/\tau_R = [(1 + \tau_R/\tau_k)(1 + \tau_R/\tau_k + b_k \tau_R \tau_k)]^{-1/2}$. In Fig. 9, these quantities are plotted as a function of the residence time τ_k for two representative b_k values and a typical biomolecular tumbling time $\tau_R = 10$ ns. For this τ_R value and a b_k value appropriate for water ^{17}O, a site with a residence time $\tau_k = 6 \, \mu$s, for example, contributes a dispersion step with a dispersion amplitude parameter b'_k that is half of the maximum value $(\omega_{Qk} S_k)^2$ and with a correlation time τ'_{Ck} that is half of τ_R. The magnitude of the dispersion step is therefore smaller by a factor 4 than the maximum value.

For residence times on the central plateau of the MRD window in Fig. 9b, only lower and upper bounds on τ_k can be established, as expressed by the inequalities (19). On the wide flanks of the MRD window, however, τ_k can be quantitatively determined. On the short-τ_k flank, this requires independent information about τ_R. If the τ_{Ck} value deduced from the MRD profile is much smaller than τ_R, as is the case for water molecules in the minor groove of B-DNA,[41,42] then it can be directly identified with the residence time τ_k without the need for an accurate estimate of τ_R or for modeling any rotational anisotropy. In contrast, determination of residence

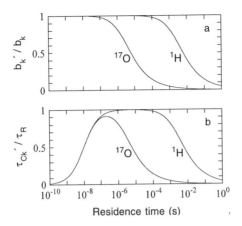

FIG. 9. The MRD window on residence times for a biomolecular tumbling time $\tau_R = 10$ ns and with a dispersion amplitude parameter $b_k = 5 \times 10^{13}$ s^{-2} (typical for water ^{17}O) or 5×10^{10} s^{-2} (typical for water ^{1}H). Deviations from fast-exchange conditions reduce the effective dispersion amplitude parameter (a) and the effective correlation time (b) over a wide range of residence times.

times comparable to τ_R requires knowledge of the precise value of τ_R. This can be obtained from ^{1}H or ^{2}H MRD data, which generally contain contributions from labile biomolecular hydrogens (with $\tau_k \gg \tau_R$). Residence times of water molecules located in pockets on proteins surfaces have been determined in this way.[49,50].

Longer residence times can be determined on the long-τ_k flank of the MRD window. According to Eqs. (13) and (14), τ_k can be obtained from either b'_k or τ'_{Ck}, since $b'_k/b_k = \tau'_{Ck}/\tau_R = (1 + b_k \tau_R \tau_k)^{-1/2}$ in this regime. To determine τ_k at a given temperature, τ_R and b_k must be known. From a variable-temperature MRD study, b_k (assumed to be temperature-independent) and the activation enthalpy of τ_k can be determined if τ_R is taken to scale as η/T (as expected for rotational diffusion). If τ_R is known at one temperature, τ_k is obtained at all investigated temperatures. As an illustration of this approach, Fig. 10 shows the temperature dependence of b'_k/b_k, deduced from water ^{2}H and ^{17}O difference dispersions isolating the contribution from a single buried water molecule in the small protein BPTI. A joint fit to the two curves in Fig. 10 yields a residence time $\tau_k = 170 \pm 20$ μs at 300 K and an apparent activation enthalpy $\Delta H^{\#} = 90 \pm 5$ kJ mol^{-1}.[38] The temperature shift and slope difference between the ^{2}H and ^{17}O curves follow quantitatively from the different quadrupole frequencies of the two nuclei.

[49] V. P. Denisov and B. Halle, *Biochemistry* **37**, 9595 (1998).
[50] U. Langhorst, R. Loris, V. P. Denisov, J. Doumen, P. Roose, D. Maes, B. Halle, and J. Steyaert, *Protein Sci.* **8**, 722 (1999).

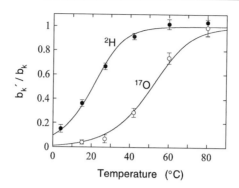

FIG. 10. Temperature variation of the relative effective dispersion amplitude parameter for a single buried water molecule in BPTI. The data were obtained from difference-MRD profiles recorded on wild-type BPTI and the G36S mutant.[38]

Complications

Hydrogen Exchange

Exchange averaging with biomolecular hydrogens is a potential pitfall in all ^{1}H and ^{2}H MRD studies of small molecules, such as water, that contain labile hydrogens. Failure to appreciate this point has led to qualitatively incorrect conclusions about hydration behavior. For a protein with $\tau_R \approx 10$ ns, most of the labile protons have local ^{1}H relaxation times, $1/R_k(0)$, around 10 ms.[20] At seen from Fig. 11, COOH and OH protons are then in fast exchange with water protons at all pH

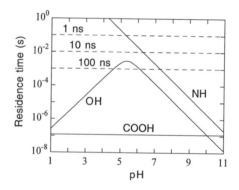

FIG. 11. Approximate residence times of labile protons in side chains of proteins in aqueous solution at 25°: COOH (Glu, Asp), OH (Ser, Thr, Tyr), and NH (Lys, Arg). The dashed lines indicate the approximate local ^{1}H relaxation time of these labile protons in proteins with the indicated tumbling times. Because of local variations in residence times and relaxation times, the curves shown here only serve as an order-of-magnitude estimate.

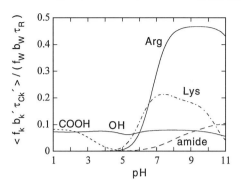

FIG. 12. Average contribution to the magnitude of the ^{1}H dispersion step from labile protons in a side chain or peptide group, relative to the contribution from a fully ordered water molecule. The curves were calculated from data for BPTI at 300 K ($\tau_R = 6.2$ ns),[20] but should be representative for most globular proteins provided that the curves (except COOH) are displaced along the pH axis in proportion to variations in τ_R (see text).

values (at room temperature), whereas NH protons in lysine and arginine side chains exchange rapidly above pH 6 (but can contribute substantially even in the intermediate-exchange regime below pH 6; see Fig. 9). Figure 12 shows the average contribution to the ^{1}H dispersion step from different types of side chains with labile protons, normalized by the contribution from a single (fully ordered) long-lived water molecule, that is, $\langle f_k b'_k \tau'_{Ck}\rangle/(f_W b_W \tau_R)$; see Eqs. (11), (13), and (14). At pH 7, for example, the labile protons in three arginine residues contribute nearly as much as the two protons in one buried water molecule. These curves refer specifically to BPTI at 27° with $\tau_R = 6.2$ ns,[20] but can serve as a rough guide also for other proteins if one notes that the base-catalyzed curves are shifted to higher pH by one unit if τ_R is increased by a factor of 10. (The pH dependence of the COOH curve reflects deprotonation, rather than exchange catalysis.) As seen from Eq. (14), the dispersion is shifted to higher frequency in the intermediate-exchange regime. Because the dispersion amplitude is simultaneously suppressed, this effect tends to show up as a high-frequency tail in the dispersion profile.

It appears that labile protons make a large, if not dominant, contribution to the water ^{1}H dispersion of small and medium-sized proteins even at neutral pH.[20] This also applies to oligonucleotides.[41,51] Before drawing conclusions about biomolecular hydration from ^{1}H MRD data, it is therefore prudent to investigate the pH dependence. Since the local relaxation times of labile hydrogens are at least an order of magnitude shorter for ^{2}H than for ^{1}H, labile hydrogens contribute less to the ^{2}H MRD profile. Nevertheless, outside the neutral pH range (particularly on the alkaline side), labile hydrogens can make large contributions.[23–26]

[51] D. Zhou and R. G. Bryant, *J. Biomol. NMR* **8,** 77 (1996).

Deviations from Motional Narrowing

For large (anisometric) or highly concentrated biomolecules in solution, in thermally or chemically induced gels, and in tissues, biomolecule rotation may become very slow or even inhibited. According to Eq. (7), the correlation time τ_{Ck} then equals the residence time τ_k. It might then appear that the distribution of residence times for all sites in a rotationally immobilized biomolecule could be determined directly from the dispersion profile. This approach is viable only if all contributing sites have residence times that are short enough to obey the motional-narrowing condition, which for longitudinal relaxation means that $b_k \tau_k^2 \ll 1 + (\omega_0 \tau_k)^2$.[52] When this inequality is violated, the frequency in the spectral density (9) must be replaced by an effective frequency $\omega_{eff} = (\omega^2 + ab_k)^{1/2}$, where a is a numerical constant ($a = 3/2$ for $I = 1$ nuclei). Provided that $f_k \ll 1$, the spectral density can still be expressed on the form of Eq. (9), but with b_k and τ_{Ck} (here equal to τ_k) replaced by effective quantities $b_{k,\mathrm{eff}}$ and $\tau_{k,\mathrm{eff}}$ defined by $b_{k,\mathrm{eff}}/b_k = \tau_{k,\mathrm{eff}}/\tau_k = (1 + ab_k\tau_k^2)^{-1/2}$.[52]

In the motional-narrowing regime, the dispersion frequency is governed by the residence time τ_k. In the limit of long residence times, however, it is governed by the effective correlation time $\tau_{k,\mathrm{eff}} = 1/(\sqrt{a}\,\omega_{Qk}S_k)$. This limiting correlation time is about 1 μs for ^{2}H. Because the magnitude of the dispersion step is proportional to $b_{k,\mathrm{eff}}\tau_{k,\mathrm{eff}}$, it follows that the dispersion will be dominated by sites with residence times in a range around $1/(\sqrt{a}\,\omega_{Qk}S_k)$.[53] These considerations account for the observation of ^{2}H dispersion profiles with an effective temperature-independent correlation time of about 1 μs in a variety of immobilized protieins[53–56] and in colloidal systems.[57] Similar effects are seen in ^{1}H MRD, but here the analysis is complicated by a distribution of intermolecular dipole couplings and by spin diffusion.

Summary

Although the MRD method has a long record in biomolecular systems, it has undergone a renaissance in the past few years as methodological developments have provided access to new types of information. In particular, MRD studies of quadrupolar nuclei such as ^{17}O and ^{23}Na have yielded valuable insights about the interactions of proteins and oligonucleotides with their solvent environment.[4,5]

[52] B. Halle, *Progr. NMR Spectrosc.* **28**, 137 (1996).
[53] B. Halle and V. P. Denisov, *Biophys. J.* **69**, 242 (1995).
[54] S. H. Koenig, R. D. Brown, and R. Ugolini, *Magn. Reson. Med.* **29**, 77 (1993).
[55] S. H. Koenig and R. D. Brown, *Magn. Reson. Med.* **30**, 685 (1993).
[56] R. Kimmich, T. Gneiting, K. Kotitschke, and G. Schnur, *Biophys. J.* **58**, 1183 (1990).
[57] P. Roose, H. Bauwin, and B. Halle, *J. Phys. Chem. B* **103**, 5167 (1999).

The biomolecular MRD literature is still dominated by hydration studies, but the method has also been used to study the interaction of organic cosolvents[58] and inorganic counterions[17] with biomolecules. The MRD method can potentially make important contributions to the understanding of the mechanisms whereby protein conformational stability is affected by nonaqueous solvent components, such as denaturants, stabilizers, and helix promoters. Residence times of water molecules and other low molecular weight species in association with biomolecules can be determined by MRD. Such residence times are of general interest for understanding the kinetics of biomolecule–ligand interactions and, when exchange is gated by the biomolecule, can be used to characterize large-scale conformational fluctuations on nanosecond–millisecond time scales.[38] By monitoring the integrity and specific internal hydration sites as well as the global solvent exposure, the MRD method can also shed light on the structure and dynamics of biomolecules in fluctuating nonnative states.[49,59] Because it does not rely on high resolution, the MRD method is also applicable to very large biomolecules and complexes and has even been used to investigate protein crystals,[60] gels,[54,55] and biological tissues.[22] In fact, dynamic studies of solids and liquid crystals were among the earliest applications of the MRD method. In many of its diverse applications, the MRD method provides unique information, complementing that available from high-resolution NMR.

Acknowledgments

We are grateful for research support from the Crafoord Foundation, the Swedish Natural Science Research Council (NFR), the Swedish Research Council for Engineering Sciences (TFR), and the Swedish Medical Research Council (MFR).

[58] H. Jóhannesson, V. P. Denisov, and B. Halle, *Protein Sci.* **6,** 1756 (1997).
[59] V. P. Denisov, B.-H. Jonsson, and B. Halle, *Nature Struct. Biol.* **6,** 253 (1999).
[60] K. Venu, L. A. Svensson, and B. Halle, *Biophys. J.* **77,** 1074 (1999).

[8] Nuclear Magnetic Resonance-Based Approaches for Lead Generation in Drug Discovery

By Jeffrey W. Peng, Christopher A. Lepre, Jasna Fejzo, Norzehan Abdul-Manan, and Jonathan M. Moore

Introduction

In the past several years, nuclear magnetic resonance (NMR) strategies in pharmaceutical research have shifted from traditional methods focused on gaining structural and mechanistic insight on drug targets, to a new set of experiments aimed at identifying and optimizing potent lead molecules for drug design.[1,2] This "new" set of methods has been known to NMR practitioners for some time, but shares the same ability to characterize the interaction of small organic molecules with larger biological macromolecules. In their present context, these experiments are implemented in a high throughput mode to screen not a single small molecule, but rather large libraries of compounds of pharmaceutical interest. This review focuses on selected theoretical, experimental, and practical aspects of performing screening experiments. We begin by summarizing the basic theory of exchanging systems on which these experiments are based, followed by discussions of the most commonly used experiments for NMR-based screening. Because the compounds used for screening are of at least as much importance as the experimental methods, a section is devoted to design of screening libraries. Finally, practical considerations such as sample preparation, automation, and throughput are discussed.

Basic Considerations of Exchanging Systems

In a typical target–ligand interaction there are three species in equilibrium: the target, E, the ligand L, and the complex EL:

$$E + L \underset{k_{off}}{\overset{k_{on}}{\rightleftharpoons}} EL \tag{1}$$

The ligand and target molecules exchange between the free and complexed states. This exchange can significantly perturb the NMR parameters of both the ligand and target. Identifying such perturbations forms the basis for all NMR screening experiments. In order to better understand these experiments, it is useful to highlight some basic aspects of exchange processes and how these can modulate NMR parameters. In the following text, we refer to the target molecule as the protein.

[1] J. M. Moore, Curr. Opin. Biotechnol. 10, 54 (1999).
[2] M. J. Shapiro and J. R. Wareing, Curr. Opin. Drug Disc. Devel. 2, 396 (1999).

However, many of the screening methods described here are equally applicable to other therapeutically interesting macromolecules such as nucleic acids.

The binding affinity for the equilibrium of Eq. (1) can be described by a dissociation constant, K_D. K_D is equal to the ratio k_{off}/k_{on} and is a function of the temperature. If the ligand concentration is disturbed, the rate constant describing its relaxation back to equilibrium is $k_{ex} = k_{on}[E] + k_{off}$. Note that k_{ex} is the essential rate constant needed for gauging the rapidity of exchange and the consequent effects on the ligand NMR parameters. It can also be cast as:

$$k_{ex} = k_{off}/(1 - P_b) \tag{2}$$

Here, P_b is the bound ligand fraction $[EL]/L_T$, where L_T is the total ligand concentration. This relation implies that k_{ex} generally depends both on the off-rate *and* the relative amounts of target and ligand. P_b can be easily determined from the K_D, the total protein concentration E_T, and the total ligand concentration, L_T. Assuming a single protein binding site, we have:

$$P_b = \frac{(E_T + L_T + K_D) - \left(\sqrt{(K_D + E_T + L_T)^2 - 4E_T L_T}\right)}{2L_T} \tag{3}$$

A useful guide is that the protein binding site is approximately half-saturated when $L_T \approx K_D$, assuming $L_T > E_T$. Merely having a huge excess of ligand is not enough to guarantee saturation of the binding site; rather, L_T must be compared to K_D. On the other hand, having a great ligand excess does force the bound ligand fraction P_b toward zero. In this case, the net exchange rate constant [cf. Eq. (2)] then approaches the off-rate, k_{off}.

Exchange toggles the ligand and protein NMR parameters between values characteristic of the free and bound states. As a consequence, we observe exchange-averaged NMR parameters for both sets of molecules. The precise form of this exchange-averaging depends on the rate of exchange relative to the magnitude difference between the free and bound NMR parameter values. Specifically, if Q_f and Q_b are the free and bound state values of some ligand NMR parameter Q, then the exchange is fast if $k_{ex} \gg |Q_f - Q_b|$, intermediate if $k_{ex} \approx |Q_f - Q_b|$, or slow if $k_{ex} \ll |Q_f - Q_b|$. Depending on which exchange limit prevails, we observe different exchange-averaging effects. The case in which Q is a ligand chemical shift provides the most familiar illustration. In the fast exchange limit, $k_{ex} \gg 2\pi|\delta_f - \delta_b|$, where δ_f and δ_b are the chemical shifts of the free and bound states in hertz. The large k_{ex} indicates that many exchange events occur during all stages of the NMR experiment. As a result, we observe a single Lorentzian line that appears at the population-weighted average resonance, $\delta_{obs} = P_b\delta_b + (1 - P_b)\delta_f$. In contrast, the slow exchange limit gives $k_{ex} \ll 2\pi|\delta_f - \delta_b|$. Nuclei in the free and bound states precess almost independently, and separate Lorentzian lines are observed for the free and bound species at their respective shifts. The most complex scenario is the intermediate exchange limit, in which $k_{ex} \approx 2\pi|\delta_f - \delta_b|$. As with the fast exchange limit, we observe only a single line. However, the line shape is not Lorentzian and is expected to be broader. Additionally, the exchange-averaged shift is not at the population-weighted average position except in the special case for $P_b = 0.5$.

Generally, Q can be a variety of NMR parameters including the longitudinal and transverse relaxation rate constants $R_1 = 1/T_1$ and $R_2 = 1/T_2$, cross-relaxation rates arising from dipole–dipole interactions between pairs of like spins (NOE and ROE), and the translational molecular diffusion coefficient, D. As in the case of the chemical shift, the fast exchange limit results in the observation of population-weighted averages. These are given below for the four parameters mentioned above:

$$R_{1,\text{obs}} = P_b R_{1,b} + (1 - P_b)R_{1,f} \tag{4}$$

$$R_{2,\text{obs}} = P_b R_{2,b} + (1 - P_b)R_{2,f} + R_{\text{ex}} \tag{5}$$

$$\sigma_{\text{noe,obs}} = P_b\sigma_{\text{noe},b} + (1 - P_b)\sigma_{\text{noe},f} \tag{6}$$

$$\sigma_{\text{roe,obs}} = P_b\sigma_{\text{roe},b} + (1 - P_b)\sigma_{\text{roe},f} \tag{7}$$

$$D_{\text{obs}} = P_b D_b + (1 - P_b)D_f \tag{8}$$

Two caveats are worth noting here. First, R_2 can have an additional contribution, R_{ex}. For the simple two-state binding exchange:

$$R_{\text{ex}} = P_b(1 - P_b)\frac{4\pi^2(\delta_f - \delta_b)^2}{k_{\text{ex}}} \tag{9}$$

The R_{ex} contribution depends not only on k_{ex} but the bound ligand fraction P_b and the difference between the free and bound chemical shifts. Note that the R_{ex} term may tend to zero if P_b is very small, the exchange is extremely fast, or the bound and free state shifts are nearly degenerate. Screening protocols using ^{1}H NMR and an excess of ligand typically meet one of these criteria, and so the R_{ex} term is often negligible. Another important caveat is that the same exchange process may be fast for one NMR parameter while being slow for another. For example, a given nucleus could hop between two sites having drastically different chemical shifts but with similar longitudinal relaxation rate constants R_1. The corresponding k_{ex} might be less than the difference in chemical shifts while simultaneously exceeding the difference in R_1 values. Accordingly, the exchange would be slow or intermediate in chemical shift space, but fast in "relaxation space." A superficial interpretation of the data based only on consideration of the chemical shifts would then be hazardous to the accuracy of the conclusions. Even if one is just concerned with chemical shifts, the exchange rate limit can vary significantly depending on what kind of nucleus is being observed and on the strength of the magnetic field.

The fast-exchange expressions of Eqs. (4)–(8) are appealing because of their simplicity and provide adequate descriptions of the NMR spectra for most screening applications. For example, typical experimental conditions include a greater than 10-fold excess of ligand over protein and ligand K_D values in the range of 10 μM to 1 mM. If we assume further that k_{on} is well approximated by a diffusion-limited value ($1 \times 10^8 \, s^{-1} M^{-1}$), then k_{ex} will fall in the range between 1000 to 100,000 s^{-1}, and thus, the fast-exchange limit seems to be a reasonable approximation.

If one is interested in carrying out more quantitative follow-up experiments beyond the simple determination of binding, for example to estimate exchange rates or to determine K_D values to rank ligand affinities, then ascertaining the correct exchange regime becomes important. For such studies, proper data interpretation demands proper data reduction, and one should then consult more in-depth reviews in the literature.[3–8]

NMR Pulse Sequences Used for Screening

The growing interest in screening mixtures of compounds against a given therapeutic target has led to the development of new NMR pulse sequences as well as the rediscovery of old ones. The goal of these experiments is to distinguish binders from nonbinders in a mixture of compounds. Toward this end, these experiments employ several broad strategies that include the following:

1. *Looking for binding-induced changes in the chemical shift.* Binding-induced changes occur for both protein and ligand chemical shifts. Popular strategies have been devised to observe effects for either species. To detect such changes obviously requires well-resolved resonances, so if protein resonances are to be followed, isotope enrichment is usually necessary.

2. *Exploiting the differential mobility of the large protein relative to the low molecular mass compounds.* This includes both overall rotational mobility as well as translational diffusion. By convention, the former is quantified by the rotational correlation time, τ_c, and the latter by a translational diffusion coefficient, D. τ_c increases with molecular weight while D decreases with molecular weight. Compounds that bind will transiently adopt the slow mobility characteristics of the protein.

3. *Exploiting differences in the relaxation mechanisms in the protein versus those in the compounds.* The relaxation mechanisms of the protein may differ in kind or abundance. An example of the former would be the existence of a

[3] S. Alexander, *J. Chem. Phys.* **37**, 974 (1962).
[4] S. Alexander, *J. Chem. Phys.* **37**, 967 (1962).
[5] G. Binsch, *J. Am. Chem. Soc.* **91**, 1304 (1969).
[6] J. Feeney, J. G. Batchelor, J. P. Albrand, and G. C. K. Roberts, *J. Magn. Reson.* **33**, 519 (1979).
[7] C. S. Johnson, Jr., *J. Magn. Reson. A* **102**, 214 (1993).
[8] J. Fejzo, C. A. Lepre, J. W. Peng, G. W. Bemis, Ajay, M. A. Murcko, and J. M. Moore, *Chem. Biol.* **6**, 755 (1999).

paramagnetic center in the protein. An example of the latter would be the greater number of potential proton–proton dipolar relaxation pathways arising from the larger proton density of the protein. Binding compounds can be distinguished from nonbinding compounds via relaxation by these additional mechanisms.

In most laboratories, the availability of large quantities of protein for screening is often limiting. Use of substoichiometric concentrations of protein ($L_T \gg E_T$) results in low bound ligand fractions, P_b. Under these conditions, P_b must lie in the range $0 \leq P_b \leq (E_T/L_T)$. The choice of E_T/L_T thus determines the maximum P_b attainable, and, correspondingly, the maximum extent to which binding manifests in the observed NMR parameters. E_T/L_T can therefore be judiciously chosen to select only for binders with affinities above a certain threshold. This fact and reference to Eqs. (4)–(8) suggest that the most sensitive methods, and thus, the most desirable, will be those measuring parameters that are *amplified* in the bound state. Another desirable feature of any screening pulse sequence is its simplicity. Even more than in structural NMR, little time is available or tolerated for elaborate spin gymnastics that yield only modest signal improvements. Rather, the pulse sequences should require little or no calibration on the part of the user and, most importantly, should be compatible with commercial automation.

Ligand-Based Pulse Methods to Screen for Binding to Large Proteins

Ligand-based NMR screening records and compares the NMR properties of a mixture of compounds in the presence and absence of target molecules. Observation of the ligand signals renders target size irrelevant, since it bypasses the lengthy time investments and problems associated with resolving and assigning the immense number of broad protein resonances. This is an important advantage since protein target selection is based on potential therapeutic value and not low molecular weight. Sample preparation is also considerably easier given that no isotope labeling is needed. Moreover, some of the methods described below require protein concentrations of only 1 μM or less, and thus the total demand on protein supplies is dramatically reduced.

Disadvantages of ligand-based approaches include the inability to directly localize the ligand binding sites. Additionally, because isotope labeling of ligands is expensive and thus impractical, the NMR experiments are usually limited to ^{1}H NMR (*vide infra*). Thus, assuming the absence of any paramagnetic labels and field strengths ≤ 11.7 tesla, the dominant relaxation mechanisms for both compounds and protein are the pairwise dipole–dipole interactions between protons.

1D Proton Relaxation Methods

Nuclear spin relaxation parameters depend on the overall rotational mobility of the molecule through their dependence on spectral density functions $J(\omega)$ that describe this same mobility. Of particular interest are those relaxation parameters that have a direct dependence on the overall rotational correlation time of the molecule, τ_c (*vide supra*) via a $J(0)$ dependence. Such parameters include the selective longitudinal relaxation rate constant $R_1 = 1/T_1$, the transverse rate constant $R_2 = 1/T_2$, and homonuclear cross-relaxation rate constants that govern both laboratory and rotating frame Overhauser effects (NOE, ROE). The direct dependence on τ_c means that these parameters are amplified in the bound state. One can exploit this amplification to screen for binders. Specifically, in Eqs. (4)–(6), one expects that bound state relaxation rate constants will exceed those of the free state. Additionally, the fact that R_1 and R_2 are actually sums over all possible dipolar couplings involving the proton of interest means that these rate constants will be further boosted in the bound state. R_2 may be further enhanced by the R_{ex} contribution [cf. Eq. (5)]. Thus, for ligands exchanging on and off the protein, we will observe enhanced relaxation rates. On the basis of these enhancements, binders can be distinguished from nonbinders. In this section, we discuss only the autorelaxation rates, $R_1 = 1/T_1$ and $R_2 = 1/T_2$. Cross-relaxation rates are discussed later in this article.

The transverse relaxation rate constant, $R_2 = 1/T_2$, is a particularly sensitive probe of binding.[8,9] Enhanced R_2 values of binding compounds may be evident by a simple inspection of the 1D proton spectrum. Because the homogeneous line width (full width half-maximum) equals R_2/π, binders may show obvious line broadening. Such broadening effects are highly pronounced with large molecular weight targets, as illustrated in Fig. 1 with the 224 kDa target enzyme inosine-5′-monophosphate dehydrogenase (IMPDH). A tenfold increase in line width is apparent for the downfield singlet shown. For lower molecular weight targets, broadening may not be as severe, and binders may be identified by filtering signals according to their R_2 values using the Hahn echo pulse sequence 90_x–delay–180_y–delay–Acquire[10] or the simple sequence 90_x–[Spin lock y]–Acquire. An example is shown in Fig. 2 for the 42 kDa p38 MAP kinase enzyme with a mixture of two ligands. The delay period or spin lock acts as the relaxation filter and is set for a fixed duration. One tries to find a duration that suppresses all protein signals while retaining compound signal. This becomes easier to achieve with increasing protein size due to the direct dependence of R_2 on τ_c. Note that for smaller targets that require longer relaxation filtering delays, a spin lock sequence should be used instead of a simple Hahn echo. Longer Hahn echo times can

[9] P. J. Hajduk, E. T. Olejniczak, and S. W. Fesik, *J. Am. Chem. Soc.* **119**, 12257 (1997).
[10] I. D. Campbell, C. M. Dobson, R. J. Williams, and P. E. Wright, *FEBS Lett.* **57**, 96 (1975).

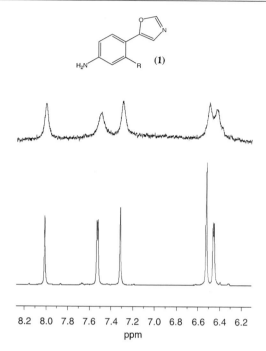

FIG. 1. 1D–^{1}H spectra of (bottom) free ligand **(1)** and (top) the ligand **(1)** in the presence of IMP dehydrogenase, a 224 kDa protein. The significant line broadening observed in the presence of the enzyme indicates binding of the small molecule. Line widths at half height for the furthest downfield singlet of **(1)** (leftmost peak) are 3 Hz (bottom) and 30 Hz (top). The sample contained 1 mM ligand + 100 μM IMPDH in 25 mM Tris-d pH 8 (D$_2$O), 300 mM KCl, 5% glycerol-d, 5 mM dithiothreitol (DTT). 1D spectra were collected at 277 K with 1 sec low power presaturation of the residual HDO signal, using a Bruker DRX-800 spectrometer.

lead to significant scalar (J_{HH}) coupling evolution that distorts the line shapes and could thus corrupt the data analysis. Potential spin lock sequences are the Carr–Purcell–Meiboom–Gill (CPMG) train, or a simple continuous wave (CW) radio frequency field. If the CPMG sequence is used, then the gap between successive 180° pulses, $2T_{cpmg}$, should satisfy $|4\pi J_{HH} T_{cpmg}| \ll 1$ to minimize scalar coupling evolution.[11] It is also conceivable to use windowless composite pulse sequences, such as CLEANEX-PM,[12] that have been optimized for 2D-exchange spectroscopy. By design, these sequences suppress scalar coupling evolution and dipolar cross-relaxation. However, since the magnetization trajectory during the sequence roams over both the rotating frame z axis and the transverse plane, the

[11] R. Freeman and H. D. W. Hill, *J. Chem. Phys.* **55**, 1985 (1971).
[12] T.-L. Hwang, S. Mori, A. J. Shaka, and P. C. M. van Zijl, *J. Am. Chem. Soc.* **119**, 6203 (1997).

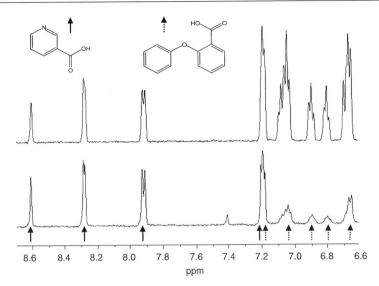

FIG. 2. 1D–^{1}H spectra of (top) a mixture of two ligands compared to (bottom) the mixture of the ligands in the presence of p38 MAP kinase, a 42 kDa protein. Resonances from nicotinic acid (top left structure) and 2-phenoxybenzoic acid (top right structure) are marked with solid and dashed arrows, respectively. The peak at 7.2 ppm consists of overlapping resonances from both compounds. Line broadening, suppression of fine structure, and attenuation of ligand resonance peak height due to relaxation filter in the bottom spectrum indicate 2-phenoxybenzoic acid binds to p38, while nicotinic acid does not. The sample contained 1 mM ligands, 0.2 mM p38 MAP kinase, 25 mM Tris-d, 10% gycerol-d, 20 mM DTT-d at pD* 8.4. Experiments were carried out at 278 K. 1D NOESY spectra were collected with 16K data points, 128 transients, and a relaxation delay of 3 s. A relaxation filter was used after the preparatory delay to attenuate broad resonances arising from the protein.

effective relaxation rate constant is $\approx 1/2(R_1 + R_2)$. An alternative approach to detecting binding with relaxation filtering uses difference methods to eliminate signals from protein and nonbinding ligands.[9] However, these methods have not been widely used due to the lack of reproducibility and reliability associated with such subtraction.[13]

In principle, relaxation filters using selective R_1 values may also probe for binding. Selective R_1 values require experiments that invert or saturate a restricted set of proton spins while leaving all others unaffected. Frequency-selective shaped pulses are often used for this purpose. In contrast, nonselective R_1 values arise if all spin populations are identically disturbed. The nonselective R_1 values lack the direct dependence on τ_c contained in the selective R_1 values and are therefore

[13] P. J. Hajduk, T. Gerfin, J.-M. Boehlen, M. Haberli, D. Marek, and S. W. Fesik, *J. Med. Chem.* **42,** 2525 (1999).

much less sensitive probes of binding.[14,15] Selective R_1 experiments have not been as common as the R_2 experiments in NMR screening because of the problems of achieving selectivity for a large library of chemically diverse compounds. Clearly, one wants to avoid repeated calibration of selective inversion/saturation pulses when screening libraries containing hundreds or thousands of compounds. What is essential is the ability to selectively invert/saturate ligand protons relative to the protein protons. Selective inversion of the ligand resonances can be approximated by two 90° pulses intercalated by a Hahn spin–echo or CPMG pulse train. The Hahn echo or CPMG train serves as an R_2 relaxation filter that saturates the protein protons but inverts the compounds' protons; this inversion strategy has been used to achieve "reverse NOE pumping"[16] (see below). After a relaxation delay, a final 90° pulse reads out the resulting magnetization. Toggling the phase of the second 90° pulse places the ligand magnetization alternately on the $\pm z$ axis. Subtraction in the receiver yields the ligand selective R_1. Binders have enhanced selective R_1 values and would be expected to have greater residual signal after the subtraction.

A potential problem for relaxation filtering methods is the fact that not all the resonances of the protein may have the same relaxation rates. Certain protons that are part of molecular fragments having significant internal motion have relaxation properties more similar to those of the low molecular weight ligands. As a result, suppression of these protein proton signals may be quite difficult and their remnants may interfere with the observation of the compounds. Examples include methyl groups and rapidly rotating aromatic groups. To minimize these problems one can work with an excess of ligand over target, or introduce thickening agents that have a strong temperature dependence (e.g., glycerol). Of course, in the latter situation, one must ensure that such additives do not interfere with binding properties.

1D Proton NOE Methods

Dipolar interactions between two protons result in mutual cross-relaxation (spin flips), leading to proton–proton NOEs in the laboratory frame and ROEs in the rotating frame. The rate constants governing the cross-relaxation depend on the distance between the two protons and the rotational mobility of the vector connecting the two proton centers. The sensitivity to rotational mobility arises via a direct dependence on the overall rotational correlation time of the molecule, τ_c. Ligands that exchange on and off the protein will experience alternately the large τ_c of the protein (e.g., 10–20 ns/rad) and the much smaller τ_c of the free state (100–500 ps/rad). Additionally, the ligand protons may pick up additional cross-relaxation with the protein protons. Both of these effects modulate the

[14] F. Ni, *Prog. Nucl. Mag. Reson. Spectrosc.* **26,** 517 (1994).
[15] G. Valensin, T. Kushnir, and G. Navon, *J. Magn. Reson.* **46,** 23 (1982).
[16] A. Chen and M. J. Shapiro, *J. Am. Chem. Soc.* **122,** 414 (2000).

ligand NOE and ROE cross-relaxation rate constants. The fast exchange expressions for these constants are the simple population-weighted averages of Eq. (6).

The possibility of ligand protons picking up additional dipole–dipole interactions with protein protons can be exploited for NMR screening. Such interactions lead to intermolecular NOE transfer, which perturbs the spin populations of the proximal ligand protons. In principle, 1D transient NOE experiments can detect these perturbations, and thus provide a vehicle for screening. What are needed are independent perturbations of the ligand proton spin populations versus those of the protein protons. To this end, the Novartis Group[16,17] has proposed two new 1D transient NOE methods. The first method has been termed NOE pumping.[17] In this experiment, a diffusion filter prefaces an NOE mixing period. The diffusion filter attenuates signal based on the translational diffusion coefficient of the molecules. Because the mixture compounds diffuse orders of magnitude faster than the target protein, their signals are destroyed. Note that the diffusion filter has an advantage over the relaxation filter in that all ligand lines are suppressed, regardless of their individual relaxation rates, which may vary according to local proton density and internal mobility. Protein magnetization survives the diffusion filter, and phase cycling of the diffusion filter ensures that the remaining magnetization is placed alternately on the $\pm z$ axis at the start of the mixing time. The receiver phase toggles with the sign of the protein magnetization such that only NOEs "pumped" from the protein to the ligand are detected. Contributions from R_1 relaxation as well as magnetization from nonbinding compounds are eliminated by phase cycling. The diffusion filter uses pulsed field gradients to spatially encode and decode the positions of molecules in the sample. Filters using either transverse spin echoes[18,19] or stimulated echoes with the LED modifications[20,21] may be used. An advantage of the latter method is that rapid R_2 relaxation is avoided. This is a critical consideration since rapid R_2 relaxation could kill the protein signal during the diffusion filter and thus undermine the NOE pumping experiment. Also, like the relaxation-based methods, the diffusion filter will perform better the greater the difference in molecular weight (and therefore, diffusion coefficients) between the protein and ligand.

Binders may also be detected using a second method called reverse NOE pumping.[16] This method involves two experiments that measure a ligand selective R_1 and a nonselective R_1, respectively. The first experiment uses a relaxation filter followed by an NOE mixing time and read pulse. The relaxation filter saturates

[17] A. Chen and M. J. Shapiro, *J. Am. Chem. Soc.* **120,** 10258 (1998).
[18] E. O. Stejskal and J. E. Tanner, *J. Chem. Phys.* **42,** 288 (1961).
[19] H. Ponstingl and G. Otting, *J. Biomol. NMR* **9,** 441 (1997).
[20] J. E. Tanner, *J. Chem. Phys.* **52,** 2523 (1970).
[21] S. J. Gibbs and C. S. Johnson, Jr., *J. Magn. Reson.* **93,** 395 (1991).

the protein protons while inverting the ligand protons. The ligand proton spin populations recover because of a ligand-selective R_1 that includes the effect of magnetization transfer to the protein protons. The nonselective R_1 comes from a second experiment that reverses the order of mixing time and relaxation filter. The nonselective R_1 lacks the effect of magnetization transfer to the protein. Thus, for binding compounds, the magnetization decays at a slower rate compared to the first experiment. The two experiments can be recorded in an interleaved fashion. Taking the difference between the spectra of the two experiments exposes the binding compounds. Clearly, both NOE pumping methods become more effective as the protein size increases, since the corresponding higher molecular weights lead to larger cross-relaxation rate constants for the bound state.

A disadvantage of the 1D methods described above are that the detected signal relies on intermolecular NOE transfers alone. It does not directly take advantage of the enhanced intramolecular NOE cross-relaxation the ligand experiences while bound to the protein. Additionally, depending on the rigidity of the ligand-binding mode, the bound state cross-relaxation rate constants may be attenuated by flexibility, thus requiring long NOE mixing delays to accumulate appreciable NOE. For the NOE-pumping experiment, one should note that the optimal intermolecular transfer occurs on the time scale of the protein proton $R_1 = 1/T_1$. Short protein T_1 values or nonuniform protein saturation can therefore also compromise the buildup of intermolecular NOE. Thus, in situations where the protein concentration is limited (e.g., $\leq 1 \ \mu M$), these methods may suffer in sensitivity.

2D NOE Methods

In contrast to the 1D methods above, the 2D transferred NOESY (tNOE) effectively inverts all protons at different times via a t_1 chemical-shift labeling period; thus, both inter- and intraligand NOE transfers can be observed. Under conditions of fast exchange, the intensity of the intraligand NOE cross peaks are governed by the population-weighted cross-relaxation rate shown in Eq. (6). The bound state cross-relaxation rate constants differ starkly from those of the free state in that (i) the cross-relaxation is of opposite sign from the free-state cross-relaxation, and (ii) the absolute magnitude of the rate constant is much larger than in the free state. Thus, the bound state NOE dominates the weighted sum in Eq. (6). These two factors result in strong positive NOESY cross peaks for binders, as opposed to weakly negative or zero cross peaks for the nonbinders. Sample tNOE spectra for the same mixture of ligands shown in Fig. 2 are given in Fig. 3. Note that the sign flip of the NOE cross peak between the free versus bound states acts as a simple binary filter to distinguish binders. If one seeks a more quantitative interpretation of these cross peaks for the purposes of structure determination, then a more rigorous approach is required, as has been suggested by several investigators.[14,22-24]

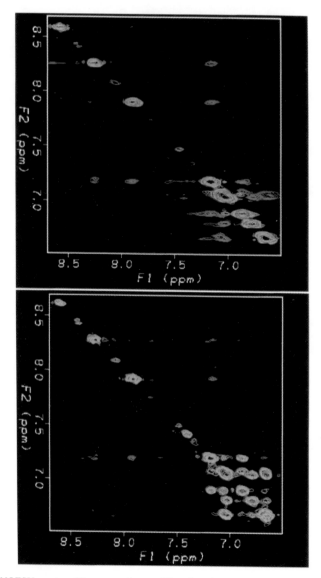

FIG. 3. 2D NOESY spectra of the same mixture of ligands and protein shown in Fig. 2. Positive contours are shown in cyan and negative contours are shown in green. Both ligands in the mixture without protein present (left) have weak negative cross peaks and positive diagonal peaks. In the presence of the protein (right) cross peaks remain opposite in phase to the diagonal peaks for the downfield resonances corresponding to nicotinic acid, indicating this compound does not bind. However, the sign of the upfield cross peaks of 2-phenoxybenzoic acid are the same sign as the diagonal peaks, indicating this compound binds to the protein. The peak at 7.2 ppm consists of overlapping resonances from both compounds (see Fig. 6 with 1D spectra), and thus both positive and negative cross peaks are found at this frequency. 2D NOESY spectra were collected with 400 t_1 increments and 2K complex points in t_2, with mixing times of 50 and 200 ms, relaxation delay of 2 s, and 16 transients per t_1 increment. A spin echo sequence after the first proton pulse was used in the NOESY experiments as a relaxation filter.

Obviously, at sufficiently large ligand excesses, the free state NOE will ultimately dominate the exchange-averaged cross-relaxation rate [cf. Eq. (6)]; the boundary where this occurs depends both on the ligand affinity and the size of the protein target. For example, we have found that 2D tNOE using a ligand–protein molar ratio of 20 : 1 permits facile detection of binders to a 42 kDa target having K_D values in the range of 10 μM to 10 mM. On the other hand, when the ligand–protein ratio reaches $\approx 1000 : 1$, ligands with K_D values ≈ 200 μM escape detection (see Fig. 6).

Pulse sequences for 2D transferred NOE experiments differ little from the canonical three-pulse sequence.[25,26] The primary modification is the insertion of a relaxation filter to suppress protein signals during detection. Inserting an R_2 filter prior to the first 90° pulse in the NOESY sequence can achieve this. The R_2 filter obviously starts from z magnetization so the transverse relaxation period should be bracketed by 90° pulses. Alternatively, one can use windowless composite pulse sequences[14] that perform cyclic rotations of magnetizations that start on the z axis. Examples of this include DIPSI-2[27] and FLOPSY-8.[28] Note that the effective relaxation during these composite-pulse sequences is the trajectory average of R_1 and R_2, and thus the filtering time required may be longer than that of a simple Hahn spin–echo, and as mentioned above, avoid scalar-coupling evolution and diffusion losses that arise with a long Hahn echo. Relaxation filters may also be used *after* the last 90° pulse. This is particularly useful when solvent suppression is also required. Solvent suppression schemes use spin–echo delays that may be exploited for their inherent relaxation filtering properties. The double spin–echo scheme used in the "excitation-sculpting" method of Hwang and Shaka[29] is an excellent choice that is easy to set up and quite robust against user error.

1D Saturation Transfer Methods

The NOE methods presented above rely on a transient perturbation of either ligand or protein spin populations. One may also impose a steady-state insult in the form of selective proton saturation. The use of a long irradiation period to selectively saturate protons goes back to earlier studies of NMR and is the basis for 1D saturation transfer experiments,[30] presaturation of solvent lines, and the original 1D truncated driven NOE experiments.

[22] J. Zheng and C. B. Post, *J. Phys. Chem.* **100**, 2675 (1996).
[23] A. P. Campbell and B. D. Sykes, *J. Magn. Reson.* **93**, 77 (1991).
[24] A. P. Campbell and B. D. Sykes, *Annu. Rev. Biophys. Biomol. Struct.* **22**, 99 (1993).
[25] J. Jeener, B. H. Meier, P. Bachmann, and R. R. Ernst, *J. Chem. Phys.* **71**, 4546 (1979).
[26] A. Kumar, G. Wagner, R. R. Ernst, and K. Wüthrich, *J. Am. Chem. Soc.* **103**, 3654 (1979).
[27] A. J. Shaka, C. J. Lee, and A. Pines, *J. Magn. Reson.* **77**, 274 (1988).
[28] M. Kadkhodaie, O. Rivas, M. Tan, A. Mohebbi, and A. J. Shaka, *J. Magn. Reson.* **91**, 437 (1991).
[29] T.-L. Hwang and A. J. Shaka, *J. Magn. Reson. A* **112**, 275 (1995).
[30] S. Forsén and R. A. Hoffman, *J. Chem. Phys.* **39**, 2892 (1963).

FIG. 4. Example of a pulse sequence using the saturation transfer difference (STD) method.[31] Thin and thick rectangles on the ^{1}H staff indicate 90° and 180° pulses, respectively. The hatched pulses within the dashed lines represent a frequency-selective saturation train lasting 1–3 s. Selective 90° pulses of 50 ms length were used in the train, with an interpulse delay of ≈10 ms. In our laboratory both Gaussian and Seduce-1[64] profiles have been used for these purposes. The arrows before and after the saturation train indicate the ^{1}H carrier jumps to the point of saturation, and the residual water line, respectively. After the saturation train, a gradient selects for z magnetization. Excitation sculpting[29] is then used as a readout. The shaped pulses preceding the high-power 180° pulses are 2 ms Seduce-1 180° pulses. Gradients are applied on the z axis with rectangular profiles of 800 μs to 1 ms length and amplitudes ranging from 2 to 22 G/cm. Two experiments are performed in which the saturation train is on- and off-resonance, respectively, in a spectral region free of ligand peaks. Phase cycling is $\phi_1 = x, \phi_2 = x, y, -x, -y, \phi_3 = 4(+x), 4(+y), 4(-x), 4(-y), \phi_{rec} = 2[(+x, -x, +x, -x)(-x, +x, -x + x)]$. Pulses with no phase label indicate application along $+x$.

One-dimensional (1D) saturation transfer methods have been adapted for screening purposes with excellent results. Selective saturation of the protein quickly results in saturation of all protein protons via the high efficiency of NOE-driven cross-relaxation within the protein. As stated, the cross-relaxation magnitude varies with τ_c, and so the saturation diffusion occurs more rapidly for larger proteins. Mayer and Meyer[31] have demonstrated that this effect serves as the basis for a powerful method for screening. Specifically, compounds that bind transiently to the target will lose some magnetization as a consequence of cross-relaxation with the saturated protein protons. The total ligand magnetization continues to decrease as more "fresh" ligand encounters saturated protein. The losses continue to accumulate over a duration of ≈1 sec because of the relatively small R_1 values of the compounds and the constant influx of protein proton saturation. In the ideal limit, one can expect all of the binding compounds to have their proton magnetization similarly decreased. In contrast, the net magnetization of the nonbinders will remain unchanged.

The 1D saturation transfer method of Mayer and Meyer,[31] a modified version of which is shown in Fig. 4, involves two experiments. In a first experiment, a selected region of the protein spectrum is saturated. Saturation can be achieved by a train of frequency-selective 90° or 120° pulses, whose total duration is 1–3 seconds. A subsequent 90° read pulse records the residual z magnetization. Next, a control experiment is performed in which the saturation is applied far off resonance. Spectra from both experiments can be recorded in an interleaved fashion and subsequently subtracted. The only resonances that survive the subtraction must

[31] M. Mayer and B. Meyer, *Angew. Chem. Int. Ed. Engl.* **38**, 1784 (1999).

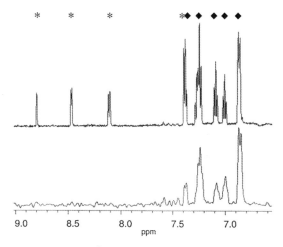

FIG. 5. Comparison of (top) the 1D ^{1}H NMR spectrum of two potential small molecule ligands in the absence of target and (bottom) the STD[31] spectrum of the same mixture in the presence of p38 MAP kinase. Resonances from nicotinic acid and 2-phenoxybenzoic acid are marked with asterisks and diamonds, respectively. Only peaks from the ligand that binds to p38, 2-phenoxybenzoic acid, appear in the STD spectrum. The sample contained 35 μM p38 MAP kinase, 1 mM ligands, 100 mM Tris-d, 10% glycerol-d, 20 mM DTT-d at pD* 8.4. Experiments were carried out at 278 K. The 1D spectrum was collected usng a standard NOESY pulse sequence with 16 K data points, 128 transients, and a relaxation delay of 3 s. The STD spectrum was collected using a pulse sequence similar to that shown in Fig. 4; 2K data points were collected with 256 transients each for the on- and off-resonance spectra. Internal subtraction was achieved via phase cycling. On-resonance irradiation was at $\delta = 0.74$ and off-resonance at $\delta = -20$ ppm. A pulse train of 60 Gaussian selective pulses of 50 ms duration separated by a 1 ms delay was used to saturate the protein. An excitation sculpting[29] sequence appended after 90° read pulse was used for water suppression.

belong to molecules that contacted the saturated protein. Accordingly, this screening method is referred to as the STD (saturation transfer difference) method. STD spectra collected on the same mixture of ligands used in Figs. 2 and 3 with p38 MAP kinase are shown for comparison in Fig. 5. As shown, an advantage of the STD experiment is that only signals from ligands that bind will appear in the spectra, thus simplifying analysis and deconvolution of the spectra, especially for mixtures with many components. Compared to line broadening or NOE-based methods, STD experiments are significantly more sensitive. Because of this sensitivity increase, it is possible to work with significantly lower concentrations of protein. Besides the obvious advantage of needing less sample material, working with lower protein (or target) concentrations provides the additional possibility of tuning the detection limits for binding. This phenomenon is illustrated in Fig. 6 for two small molecule ligands that bind the p38 MAP Kinase with different affinities. At high concentrations of target enzyme (bottom traces in Fig. 6), binders with affinities in the range from low micromolar to millimolar will be detected.

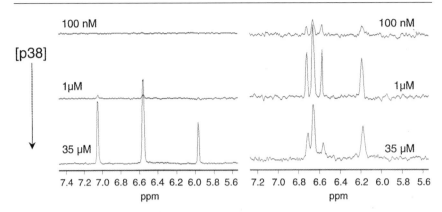

FIG. 6. Comparison of the STD[31] spectra collected using various concentrations of the p38 MAP kinase target for two compounds with different binding affinities: (right) 4-[4-(4-fluorophenyl)thiazoyl] benzoic acid with $K_d = 200 \, \mu M$ and (left) 3-(4-pyridyl)pyrazole with $K_d = 2$ mM. Sample conditions: 0.5 mM ligands, protein concentration as marked, 100 mM Tris-d/D$_2$O, 10% glycerol-d, 20 mM DTT-d at pD* 8.4. STD experiments were collected as described in Fig. 5.

However, with the use of lower concentrations of protein, e.g., 1 μM or 100 nM, only compounds that bind at several hundred micromolar or tighter will be detected (middle and top traces, Fig. 6). Therefore one can tune the sensitivity of the experiment to correspond to the desired results. For example, if large libraries are used, one might choose to screen at lower concentrations in order to keep the hit rate at a level sufficiently low that all interesting hits may be followed up on with regard to specificity. Alternatively, if one screens a smaller library, higher concentrations of protein may be used, allowing detection of weaker binding ligands that can then be pursued using, for example, linking strategies.[8,32–34]

In the event that protein and solvent signals need to be suppressed, spin–echo based solvent-suppression schemes, such as excitation sculpting,[29] may be appended after the 90° read pulse (Fig. 4). Care must be taken to ensure that the saturation train does not perturb ligand resonances. Depending on the nature of the mixtures, certain spectral regions will be richer in ligand resonances than others. These are obviously regions to avoid placing the saturation carrier frequency.

Saturation and inversion of the water resonance can provide for a very efficient saturation or inversion-transfer experiment.[35] In effect, protein saturation/inversion enters the protein through multiple avenues, including saturation/inversion of

[32] P. J. Hajduk, G. Sheppard, D. G. Nettesheim, E. T. Olejniczak, S. B. Shuker, R. P. Meadows, D. H. Steinman, G. M. Carrera, Jr., P. A. Marcotte, J. Severin, K. Walter, H. Smith, E. Buggins, R. Simmer, T. F. Holzman, D. W. Morgan, S. K. Davidsen, J. B. Summers, and S. W. Fesik, *J. Am. Chem. Soc.* **119**, 5818 (1997).
[33] S. B. Shuker, P. J. Hajduk, R. P. Meadows, and S. W. Fesik, *Science* **274**, 1531 (1996).
[34] D. J. Maly, I. C. Choong, and J. A. Ellman, *Proc. Natl. Acad. Sci. U.S.A.* **97**, 2419 (2000).
[35] C. Dalvit, P. Pevarello, M. Tato, A. Vulpetti, and M. Sundstrom, *J. Biomol. NMR,* in press (2000).

α protons, exchange-transferred saturation/inversion from water to labile protein protons, and direct intermolecular NOE between the water molecules and the protein. Provided the water has a long T_1, saturated or inverted bulk water acts as a vast reservoir of nonequilibrium magnetization available to the protein target. Depending on the amount of water present, radiation-damping considerations come into effect and care must taken to use selective pulses that do not involve long periods of transverse water magnetization.

Diffusion Methods using Pulsed Field Gradients

As mentioned earlier, the translational mobilities of low molecular weight compounds are much greater than those of the protein target. Accordingly, their translational diffusion coefficients are as much as an order of magnitude larger. However, compounds that bind to the protein will transiently adopt the slow diffusion characteristics of the protein. In the fast exchange limit, the exchange-averaged coefficients for the binding ligands are shown in Eq. (7). Because $D_b \ll D_f$, the ligand diffusion coefficient experiences an apparent decrease. Accordingly, diffusion filtered experiments may be used to screen for binders.

A popular diffusion filter is the LED version of the Tanner stimulated echo.[20,21] The use of bipolar gradients to phase encode and decode the spatial locations of the protein and compound molecules reduces artifacts from exchange and eddy–current recovery times.[36–38] An advantage of the stimulated echo experiment is that the magnetization is placed along the z axis during the potentially long diffusion delay (e.g., 50–500 ms), thus avoiding the possibility of complete signal loss due to rapid transverse relaxation. However, the obligatory toll paid for avoiding the transverse plane is a twofold loss in signal.

In contrast with relaxation rates and NOEs, which are amplified in the bound state, the bound-state diffusion coefficient is *smaller* than the free-state coefficient. In Eq. (7), perturbations to the apparent diffusion coefficient enter as the product of P_b and D_b. If P_b is small because of low concentrations of protein, the binding-induced perturbation may escape the limits of detection. To avoid this situation, the ligand-to-protein ratio should be $\approx <2$. Such ratios, however, make suppression of the protein signal crucial. This can be achieved by the R_2 relaxation filtering built in to the stimulated echo sequence as a consequence of the spin–echo periods. If such periods are unable to provide suppression, then difference spectra need to be recorded in which the protein signal can be subtracted out.[9] A third alternative is to turn the diffusion delay into a relaxation filter itself. In our experience, a TOCSY spin lock bracketed by bipolar phase encode and decode gradients works well.[8,39]

[36] G. Wider, V. Dötsch, and K. Wüthrich, *J. Magn. Reson. Ser. A* **108**, 255 (1994).
[37] V. Dötsch, G. Wider, and K. Wüthrich, *J. Magn. Reson. A* **109**, 263 (1994).
[38] A. Chen, C. S. Johnson, M. Lin, and M. J. Shapiro, *J. Am. Chem. Soc.* **120**, 9094 (1998).
[39] N. Birlirakis and E. Guittet, *J. Am. Chem. Soc.* **118**, 13083 (1996).

Given the above considerations, diffusion-filtered methods are not sensitive probes of binding when dealing with an excess of ligand over protein. On the other hand, diffusion measurements may be more useful as "titrationless" techniques for estimating ligand K_D.[8,40] In order to calculate K_D, the parameters D_{obs}, D_f, and D_b must be determined. An advantage of NMR diffusion measurements is that the bound-state parameters can be measured easily by measuring the diffusion coefficient of the protein itself, whereas the analogous relaxation measurements are either not feasible or impractical. D_f can be obtained directly from measurements for the free ligand. Provided one can detect significant changes in D_{obs}, then P_b can be determined with sufficient accuracy using Eq. (7). Equation (2) may then be inverted to obtain K_D. A significant advantage of this method is that it obviates the need for ligand titrations and the associated problems.[8] To avoid problems associated with protein background, ^{19}F NMR may also offer significant advantages.

1D ^{19}F NMR Methods

Given the ubiquity of protons and the low levels of isotopes at natural abundance, ligand-based NMR screening methods have mainly focused on ^{1}H NMR. However, many pharmaceutically interesting compounds also have fluorine moieties (e.g., fluorinated aromatics or trifluoromethyl groups). Fluorine (^{19}F) is also a spin-1/2 nucleus at 100% natural abundance. Its gyromagnetic ratio is about 0.94 that of proton. [19]F observation holds several key advantages over [1]H. First, because there is no background signal from the protein, additional pulse sequence elements such as relaxation filters are not necessary. Second, the chemical shift range of fluorine is much larger than for ^{1}H, ≈ 400 ppm, thus leading to well-separated lines.[41] Third, the chemical shift is much more sensitive to local environmental effects than for [1]H. This translates not only into larger binding-induced chemical shift perturbations, but also more extensive line broadening (R_2 relaxation enhancement) due to significant R_{ex} contributions [cf. Eq. (8)]. In particular, for a given ligand, the chemical shift differences between the free and bound states are likely to be much greater for the fluorines than for the protons. Thus, in contrast to ^{1}H, even small P_b values can have significant R_{ex} terms because of large ^{19}F shift differences $|\delta_f - \delta_b|$. Another possibility is that the binding exchange could be fast with respect to the proton shifts, but intermediate or slow with respect to the fluorine shifts. Because of the unique chemical shift properties of ^{19}F, a simple one-pulse 1D spectrum of a fluorinated compound in the presence and absence of a biomolecular target constitutes a sufficient diagnostic for binding. These same properties also make ^{19}F an attractive candidate for the aforementioned "follow-up" experiments

[40] D. J. Detlefsen, F. Xu, S. E. Hill, and M. E. Hail, *Am. Pharm. Rev.* **3,** 43 (2000).
[41] J. T. Gerig, *Prog. Nucl. Mag. Reson. Spectrosc.* **26,** 293 (1994).

aimed at estimating exchange rates and binding dissociation constants. The former can be addressed by a variety of transverse and rotating-frame relaxation experiments, while diffusion experiments can address the latter (J. Peng, unpublished results, 2000).

The obvious disadvantage of ^{19}F NMR is clearly that ^{19}F is not as ubiquitous in screening libraries as ^{1}H. However, acknowledging the advantages described above, it becomes reasonable to contemplate screening libraries biased toward fluorine. Work along these lines has already begun (L. Poppe, personal communication, 2000).

Protein-Based Pulse Methods to Screen for Binders to Small Proteins

Protein-based methods follow protein resonances to probe for binding. The motivation behind these methods is not only to detect small molecules that bind to a target, but also to identify the site or sites of interactions. These types of studies require previous knowledge of the sequential resonance assignments for the protein target, and either a three-dimensional structure of the target or a suitable model. Carrying out these studies in a cost-effective manner requires high expression levels of the target protein in bacterial cells in order to achieve uniformly ^{15}N,^{13}C- or ^{15}N,^{13}C,^{2}H-labeled protein, folowed by a battery of triple-resonance 2D and 3D NMR experiments for assignment and structure determination. At present, backbone resonance assignments are currently feasible for proteins with $M_r < 30,000$. Although this size seems large from a protein NMR perspective, it constitutes only a very small subset of proteins of biological or therapeutic importance.

Protein-based screening has been developed most extensively by the group at Abbott Laboratories.[13,32,33,42–45] In their approach, a given protein target is first assigned using the arsenal of standard techniques described above. Afterward the core screening experiment is a heteronuclear 2D spectrum, more frequently a backbone amide ^{15}N–^{1}H HSQC.[46] One compares the HSQC in the presence and absence of a mixture of compounds. Changes in the amide ^{15}N and/or ^{1}H chemical shifts induced by the presence of the compounds reveals binding. Using

[42] P. J. Hajduk, J. Dinges, G. F. Miknis, M. Merlock, T. Middleton, D. J. Kempf, D. A. Egan, K. A. Walter, T. S. Robins, S. B. Shuker, T. F. Holzman, and S. W. Fesik, *J. Med. Chem.* **40**, 3144 (1997).

[43] P. J. Hajduk, M.-M. Zhou, and S. W. Fesik, *Biorg. Med. Chem. Lett.* **9**, 2403 (1999).

[44] P. J. Hajduk, J. Dinges, J. M. Schkeryantz, D. Janowick, M. Kaminski, M. Tufano, D. J. Augeri, A. Petros, V. Nienaber, P. Zhong, R. Hammond, M. Coen, B. Beutel, L. Katz, and S. W. Fesik, *J. Med. Chem.* **42** (1999).

[45] E. T. Olejniczak, P. J. Hajduk, P. A. Marcotte, D. G. Nettesheim, R. P. Meadows, R. Edalji, T. F. Holzman, and S. W. Fesik, *J. Am. Chem. Soc.* **119**, 5828 (1997).

[46] G. Bodenhausen and D. J. Ruben, *Chem. Phys. Lett.* **69**, 185 (1980).

the assignments, the perturbed shifts can be correlated with the primary sequence and thus the binding site(s) can be localized. Working together with binding assays and structural studies, subsequent rounds of NMR screening can suggest linking strategies to improve binding specificity and affinity. The ability to observe the protein also suggests that dissociation binding constants may be estimated by monitoring the shift changes under a ligand titration. In the case of targets with multiple cooperative binding sites, this may be particularly useful.

Of primary interest in this approach are 2D heteronuclear correlation pulse sequences that offer the maximum sensitivity for samples often containing as little as 100 μM protein.[47] If the sample is deuterated and one is working at high fields, TROSY techniques[48] with PEP sensitivity enhancement[49,50] may offer some improvements. However, the largest gains appear to be found not in pulse-sequence development, but via improvements in instrumentation. It has been demonstrated that use of cryogenically cooled probes in these screening experiments can offer significant improvements in sensitivity and, consequently, throughput. Improvements of ~2.5-fold in sensitivity have enabled screens to be run with less protein and with lower concentrations of small molecules (~0.05 mM protein and ligand).[51] These advantages offer the possibility that mixtures of 100 compounds may be screened simultaneously,[13] although questions regarding the integrity of such large mixtures have made this a somewhat controversial topic.[52]

Problems with this method, beyond the large prerequisite investments in the initial protein assignment, are situations in which ligand binding induces gross changes in the 2D heteronuclear spectrum. For example, proteins may be inherently more plastic in the absence of a ligand. Binding then shifts so many cross peaks that the initial assignments are useless and localizing the binding sites becomes very complex. A solution to this problem is to compare the shift changes for a series of related ligands. This approach can also reveal the coarse "orientation" of the ligand in the binding site if the ligands differ in the appropriate systematic manner.[53]

On the other hand, amide bonds are not the most prevalent moieties in ligand binding sites. Thus, binding may not induce significant shift changes, or it may induce remote changes that are difficult to interpret. Alternative approaches to this problem have been proposed by the Abbott group in which ^{13}C protein methyl resonances are observed, but these methods are still under development.

[47] Or 50 nmol, assuming a 20 kDa protein at a concentration of 100 μM in 0.5 ml.
[48] K. Pervushin, R. Riek, G. Wider, and K. Wüthrich, *Proc. Natl. Acad. Sci. U.S.A.* **94,** 12366 (1997).
[49] L. E. Kay, P. Keifer, and T. Saarinen, *J. Am. Chem. Soc.* **114,** 10663 (1992).
[50] M. Rance, P. J. Loria, and A. G. Palmer III, *J. Magn. Reson.* **136,** 92 (1999).
[51] Or 100 nmol, assuming a 20 kDa protein at a concentration of 50 μM in 0.5 ml.
[52] A. Ross, G. Schlotterbeck, W. Klaus, and H. Senn, *J. Biomol. NMR* **16,** 139 (2000).
[53] A. Medek, P. J. Hajduk, J. Mack, and S. W. Fesik, *J. Am. Chem. Soc.* **122,** 1241 (2000).

Design of Compound Libraries for NMR Screening

Most of the principles for constructing libraries of compounds for NMR screening are also relevant for more conventional screening methods. For the purposes of NMR screening, the goal of library design is to simultaneously optimize a large number of factors relating to diversity, solubility, cost, synthetic accessibility, and NMR spectral parameters. The strength of such an approach is that a set of simple molecules provides an array of binding information for almost any target. These molecules and their derivatives translate into hits whose chemistry tends to be well understood and whose physical properties tend to be desirable for lead compounds.

Druglike Character

An approach we have used for designing screening libraries is to bias the selection of compounds toward druglike molecules. This approach is based on the assumption that compounds resembling known drugs are more likely to possess desirable physicochemical properties, such as low toxicity, high oral absorption and permeability, resistance to metabolic degradation, and the absence of rapid excretion.

A very simple rule for choosing druglike compounds was devised by Lipinski et al.[54] According to the "Rule of 5," the majority of orally available compounds possess 5 or fewer hydrogen bond donors, 10 or fewer hydrogen bond acceptors, molecular weight of no more than 500, and a log P of 5 or less. Compound classes that are substrates for natural transporters (e.g., antibiotics, antifungals, vitamins, and cardiac glycosides) are exceptions to this rule. In a similar fashion, Ghose et al.[55] used calculated physicochemical property profiles to determine the preferred ranges for molecular weight, c log P, and other properties.

Another approach uses molecular shapes that occur frequently in known drugs. Shape description methods have been used to show that surprisingly few molecular frameworks describe a large percentage of known therapeutics.[56] The most common drug side chains were identified in a similar manner,[57] and these results suggest that a small number of common, druglike scaffolds and side chains serve as building blocks for a large number of therapeutics over a broad range of indications. At Vertex, our "SHAPES" NMR screening library was designed using commercially available compounds that contain the most common drug scaffolds combined with the most common drug side chains (Fig. 7).[8] Finally, it is desirable to avoid molecules that contain functional groups that are chemically

[54] C. A. Lipinski, F. Lombardo, and B. Dominy, *Adv. Drug Deliv. Rev.* **23**, 3 (1997).
[55] A. K. Ghose, V. N. Viswanadhan, and J. J. Wendoloski, *J. Comb. Chem.* **1**, 55 (1999).
[56] G. W. Bemis and M. A. Murcko, *J. Med. Chem.* **39**, 2887 (1996).
[57] G. W. Bemis and M. A. Murcko, *J. Med. Chem.* **42**, 5095 (1999).

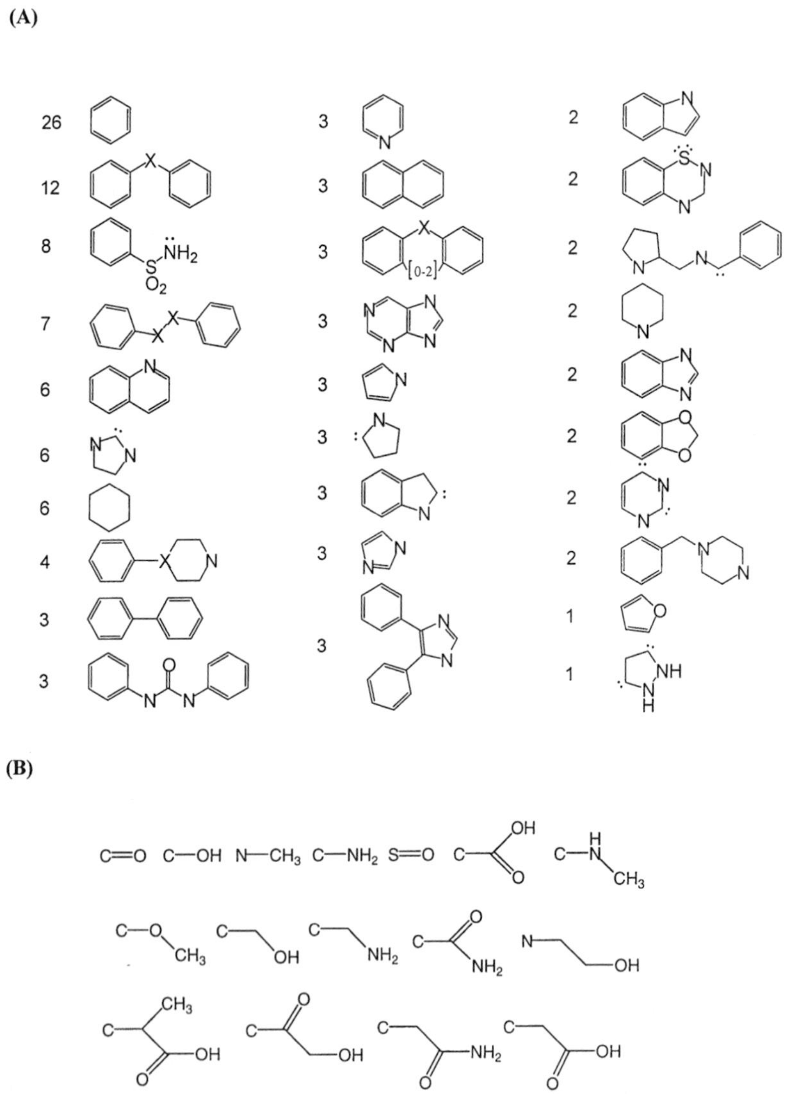

(A)

(B)

FIG. 7. (A) Molecular frameworks used in the SHAPES library. The numbers denote the frequency of occurrence of each framework. Side-chain positions are indicated by lone pairs. X denotes a C, N, O, or S atom. Frameworks representing less common classes of drug scaffolds are not shown. (B) Common drug side chains used in the SHAPES library. The leftmost atom indicates the point of attachment to the framework.

reactive. Rishton[58] presents a list of 25 functional groups that are likely to undergo hydrolysis or react with biological nucleophiles. We apply an automated filter that identifies these and other undesirable functional groups.[59]

NMR-Specific Criteria

NMR screening imposes certain requirements that do not exist for other screening methods. The most stringent requirement is that the compounds must have high aqueous solubility. Reported concentrations for compounds used in screening experiments range from 1–10 mM for most ligand and protein-based methods to 50 μM for experiments using a cryoprobe.[13]

Aqueous solubility of screening compounds is crucial because precipitated or aggregated compounds may produce false positives (spurious hits) in NMR screens based on NOE, relaxation, or diffusion methods, as well as false negatives due to removal of the compounds from solution (C. Lepre, unpublished, 1997). Although solubility problems are less likely to produce false positives in HSQC- or STD-based screens, false negatives can still occur. Solubility and pH control problems have been diagnosed in HSQC-based screens using principal component analysis of the amide chemical shifts.[52] To avoid these problems, we test the solubility of all compounds by diluting concentrated dimethyl sulfoxide (DMSO) stock solutions into NMR buffer at the appropriate pH. In our experience, approximately 10% of the screening compounds are eliminated by this procedure. The practice of diluting DMSO stocks into aqueous buffer can be deceptive, however, since the apparent solubility is largely kinetically driven,[54] and compounds may precipitate or aggregate slowly over many hours. Furthermore, we have found that optically clear solutions often conceal high molecular weight aggregates of compounds (most common for fused aromatic compounds). Even if the individual compounds are highly soluble, chemical interactions between molecules in a mixture (such as ion pairing between positively and negatively charged molecules) could prevent them from binding. This could be particularly problematic for HSQC-based screens using large (100 compound) mixtures,[13] since the ligand resonances are not directly observed. For these reasons, we also collect NOESY spectra of every mixture in the absence of protein to check for NOEs indicative of high molecular weight species, although for larger libraries, this may not be practical. Because solubility varies with pH, we use slightly different libraries for screening at pH 7 and 8. It is worth noting that at least one group has bypassed this problem by screening individual compounds rather than mixtures.[52]

In addition to solubility, there are several other practical considerations. Chemical and isomeric purity of the compounds is important, although more so for

[58] G. M. Rishton, *Drug. Discov. Today* **2**, 382 (1997).
[59] W. P. Walters, M. T. Stahl, and M. A. Murcko, *Drug. Discov. Today* **3**, 160 (1998).

HSQC-based screens than for methods that detect binding using the ligand resonances. In the latter case, false positives arising from minor species are less of a problem because their resonances can be directly observed. For tNOE-based screens, it is obviously essential that every compound possess at least two protons within 5 Å of one another that give resolved ^{1}H resonances. Lastly, the compounds must not react with biomolecules or one another.

Synthetic Accessibility

When selecting compounds for NMR screening libraries, it is essential to consider the needs of the end users of the information that will be obtained. In our experience, it is not enough for compounds to be druglike; unless they are also easy to modify, the synthesis of analogs is unlikely to be attempted. Particularly in cases where NMR and HTS screening are run against the same target, relatively weakly binding NMR hits will elicit little interest for follow-up chemistry unless they are synthetically accessible.

In order to make the leads produced by our screening library both druglike and more synthetically accessible, we expanded the SHAPES library using compounds containing linkers and side chains that are amenable to high-yield, parallel chemistry. This strategy is similar to the "combinatorial target-guided ligand assembly" method described by Ellman and co-workers,[34] in which a library of compounds containing different binding elements but common linking groups is screened against a biological target at high (mM) concentration. From those compounds that bind to the target, a combinatorial library is prepared in which the binding elements are connected via the common linking groups using flexible linkers of varying lengths. The combinatorial library is then screened to identify the most tightly binding ligands.

The compounds in our "linking library" contain either the scaffolds used in the original SHAPES library or proprietary scaffolds, along with either a preferred linker or a side chain. Therefore, any hits produced by this library are highly amenable to follow-up using conventional chemistry or combinatorial methods. If the structure of the active site is known, computational methods such as virtual screening[59] can be used, e.g., to dock virtual compound libraries to the target, and to guide the design of additional follow-up compounds for synthesis.

Practical Aspects

Prescreening Protein Stability

The NMR screening process exposes the target molecule to what most biologists would consider harsh conditions. Typically, small aliquots of concentrated ligand stocks in deuterated dimethyl sulfoxide (DMSO) solution are added to achieve the desired concentration, e.g., 1 mM. With this protocol, DMSO concentrations may be in the range of 1% or more, depending on how many ligands are used in a mixture. Furthermore, adding acidic or basic ligand at 1 mM concentrations can significantly alter the pH of the sample. Finally, the NMR samples are placed at room temperature for significant periods of time, both in the sample changer rack (or sample mixing apparatus) and in the magnet. For these reasons, it is important to examine target stability and activity as a function of organic solvent concentration, pH (buffer type and buffer concentration), temperature, and time before a screen is executed. These preliminary experiments will ensure that protein losses are minimal, and that the protein under these experimental conditions will represent a viable target for lead generation.

For most targets, we routinely employ the microdrop protein screening method,[60] which uses submilligram quantities of protein to optimize sample conditions before NMR screens are carried out. To achieve critical pH control, it is necessary to first measure pH for a proposed set of ligand mixtures in aqueous buffer in the absence of the target. We have found empirically that use of buffer concentrations much higher than needed to maintain the pH of the protein (e.g., 100 mM) are required to maintain pH of the solution in the presence of a mixture of 1 mM organic acids or bases.

Sample Preparation

NMR screening may be carried out in standard 5 mm sample tubes or may be mixed immediately prior to spectroscopic measurement using a liquid handling apparatus such as a Gilson 215, followed by direct injection in a NMR flow probe. Several NMR vendors market these setups. The use of flow techniques is the subject of another review in this volume and so will not be discussed in detail here. However, since many groups appear to be adopting these method,[52] it is clear that in these settings ease of use outweighs the lower sensitivity of current flow probes versus conventional NMR probes.

Another alternative is to use an automated liquid handler to prepare solutions and inject them into 5 mm NMR tubes, then use a sample changing robot to collect the appropriate NMR experiments for each sample. We currently use this method in our laboratory for preparing samples and screening relatively small libraries of compounds. Turnkey systems for NMR sample tube preparation for biological systems are currently unavailable; however, such systems are straightforward to set up using commercially available programmable liquid handlers with custom racks designed to accommodate NMR tubes.

Optimal Mixture Design

When screening a library of compounds by NMR, is it better to screen mixtures or individual compounds? This topic has been widely discussed in the field.

---

[60] C. A. Lepre and J. M. Moore, *J. Biomol. NMR* **12,** 493 (1998).

Screening mixtures of compounds provides an advantage in throughput; however, there are many factors that can complicate these experiments.[8] Clearly, a trade-off exists between conservation of limited resources (protein and instrument time) and ease of data analysis and reliability. Unfortunately, little has appeared in the literature regarding optimal mixture design.[8,52]

Based on their hit rate, Ross *et al.*[52] have suggested that the optimal number of compounds in mixtures (for their screening library) was three. Using this number of compounds, deconvolution would be minimized. However, rather than screening in mixtures of three, these investigators felt the ease of analysis and absence of a deconvolution step compensated for the additional resources required, and opted to screen single compounds. On the other extreme, to demonstrate a potential high-throughput implementation of SAR by NMR, the Abbott group used a mixture of 100 compounds.[13] Using as large a mixture as possible makes sense with protein-based methods, since the ligands are not observed (or with a ligand-based method such as STD, since only binding ligands are observed). However, this approach is somewhat controversial, as no work has appeared in the literature to carefully examine the effects of using such a large number of compounds together in a single sample. It is highly likely that within a mixture of 100 randomly chosen compounds, several components will react or interact in some manner. These interactions will result in "false negatives," which may go undetected since the HSQC-based method observes protein rather than ligand signals. Even if such large mixtures could be designed to eliminate compound interactions, the deconvolution process involved with such a task is extremely daunting. However, given the possibility of screening 200,000 compounds/month,[13] some investigators might be willing accept a very high incidence of false negatives to achieve this level of throughput.

For smaller libraries, mixtures of 4–10 compounds are common.[8,9,32,33,42,43] In our experience, using ligand-based methods, design of noninteracting mixtures with nonoverlapping spectral components is straightforward. Details of procedures used to design such mixtures are given in Fejzo *et al.*[8]

Protein Requirements and Sensitivity

As illustrated by Eqs. (2)–(8), the sensitivities of the various experimental NMR methods discussed in this review are all highly dependent on the percent of ligand bound, P_b, and thus protein concentration. However, the protein and ligand concentrations necessary to provide satisfactory results for both protein-based and ligand-based methods vary considerably, as do the detection limits for binding. Some typical parameters used in these experiments are shown in Table I. It is important to note that the guidelines in Table I are given as approximate values and can vary significantly depending on the molecular mass of protein. Parameters shown in Table I assume a protein of molecular mass 20 kDa for

TABLE I

DETECTION LIMITS, LIGAND : PROTEIN RATIOS, MINIMUM PROTEIN CONCENTRATIONS, AND ESTIMATED
PROTEIN CONSUMPTION FOR VARIOUS APPROACHES USED IN NMR SCREENING[a]

Method	Range of affinities detected	Ligand : protein ratio	Minimum protein concentration (μM)	Protein consumption per 100 compounds (mg)
Protein-based HSQC	$< nM–mM$	1–2	300	30
Ligand-based Relaxation/LB	$< nM–\mu M$	1–10	50	25
Diffusion	μM	< 2	100	50
2D transferred NOE	$\mu M–mM$	5–50	20	10
NOE pumping	$\mu M–mM$	50	20	10
Saturation transfer difference (STD)	$\mu M–mM$	10,000	0.1	0.05

[a] The values for range of affinities detected represents ranges detected under the given conditions. Values for typical ligand : protein concentrations and minimum protein concentrations come from the literature references indicated or, if no reference is given, reflect conditions used in our own laboratory. For HSQC-based methods, protein consumption is calculated assuming a molecular mass of 20 kDa, 10 compounds per mixture, and 500 μl sample volumes. For the ligand-based methods, protein consumption is calculated assuming a molecular mass of 40 kDa, 4 compounds per mixture, and 500 μl sample volumes.

HSQC-based methods, while a molecular mass of 40 kDa is used for all ligand-based methods. Furthermore, the "Range of affinities detected" is meant to indicate the range of affinities that may be detected under the given conditions. For example, HSQC-based techniques are most useful for detecting binders in the micromolar to millimolar range; however, under the conditions depicted, nanomolar or even tighter binders may be detected. This is because these methods observe *protein* resonances. So long as the fraction of bound protein molecules remains significant, protein resonances will be perturbed and binders will be revealed. On the other hand, while STD methods are capable of detecting ligands that bind in the high nanomolar range, *under the conditions depicted,* STD and other methods that employ large excesses of ligands over protein cannot reliably detect nanomolar affinity binders. For such high-affinity binders, k_{ex} is comparatively small, and the efficiency with which bound state information is transferred to the free ligand excess is significantly reduced. In such cases, the free ligand signal contains little or no bound-state information and the large ligand excess serves only to obscure the bound-state signals.

Suitable choices of the ligand : protein ratio should also take into account the molecular weight of the target. As stated, ligand-based methods exploit NMR parameters that increase with molecular weight. Thus, for the same ligand affinity, higher ligand: protein ratios can be used when screening larger targets than for smaller targets. For example, for ligands that bind to a target of $\sim$42 kDa with binding affinity 10 $\mu M < K_D <$ 1 mM, one must keep the ligand: protein ratio less than 10 : 1 in order to reliably observe line broadening. However, for binders to a target such as IMP dehydrogenase (molecular mass of 224 kDa), larger excesses can be used to detect weaker millimolar binders. In this regard, larger proteins represent the best targets for ligand-based NMR screening methods.

The choice of ligand: protein ratio is also affected by the type of NMR experiment chosen. For example, when 2D transferred NOESY is used, discernible binding effects are observed for ligand: protein ratios as large as 50 : 1. At larger excesses of ligand (ligand: protein ratio > 100) P_b becomes very small, and the free state NOE may exceed the bound state NOE in Eq. (6). In contrast, STD methods provide for a longer exchange period, and thus, a longer period for the exchange-transferred signal to accrue. Thus, STD experiments have successfully detected binding even for ligand: protein ratios in the range of 100–10,000. These high ratios will occur in situations where one uses higher ligand concentrations (on the order of 1 mM) and how concentrations of protein ($\sim$100 nM) to tune for higher affinity binders (Fig. 6).

If methods are chosen that use significant amounts of protein, one can opt to recycle the material providing no deterioration of the protein has been observed during the course of the NMR measurements; however, we have never found this necessary. Developments in NMR hardware have made it possible to reduce protein consumption significantly. These improvements, such as "tubeless" NMR[61] and cryoprobes,[62] are discussed below.

Automation and Throughput

One attraction of NMR-based screening is the flexibility of available methods to detect small molecule binders of varying affinities, from low micromolar to several millimolar. To detect compounds that bind more weakly, a small, diverse library can provide an adequate sampling of druglike compound space.[8] However, if one chooses to screen only for high-affinity binders (e.g., $K_D <$ 100 μM), the experiments can easily be tuned to detect only these binders by using lower protein concentrations. Subsequently, larger libraries are necessary to achieve satisfactory hit rates.[13] Many of the standard experiments described in this review may be implemented in a high throughput manner using vendor supplied hardware and software.[8,13,52] Standard methods using a robotic sample changer are straightforward, but are limited by the capacity of the autosampler (e.g., 120 tubes

[61] P. A. Keifer, *Drug. Discov. Today* **2**, 468 (1997).
[62] M. E. Busse-Grawitz and W. Roeck, "NMR Probe Head with Cryogenically Cooled Preamplifiers." U.S. Patent No. 5,814,992 (1998).

for our system). Screening more than 120 tubes requires halting the run and manually loading more sample tubes, then reprogramming the autosampler software accordingly. This approach would become very tedious for large libraries.

To bypass the above limitations, many groups are turning toward more promising technologies such as "tubeless" NMR and cryoprobe NMR. Tubeless NMR incorporates a commercially available liquid handling apparatus with an LC-NMR probe.[52,61] These systems allow samples to be prepared and transferred to the LC-NMR probe immediately prior to measurement, although this creates a risk of clogging if the protein or compounds precipitate. Despite the ease of sample preparation and transfer, LC-NMR probes currently suffer from lower sensitivity than standard probes, and thus place a much greater demand on available instrument time. Use of these systems to screen very large libraries of compounds at present is impractical.

One solution to the sensitivity problem lies in the emerging technology of cryoprobes.[62] These cryogenically cooled probe/preamplifier setups are now available from several vendors, albeit at a significant cost. Since cryoprobes are approximately 2.5-fold more sensitive than standard 5 mm probes,[63] they potentially provide a 6.5-fold increase in throughput when screening at a given concentration of protein and ligand, or provide the ability to screen at 2.5-fold lower concentrations of protein or ligand in the same amount of time. Reducing the amount of protein has obvious benefits, but reducing both protein and ligand concentrations allows one to detect higher affinity binders, as well as use large numbers of ligands in a single mixture.[13] Although the caveats of using such large mixtures have already been discussed, it is clear that the increased sensitivity available from the cryoprobe platform provides significant advantages regardless of the screening strategy used if the additional costs can be justified. Further technology advances incorporating the sensitivity of cryoprobes with the efficiency and ease of use of tubeless NMR systems are currently under development and should provide an even greater boost to throughput and productivity as they become available.

Conclusions

This review has provided an overview of the many experimental approaches used in NMR-based screening in drug discovery. Clearly, there is no "best" approach, and the approach selected will depend on many factors, including the molecular mass of the target of interest, the available instrumentation and resources, and the expected time frame for results. The shift in focus away from

[63] Although sensitivity for samples in organic solvent is fourfold higher than for a standard 5 mm probe, sensitivity with high dielectric constant samples used in biological NMR is closer to 2.5-fold higher.

[64] M. A. McCoy and L. Mueller, *J. Am. Chem. Soc.* **114**, 2108 (1992).

structural characterization and in the direction of lead discovery has led to a resurgence in the field of pharmaceutical NMR. As more groups become involved and technologies advance, NMR-generated data will become increasingly critical in accelerating the drug discovery process. Because of its ability to detect binding regardless of target function, NMR may be used very early in the drug discovery process to assess the drug design potential for targets before conventional assays need to be developed. These methods should prove highly valuable even for targets of unknown function generated from genomics initiatives.[13]

[9] Screening of Compound Libraries for Protein Binding Using Flow-Injection Nuclear Magnetic Resonance Spectroscopy

By BRIAN J. STOCKMAN, KATHLEEN A. FARLEY, and DANEEN T. ANGWIN

Introduction

Flow nuclear magnetic resonance (NMR) spectroscopy techniques are becoming increasingly utilized in drug discovery and development.[1] The technique was first applied to couple the separation characteristics of liquid chromatography (LC) with the analytical capabilities of NMR spectroscopy.[2] Since then, high-performance liquid chromatography (HPLC)-NMR, or LC-NMR as it is more commonly referred to, has been broadly applied to natural products biochemistry, drug metabolism and drug toxicology studies.[3-6] The wealth and complexity of data made available from the latter two applications have created the potential for NMR-based metabonomics to complement genomics and proteomics.[7] Stopped-flow analysis in LC-NMR, where the chromatographic flow is halted to obtain an NMR spectrum with higher signal-to-noise and then restarted when the spectrum has finished collecting, was the forerunner to the flow-injection systems that will be described here. The largest difference between the two systems is that one includes a separation component (LC column) and the other does not. The

[1] B. J. Stockman, *Curr. Opin. Drug Disc. Dev.* **3,** 269 (2000).

[2] N. Watanabe and E. Nike, *Proc. Jpn. Acad. Ser B* **54,** 194 (1978).

[3] J. C. Lindon, J. K. Nicholson, and I. D. Wilson, *Prog. NMR Spectr.* **29,** 1 (1996).

[4] J. C. Lindon, J. K. Nicholson, U. G. Sidelmann, and I. D. Wilson, *Drug. Met. Rev.* **29,** 705 (1997).

[5] B. Vogler, I. Klaiber, G. Roos, C. U. Walter, W. Hiller, P. Sandor, and W. Kraus, *J. Nat. Prod.* **61,** 175 (1998).

[6] J.-L. Wolfender, K. Ndjoko, and K. Hostettmann, *Curr. Org. Chem.* **2,** 575 (1998).

[7] J. K. Nicholson, J. C. Lindon, and E. Holmes, *Xenobiotica* **29,** 1181 (1999).

rapid throughput possible for combinatorial chemistry samples and protein/small molecule mixtures has allowed flow-injection NMR methods to impact medicinal chemistry and protein screening.[8-12]

Changes in chemical shifts, relaxation properties, or diffusion coefficients that occur on the interaction between a protein and a small molecule have been documented for many years (for recent reviews see Refs. 13–15). Observables typically used to detect or monitor the interactions are chemical shift changes for the ligand or isotopically enriched protein resonances,[16] or line broadening,[17,18] change in sign of the nuclear overhauser effect (NOE) from positive to negative,[19,20] or restricted diffusion[21] for the ligand. For the most part, these studies have focused on protein/ligand systems where the small molecule was already known to be a ligand or was assumed to be one. In the last several years, however, the elegant work of the Fesik,[22,23] Meyer,[24] Moore,[25] and Shapiro[26] laboratories has demonstrated the applicability of these same general methods as a screening tool to identify ligands from mixtures of small molecules.

These screening protocols typically involve the preparation of a series of individual samples in glass NMR tubes and the use of an autosampler to achieve reasonable throughput. Variations in volume or positioning that occur during sample preparation or tube insertion can necessitate tuning and calibration of the probe between each sample, thereby reducing throughput of data collection.

By contrast, flow-injection NMR has several advantages. The stationary flow cell provides uniform locking and shimming from one sample to the next, and, with the radio frequency coils mounted directly onto the glass surface of the flow

[8] P. A. Keifer, *Drugs Fut.* **23,** 301 (1998).

[9] P. A. Keifer, *Drug Disc. Today* **2,** 468 (1997).

[10] P. A. Keifer, *Curr. Opin. Biotech.* **10,** 34 (1999).

[11] K. A. Farley, C. A. Schering, E. A. Steinbrecher, C. Zhou, and B. J. Stockman, SMASH'99, Argonne, IL, 15–18 August 1999.

[12] A. Ross, G. Schlotterbeck, W. Klaus, and H. Senn, *J. Biomol. NMR* **16,** 139 (2000).

[13] M. J. Shapiro and J. R. Wareing, *Curr. Opin. Drug. Disc. Dev.* **2,** 396 (1999).

[14] J. M. Moore, *Biopolymers* **51,** 221 (1999).

[15] B. J. Stockman, *Prog. NMR Spectr.* **33,** 109 (1998).

[16] J. Wang, A. P. Hinck, S. N. Loh, D. M. LeMaster, and J. L. Markley, *Biochemistry* **31,** 921 (1992).

[17] D. L. Rabenstein, T. Nakashima, and G. Bigam, *J. Magn. Reson.* **34,** 669 (1979).

[18] T. Scherf and J. Anglister, *Biophys. J.* **64,** 754 (1993).

[19] P. Balaram, A. A. Bothner-By, and E. Breslow, *J. Am. Chem. Soc.* **94,** 4017 (1972).

[20] A. A. Bothner-By and R. Gassend, *Ann. NY Acad. Sci.* **222,** 668 (1972).

[21] A. J. Lennon, N. R. Scott, B. E. Chapman, and P. W. Kuchel, *Biophys. J.* **67,** 2096 (1994).

[22] S. B. Shuker, P. J. Hajduk, R. P. Meadows, and S. W. Fesik, *Science* **274,** 1531 (1996).

[23] P. J. Hajduk, E. T. Olejniczak, and S. W. Fesik, *J. Am. Chem. Soc.* **119,** 12257 (1997).

[24] B. Meyer, T. Weimar, and T. Peters, *Eur. J. Biochem.* **246,** 705 (1997).

[25] J. Fejzo, C. A. Lepre, J. W. Peng, G. W. Bemis, Ajay, M. A. Murcko, and J. M. Moore, *Chem. Biol.* **6,** 755 (1999).

[26] M. Lin, M. J. Shapiro, and J. R. Wareing, *J. Org. Chem.* **62,** 8930 (1997).

cell, high sensitivity. Fast throughput of data collection is thus possible. Use of a liquid handler to prepare and inject samples, such as the Gilson 215 liquid handler used on Bruker and Varian systems, allows the potential for on-the-fly sample preparation,[12] thus maximizing sample integrity and uniformity. Because the use and/or reuse of glass NMR tubes is avoided, costs are minimized.

This review will focus on the use of flow-injection NMR spectroscopy to screen compound libraries for protein binding. Topics to be covered are: data acquisition hardware and software; flow probe calibration and system optimization; design of small molecule screening libraries; relaxation-edited flow-injection NMR screening; data analysis; and comparisons between flow and traditional methods. Practical enhancements based on our own observations and those of others will be discussed.

Data Acquisition Hardware and Software

A typical flow NMR system consists of a magnet, an NMR console, a computer workstation, a Gilson sample handler, and a flow-injection probe. Two vendors currently offer complete flow-injection systems: Bruker Instruments (Rheinstetten) and Varian Instruments (Palo Alto, CA). In addition, the Nalorac Corporation (Martinez, CA) manufactures an LC probe that can also be used for flow-injection NMR screening. A schematic of the Bruker Efficient Transport System (BEST) manufactured by Bruker Instruments is shown in Fig. 1. The Gilson 215 sample handler supplied by Bruker is equipped with two Rheodyne 819 valves. The first valve is attached to a 5 ml syringe, the needle capillary in the sample handler injection arm, the bridge capillary, the waste reservoir, and the second valve. The second Rheodyne valve is attached to the input and output of the probe, the source of nitrogen gas, the first valve, and the injection port. FEP (fluorinated ethylene propylene) Teflon tubing is used in each of the connections with the exception of the gas connection, which uses PEEK (polyetherether ketone) tubing.

A sample is injected into the Bruker probe by filling the needle capillary and transferring the sample into the inlet tubing for the probe using the second Rheodyne valve. In quick mode, the next sample is loaded into the tubing during the spectral acquisition of the previous sample. When the spectral acquisition has completed, the first sample exits the probe through the outlet capillary. This action pulls the next sample into the probe through the inlet port and spectral acquisition can immediately begin. Quick mode acquisition can save approximately 1 min per sample from the time it would take to load each sample individually. However, sample recovery is not currently an option with this method. In order to recover a sample, each sample must be injected individually using normal mode acquisition. The sample is recovered by selecting either nitrogen gas or the syringe to pull the sample back from the probe through the inlet tube. The sample can then be returned to the Gilson liquid handler into its original well or into a new 96-well

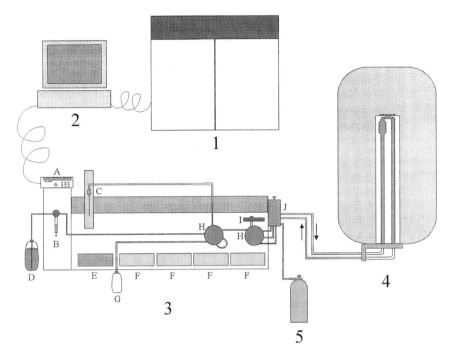

FIG. 1. Schematic of the BEST flow system: 1, NMR console; 2, computer workstation; 3, Gilson sample handler; 4, flow probe in the magnet; and 5, nitrogen gas. The Gilson sample handler is labeled as follows: A, keypad; B, syringe; C, injector; D, solvent reservoir; E, solvent rack; F, sample racks; G, waste reservoir; H, Rheodyne valves; I, injection port; and J, recovery unit.

plate. A recovery unit has been added to the BEST system to improve the efficiency of recovery of the syringe by using the nitrogen gas to create a back pressure on the sample. We currently have only limited experience with this unit.

Two useful accessories available for the BEST system are a Valvemate solvent switcher and a heated transfer line. The solvent switcher was added to the flow system for the combinatorial chemist who may want to analyze samples in various organic solvents, but it can also be used for a library screen to vary buffer conditions or to clean out the probe with an acid or a base. The heated transfer line is used to equilibrate the sample temperature to the probe temperature during sample transfer. Both the inlet and output capillary transfer lines are threaded through the heated transfer line. This feature is desirable when the spectral analysis time is short and a high throughput of samples is required. In the ideal case, data acquisition using this accessory can begin immediately after the sample enters the probe. In our experience, some samples may still require a temperature equilibration period after entering the probe.

The setup of the Versatile Automated Sample Transport (VAST) system produced by Varian is similar to the Bruker system. The VAST system consists of a Gilson 215 liquid handler, a Varian NMR flow probe, an NMR console, and a Sun workstation. The Gilson liquid handler supplied by Varian is equipped with a single Rheodyne 819 valve and is connected to the NMR flow probe with 0.010 inch inside diameter PEEK tubing.[27] In the Varian system design, the sample handler injects a specified volume of sample into the probe, the data are acquired, and then the flow of liquid through the tubing is reversed and the sample is returned to its original vial or well. The return of the sample to the Gilson by the syringe pump is assisted by a Valco valve and nitrogen gas, which supply some back pressure on the outlet portion of the Varian flow probe. With the VAST system setup, the probe is rinsed just prior to sample injection and then is dried with nitrogen gas to minimize dilution of the sample during injection. The Varian design gives excellent sample recovery without dilution, but it is strongly recommended that samples be filtered to prevent clogging of the capillary transfer lines.[27]

Flow NMR systems are ideally suited for use with the shielded magnets manufactured by Bruker Instruments or Oxford Magnets. Actively shielding a 600 MHz magnet reduces the radial 5 gauss line from approximately 4 m to less than 2 m, which allows the Gilson liquid handler to be placed significantly closer to the magnet. This reduces the length of tubing needed between the Rheodyne valve and the flow-injection probe and minimizes the sample transfer time. The potential for clogging and sample dilution are concomitantly reduced.

Bruker uses two software packages to run the BEST system: BEST Administrator and ICONNMR.[28] The BEST administrator is activated by typing the command "BESTADM" in XWINNMR. This portion of the software is used during method generation and optimization. Samples are injected into the probe one at a time and data are collected under XWINNMR. Early versions of the BEST software utilized three separate programs: CFBEST, SUBEST, and OTBEST. These functions were recently combined under the single software package, BEST Administrator. In addition, the parameters available for customization have been greatly expanded to include automated solvent switching and method switching, which were not available in earlier versions of the software. The software package ICONNMR is used after a flow method has been optimized with the BEST administrator. This package is set up for full automation and is the same software used with automated NMR tube sample changers. In a similar fashion, Varian software uses the command "Gilson" to generate a method before sample injection and data acquisition is initiated using Enter/Autogo in VNMR.[29]

[27] P. A. Keifer, S. H. Smallcombe, E. H. Williams, K. E. Salomon, G. Mendez, J. L. Belletire, and C. D. Moore, *J. Comb. Chem.* **2,** 151 (2000).

[28] Bruker Instruments, AMIX, BEST and ICONNMR software packages.

[29] Varian NMR Systems, VNMR software package.

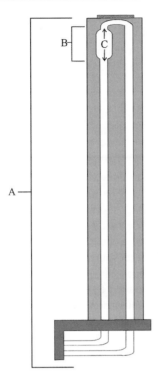

FIG. 2. Schematic of a Bruker flow probe showing (A) the total probe volume, (B) the flow cell volume, and (C) the positioning volume.

Flow Probe Calibration and System Optimization

In addition to the normal 90° pulse lengths and power levels that must be calibrated for any NMR probe, several additional calibrations are required for a flow probe. The three additional volumes required to calibrate a Bruker flow probe are shown schematically in Fig. 2.[28] The first volume calibrated is the total probe volume. This can be accomplished by injecting a colored liquid into the inlet of a dry probe with a syringe and watching for the liquid to appear in the outlet port (approximately 700–800 μl). With the Varian system, the system filling volume also includes the capillary tubing that connects the injector port to the flow probe.[27] This volume is used to calculate the distance required to reposition a sample from the Gilson sample handler to the center of the flow cell in the probe.

The second volume calibrated is the flow cell volume. This is the volume of liquid required to fully fill the coil around the flow cell. The three flow probe vendors (Bruker, Varian, and Nalorac) have probes available with active volumes ranging from 30 to 250 μl. The stated volume of the flow cell in a 5 mm Bruker flow probe is 250 μl, but we calibrated the sample volume for our probe to be approximately 300 μl. This volume can be calibrated by making repeated injections of a standard sample, starting with a volume less than the stated active volume of the probe, and collecting a 1D ^{1}H NMR spectrum. The injection volume can then be increased incrementally until no further improvement in signal-to-noise is observed.

In addition to the two probe volume calibrations already discussed, Bruker software also includes a third volume for calibration. This volume, referred to as the positioning volume, is used to optimize the centering of a sample in the flow cell. Early versions of ICONNMR software (prior to 3.0.a.9) did not include the ability to set the positioning volume. Rather, Bruker literature suggested that the flow cell volume should be roughly doubled to ensure that the sample would completely fill the coil.[28] Fortunately, this is no longer necessary. The positioning volume can now be used to optimize the sample position. This calibration reduced the sample size required for injection from 450 μl in our first few protein screens to 300 μl for our current screens using a Bruker 5 mm flow probe with an active volume of 250 μl. Optimization of this parameter minimized the sample volume required for each spectrum. Importantly, this significantly reduced the total amount of protein (or other target) at a given concentration needed to screen our small-molecule library. The positioning volume can be optimized by collecting a series of spectra on a standard sample. In each spectrum collected, the positioning volume can first be varied by large increments (50–100 μl) to get a rough estimate of the volume. An example of three such spectra is shown in Fig. 3. The positioning volume can then be varied in smaller increments (10–25 μl) to identify the best volume for this parameter. The best signal-to-noise was obtained for our 5 mm Bruker flow probe on a DRX-600 when the positioning volume was set to +25 μl, but this volume is probe specific and must be calibrated for each flow probe.

The optimization of a flow-injection system for screening has three main objectives. The first objective is to transfer an aqueous sample to the center of the flow cell for analysis using the parameters determined during the flow probe calibration described above. The second objective is to reposition a sample from the Gilson liquid handler into the flow-injection probe without bubbles and with minimal sample dilution. This can be achieved by using nitrogen as a transfer gas (which keeps the system under pressure) and by using a series of leading and trailing solvents. In our experiments, we typically use 150 μl of ^{2}H$_2$O as a leading solvent, 20 μl of nitrogen gas, 300 μl of sample, 20 μl of nitrogen gas, and 100 μl of ^{2}H$_2$O as a trailing solvent. Alternatively, a larger volume of sample can be used in place of the push solvents. The third objective is to determine a cleaning procedure that would reduce sample carryover to less than 0.1%. Typically, this involves rinsing the probe with a predetermined volume of water. The rinse cycle can also be followed by a dry cycle, in which the capillary lines and flow probe are dried with

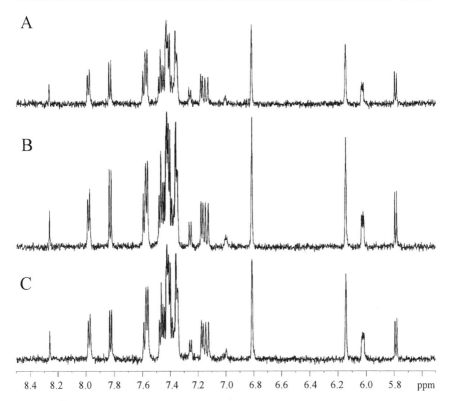

FIG. 3. ^{1}H NMR spectra, 600.13 MHz, of a 100 μM NMR library sample with the positioning volume set to (A) -100 μl, (B) 0 μl, and (C) 100 μl.

nitrogen gas to further minimize sample dilution. In our experiments, we typically use a 1 ml wash volume followed by a 30-s drying time with nitrogen gas.

Design of Small Molecule Screening Libraries

With the increasing prevalence of extremely high throughput screening equipment in the pharmaceutical industry, it may seem counterintuitive to suggest screening smaller collections of compounds in an NMR-based assay. However, a correlation between the quality of hits obtained and the number of compounds screened has not been well documented. In fact, compounds are typically added to screening collections not simply to increase their numbers, but to increase the diversity and quality of the compound collection. Thus, if one could find suitable hits from a smaller collection of well-chosen compounds, it may not be necessary to expend the time and chemical resources to screen the entire compound library

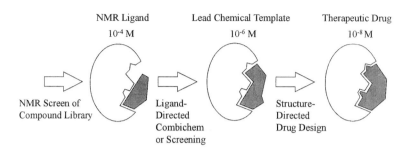

NMR Ligand   Lead Chemical Template   Therapeutic Drug
10^{-4} M 10^{-6} M 10^{-8} M

NMR Screen of   Ligand-   Structure-
Compound Library   Directed   Directed
                       Combichem   Drug Design
                       or Screening

FIG. 4. Schematic diagram illustrating the use of NMR to discover a $\sim$1.0 $\times$ 10^{-4} M affinity ligand (left), to use the discovered ligand to direct the discovery of a $\sim$1.0 $\times$ 10^{-6} M affinity lead chemical template (middle), and then via synthetic chemistry and structure-directed drug design arrive at the ultimate $\sim$1.0 $\times$ 10^{-8} M affinity drug candidate (right).

against every single target. Hits so identified could then be used to focus further screening efforts or to direct combinatorial syntheses, thus saving both time and chemical resources, as shown schematically in Fig. 4. An NMR-based screen, like other binding assays, has the advantage in that a high throughput functional assay does not need to be developed. This will become increasingly important as more and more targets of interest to pharmaceutical research are derived from genomics efforts and thus may not have a known function that can be assayed. Having said this, developments in cryoprobe technology and improvements in NMR method-ologies may yet increase the sensitivity of NMR-based screening techniques to the point of being competitive with existing technologies.[30] Screening of entire compound collections could then become routine.

Until that time, however, the design of NMR screening libraries is an impor-tant component of the overall process. Several types of libraries are possible: broad screening libraries applicable to many types of target proteins, directed libraries that are designed with the common features of an active site in mind that might be useful for screening a series of targets from the same protein class, such as protease enzymes, and "functional genomics" libraries composed of known sub-strates, cofactors, and inhibitors for a diverse array of enzymes that might be useful for defining the function of genomics-identified targets.

We have focused on developing a broad screening library since our screen-ing targets have included enzymes with diverse functions, known receptors, and proteins of unknown function.[11] Ideally, the size and content of a broad screening library should be such that screening can be accomplished in a day or two with a favorable chance of identifying several hits for each of the target proteins to be

[30] P. J. Hajduk, T. Gerfin, J.-M. Boehlen, M. Häberli, D. Marek, and S. W. Fesik, *J. Med. Chem.* **42,** 2315 (1999).

screened. Rather than just randomly choosing a subset library, several rationale approaches have been implemented. These include the SHAPES library developed by Fejzo and co-workers that is composed largely of molecules that represent frameworks commonly found in known drug molecules,[25] druglike or leadlike libraries, and diversity-based libraries. A number of studies have appeared that discuss the properties of known drugs and methods to distinguish between druglike and nondruglike compounds.[31–37] Superimposing druglike[38] or leadlike[39] properties on a diversity-selected compound set may yield the best library of compounds. Leadlike properties of low molecular weight and low lipophilicity are important considerations since the NMR-based assay is designed to identify weak-affinity compounds that will most likely gain molecular weight and lipophilicity as they evolve into drug candidates or even lead chemical templates.[39]

Development and expansion of our leadlike NMR screening library to mimic the structural diversity of our larger compound collection has made use of the DiverseSolutions software for chemical diversity.[40] In this approach, each compound is described by a set of descriptors referred to as BCUTs, which are metrics of chemistry space. Six orthogonal descriptors, related to substructures as opposed to the entire molecule, are often used. Although the BCUTs to use can be automatically chosen to maximize diversity, typically there are two each corresponding to charge, polarizability, and hydrogen bonding. A cell-based diversity algorithm is employed to divide the descriptor axes into bins and thus into a lattice of multidimensional hypercubes. As an example of how this can be used to construct or expand a small screening library, consider the selection of 1000 compounds from a compound library of 250,000 compounds. First, the cell-based algorithm is used to partition the 250,000 compounds into approximately 1000 cells. The number of compounds per cell will vary and some will be empty. Maximum structural diversity will be obtained by taking one compound from each occupied cell (and as close to the center as possible). The actual compounds chosen are based on desirable leadlike properties such as low molecular weight and hydrophilicity as well as availability and chemical nonreactivity as explained below. Diversity voids, as exemplified by empty cells, can be filled from external sources or by chemical syntheses if desired. Identifying and filling diversity voids is critical since larger

[31] G. W. Bemis and M. A. Murcko, *J. Med. Chem.* **39**, 2887 (1996).
[32] C. A. Lipinski, F. Lombardo, B. W. Dominy, and P. J. Feeney, *Adv. Drug Del. Rev.* **23**, 3 (1997).
[33] Ajay, W. P. Walters, and M. A. Murcko, *J. Med. Chem.* **41**, 3314 (1998).
[34] J. Sadowski and H. Kubinyi, *J. Med. Chem.* **41**, 3325 (1998).
[35] A. K. Ghose, V. N. Viswanadhan, and J. J. Wendoloski, *J. Comb. Chem.* **1**, 55 (1999).
[36] J. Wang and K. Ramnarayan, *J. Comb. Chem.* **1**, 524 (1999).
[37] G. W. Bemis and M. A. Murcko, *J. Med. Chem.* **42**, 5095 (1999).
[38] E. J. Martin and R. E. Critchlow, *J. Comb. Chem.* **1**, 32 (1999).
[39] S. J. Teague, A. M. Davis, P. D. Leeson, and T. Oprea, *Angew. Chem. Int. Ed.* **38**, 3743 (1999).
[40] R. S. Pearlman and K. M. Smith, *Persp. Drug Disc. Des.* **9/10/11**, 339 (1998).

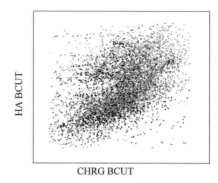

CHRG BCUT

FIG. 5. Overlay of the two-dimensional HA (hydrogen-bond acceptor) vs CHRG (charge) BCUT plots for the compounds in the CMC index (gray) and the leadlike compounds contained therein (black).

compound collections are often heavily weighted in certain classes of compounds stemming from earlier research projects.

An example of diversity-based subset selection using these methods is shown in Fig. 5. Here, the 6436 compounds from the Comprehensive Medicinal Chemistry index[41] have been divided into 2012 cells to maximize diversity using five BCUTs. This five-dimensional chemistry space projected onto its hydrogen bond acceptor (HA) and charge (CHRG) BCUT axes is shown in gray. The black squares correspond to the 1474 leadlike compounds (molecular weight less than 350 and $1 < c \log P < 3$) contained in the CMC index. A total of 806 of the 2012 cells were occupied by leadlike compounds. A similar approach could be used to select diverse, leadlike compounds from a large corporate compound collection.

The cell concept of structural space is quite useful after the screening is complete. When a hit is identified, other compounds from the same or nearby cells are obvious candidates for secondary assays. One can think of this as the gold mine analogy: when gold is struck, the search is best continued in close proximity.

In addition to structural diversity, there are other important characteristics to consider when selecting the subset molecules. These include purity, identity, reactivity, toxicological properties, molecular weight, water solubility, and suitability for chemical elaboration by traditional or combinatorial methods. It makes sense to populate the screening library with compounds of high integrity that are not destined for failure down the road. Time spent up front to ensure purity and identity with LC-MS or LC-NMR analyses will save resources downstream. Filtering tools should be used to avoid compounds that are known to be highly reactive or toxic or to have poor metabolic properties. Lack of reactivity is critical since compounds

---

[41] Integrated Scientific Information System (ISIS); available from MDL Information Systems, Inc., San Leandro, CA (1998).

can be screened more efficiently as mixtures. Like other laboratories[22,24–26] we typically pool our selected small molecules into mixtures of 6–10 compounds for screening.[11]

Compounds chosen for our diversity library are leadlike as opposed to drug-like. It is often the case that chemical elaborations to improve affinity also increase molecular weight and decrease solubility.[39] The molecular weight of the compounds therefore should not exceed 350. Since most hits obtained will have affinities for their target in the ~100 μM range, low molecular weight will leave room for chemical elaboration to build in more affinity and selectivity. Using larger molecular weight druglike compounds would not substantially improve affinity of the hits and could easily preclude obtaining lead chemical templates of reasonable size. Leadlike hits that are reasonably water soluble allow for chemical elaboration that results in modest increased lipophilicity of the final therapeutic entity.[39] Water solubility is also important since it enhances the potential success of downstream studies such as calorimetry, enzymology, cocrystallization, and NMR structural studies. Compound solubility is especially important for flow-injection NMR methods in order to prevent clogging of the capillary lines.

Compounds should also be chosen with their suitability for chemical elaboration by traditional or combinatorial chemistry methods in mind. Hits with facile handles for synthetic chemistry will be of more interest and will allow more efficient use of often limited medicinal chemistry resources. Input from medicinal chemists regarding which compounds to include is critical. From an extreme perspective, favorite compounds suggested by chemists should always be included in the screening library. With hundreds of targets to screen, there is a good chance of finding a target for these so-called orphan compounds.[42]

Relaxation-Edited Flow Injection NMR Screening Methods

Calibration and validation of the flow system and creation of a small-molecule screening library yields an automated system that is ready to screen new targets. A protein target can be analyzed for protein–ligand interactions by adding sufficient protein to each well of the 96-well library plate to give a 1 : 1 (protein : ligand) ratio at a concentration of approximately 50 μM. In our case, the total volume in each well is 350 μl in order to provide a 300 μl injection volume. Homogeneous sample dispersion throughout the well can be facilitated by agitating the plate on a flatbed shaker. Screening at this concentration allows a decent 1D ^{1}H NMR spectrum to be acquired in about 10 min. In our experience, this concentration of target and small molecule requires identified ligands to have affinities on the order of ~200 μM or tighter.

---

[42] M. S. Lajiness, personal communication (1999).

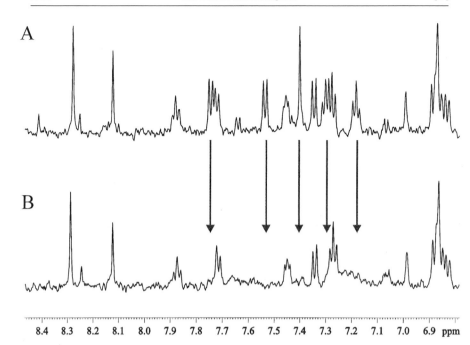

8.4   8.3   8.2   8.1   8.0   7.9   7.8   7.7   7.6   7.5   7.4   7.3   7.2   7.1   7.0   6.9 ppm

FIG. 6. Regions of the 600.13 MHz relaxation-edited ^{1}H NMR spectra of a nine compound mixture (A) without and (B) with added target protein. Protein and each ligand were 50 μM. Spectra were acquired on a Bruker 5 mm flow-injection probe at 27°. A total of 1K scans were collected, resulting in a total acquisition time of about 60 min per spectrum. A relaxation filter of 174 ms was used. Arrows identify resonances that disappear in the presence of protein.

Once the screening plate has been prepared, the Gilson liquid sample handler transfers samples from 96-well plates into the flow-injection probe and, if desired, returns the samples back into either the original 96-well plate or a new plate. Once the sample is in the magnet, spectra that can detect changes in chemical shifts, relaxation properties, or diffusion properties can be collected. Our NMR screening assay comprises two 1D relaxation-edited ^{1}H NMR spectra: one spectrum is collected on the ligand mixture in the presence of protein, and the second, control spectrum is collected on the ligand mixture in the absence of protein. Ligands are identified as binding to a target when their resonances are greatly reduced when compared to a relaxation-edited spectrum collected in the absence of protein, as illustrated in Fig. 6. In this example, the target protein was a genomics-derived protein of unknown function.

Ligand binding can be confirmed by collecting a 1D relaxation-edited ^{1}H NMR spectrum of each individual ligand that was identified as binding to the protein

A

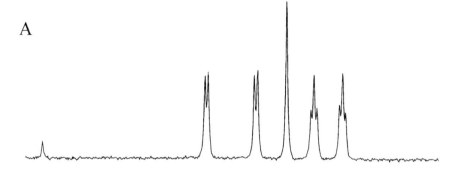

B

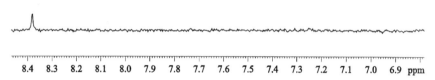

8.4   8.3   8.2   8.1   8.0   7.9   7.8   7.7   7.6   7.5   7.4   7.3   7.2   7.1   7.0   6.9 ppm

FIG. 7. Regions of the 600.13 MHz relaxation-edited ¹H NMR spectra of a single compound (A) without and (B) with added target protein. Protein and ligand were 50 μM. Spectra were acquired on a regular Bruker 5 mm TXI probe at 27°. A total of 512 scans were collected resulting in a total acquisition time of about 30 minutes per spectrum. A relaxation filter of 174 ms was used.

in a given mixture as shown in Fig. 7. In addition, the binding constant of the protein/ligand interaction can be estimated using 1D diffusion-edited spectra of the ligand in the presence and absence of protein.[21] If labeled protein is available, a 2D ¹H–¹⁵N HSQC spectrum can also be obtained to locate the ligand binding site on the protein.[16,22] In cases where the protein is small enough and structural characterization of the binding interaction is desired, further experiments can be carried out using ¹⁵N and/or ¹³C/¹⁵N protein/ligand complexes.

Data Analysis

The development of flow probes has facilitated the transition to high-throughput NMR and has made possible the routine collection of tremendous volumes of data. However, a bottleneck still remains at the data analysis stage.

Software capabilities do exist, in the form of databases and statistical analysis, to aid in the interpretation and data reduction of large screening data sets, although optimal use of these still requires manual manipulation.

Recent software developments have advanced the automated handling of large data sets collected on combinatorial chemistry libraries.[27–29,43] Visualization of results in a 96-well format allows rapid evaluation of the data sets. The integration of features such as this into a software package tailored more for data reduction and evaluation of library screening data sets parallels the combinatorial chemistry software development but remains slightly behind. However, advancements that have been made for combinatorial chemistry data analyses portend similar developments for the automation of protein binding screening data.

In our 1D relaxation-edited ^{1}H NMR data sets, one can simply identify the ligand resonances by inspection since their intensity is reduced in the presence of protein as shown in Fig. 6. Comparison to an assigned small molecule control spectrum must be made to identify the compound associated with the indicated resonances. Although we currently carry out this analysis manually, the use of automated peak picking coupled with spectral databases could facilitate the ligand identification process provided that the resonances of the individual compounds have been previously assigned.

Other laboratories have relied on difference spectra to analyze relaxation- or diffusion-edited 1D ^{1}H NMR data sets.[23,44,45] After a series of spectral subtractions, the resulting spectrum represents the resonances of the compounds that bind to the protein. Two factors that pose problems are line broadening and shifting resonances, both of which can lead to subtraction artifacts. Changes in intensity can also add the need for a scaling factor in the data analysis step. These additional steps, which can vary from one spectra to the next, make strategies for automated data analysis complex.

Data analysis for 2D screening methods typically involves either the analysis of protein chemical shift perturbations indicative of ligand binding,[12,22] or the analysis of changes in signals from the small molecules in NOE or DECODES spectra indicative of binding.[24,25,46] Although a series of 2D ^{1}H–^{15}N HSQC spectra can be compared manually, automated analysis using both nonstatistical and statistical approaches of a series of ^{1}H–^{15}N HSQC spectra acquired with flow-injection NMR methods has been demonstrated.[12] AMIX was used for the nonstatistical analysis by comparing spectra collected in the presence of single compounds to the reference spectrum of the protein alone. Bucketing calculations, which divide spectra into small regions (buckets) and extract the information content, were used

[43] A. Williams, Book of Abstracts, 218th ACS National Meeting (1999).
[44] N. Gonnella, M. Lin, M. J. Shapiro, J. R. Wareing, and X. Zhang, *J. Magn. Reson.* **131**, 336 (1998).
[45] A. Chen and M. J. Shapiro, *J. Am. Chem. Soc.* **122**, 414 (2000).
[46] M. Lin, M. J. Shapiro, and J. R. Wareing, *J. Am. Chem. Soc.* **119**, 5249 (1997).

for data reduction. A table ranked by the correlation coefficients was then generated, but no clear correlations were observed using this method. Subsequently, integration patterns for all 300 small-molecule spectra were analyzed by AMIX to generate a data matrix of N integration regions times 300. A statistical software package, UNSCRAMBLER 6.0, was then used to analyze this data matrix using principal components analysis. Two classes of spectral changes were observed. Ultimately, one class was found to correspond to pH changes caused by certain small molecules while the other class corresponded to small molecules binding to the target protein.[12] A description of the analysis of changes in signals from small molecules in 2D spectra indicative of binding can be found in the accompanying chapter by Moore and co-workers.[47]

Data reduction is an important aspect for handling the amounts of data generated if high-throughput screening by NMR is to be successful. Nonstatistical methods such as the bucketing calculations of AMIX[28] or the database comparisons of ACD[43] compare chemical shift, multiplicity, integration regions, and patterns to give correlation factors between spectra. These software packages can be used for data reduction of both one- and two-dimensional data. Prediction software is also available to help aid in interpretation of data sets. Statistical methods such as principal components analysis can be used to analyze data for other correlations that are not apparent using nonstatistical methods alone. In the case of 2D ^{1}H–^{15}N HSQC data, an adaptive, multivariate method that incorporates a weighted mapping of perturbations to correlate information within a spectrum or across many spectra has also been described.[48]

Comparison of Flow vs Traditional Methods

The advantage of working with samples in the flow NMR screening environment is that each set of spectra are collected on samples that are at the same concentration. This accelerates spectral acquisition considerably. Since the samples are fairly homogenous, many of the routine tasks need to be completed on only the first sample: probe tuning, ^{1}H 90° pulse calibration, receiver gain, number of transients, locking, and gradient shimming. On subsequent samples, these steps can be omitted, although simplex shimming of Z_1 and Z_2 can still be used with multiday acquisitions.

Prerequisites for a high-throughput assay include rapid data collection, sample-to-sample integrity, and minimal costs. Flow NMR techniques have been developed with each in mind. For 1D ^{1}H NMR screening experiments, the process of

[47] J. W. Peng, C. A. Lepre, J. Fejzo, N. Abdul-Manan, and J. M. Moore, *Methods Enzymol.* **338,** [8] 2001 (this volume).

[48] F. Delaglio, CHI Conference on NMR Technologies: Development and Applications for Drug Discovery, Baltimore, MD, 4–5 November 1999.

removing the previous sample from the flow cell, rinsing the flow cell, injecting the next sample, allowing for thermal equilibration, automating solvent suppression, and acquiring the data can take less than 10 min. In practice, the use of this procedure is two to three times faster than a sample changer with conventional NMR tubes. If compounds were screened in mixtures of 10, this results in a throughput of about 1500 compounds per day. Use of a liquid handler, such as the Gilson 215 typically employed by Bruker and Varian flow NMR systems, can simplify the preparation of samples as well. Ross and co-workers have demonstrated on-the-fly sample preparation by using the liquid handler to mix the protein to be screened with the small molecule immediately prior to injection.[12] Sample conditions can thus be highly standardized with the resulting spectra very consistent and reproducible. Even if target protein is added manually to preplated screening libraries, the amount of pipetting is still less than if NMR tubes were used. Recurring expenses associated with purchasing and/or cleaning NMR tubes are eliminated with flow-injection NMR methods. The cost of the 96-well microtiter plates is insignificant compared to that of NMR tubes.

Future Directions

The rapid throughput of flow-injection NMR as a tool to screen protein/small molecule mixtures has only begun to be appreciated. Continuing advances in instrumentation will undoubtedly occur as these techniques become more common in the field. For instance, a cryogenic NMR flow probe with increased sensitivity would be expected to enhance the throughput of screening experiments much like the traditional cryogenic NMR probe.[30] As flow NMR spectroscopy becomes increasingly important in the field of high-throughput screening, continued emphasis will need to be placed on data analysis tools to extract the key bits of information present in this vastness of data space.

Acknowledgments

We thank V. Shanmugasundaram, M. S. Lajiness, and G. M. Maggiora for preparing Fig. 5 and for continued collaborations regarding the design of small molecule libraries.

[10] The Internet for Nuclear Magnetic Resonance Spectroscopists

By DAINA AVIZONIS and SHAUNA FARR-JONES

The Internet has had an enormous impact on how scientists work. Access to databases, programs, opportunities for collaboration, and teaching material has never been easier. We present a small selection of Internet sites that we identified as the most useful, and do not attempt to provide an exhaustive list. The scope of this article includes sample preparation, nuclear magnetic resonance (NMR) experiments, structure calculations, and analysis, but we do not delve into structure prediction and other bioinformatics topics.

Expression Systems and Purification

In meeting the challenge of developing protein expression or purification protocols, the Internet has numerous resources. The ATCC (American Type Cutting Collection, Rockville, MD) Bacteriology Collection is a good place to obtain appropriate expression cell lines. (http://www.atcc.org) While expressing proteins in a line of bacteria may be convenient, *Pichia* expression systems for producing high levels of functionally active recombinant protein can make an excellent alternative since methanol can be used as a carbon source. Baculovirus systems should also be considered. Invitrogen provides a large number of choices when it comes to finding a workable expression system, and provides numerous technical application notes on optimizing expression levels. (http://www.invitrogen.com)

Several Web sites are collecting molecular biology protocols. One such site, the Dartmouth College Comprehensive Protocol Collection (http://www.dartmouth. edu/artsci/bio/ambros/protocols/molbio.html), has molecular biology protocols from RNA extraction to plasmid preparation. Two other protocol collections are at http://www.protocol-online.net and http://www.metazoa.com; the former is maintained by Longcheng Li at the University of California, San Francisco (UCSF), and the latter is a recently created commercial initiative. More resources for protein expression are available at vendor sites such as Novagen (http://www.novagen.com/ html/vectfram.html) and New England Biolabs (http://www.neb.com). Both sites provide restriction enzymes and an extensive array of protein expression vectors along with detailed protocols on how to use them. For protein purification materials and protocols Amersham Pharmacia (http://www.apbiotech.com/), PerSeptive Biosystems (http://www.pbio.com), and Bio-Rad Laboratories (Hercules, CA) (http://www.bio-rad.com/) are good places to start.

Before you start to make your protein, peptide, or nucleic acid, you should visit the ExPASy (Expert Protein Analysis System) proteomics server of the Swiss Institute of Bioinformatics. ExPASy is dedicated to the analysis of protein sequences and structures. (http://www.expasy.ch) Of particular use to NMR spectroscopists are the secondary structure prediction programs and the structure databases.

Isotopes and Labeling

Stable isotope suppliers with information about their products on line are Isotec Inc. (http://www.isotec.com/), Martek Biosciences Corporation (http://www.martekbio.com/), and NEN (http://www.nen.com/products/custom/custom.htm).

Some of the isotope suppliers allow for online ordering such as SciQuest (http://www.sciquest.com) and Sigma-Aldrich, St. Louis, MO (http://www.sigma-aldrich.com).

For more exotic isotopes and less common labeled compounds, explore the National Tritium Labeling Facility (NTLF). (http://www.lbl.gov/LBL-Programs/NTLF/Directory.html) NTLF can synthesize tritium-labeled compounds with techniques such as hydrogenation, dehalogenation, methylation, acetylation, hydride reduction, and exchange. Similarly, the National Stable Isotope Resource at Los Alamos advances biomedical applications by offering training in the synthesis and applications of compounds labeled with stable isotopes ^{13}C, ^{15}N, ^{17}O, ^{18}O, ^{33}S, ^{34}S, and ^{77}Se. (http://pearl1.lanl.gov/sir/) They will also distribute labeled compounds not available from commercial resources to accredited investigators.

NMR Theory and Tutorials

Our favorite Web site is educational, multimedia, and fun! Walter Bauer at the Institute of Organic Chemistry, University of Erlangen-Nuremberg, Germany, cleverly demonstrates that audio can be used to understand and monitor NMR phenomena. Sample sound files, available in either *.wav* or *.aiff* format, demonstrate the use of sound for tuning, shimming, and spectroscopy of solids and liquids. Although listening to NMR spectrometers has a long history, this site brings the use of sound for data analysis to new heights. (http://www.organik.uni-erlangen.de/research/NMR/music11.html)

For a complete Web-based class on basic NMR theory, look up Joseph Hornak's "The Basics of NMR". Hornak makes excellent use of hypertext and the Web's animation capabilities to illustrate complex nuclear spin behavior and the mathematics behind it. (http://www.cis.rit.edu/htbooks/nmr)

Françoise Sauriol at Queens' University has put together a fairly complete practical NMR course covering simple 1D up to 3D heteronuclear experiments. He has made extensive use of *Shockwave* to make slide shows that walk you through the experiments. Some of the vector diagrams are animated. (http://web.chem.queensu.ca/FACILITIES/NMR/nmr/webcourse/index1.htm)

A site with lucid explanations of more advanced NMR concepts is maintained by James Keeler. It has pdf files on topics such as coherence selection and product operator formalism. (http://www.nmr.de/html/library/keelers.htm)

Another use of the Web's capabilities is an evolving book edited by David Gorenstein. Although not yet complete, entire chapters, written by various NMR luminaries, are available for download in pdf format. (http://biosci.umn.edu/biophys/OLTB/NMR.html)

Webelements is an extremely useful interactive periodic table. It provides access to NMR information, as well as other physical and chemical data. This site, started at the University of Sheffield by Mark Winter, makes excellent use of Internet multimedia technology. You can even download the "Palmelements" periodic table for your Palm. (http://www.webelements.com)

Many other academics have put their lecture notes and tutorials online; some of these are linked in the directory sites listed below.

Virtual NMR Laboratories

One NMR facility that has made exceptional use of Internet technologies for using NMR facilities remotely is the Pacific Northwest National Laboratory's Environmental Molecular Sciences Laboratory (EMSL), in Richland, WA. It has a number of spectrometers ranging from 300 to 900 MHz that can be run remotely by users around the world. The EMSL has set up a Virtual Facilities Collaboratory that gives users of the facility remote desktop access, wherever they might be, to the EMSL's researchers, data management systems, computers, and scientific instruments. Virtual Facilities use Internet technologies to provide secure, direct acquisition control of instruments. ESML also provides video conferencing tools allowing users to see and talk with colleagues, and to look around the NMR laboratory. A Web whiteboard allows remote interactive sketching. Another example of the collaboratory tools is the electronic laboratory notebook, a Web-based tool for sharing, viewing, and analyzing data among a group of researchers. The electronic lab notebook runs in the browser and is platform independent. Any type of electronic data may be shared this way. The collaboratory and the electronic notebook software are publicly available from the EMSL and use Java technology. (http://www.emsl.pnl.gov:2080/docs/collab/virtual/EMSLVNMRF.html)

NMR Pulse Sequences

Before you start to write your own pulse sequences from scratch, you might want to look at the numerous pulse sequence archives.

One of the best archives of Bruker pulse sequences for solution NMR can be found at the EMBL Heidelberg. (http://www.nmr.EMBL-Heidelberg.DE/sattler/PP/pulseprograms.html) Some solids experiments can be found at the Ecole Normale Superieure in Lyon, at a site hosted by Lyndon Emsley's Laboratory. (http://www.ens-lyon.fr/STIM/NMR/pp.html)

More choices are available to access Varian pulse sequences. The best place to start is with the Varian NMR Corporate site user pages (http://www.varianinc.com/nmr/apps/userpages.html). Here you can download a large collection of multidimensional protein or RNA pulse sequences. The most useful programs to get are ProteinPack and RNAPack. These two programs will help you set up complex 2D or 3D experiments in a semiautomatic way. The programs will detect the field strength and hardware capabilities and perform necessary calibrations. The advantage of these programs is that setup errors are reduced to the extent that even novice NMR users can collect data efficiently.

Gaetano Montelione at Rutgers University maintains an extensive list of solution pulse sequences and also pulse sequence diagrams of those sequences in pdf format. (http://www-nmr.cabm.rutgers.edu/nmr) Lewis Kay's group at the University of Toronto has made most of the pulse sequences that they have published available via ftp. (http://abragam.med.utoronto.ca/sequences.html)

NMR Relaxation

Numerous programs that analyze NMR relaxation data are available online either for free or for a nominal licensing fee.

Arthur Palmer at Columbia University provides the program ModelFree, which fits the extended model free spectral density function to NMR spin relaxation data. ModelFree optimizes "Lipari-Szabo model free" parameters to heteronuclear relaxation data. Another program, CurveFit, can fit R1 inversion recovery and R2 CPMG relaxation data for subsequent use in ModelFree. (http://cpmcnet.columbia.edu/dept/gsas/biochem/labs/palmer/)

At least three programs are available to do relaxation matrix analysis: IRMA, MARDIGRAS and MORASS. IRMA can be found at Bijovoet Centre Utrecht (http://www.nmr.chem.uu.nl/~abonvin/irma.html), CORMA/MARDIGRAS at UCSF (http://picasso.nmr.ucsf.edu/New/download.html), and MORASS/GRASP at University of Texas Medical Branch. (http://www.nmr.utmb.edu/#mrass)

NMR Processing and Display Software

Many suites of processing and display software are available online. We mention only few of those available for nominal licensing fees. For a fairly complete list, see the directories below.

ANSIG, by Per Kraulis, is an interactive graphics package that assists in assigning spectra. One can easily scroll through 2D planes of multidimensional homo- and heteronuclear data. Multiple experiments and screens can be opened simultaneously in a correlated fashion. (http://www-ccmr-nmr.bioc.cam.ac.uk/public/ANSIG/ansig.html)

NMRPipe is a comprehensive package for Fourier processing of spectra in one to four dimensions, and spectral display and analysis. It is widely used and runs under Solaris, Linux, Unix, and Windows NT. Its companion program, NMRWish, has a general purpose database engine for manipulating peak tables, assignments, and molecular coordinates. Written by Frank Delaglio, it can be obtained from the NIH. (http://spin.niddk.nih.gov/bax/software/NMRPipe/index.html)

For a spectral display and analysis program with excellent database management, assisted assignment, and spectral display, you can get Sparky from UCSF. (http://www.cgl.ucsf.edu/home/sparky) The database is integrated with the molecular display program MIDASPlus. Sparky runs on Unix platforms and is somewhat unusual in that it is also available for MS Windows platforms. You can write extensions that access peak and spectrum data and use the display capabilities of Sparky. Extensions are short programs written in the free and simple-to-learn scripting language, Python.

Few options exist for Mac Platforms; however, SWAN-MR by Giuseppe Balacco at Menarini Ricerche (Florence, Italy) does run on Mac platforms. SWAN-MR will do spectral processing, but does not have database management required for molecular assignment. (http://qobrue.usc.es/jsgroup/swan/index.html)

XEASY from the ETH in Zurich is very popular and runs under X Window Systems with Motif libraries. It facilitates spectral display, analysis, and molecular assignment. XEASY has facile interfaces to programs for spectral processing and molecular structure calculations, including DYANA. (http://www.mol.biol.ethz.ch/wuthrich/software/xeasy)

Pronto 3D^2 can analyze multidimensional spectra as well as display 3D protein structures. Pronto began as a joint project between CRI, Bruker, and Carlsberg A/S and is freely available from Carlsberg Laboratory to both academic and for-profit institutions. (http://mail.crc.dk/chem/pronto)

Structure Calculations

A place to start learning about the numerous programs that are available for molecular modeling is the NIH Center of Molecular Modeling. Here you will find a number of links to tutorials for various molecular modeling suites. (http://cmm.info.nih.gov/modeling/tutorials.html)

What follows is a sample of a few of the most popular academic molecular modeling software packages.

Amber and Sander can be obtained from the Kollman laboratory at UCSF. Features of the latest release of this widely used program are inclusion of Locally Enhanced Sampling, Particle Mesh Ewald, and One Window Free Energy Grid that allows generation of free energy grids from a single MD trajectory. Another interesting addition to this version is Chemical Monte Carlo/Molecular Dynamics (CMC/MD), a new approach to ligand design, which can consider many molecules at once. Also available is the generalized Born surface area (GB/SA), which is a way to put solvation effects implicitly into calculations of complex systems for use in minimization and molecular dynamics calculations. (http://www.amber.ucsf.edu)

DYANA ("Dynamics algorithm for NMR applications") by Güntert and Mumenthaler uses simulated annealing by MD in torsion angle space. Recently added features include grid searches for arbitrary fragments of a molecule and a module to determine and check NOESY assignments. (http://www.mol.biol.ethz.ch/dyana/intro.html)

GROMOS, from the van Gunsteren group at the ETH in Zurich, is a general-purpose MD simulation package for biomolecules. GROMOS allows simulation of any molecules in solution or crystalline states by MD, stochastic dynamics, or the path-integral method. The latest version of GROMOS includes prediction of molecular structure in different solvents, calculation of binding constants, prediction of energetic or structural changes upon "virtual" mutation, and derivation of 3D molecular structure from NMR restraints (time averaged distance and coupling constant restraints). (http://igc.ethz.ch/gromos/gromos.html)

Michael Nilges' group, at the EMBL in Heidelberg, offers ARIA and XPLOR software along with on-line manuals, refinement protocols, and tutorials. ARIA is a user-friendly program for automated NOE assignment and NMR structure calculation. It makes use of ambiguous distance restraints and iterative calculation scheme to speed up the NOE assignment process. XPLOR (eEXPLORation of conformational macromolecules restrained to regions allowed by combinations of empirical energy functions and experimental data) is a program used to calculate protein and nucleic acid structures based on either crystallographic or NMR data. (http://www.embl-heidelberg.de/nmr/nilges/)

The program Crystallography & NMR System (CNS) is the collaborative effort of several research groups. CNS provides a flexible multilevel hierarchical approach for the most commmonly used algorithms in macromolecular structure determination. Calculations based on NMR include NOEs, J coupling, chemical shift, and dipolar coupling data. As the name of this program suggests, it is also used for crystallographic data refinement. (http://cns.csb.yale.edu/v1.0/)

Molecular Graphics

MOLMOL (MOLecule Analysis and MOLecule Display) software, by Reto Koradi in Kurt Würthrich's group; is available from the ETH in Zurich.

MOLMOL is designed primarily for the visualization of biomolecular structures derived from NMR data. It is especially efficient at displaying a number of super-imposed structures (20–40), identifying hydrogen bonds, checking and displaying NOE constraints and violations, and identifying short distances between hydrogen atom pairs. It also has a graphical user interface with menus dialogue boxes and on-line help. (http://www.mol.biol.ethz.ch/wuthrich/software/molmol/)

The program Molscript, by Per Kraulis, can take the atomic coordinates of a biomolecule and render 3D structures in both schematic and detailed representa-tions. The output can be saved as a PostScript file, Raster3D, virtual reality model-ing language (VRML), GIF, and others. They have put great effort in trying to make different output modes give the same image quality as the input. This program is very useful in reading PDB files. (http://www.avatar.se/molscript/molscript.html)

PDB3D is a great tool for making the best use of the Web. It is a real-time 3D molecular rendering program specifically designed for the Web running as an interpreted Java applet. The Java applet allow Web site visitors to rotate and manipulate structures viewed on the Web page, without installing any other molec-ular graphics software. To see it in action, visit the University of California at Los Angeles—Department of Energy (UCLA-DOE), Laboratory of Structural Biology and Molecular Medicine. (http://www.doe-mbi.ucla.edu/Services/PDB3D/)

PDB3D and Molscript make use of a language called Virtual Reality Modelling Language (VRML). It has an ISO standard format for representing 3D objects and scenes and is especially useful for displaying molecules, since VRML is platform independent and interactive. You will need a VRML browser or plug-in for your current browser. Several different plug-ins are listed in an archive by David Blauch at Davidson College. (http://www.chm.davidson.edu/VRML/pdb2vrml.html)

The Resource for Biomolecular Graphics, also known as the Computer Graph-ics Laboratory (CGL) at the UCSF, provides state-of-the-art graphics hardware and software for research on biomolecular structure. The program, MidasPlus, is a molecular graphics display system used to display and manipulate proteins, peptides, and nucleic acids. Additional features include computation of molec-ular surfaces and electrostatic potentials and generation of publication-quality space-filling images with multiple light sources and shadows. UCSF CGL dis-tributes documented source code, so that an interested researcher can alter pro-grams for his or her own needs. The UCSF site also provides step-by-step tuto-rials on how to create striking graphics. Chimera is the successor to MidasPlus. (http://www.cgl.ucsf.edu/Outreach/midasplus/)

Two more useful visualization systems are RasMol and Chime. RasMol can be used to look at PDB files on PC, Mac, and SGI platforms. It gives very fast 3D ren-derings of complex molecules that can be rotated in real time. (http://www.umass.edu/microbio/rasmol/) Chime is an applet used to visualize molecules from within a browser. The Chime plug-in is freely available from MDL Information Systems

Inc. (http://www.mdli.com/cgi/dynamic/downloadsect.html?uid=$uid&key=$key &id=1)

Nucleic Acids Structure

There is less information for nucleic acids on the Internet than can be found for proteins. One of the first places to start with nucleic acids is the list of links found at RNA World (http://www.imb-jena.de/RNA.html) Rutgers also hosts the Nucleic Acid Database Project, which uses Web-based interactive forms for structure deposition. This site contains not only a database of DNA and RNA structures, but also tutorials on how to find a particular structure. (http://ndbserver.rutgers.edu/ NDB/general/index.html)

Curves is an algorithm by Lavery and Sklenar for the analysis of nucleic acid structures from strands in multistranded complexes to fragments with mispairs, bulges open base pairs, abasic sites, or bulges. Curves calculates the optimal helical axes and base tilts. (http://www.ibpc.fr/UPR9080/Curindex.html)

Rnadraw and rRNA server are RNA secondary structure prediction and analysis programs offered by Ole Matzura or Peter De Rijk's group at the University of Antwerp.(http://rnadraw.base8.se/index.html, http://www-rrna.uia.ac.be/index. html)

Protein Structure Analysis and Validation

For a number of programs to examine and validate your newly determined NMR structure, try the UCLA-DOE site (http://www.doe-mbi.ucla.edu/Services). The following programs are included here:

Continuous Profile method (CPROF)
Structural analysis of DNA (NEWHEL)
Calculation of a protein's hydrophobic folding energy and structural hydrophobic moments (ASP)
Statistical analysis of interactions between different atom types (ERRAT)

The EMBL in Heidelberg offers automated servers to examine your structures. The Biotech Validation Suite for Protein Structures (http://biotech.embl-heidelberg.de:8400) is biased toward NMR structures, while the UCLA-DOE site is targeted more toward crystallographers.

Some calculations and analyses that the EMBL server will run are as follows:

Volume calculations (PROVE)
Programs to check the stereochemical quality of protein structures
Solvent accessibility (ACCESS)

Buried surface differences upon drug binding (DIFACC)
Comparison of surfaces to database averages (ANASRF)
Atomic distances (DIST)
Steric clashes (BUMPS)
Contacts with waters (WATER)
Distance of atoms to bulk water (CABWAT)
Hydrogen bonds (HBONDS)
Residue-based contact analysis (CONRES)
Salt bridges (SHOSBR)
Superposition possibilities (SUPPOS)

NMR spectroscopists always deal with structural ensembles, and the Sutcliffe group at University of Leicester provides a server where you can upload your coordinate ensemble for evaluation by both nmrclust and nmrcore. (http://neon.chem.le.ac.uk/olderado) The nmrclust and nmrcore programs cluster an ensemble of structures to determine the core atoms.

Part of the power of NMR is its ability to provide information on molecular motions. Mark Gerstein at Yale University (New Haven, CT) has set up a fascinating database of macromolecular movements. You can watch movies of protein motions in MPEG, QuickTime, GIF, or VRML formats. The protein can be displayed as CPK, sticks, or ribbons and can be displayed in stereo. Most importantly, you can submit your own data for inclusion in the database. (http://bioinfo.mbb.yale.edu/MolMovDB/)

At least two Web sites provide programs for automated protein chemical shift prediction. One is SHIFTY by David Wishart at the BioMagResBank (BMRB). (http://www.bmrb.wisc.edu:8999/shifty.html)

Another site that provides programs for calculating ^{13}C and proton chemical and ring current shifts of proteins can be found at the University of Sheffield. (http://www.shef.ac.uk/uni/projects/nmr/home.html)

NMR Conferences

Most NMR conferences can be found online. Here are a few locations for the most popular NMR conferences where one can examine the program, submit an abstract, or register:

The Experimental Nuclear Magnetic Resonance Conference (ENC) (http://www.enc-conference.org) and European ENC (http://www.uni-leipzig.de/~eenc) are held once a year at various locations.
The Gordon Research Conferences often offer symposia related to NMR. (http://www.grc.uri.edu/)

Every other year there is a Keystone Symposium in Structural biology and NMR methodology. (http://www.symposia.com)

Well-maintained listings for all types of NMR meetings can be found at the following sites:

http://www.varianinc.com/nmr/events/index.html
http://micro.ifas.ufl.edu/meetings.html

NMR Directories

This article barely scratches the surface of NMR information available on the Web, and several sites will direct you to many others.

Of the NMR Web directories, one of the most comprehensive and searchable sites is hosted by the Open Directory Project and maintained by volunteer contributors. (http://dmoz.org/Science/Chemistry/Nuclear_Magnetic_Resonance). Sites maintained by Prof. Kessler's group, at the Institut für Organische Chemie und Biochemie, TU München, Germany (http://www.nmr.de), and Marian Lech Buszko at the University of Florida (http://micro.ifas.ufl.edu) contain a huge list of NMR links. Peter Lundberg at Linköpings University in Sweden has compiled an extensive listing of educational NMR software. (http://huweb.hu.liu.se/inst/imv/radiofysik/info/edunmrsoft.html)

The BMRB has a large repository of NMR results from proteins, peptides, and nucleic acids. In collaboration with the Research Collaboratory for Structural Biology and the Nucleic Acids Data Bank, BMRB is a repository of biomolecular information not available in a PDB type file. Its goal is to archive NMR-specific data such as chemical shifts of assignments, J couplings, relaxation rates, and other chemical information derived by NMR such as hydrogen exchange rates and pK_a values. Also available on this Web site are pulse sequences for both Bruker and Varian spectrometers, and links to other NMR and bioinformatics sites. (http://www.bmrb.wisc.edu/)

The Future

The Web is full of NMR sites from the high-tech multimedia to the simplest text documents. However, we did not find many academic sites that made use of advanced Internet technologies. We suspect that the time and effort required to set up Web sites with multimedia, databases and e-commerce are greater than academic resources will allow. Nevertheless, a few academic sites have exploited the full potential of the Internet with animation, sound, and VRML. We would like to see more e-commerce use by academics to pay licensing fees so that software could be downloaded immediately. We also advocate provision of servers that will run calculations or database searches remotely. These are a boon to busy scientists, who are reluctant to download and install every tool they would like to explore.

Web authoring tools are available from Macromedia (http://www. macromedia. com) and many other companies. Wired has a good collection of tutorials and resources for Web authoring. (http://hotwired.lycos.com/webmonkey).

Although almost every graduate student is fluent in HTML, XML (Extensible markup language) is a streamlined metalanguage that is much more powerful and provides more facile database integration. Whereas HTML was design to provide information about a page's appearance, XML was designed for document exchange. XML can handle any link between any data. It is powerful and portable because all a browser needs to view XML is the data itself and the style sheet controlling the layout.

As XML becomes more widely used, the Web should become faster. One of the most useful features of XML is that it enables authors to use indirect links that point to a database, rather than to pages directly. This will help eliminate "404 File Not Found" errors.

In the future, as researchers become more computer savvy, and tools for building complex Web sites become better known, we hope to see more interactive Web sites that fully exploit the tremendous potential of the Internet for teaching, research, data sharing, and collaboration.

The Web is ever changing, and the links in this site may well be different by the time you read this book. We will maintain a list of sites mentioned in the article at UCSF. (http://picasso.nmr.ucsf.edu/MiE_nmr.html)

Section II

Nucleic Acids and Carbohydrates

[11] Solid-Phase Synthesis of Selectively Labeled DNA: Applications for Multidimensional Nuclear Magnetic Resonance Spectroscopy

By CHOJIRO KOJIMA, AKIRA "MEI" ONO, AKIRA ONO, and
MASATSUNE KAINOSHO

Introduction

Various methods for preparing the isotopically labeled DNA oligomers have been reported during the past several years,[1-5] and the use of labeled DNAs for nuclear magnetic resonance (NMR) studies has become more common in the literature. The situation is, however, still quite different compared to proteins and RNAs, for which isotope-assisted NMR methods have already become absolutely essential.[1,6] This is partly due to the lack of versatile methods for preparing isotopically labeled DNA oligomers sufficient for NMR studies. Even at this moment, the preparation of labeled DNA oligomers, by either enzymatic or synthetic methods, is relatively troublesome. One might also argue that isotope-assisted NMR approaches may not be necessary, since the conformational varieties of DNA, especially for DNA duplexes, are rather limited compared to those of proteins and RNAs. The situation may be totally different for noncanonical DNA structures or DNA complexed with proteins. In this review, we will focus on the chemical synthesis of isotopically labeled nucleosides and nucleotides, and some recent applications involving labeled DNA.

Although solid-phase phosphoamidite chemistry[7] has been widely used to prepare sufficient amounts of unlabeled DNA oligomers for NMR measurements, the

[1] M. Kainosho, *Nature Struct. Biol.* **4**, 858 (1997).

[2a] R. A. Jones, *in* "Protocols for Oligonucleotide Conjugates" (S. Agrawal, ed.), p. 207. Humana Press Inc., Totowa, NJ, 1994.

[2b] R. A. Jones, *in* "Stable Isotope Applications in Biomolecular Structure and Mechanisms" (J. Trewhella, T. A. Cross, and C. J. Unkefer, eds.), p. 105. Los Alamos National Laboratory, Los Alamos, NM, 1994.

[3] A. Ono, S. Tate, and M. Kainosho, *in* "Stable Isotope Applications in Biomolecular Structure and Mechanisms" (J. Trewhella, T. A. Cross, and C. J. Unkefer, eds.), p. 127. Los Alamos National Laboratory, Los Alamos, NM, 1994.

[4] A. S. Serianni, *in* "Stable Isotope Applications in Biomolecular Structure and Mechanisms" (J. Trewhella, T. A. Cross, and C. J. Unkefer, eds.), p. 209. Los Alamos National Laboratory, Los Alamos, NM, 1994.

[5] A. Földesi, T. V. Maltseva, Z. Dinya, and J. Chattopadhyaya, *Tetrahedron* **54**, 14487 (1998).

[6] G. M. Clore and A. M. Gronenborn, *Nature Struct. Biol.* **4**, 849 (1997).

[7a] S. L. Beaucage and M. H. Caruthers, *Tetrahedron Lett.* **22**, 1859 (1981).

[7b] S. L. Beaucage, *in* "Protocols for Oligonucleotides and Analogs" (S. Agrawal, ed.), p. 33. Humana Press Inc., Totowa, NJ, 1993.

application of this robust method for the synthesis of isotopically labeled oligomers has been hampered by the lack of efficient synthetic methods for isotopically labeled nucleoside phosphoamidites. For RNA, the enzymatic synthesis method is available and applicable for the preparation of isotopically labeled ($^{13}C/^{15}N/^{2}H$) oligomers.[8] The details, including the NMR studies of RNAs and RNA–protein complexes, have appeared in the literature.[8–10] With this method, labeled NTPs extracted from bacteria grown on labeled media are incorporated into RNA strands by T7 RNA polymerase.

For DNA oligomers, a similar enzymatic method that uses DNA polymerases instead of RNA polymerase is applicable and has been used for the incorporation of labeled dNTPs into DNA strands.[11] Enzymatic synthesis is effective for preparing uniformly or nucleotide-type specifically labeled DNA/RNA oligomers. In contrast, the solid-phase phosphoamidite chemistry for preparing DNA/RNA oligomers,[7] which was developed in the 1980s, is considered to be useful for preparing large amounts of DNA/RNA oligomers labeled at specific residues. Solid-phase chemical synthesis is not used widely to prepare the isotopically labeled oligomers, since there is no familiar, efficient synthetic route that enables the economic preparation of fully labeled monomer units with routinely available isotope precursors. Here we review this solid-phase synthesis method for isotopically labeled DNA oligomers, which was developed recently.[3]

Synthetic Methods

Overview: Chemical and Enzymatic Synthesis of DNA Oligomers

Before describing the details of solid-phase DNA synthesis, we compare the chemical and enzymatic methods to clarify the scope and limitation of their applications. Overviews of both the enzymatic preparation and the chemical preparation are shown in Fig. 1. The enzymatic approach requires isotopically labeled dNTPs, which are generally prepared from the DNA components extracted from bacteria grown on labeled media. Once the dNTPs are prepared, as a mixture or as isolated

[8a] R. T. Batey, J. L. Battiste, and J. A. Williamson, *Methods Enzymol.* **261**, 300 (1995).
[8b] E. P. Nikonowicz, *Methods Enzymol.* **338** [14] 2001 (this issue).
[9] A. Pardi, *Methods Enzymol.* **261**, 350 (1995).
[10] M. Allen, L. Varani, and G. Varani, *Methods Enzymol.* **339** [17] 2001.
[11a] D. P. Zimmer and D. M. Crother, *Proc. Natl. Acad. Sci. U.S.A.* **92**, 3091 (1995).
[11b] J. M. Louis, R. G. Martin, G. M. Clore, and A. M. Gronenborn, *J. Biol. Chem.* **273**, 2374 (1998).
[11c] G. Mer and W. J. Chazin, *J. Am. Chem. Soc.* **120**, 607 (1998).
[11d] D. E. Smith, J. Y. Su, and F. M. Jucker, *J. Biomol. NMR* **10**, 245 (1997).
[11e] J. E. Masse, P. Bortmann, T. Dieckmann, and J. Feigon, *Nucleic Acids Res.* **26**, 2618 (1998).
[11f] M. H. Werner, V. Gupta, L. A. Lambert, and T. Nagata, *Methods Enzymol.* **338** [12] 2001 (this issue).

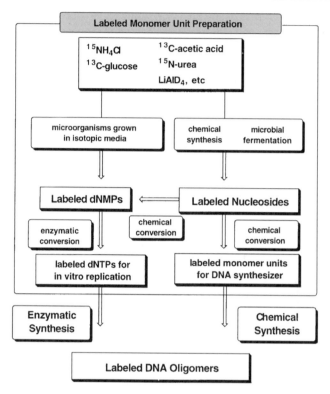

FIG. 1. A scheme for the preparation of labeled monomer units.

forms, the DNA extension reaction is performed, using the 3'–5' exo⁻ Klenow fragment, in many cases, with moderate yields.[11] This method is suitable for preparing fully labeled or nucleotide-type selectively labeled oligomers and can be used to prepare a longer DNA than the chemical synthesis method. Additionally, the DNA extension reaction works efficiently for small amounts of commercially available isotopically labeled dNTPs. The yields of the labeled oligomers, however, depend on their sequences, and the length of the DNA thus produced is not unique, which is known as the 3' nontemplated addition.[11e] On the other hand, the solid-phase chemical synthesis method uses isotopically labeled nucleoside phosphoamidites, which are synthesized chemically from the labeled nucleosides. The labeled nucleosides are prepared by chemical synthesis and/or microbial fermentation.[3] The nucleoside phosphoamidites are used with an automated DNA synthesizer, and the pure DNA oligomers are easily obtained.[5,12] This method is generally applicable for

[12] W. J. Warren and G. Vella, in "Protocols for Oligonucleotide Conjugates" (S. Agrawal, ed.), p. 233. Humana Press Inc., Totowa, NJ, 1994.

preparing DNA oligomers of any kind or labeling pattern, and the yield and purity of the labeled oligomer are almost independent of the sequence.

The monomer units for the chemical and enzymatic oligomer synthesis are the isotopically labeled nucleoside phosphoamidites and dNTPs, respectively. As shown in Fig. 1, both monomer units can be chemically prepared from labeled nucleosides. The labeled nucleosides are chemically converted to dNMPs, and then chemically or enzymatically converted to dNTPs.[13] Chemical synthesis methods allow us to prepare the monomer units having site- and stereoselective labels with ^{2}H, ^{13}C, and ^{15}N, at any desired sites. The selective labeling methods open up various possibilities to develop advanced NMR technologies, as described later. We will focus on the preparation of isotopically labeled nucleosides as it is the most crucial step to prepare selectively labeled DNA oligomers.

Synthetic Route for Various Isotopomers of Labeled Nucleosides

Synthetic routes for nucleosides have been developed by many groups.[14,15] Stable isotope labeling of nucleosides requires an efficient synthetic route starting from inexpensive labeled materials.[16] For various NMR measurements, a wide variety of labeled samples is required, and thus the synthetic routes are also expected to be applicable for the preparation of fully labeled, atom-specifically labeled, and stereoselectively labeled nucleosides. One example of such a synthetic route is shown in Fig. 2. Three components, ribose, pyrimidine bases, and purine bases, are prepared separately from the proper isotope precursors. Labeled ribose and labeled bases are chemically or enzymatically coupled to yield labeled nucleosides, so the isotope is introduced at the steps for synthesizing the three components.

The ^{13}C-labeled sugar moiety of the nucleosides and 2′-deoxynucleosides is synthesized from ^{13}C-labeled glucose. As various ^{13}C-labeled isotopomers of glucose are commercially available, nucleosides with variously labeled sugar residues can be prepared. Fully and/or specifically labeled pyrimidine bases, uracil and thymine, are chemically synthesized starting from various isotope precursors. ^{13}C-labeled glucose is chemically converted into the labeled ribose derivative,[17]

[13a] J. G. Moffatt, *Can. J. Chem.* **42,** 599 (1964).

[13b] M. Yoshikawa, T. Kato, and T. Takenishi, *Bull. Chem. Soc. Jpn.* **42,** 3505 (1969).

[13c] N. C. Mishra and A. D. Broom, *J. Chem. Soc. Chem. Commun.* 1276 (1991).

[14] T. Ueda, *in* "Chemistry of Nucleosides and Nucleotides" (L. B. Townsend, ed.), p. 1. Plenum Press, New York, 1988.

[15] P. C. Srivastava, R. K. Robins, and R. B. Meyer, Jr., *in* "Chemistry of Nucleosides and Nucleotides" (L. B. Townsend, ed.), p. 113. Plenum Press, New York, 1988.

[16] A. T. Balaban and I. Bally, *in* "Isotopes in the Physical and Biomedical Sciences" (E. Buncel and J. R. Jones, eds.), p. 288. Elsevier Science Publishers, Amsterdam, 1987.

[17a] S. Quant, W. Wechselberger, M. A. Wolter, K.-H. Worner, P. Schell, J. W. Engels, C. Griesinger, and H. Schwalbe, *Tetrahedron Lett.* **36,** 6649 (1994).

[17b] L. A. Agrofoglio, J.-C. Jacquinet, and G. Lancelot, *Tetrahedron Lett.* **38,** 1411 (1997).

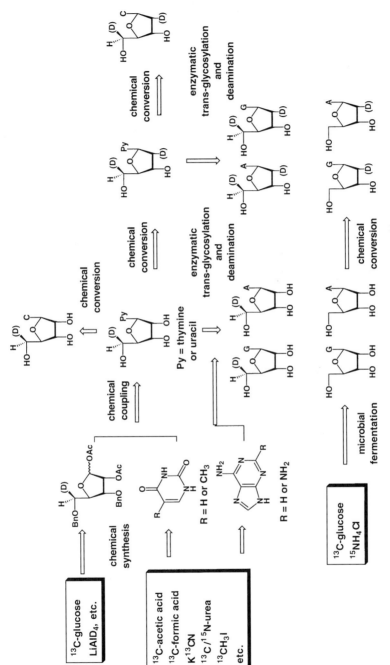

FIG. 2. A scheme for the preparation of labeled nucleosides.

and then the protected ribose is coupled[18] with labeled thymine or uracil to give labeled ribosylthymine or uridine in good yields. These pyrimidine nucleosides are chemically converted into the corresponding 2'-deoxynucleosides by established procedures.[19] If necessary, deuterium is stereoselectively introduced at the 2' position using deuterated reagents.[20] Uridine, 2'-deoxyuridine, and thymidine are chemically converted into cytidine, 2'-deoxycytidine, and 5-methyl-2'-deoxycytidine, respectively.[21] The labeled sugar moiety of the pyrimidine nucleosides is efficiently transferred into that of the purine nucleosides by the enzymatic transglycosylation reactions.[22] Fully labeled pyrimidine nucleosides can be chemically synthesized. However, the efficient chemical preparation of fully labeled purine bases and purine nucleosides is difficult, so the chemical method has been used for the labeling at limited atom positions of the purine bases.[2] On the other hand, microbial fermentation is efficiently used for preparing fully labeled purine nucleosides, adenosine and guanosine.[3] These purine nucleosides are chemically converted into the corresponding 2'-deoxynucleosides.

In the following sections some details of the synthetic route are explained. The stereoselective ^{2}H labeling at the 2' and 5' positions of ^{13}C-labeled nucleosides and the atom-specific ^{13}C/^{15}N labeling are highlighted, since chemical synthesis has a strong advantage over enzymatic synthesis in such specifically labeled nucleosides.

Stereoselective [2]H Labeling of ^{13}C-Labeled Sugars

Stereoselective ^{2}H labeling is important not only for the unambiguous stereospecific assignment of the methylene proton signals, but also for simplification and sharpening of the spectrum by eliminating the largest proton–proton dipolar and scalar interactions, i.e., between methylene protons.[23] The ^{13}C labeling is used for avoiding the spectral overlap and useful for the assignments; thus, stereoselectively ^{2}H and ^{13}C doubly labeled nucleosides are required. Many procedures are available for preparing deuterated sugars and nucleosides from various compounds; however, they are not efficient enough for preparing ^{2}H/^{13}C doubly labeled nucleosides.

[18] H. Vorbrüggen, K. Krolikiewiez, and B. Bennua, *Chem. Ber.* **114**, 1234 (1981).

[19] M. J. Robins, J. S. Wilson, and F. Hansske, *J. Am. Chem. Soc.* **105**, 4059 (1983).

[20a] E. Kawashima, Y. Aoyama, T. Sekine, E. Nakamura, M. Kainosho, Y. Kyogoku, and Y. Ishido, *Tetrahedron Lett.* **34**, 1317 (1993).

[20b] E. Kawashima, Y. Aoyama, T. Sekine, M. Miyahara, M. F. Radwan, E. Nakamura, M. Kainosho, Y. Kyogoku, and Y. Ishido, *J. Org. Chem.* **60**, 6980 (1995).

[20c] E. Kawashima, Y. Aoyama, M. F. Radwan, M. Miyahara, T. Sekine, M. Kainosho, Y. Kyogoku, and Y. Ishido, *Nucleosid. Nucleotid.* **14**, 333 (1995).

[20d] E. Kawashima, S. Uchida, M. Miyahara, and Y. Ishido, *Tetrahedron Lett.* **38**, 7369 (1997).

[21] C. B. Reese and A. Ubasawa, *Tetrahedron Lett.* **21**, 2265 (1980).

[22] T. A. Krenitsky, G. W. Koszalka, and J. V. Tuttle, *Biochemistry* **20**, 3615 (1981).

[23] D. M. LeMaster, *Annu. Rev. Biophys. Biophys. Chem.* **19**, 243 (1990).

Ishido and associates have developed some efficient stereoselective deuteration methods for both the 2' and 5' sites.[20,24] Their methods are applicable for ^{2}H/^{13}C double labeling, although, they need ^{13}C-labeled nucleosides as the starting materials. As a starting material, glucose is better than nucleosides since various ^{13}C-labeled isotopomers are commercially available with reasonable prices. Quant *et al.* and Agrofoglio *et al.* have reported the syntheses of ^{13}C-labeled nucleosides starting from ^{13}C-labeled glucose.[17] Kainosho and his collaborators have developed this method to incorporate deuteriums at the 3', and/or 4', and/or 5' sites of ^{13}C-labeled nucleosides.[25] The synthetic route of the stereoselectively ^{2}H/^{13}C doubly labeled sugars, which are the key intermediate for nucleoside synthesis, is shown in Fig. 3.

The [1-^{13}C]-, [2-^{13}C]-, [3-^{13}C]-, [4-^{13}C]-, and [5-^{13}C]-labeled isotopomers of labeled glucose, as the starting materials, are converted into the [1-^{13}C]-, [2-^{13}C]-, [3-^{13}C]-, [4-^{13}C]-, and [5-^{13}C]-labeled isotopomers of the ribose derivative **1**, and finally are converted into the [1'-^{13}C]-, [2'-^{13}C]-, [3'-^{13}C]-, [4'-^{13}C]-, and [5'-^{13}C]-labeled nucleosides, respectively. Using proper deuterated reagents for appropriate reactions in the synthetic route, deuterium is introduced into 3, and/or 4, and/or 5 positions of the ribose derivative in combination with ^{13}C label. The deuteration at the 3 and 4 positions is done in steps **2** to **2'** and **2** to **3** using D$_2$O–pyridine and NaBD$_4$, respectively.[26,27] The deuteration at the 5 position is described in the following paragraph. At the 2' position of 2'-deoxynucleosides, deuterium is stereoselectively introduced using the deuterated reagent in the step of the radical deoxygenation of the ribonucleosides. Using the methods of Kawashima *et al.*, the highly stereoselective (2'R)- or (2'S)-^{2}H$_1$ isotopomer is obtained.[20] Consequently, deuterium is introduced at any position among the 2' to 5' sites in any combination, and thus almost all of the isotopomers of ^{2}H/^{13}C doubly labeled sugars can be prepared.

For the 5'-position stereoselective deuterium labeling of the ^{2}H and/or ^{13}C labeled nucleosides, two efficient methods are available, which use the reductive deuteration of 5'-phenylselenonucleosides or 5-oxopentose.[24,28] By the 5'-phenylselenonucleosides method, the (5'R)/(5'S) deuteration ratios are in the

[24a] E. Kawashima, K. Toyama, K. Ohshima, M. Kainosho, Y. Kyogoku, and Y. Ishido, *Tetrahedron Lett.* **36,** 6699 (1995).

[24b] E. Kawashima, K. Toyama, K. Ohshima, M. Kainosho, Y. Kyogoku, and Y. Ishido, *Chirality* **9,** 435 (1995).

[25] A. M. Ono, T. Shiina, A. Ono, and M. Kainosho, *Tetrahedron Lett.* **39,** 2793 (1998).

[26] R. G. S. Ritchie, N. Cyr, B. Korsch, H. J. Koch, and A. Perlin, *Can. J. Chem.* **53,** 1424 (1975).

[27] A. K. Sinhababu, R. L. Bartel, N. Pochopin, and R. T. Borchardt, *J. Am. Chem. Soc.* **107,** 7628 (1985).

[28a] A. M. Ono, A. Ono, and M. Kainosho, *Tetrahedron Lett.* **38,** 395 (1997).

[28b] Y. Oogo, A. M. Ono, A. Ono, and M. Kainosho, *Tetrahedron Lett.* **39,** 2873 (1998).

[28c] A. Trifonova, A. Földesi, Z. Dinya, and J. Chattopadhyaya, *Tetrahedron* **55,** 4747 (1999).

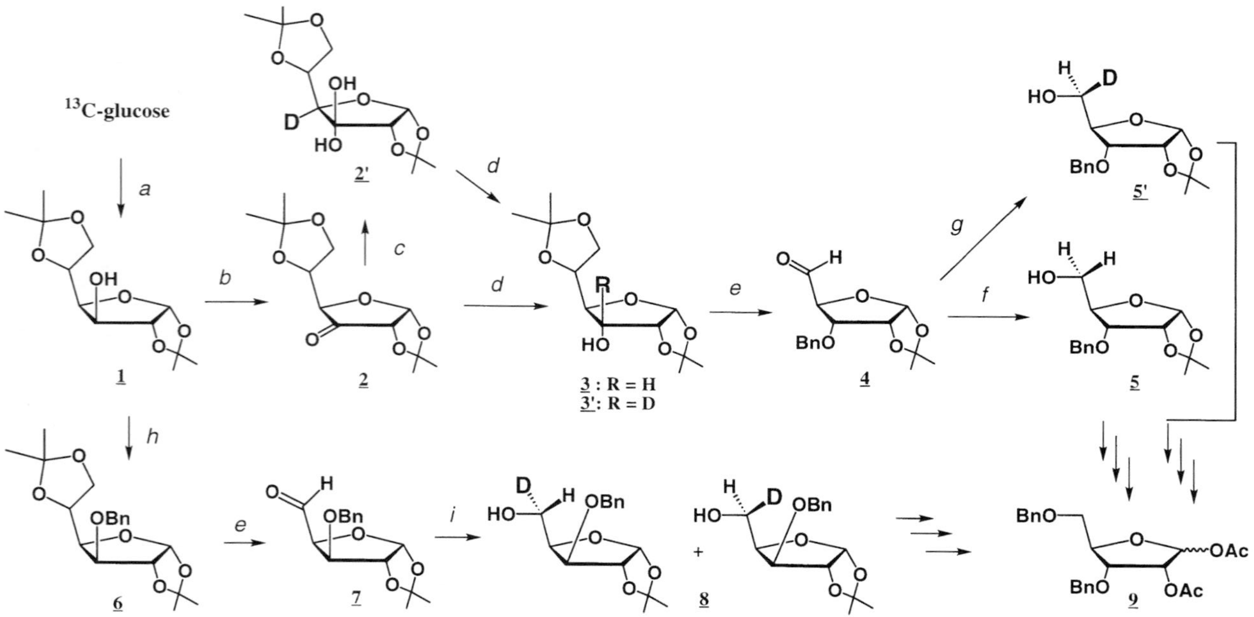

FIG. 3. A scheme for the systematic synthesis of site and/or stereoselectively deuterated ribose derivatives starting from glucose. (*a*) I$_2$, acetone; (*b*) pyridinium dichromate, molecular sieves 4A, acetic acid, CH$_2$Cl$_2$; (*c*) D$_2$O, pyridine; (*d*) NaBH$_4$ or NaBD$_4$, ethanol H$_2$O; (*e*) (1) BnBr, NaH, DMF; (2) 80% acetic acid, NaIO$_4$, ethanol, H$_2$O; (*f*) NaBH$_4$, ethanol, H$_2$O; (*g*) (−)-[2-^{2}H$_1$]isobornyloxmagnesium bromide, benzene; (*h*) BnBr, NaH, dimethylformamide (DMF); (*i*) LiAlD$_4$, LiI, *tert*-amyl alcohol.

range of 20/80 to 40/60, depending on the kind of nucleoside.[24] The 5-oxopentose method is shown in steps **4** to **5'** and **7** to **8** (Fig. 3). Treatment of the 5-oxoribose derivative **4** with $(-)$-[$^2H_1$]isobornyloxymagnesium bromide preferentially yields the (5'R)-deuterated ribose derivative **5'**.[28a] The deuteration with moderate stereoselectivity is accomplished using LiAlD$_4$ in the presence of a bulky alcohol and LiI, and the (5'R)/(5'S) deuteration ratios are in the range of 20/80 to 62/38.[28b] Using the 5-oxoxylose **7** instead of the 5-oxoribose derivative **4**, the (5'S)-deuterated xylose derivative **8** is obtained highly stereoselectively, and the (5'R)/(5'S) deuteration ratios are in the range of 55/45 to 88/12. As shown above the degree of the stereoselectivity is controlled by changing the reaction conditions, and thus the (5'R)/(5'S) deuteration ratios can be controlled in the range of 20/80 to 88/12.[28b]

Atom-Specific $^{13}C/^{15}N$ Labeling

One of the main advantages of the chemical preparation method is the incorporation into a DNA oligomer of atom-specifically $^{13}C/^{15}N$-labeled nucleosides, which are useful for NMR studies as described below. The synthetic procedures for atom-specifically labeled nucleosides are divided into two categories. One is the synthesis of labeled nucleosides from the appropriate isotope precursors, and the other is the chemical conversion of a nucleoside into a labeled nucleoside.

As mentioned above, each carbon atom in the sugar residue of a nucleoside is independently labeled with ^{13}C by the synthesis starting from specifically labeled glucose. Similarly, each atom in the base residue is labeled with ^{13}C and/or ^{15}N by employing appropriate isotope precursors. As shown in Fig. 4, all of the atoms in uracil and thymine can be labeled independently by using the proper isotope precursors as starting materials.[16] The reactions used to prepare labeled uracil and thymine have been well examined, except for the methylation of the 5-position of uracil using [^{13}C]methyl iodide, which gives a moderate yield.[29,30] N1 and N3 are not labeled separately, since both nitrogen atoms come from labeled urea. Coupling of the labeled uracil and thymine with the labeled ribose yields multiply labeled uridine and ribosylthymine, which are chemically converted into the corresponding 2'-deoxynucleosides.

The labeled 2'-deoxyuridine and thymidine, obtained as above, are further labeled by attempting the proper chemical reactions. Using the procedure for exchanging the N^3 nitrogen with $^{15}NH_3$, the N^3 of the pyrimidine ring can be labeled separately from the N^1.[31] When the [N^1, N^3-$^{15}N_2$]-derivatives are treated with

[29] J. SantaLucia, Jr., L. X. Shen, Z. Cai, H. Lewis, and I. Tinoco, Jr., *Nucleic Acids Res.* **23**, 4913 (1995).

[30] J. A. Lawson, J. I. DeGraw, and M. Anber, *J. Labell. Compd.* **11**, 489 (1975).

[31a] X. Ariza, V. Bou, J. Vilarrasa, V. Tereshko, and J. L. Campos, *Angew. Chem., Int. Ed. Engl.* **33**, 2454 (1994).

[31b] X. Ariza, V. Bou, and J. Vilarrasa, *J. Am. Chem. Soc.* **117**, 3665 (1995).

[31c] X. Ariza, J. Farras, C. Serra, and J. Vilarrasa, *J. Org. Chem.* **62**, 1547 (1997).

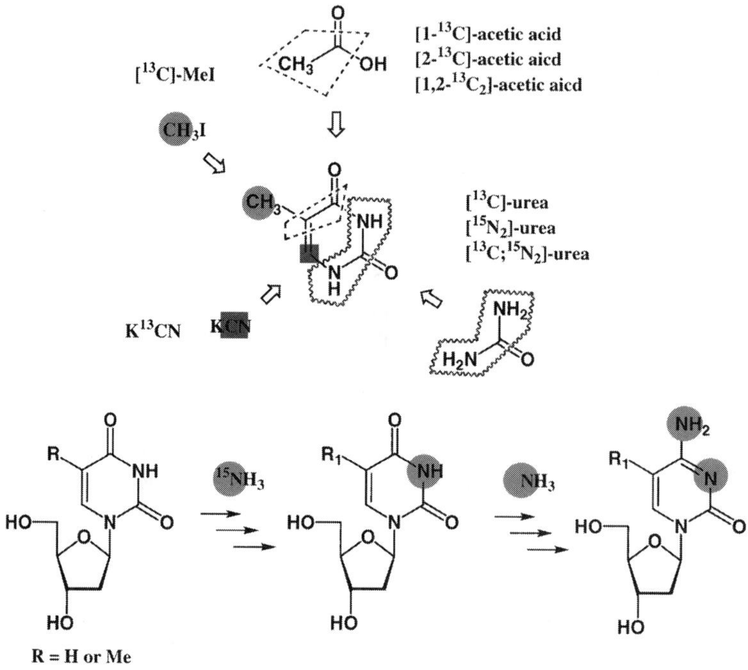

FIG. 4. Schemes for preparation of labeled pyrimidine bases and pyrimidine nucleosides.

$^{14}NH_3$, $[N^1\text{-}^{15}N]$-derivatives are obtained. Consequently, $[N^1\text{-}^{15}N]$-, $[N^3\text{-}^{15}N]$-, and $[N^1, N^3\text{-}^{15}N_2]$-derivatives with the appropriate ^{13}C labeling are prepared. Chemical conversions of the labeled 2′-deoxyuridine and thymidine into corresponding labeled cytidine derivatives with $^{15}NH_3$ give 2′-deoxycytidine and 5-methyl-2′-deoxycytidine with ^{15}N-labeled exocyclic amine groups.[32]

The preparation methods for DNA oligomers containing atom-specifically labeled purine bases have been vigorously studied by Jones and co-workers. 2′-Deoxyadenosine is converted into the atom-specifically labeled purine nucleosides, such as 2′-$[N^6\text{-}^{15}N]$deoxyadenosine, 2′-$[N^1\text{-}^{15}N]$deoxyadenosine, 2′-$[N^2\text{-}^{15}N]$deoxyguanosine, 2′-$[2\text{-}^{13}C]$deoxyguanosine, and 2′-$[2\text{-}^{13}C; N^3\text{-}^{15}N]$deoxyguanosine.[2,33] These nucleosides are chemically incorporated into the DNA oligomers used for NMR studies. Alternative methods are also available. Namely,

---

[32] E. R. Kellenbach, M. L. Remerowski, D. Ebi, R. Boelens, G. A. van der Marel, H. van den Elst, J. H. van Boom, and R. Kaptein, *Nucleic Acids Res.* **20,** 653 (1992).

[33a] A. R. Pagano, H. Zhao, A. Shallop, and R. A. Jones, *J. Org. Chem.* **63,** 3213 (1998).

[33b] H. Zhao, A. R. Pagano, W. Wang, A. Shallop, B. L. Gaffney, and R. A. Jones, *J. Org. Chem.* **62,** 7832 (1997).

appropriate isotope precursors are converted into the labeled purine nucleosides, and then the labeled bases are coupled with 2′-deoxyribose by enzymatic trans-glycosylation. [N^7-^{15}N]Adenine, [N^3-^{15}N]adenine, [N^7-^{15}N]diaminopurine, and the corresponding 2′-deoxynucleosides have been prepared.[2,34] [8-^{13}C]Adenine and [8-^{13}C]diaminopurine also can be prepared and converted into the 2′-deoxynucleo-sides.[16,29]

NMR Applications

Labeled nucleosides reported to be available by chemical synthesis are listed in Table I. These nucleosides are easily incorporated into the DNA oligomer by solid-phase phosphoamidite chemistry or the DNA extension reaction using DNA polymerase, as described before. When solid-phase synthesis is selected, site-specific labeling is available, which is useful for NMR studies. The chemical synthesis of DNA oligomers has a strong advantage in atom- and site-specific labeling and in stereoselective labeling. Such specifically and stereoselectively labeled DNA is necessary for the simplification of the complex spectra and unambiguous assignments. Once labeled DNA oligomers are prepared, simple NMR experiments, such as ^{1}H-^{13}C/^{1}H-^{15}N HSQC and ^{13}C/^{15}N direct observation, give the simple spectra where all of the resonances can be immediately assigned. Many NMR applications with labeled DNA oligomers use this advantage. NMR studies of isotopically labeled DNA oligomers, in which the chemical synthesis method was used for sample preparation, are listed in Table II. Only recent reports describing the atom- and site-specifically and the stereoselectively labeled DNA oligomers are reviewed in this chapter.

Atom- and Site-Specifically ^{13}C/^{15}N-Labeled DNA

DNA oligomers that are atom- and site-specifically ^{15}N-labeled at the N^4 amino nitrogen of 2′-deoxycytosine are widely used to study triplex formation and protein–DNA and drug–DNA interactions.[32,35] Jones and associates have used other atom- and site-specifically ^{15}N-labeled DNA oligomers to study base pairs and hydrogen bonding as follows: N^1 or N^2 atom-specifically labeled purine nucleosides for the AC and AG mispair and O^6MeG•T base pair formation,[36] N^3 atom-specifically labeled 2′-deoxyadenosine for the drug–DNA interaction,[37]

[34] A. R. Pagano, W. M. Lajewski, and R. A. Jones, *J. Am. Chem. Soc.* **117,** 11669 (1995).
[35a] A. M. MacMillan, R. J. Lee, and G. L. Verdine, *J. Am. Chem. Soc.* **115,** 4921 (1993).
[35b] V. Ramesh, Y.-Z. Xu, and G. C. K. Roberts, *FEBS Lett.* **363,** 61 (1995).
[35c] D. Leitner, W. Schröder, and K. Weisz, *J. Am. Chem. Soc.* **120,** 7123 (1998).
[36a] C. Wang, H. Gao, B. L. Gaffney, and R. A. Jones, *J. Am. Chem. Soc.* **113,** 5486 (1991).
[36b] B. L. Gaffney, B. Goswami, and R. A. Jones, *J. Am. Chem. Soc.* **115,** 12607 (1993).
[36c] B. Goswami, B. L. Gaffney, and R. A. Jones, *J. Am. Chem. Soc.* **115,** 3832 (1993).
[37] Y. Rhee, C. Wang, B. L. Gaffney, and R. A. Jones, *J. Am. Chem. Soc.* **115,** 8742 (1993).

TABLE I
ISOTOPICALLY LABELED 2'-DEOXYNUCLEOSIDE

Labeled 2'-deoxynucleoside	Starting materials, reagents	Refs.[a]
$^2H/^{13}C/^{15}N$ labeling		
$(2'R)$-[ul-$^{13}C/^{15}N;2'$-$^2H_1$]dA	[U-$^{13}C/^{15}N$]Adenosine,[b] $Bu_3Sn^2H_1$	(1)
$^2H/^{13}C$ labeling		
$(5'S)$-[5'-$^2H_1$;1',2',3',4',5'-$^{13}C_5$]T	[$^{13}C_6$]Glucose, $LiAl^2H_4$	(2)
$(2'R)$-[2',5'-$^2H_2$;1',2 ',3',4',5'-$^{13}C_5$]dA,T	[$^{13}C_6$]Glucose, NaB^2H_4, $Bu_3Sn^2H_1$	(3)
$(2'S, 5'S)$-[2',5'-$^2H_2$;1',2',3',4',5'-$^{13}C_5$]dA,dG,dC,T	[$^{13}C_6$]Glucose, $LiAl^2H_4$, $Bu_3Sn^2H_1$	(4)
^2H labeling		
[1'-$^2H_1$]dA,dG,dC,T	Ribonoractone, DIAL-$^2H_1$	(5)
$(2'R)$-[2'-$^2H_1$]dA,dG,dC,T	Ribonucleoside, Bu_3Sn^2H, $Me_3Si^2H_1$	(6, 7)
$(2'S)$-[2'-$^2H_1$]dA,dG,dC,T	Ribonucleoside, NaB^2H_4, Me_3SiH	(7, 8)
[3'-$^2H_1$]T	Xylose, NaB^2H_4	(9)
[4'-$^2H_1$]dC	Ribose derivative, $Bu_3Sn^2H_1$	(10)
[5'-$^2H_1$]T	Glucose, $LiAl^2H_4$	(11)
[5'-$^2H_1$]dA,dG,dC,T	Deoxyribonucleoside, $Bu_3Sn^2H_1$	(12)
$(5'R)$-[5'-$^2H_1$]dC	Glucose, $Ph_3Sn^2H_1$	(13)
$(5'S)$-[5'-$^2H_1$]dC	$(5'R)$-[5'-$^2H_1$]dC	(10)
[5',5''-$^2H_2$]dC	dC	(10)
[1',2',2''-$^2H_3$]dA,dG,dC,T	Ribonoractone, DIAL-$^2H_1$, MeO^2H_1	(5)
[2',3',5'-$^2H_3$]dA,dG,dC,T	Glucose derivative, NaB^2H_4	(14)
[1',2',3',4',5',5''-$^2H_6$]dA,dG,dC,T	Ribose derivative, ^2H_2O	(15)
[8-$^2H_1$]dA,dG	dA,dG, ^2H_2O	(16)
[6-$^2H_1$]dC,T	dC,T, $C^2H_3O^2H_1$, C^2H_3ONa, NaO^2H_1	(16)
^{13}C labeling		
[1',2',3',4',5'-$^{13}C_5$]dA,dG,dC,T	[$^{13}C_6$]Glucose	(17, 18)
[5'-$^{13}C_1$]dA,dG,dC,T	Ribose, $^{13}CH_3I$	(19)
[6-$^{13}C_1$]T	2-Bromopropyl bromide, $K^{13}CN$	(20)
[Me-$^{13}C_1$]T	5-BrT, $^{13}CH_3I$	(21)
^{15}N labeling		
[1-$^{15}N_1$]dA	Deoxyinosine, $^{15}NH_4OH$	(22)
[1-$^{15}N_1$]dG	AICA-deoxyriboside, $^{15}NH_4Cl$	(23)
[2-$^{15}N_1$]dG	[6-$^{15}N_1$]dA	(24)
[6-$^{15}N_1$]dA	dA, $^{15}NH_4Cl$	(25)
[7-$^{15}N_1$]dG	4-Hydroxy-2,5,6-triaminopyrimidine, $Na^{15}NO_2$	(26)

[a] *Key to References:* (1) J. Kurita, M. Kawaguchi, T. Shiina, S. Tate, A. M. Ono, A. Ono, and M. Kainosho, *Nucleic Acids Symp. Ser.* **34**, 49 (1995); (2) A. M. Ono, A. Ono, and M. Kainosho, *Tetrahedron Lett.* **38**, 395 (1997); (3) A. Földesi, T. V. Maltseva, and J. Chattopadhyaya, *Nucleosid. Nucleotid.* **18**, 1377 (1999); (4) A. M. Ono, Y. Oogo, S. Tate, A. Ono, and M. Kainosho, *Nucleic Acids Symp. Ser.* **37**, 73 (1997); (5) R. P. Hodge, C. K. Brush, C. M. Harris, and T. M. Harris, *J. Org. Chem.* **56**, 1553 (1991);

and N^7 atom-specifically labeled purine nucleosides for the triplex and tetraplex formation.[38]

Kainosho and associates have used chemically synthesized DNA oligomers containing ^{13}C or $^{13}C/^{15}N$ fully labeled nucleosides at specific sites for NMR technical developments and analyses as follows; through bond intraresidue correlation between sugar protons,[3] through bond intraresidue correlation between a sugar and a base proton,[39] through bond sequential correlation between sugar protons,[40] and through vicinal coupling constants between sugar protons.[41] A collaborative team of the Wüthrich and Kainosho groups has successfully used such labeled oligomers to measure the conformational changes in the sugar–phosphate backbone induced by protein binding,[42] the precise structures of DNA–protein complexes,[43] and the

[38a] B. L. Gaffney, C. Wang, and R. A. Jones, *J. Am. Chem. Soc.* **114**, 4047 (1992).

[38b] B. L. Gaffney, P.-P. Kung, C. Wang, and R. A. Jones, *J. Am. Chem. Soc.* **117**, 12281 (1995).

[39] S. Tate, A. Ono, and M. Kainosho, *J. Am. Chem. Soc.* **116**, 5977 (1994).

[40] S. Tate, A. Ono, and M. Kainosho, *J. Magn. Res. Ser. B* **106**, 89 (1995).

[41] A. Ono, S. Tate, Y. Ishido, and M. Kainosho, *J. Biomol. NMR* **4**, 581 (1994).

[42a] T. Szyperski, A. Ono, C. Fernández, H. Iwai, S. Tate, K. Wüthrich, and M. Kainosho, *J. Am. Chem. Soc.* **119**, 9901 (1997).

(6) E. Kawashima, Y. Aoyama, T. Sekine, M. Miyahara, M. F. Radwan, E. Nakamura, M. Kainosho, Y. Kyogoku, and Y. Ishido, *J. Org. Chem.* **60**, 6980 (1995); (7) E. Kawashima, S. Uchida, M. Miyahara, and Y. Ishido, *Tetrahedron Lett.* **38**, 7369 (1997); (8) E. Kawashima, Y. Aoyama, M. F. Radwan, M. Miyahara, T. Sekine, M. Kainosho, Y. Kyogoku, and Y. Ishido, *Nucleoside Nucleotides* **14**, 333 (1995); (9) T. Chen and M. M. Greenberg, *Tetrahedron Lett.* **39**, 1103 (1998); (10) J. J. De Voss, J. J. Hangeland, and C. A Townsend, *J. Org. Chem.* **59**, 2715 (1994); (11) Y. Oogo, A. M. Ono, A. Ono, and M. Kainosho, *Tetrahedron Lett.* **39**, 2873 (1998); (12) E. Kawashima, K. Toyama, K. Ohshima, M. Kainosho, Y. Kyogoku, and Y. Ishido, *Tetrahedron Lett.* **36**, 6699 (1995); (13) J. J. Hangeland, J. J. De Voss, J. A. Heath, and C. A. Townsend, *J. Am. Chem. Soc.* **114**, 9200 (1992); (14) A. Földesi, T. V. Maltseva, Z. Dinya, and J. Chattopadhyaya, *Tetrahedron* **54**, 14487 (1998); (15) A. Földesi, F. P. R. Nilson, C. Glemarec, C. Gioeli, and J. Chattopadhyaya, *Tetrahedron* **48**, 9033 (1992); (16) X. Huang, P. Yu, E. LeProust, and X. Gao, *Nucleic Acids Res.* **25**, 4758 (1997); (17) S. Quant, R. W. Wechselberger, M. A. Wolter, K.-H. Wörner, P. Schell, J. W. Engels, C. Griesinger, and H. Schwalbe, *Tetrahedron Lett.* **35**, 6649 (1994); (18) L. A. Agrofoglio, J.-C. Jacquinet, and G. Lancelot, *Tetrahedron Lett.* **38**, 1411 (1997). (19) E. Kawashima, T. Sekine, K. Umabe, Y. Naito, K. Kamaike, C. Kojima, T. Mizukoshi, E. Suzuki, and Y. Ishido, *Nucleosid. Nucleotid.* **18**, 1597 (1999); (20) J. R. Williamson, and S. G. Boxer, *Nucleic Acids Res.* **16**, 1529 (1988); (21) E. R. Kellenbach, M. L. Remerowski, D. Eib, R. Boelens, G. A. van der Marel, H. van den Elst, J. H. van Boom, and R. Kaptein, *Nucleic Acids Res.* **20**, 653 (1992); (22) L. De Napoli, A. Messere, D. Montesarchio, and G. Piccalli, *J. Org. Chem.* **60**, 2251 (1995); (23) C. Bleasdale, S. B. Ellwood, B. T. Golding, P. K. Slaich, O. J. Taylor, and W. P. Watson, *J. Chem. Soc. Perlin Trans.* **1**, 2859 (1994); (24) B. Goswami and R. A. Jones, *J. Am. Chem. Soc.* **113**, 644 (1991); (25) X. Gao and R. A. Jones, *J. Am. Chem. Soc.* **109**, 1275, (1987); (26) W. Massefski, Jr. and A. Redfield, *J. Am. Chem. Soc.* **112**, 5350 (1990). [b] M. Kainosho, *J. Am. Chem. Soc.* **101**, 1031 (1979).

TABLE II

ISOTOPICALLY LABELED DNA OLIGOMERS STUDIED BY NMR[a]

DNA sequence[b]	Labeled 2'-deoxynucleotides	Experiments	Refs.[c]
²H/¹³C and ²H/¹³C/¹⁵N labeling			
d(CGCGAA**T**TCGCG)₂	[1'2'3'4'5'-¹³C₅; 5'-²H₁]T	¹H–¹³C dual CT-HSQC	(1)
d(CGCGA**A**TTCGCG)₂	(2'R)-[U-¹³C/¹⁵N; 2'-²H₁]dA, (2'S)-[U-¹³C/¹⁵N; 2'-²H₁]dA	HCCH-E.COSY	(2)
d(CGA**TT**AA**T**CG)₂	[1'2'3'4'5'-¹³C₅; 2'5'-²H₂]T, [1'2'3'4'5'-¹³C₅; 2'5'-²H₂]dA	¹H, ¹³C 1D	(3)
d(CGCGA**A**TTCGCG)₂, d(CGCGA**A**TTCGCG)₂	(2'R)-(5'S)-[1'2'3'4'5'-¹³C₅; 2'5'-²H₂]dA,dG,dC,T	¹H–¹³C HSQC	(4)
¹³C/¹⁵N labeling			
d(GC**T**G**TT**C**T**GC) • d(GCAGAA**C**AGC)	[5-¹³CH₃]T, [4-¹⁵NH₂]dC	¹H–¹³C HMQC, ¹H–¹⁵N HMQC, ¹³C HMQC-NOESY	(5)
d(CGCGA**A**TTCGCG)₂	[U-¹³C, ¹⁵N]dA	HCNNH, HCP-CCH-COSY	(6, 7)
d(**C**GCCATTAGAG) • d(**CTCT**AATGG**C**T**TTC**), d(GAAAGC**C**ATTAGAG) • d(CTC**T**AATGGCTTTC)	U-¹³C/¹⁵N	[¹⁵N,¹H]-COSY, [¹³C,¹H]-COSY, HCCH-COSY, ¹³C-edited NOESY, HCH-COSY, {³¹P}-sedct-[¹³C,¹H]-HSQC, {³¹P}-sedct-[¹³C,¹H]-HMQC, ʰJ_NN-correlated-[¹⁵N,¹H]-TROSY	(8–14)
d(**C**A**T**TTG**C**ATC) • d(**G**A**T**G**C**A**AA**TG)	[U-¹³C/¹⁵N]dA,dG (¹³C 15%)	¹³C-(T1, T2, NOE), DD-CSA cross-correlation	(15, 16)
¹³C labeling			
d(CGCACGC) • d(GCG**U**GCG), d(GGCACGG) • d(CCG**U**GCC), d(CGCNCGC) • d(GCG-**Aba**-GCG)	[1',3'-¹³C₂]dU,dAba	¹³C 1D	(17, 18)
d(CGCG**TT**GTTCGCG)	[6-¹³C]T	¹³C-(T1, NOE)	(19, 20)
d(**C**GC**T**CAC**AATT**) • d(**AATT**GTGAGC**G**)	1'-¹³C	¹H–¹³C HMQC, HMQC-TOCSY, HMQC-NOESY	(21)

274

Sequence	Labeling	NMR method	Ref.
d(**TCCTTCC**—CCTTCCTAG—CTAGGAAGG), d(TCCTTCC—**CCTTCCTAG**—**CT**AGGAAGG)	[1'-13C]dC,T	1H–13C HMQC, HMQC-TOCSY, HMQC-NOESY	(22, 23)
d(CGCGAA**TT**CGCG)2		HCCH-E.COSY	(24)
d(**AGCCAATA**), d(AGCC**A**ATA)	[1'2'3'4'5'-13C5]T	1H–13C HMQC, COSY, TOCSY, ROESY	(25)
d(CGCAAA**TTT**GCG)2	[1'-13C]dA	13C-(T1, T2, NOE)	(26)
d(TATCACCGCAAGGGATA) •	[1'-13C]T	1H–13C HSQC	(27)
d(**T**ATCCCTTGCGGTGATA),	[1'2'3'4'5'-13C5]T		
d(TATCACCGCAAGGGATA) •			
d(TA**T**CCCTTGCGGTGATA),			
d(TATCACCGCAAGGGATA) •			
d(TATCCC**T**TGCGGTGATA),			
d(TATCCC**T**TGCGGTGATA),			
d(TATCACCGCAAGGGATA) •			
d(TATCCC**T**TGCGGTGATA),			
d(TATCACCGCAAGGGATA) •			
d(TATCCCTTGCGGTGATA),			
d(TATCACCGCAAGGGATA) •			
d(TATCCCTTGCGGTGA**T**A)			
d(**CGCTCACAAT**) • d(**AATTGTGAGCG**)	1'-13C	13C-(T1, T2, NOE)	(28)
d(**CGCGAATTCGCG**)2	5-13C	1H–13C HMQC, HMQC-NOESY	(29)

15N labeling

Sequence	Labeling	NMR method	Ref.
d(CG**T**ACG)2	[1-15N]dA, [6-15NH2]dA	15N 1D	(30, 31)
d(ATTGTGA**G**CGCTCACAT)2, d(ATTGTGAGCGCT**C**ACAAT)2	[6-15NH2]dA, [4-15NH2]dC	15N 1D, 1H-15N HMQC	(32)
d(CGCGAATTCGCG)2	[6-15NH2]dA	15N 1D	(33)
d(GGCGGAGTT**A**GG) • d(CCTAACTCCGCC)	[6-15NH2]dA	15N 1D	(34)
d(TACCACTGGCGGTGATA) •	[7-15N]dG	15N-edited 1H 1D	(35)
d(TATCACCGCCAG**TGG**TA)			

(continued)

275

TABLE II (continued)

DNA sequence[b]	Labeled 2′-deoxynucleotides	Experiments	Refs.[c]
d(CGAGAATTCCCG)₂, d(CGGGAATTCACG)₂	[1-¹⁵N]dA	¹H–¹⁵N HMQC	(36)
d(GGTTTTTGG)₄, d(TGGGT)₄, d(GGTTTTTGG) • d(CCAAAAACC), d(TGGGT) • d(ACCCA)	[7-¹⁵N]dG	¹H–¹⁵N HSQC	(37)
d(CGCGAATTCC-**O⁶MeG**-CG)₂	[1-¹⁵N]O⁶Me-dG, [2-¹⁵NH₂]O⁶Me-dG	¹⁵N 1D	(38)
d(CGTGAATTC-**O⁶MeG**-CG)₂	[1-¹⁵N]O⁶Me-dG, [2-¹⁵NH₂]O⁶Me-dG	¹⁵N 1D	(39)
d(AATACCACTGGCGGTGATATA) • d(TATATCACCGCCAGTGGTATT), d(AATACCACTGGCGGTGATATA) • d(TATATCACCGCCAGTGGTATT)	[4-¹⁵NH₂]dC	¹⁵N-Edited ¹H 1D	(40)
d(CGCGAATTCGCG)₂, d(CGCGAATTCGCG)₂	[3-¹⁵N]dA	¹H–¹⁵N HSQC	(41)
d(TTTTTCTCTCTCTCT) • d(GCTAAAAAGAGAGAGAGATCG) • d(CGATCTCTCTCTTTTTAGC), d(TTTTTCTCTCTCT) • d(GCTAAAAAGAGAGAGAGATCG) • d(CGATCTCTCTCTCTTTTTAGC)	[7-¹⁵N]dA,dG	¹⁵N 1D	(42)
d(CGAACTAGTTAACTAGTTCG)₂	[4-¹⁵NH₂]dC	¹⁵N-Edited NOESY	(43)
d(CGCAATCATGAGCACG) • d(CGTGCTCAGCGAATGC) • d(GCATTCGCGCTATGGC) • d(GCCATAGCTGATTGCG)	[1,3-¹⁵N₂]T	¹⁵N-Edited ¹H 1D	(44,45)
d(CGCAAATTTGCG)₂, d(CGCGAGCTCGCG)₂	[6-¹⁵NH₂]dA	¹⁵N-(T1, NOE)	(46)
d(GAAGAGG-TTTT-CCTCTTC-TTTT-CTTCTCC)	[4-¹⁵NH₂]dC	¹H–¹⁵N HSQC	(47)

²H labeling (solution NMR)

Sequence	Labeling	Method	Ref.
d(CGCGAATTCGCG)₂	[8-²H]dA,dG	¹H 1D	(48, 49)
d(CGTTATAATGCG) • d(CGCATTATAACG)	[8-²H]dA,dG, [5-²H]dC	NOESY	(50)
d(TTTT)	[5-C²H₃]T	¹H 1D	(51)
d(CGCGCGCGAATTCGCGCGCG)₂	[1',2',3',4',5'-²H₇], [1',2',3',4'-²H₅]	DQF-COSY, NOESY, HAL-NOESY	(52–54)
d(CGCGAATTCGCG)₂	5'-²H₁	¹H–³¹P HSQC	(55)
d(CGGCGGCGCGGCGG)	[8-²H]dG	NOESY	(56)
d(CGCGAATTCGCG)₂	[2'R-²H₁]dA	stripe-COSY	(57)
d(CGCGAATTCGCG)₂	2',3',5'-²H₃	DQF-COSY, NOESY, ¹H–¹³C HSQC	(58, 59)
d(GCATTAATGC)₂	5'-²H₁	DQF-COSY, NOESY	(60)
d(GCTGCCATGGTTTTTGTGCACCAGC)₂	[8-²H]dA,dG	NOESY	(61)

²H labeling (solid-state NMR)

Sequence	Labeling	Method	Ref.
d(CGCGAATTCGCG)₂, d(CGCGAATTCGCG)₂	[6-²H]T, [8-²H]dA,dG	²H 1D	(62)
d(CGCGAATTCGCG)₂	[5-C²H₃]T	²H 1D	(63, 64)
d(CGCGAATTCGCG)₂, d(CGCGAATTCGCG)₂, d(CGCGAATTCGCG)₂	[5-C²H₃]T, [8-²H]dA,dG, [2'R-²H₁]dA	²H 1D	(65)
d(CGCGAATTCGCG)₂	[2'R-²H₁]T	²H 1D	(66)
d(CGCGAATTCGCG)₂	[5'-²H₂]T	²H 1D	(67)
d(CGCGAATTCGCG)₂	[8-²H]dA,dG	²H 1D	(68)
d(CGCGAATTCGCG)₂	[2'R-²H₁]dC, [5'-²H₂]dC, [5,6-²H₂]dC	²H 1D	(69)

[a] Limited for chemically synthesized DNA oligomers. Enzymatically prepared oligomers are excluded.

[b] Bold–underlined letters indicate isotopically labeled residues.

[c] Key to References: (1) S. Tate, Y. Kubo, A. Ono, and M. Kainosho, J. Am. Chem. Soc. 117, 7277 (1995); (2) J. Kurita, M. Kawaguchi, T. Shiina, S. Tate, A. M. Ono, A. Ono, and M. Kainosho, Nucleic Acids Symp. Ser. 32, 49 (1995); (3) A. Földesi, T. V. Maltseva, and J. Chattopadhyaya, Nucleosid. Nucleotid. 18, 1377 (1999); (4) N. Tjandra, S. Tate, A. Ono, M. Kainosho, and A. Bax, J. Am. Chem. Soc. 122, 6190 (2000); (5) E. R. Kellenbach, M. L. Remerowski, D. Ebi, R. Boelens, G. A. van der Marel, H. van den Elst, J. H. van Boom, and R. Kaptein, Nucleic Acids Res. 20, 653 (1992); (6) S. Tate, A. Ono, and M. Kainosho, J. Am. Chem. Soc. 116, 5977 (1994); (7) S. Tate, A. Ono, and M. Kainosho, J. Magn. Res. Ser. B 106, 89 (1995); (8) T. Szyperski, A. Ono, C. Fernández, H. Iwai, S. Tate, K. Wüthrich, and M. Kainosho, J. Am. Chem. Soc. 119, 9901 (1997); (9) C. Fernández, T. Szyperski, H. Iwai, S. Tate, A. Ono, K. Wüthrich, and M. Kainosho, J. Biomol. NMR 12, 25 (1998); (10) K. Pervushin, A. Ono, C. Fernández, T. Szyperski, M. Kainosho, and K. Wüthrich, Proc. Natl. Acad. Sci. U.S.A. 95, 14147 (1998); (11) T. Szyperski, C. Fernández, A. Ono, M. Kainosho, and K. Wüthrich, J. Am. Chem. Soc. 120, 821 (1998); (12) C. Fernández, T. Szyperski, *(continued)*

M. Billter, H. Iwai, A. Ono, M. Kainosho, and K. Wüthrich, *J. Mol. Biol.* **292**, 609 (1999); (13) T. Szyperski, C. Fernández, A. Ono, K. Wüthrich, and M. Kainosho, *J. Magn. Reson.* **140**, 491 (1999); (14) K. Pervushin, C. Fernández, R. Riek, A. Ono, M. Kainosho, and K. Wüthrich, *J. Biomol. NMR* **16**, 39 (2000); (15) C. Kojima, A. Ono, M. Kainosho, and T. L. James, *J. Magn. Reson.* **135**, 310 (1998); (16) C. Kojima, A. Ono, M. Kainosho, and T. L. James, *J. Magn. Reson.* **136**, 169 (1999); (17) M. Manoharan, J. A. Gerlt, J. A. Wilde, J. M. Withka, and P. H. Bolton, *J. Am. Chem. Soc.* **109**, 7217 (1987); (18) M. Manoharan, S. C. Ransom, A. Mazumder, J. A. Gerlt, J. A. Wilde, J. M. Withka, and P. H. Bolton, *J. Am. Chem. Soc.* **110**, 1620 (1988); (19) J. R. Williamson and S. G. Boxer, *Nucleic Acids Res.* **16**, 1529 (1988); (20) J. R. Williamson and S. G. Boxer, *Biochemistry* **28**, 2819 (1989); (21) G. Lancelot, L. Chanteloup, J.-M. Beau, and N. T. Thuong, *J. Am. Chem. Soc.* **115**, 1599 (1993); (22) O. Bornet, G. Lancelot, L. Chanteloup, N. T. Thuong, and J.-M. Beau, *J. Biomol. NMR* **4**, 575 (1994); (23) O. Bornet and G. Lancelot, *J. Biomol. Struct. Dynam.* **12**, 803 (1995); (24) A. Ono, S. Tate, Y. Ishido, and M. Kainosho, *J. Biomol. NMR* **4**, 581 (1994); (25) J. Wu and A. S. Serianni, *Biopolymers* **34**, 1175 (1994); (26) F. Gaudin, F. Paquet, L. Chanteloup, J.-M. Beau, N. T. Thuong, and G. Lancelot, *J. Biomol. NMR* **5**, 49 (1995); (27) Y. Kyogoku, C. Kojima, S. J. Lee, H. Tochio, N. Suzuki, H. Matsuo, and M. Shirakawa, *Methods Enzymol.* **261**, 524 (1995); (28) F. Paquet, F. Gaudin, and Lancelot, *J. Biomol. NMR* **8**, 252 (1996); (29) E. Kawashima, T. Sekine, K. Umabe, Y. Naito, K. Kamaike, C. Kojima, T. Mizukoshi, E. Suzuki, and Y. Ishido, *Nucleosid. Nucleotid.* **18**, 1597 (1999); (30) X. Gao and R. A. Jons, *J. Am. Chem. Soc.* **109**, 3169 (1987); (31) C. Wang, X. Gao, and R. A. Jones, *J. Am. Chem. Soc.* **113**, 1448 (1991); (32) G. Kupferschmitt, J. Schmidt, Th. Schmidt, B. Fera, F. Buck, and H. Rüterjans, *Nucleic Acids Res.* **15**, 6225 (1987); (33) A. L. Pogolotti Jr., A. Ono, R. Subramaniam, and D. V. Santi, *J. Biol. Chem.* **263**, 7461 (1988); (34) C. H. Lin and L. H. Hurley, *Biochemistry* **29**, 9503 (1990); (35) W. Massefski, Jr., A. Redfield, U. D. Sarma, A. Bannerji, and S. Roy, *J. Am. Chem. Soc.* **112**, 5350 (1990); (36) C. Wang, H. Gao, B. L. Gaffney, and R. A. Jones, *J. Am. Chem. Soc.* **113**, 5486 (1991); (37) B. L. Gaffney, C. Wang, and R. A. Jones, *J. Am. Chem. Soc.* **114**, 4047 (1992); (38) B. L. Gaffney, B. Goswami, and R. A. Jones, *J. Am. Chem. Soc.* **115**, 12607 (1993); (39) B. Goswami, B. L. Gaffney, and R. A. Jones, *J. Am. Chem. Soc.* **115**, 3832 (1993); (40) A. M. MacMillan, R. J. Lee, and G. L. Verdine, *J. Am. Chem. Soc.* **115**, 4921 (1993); (41) Y. Rhee, C. Wang, B. L. Gaffney, and R. A. Jones, *J. Am. Chem. Soc.* **115**, 8742 (1993); (42) B. L. Gaffney, P.-P. Kung, C. Wang, and R. A. Jones, *J. Am. Chem. Soc.* **117**, 12281 (1995); (43) V. Ramesh, Y.-Z. Xu, and G. C. K. Roberts, *FEBS Lett.* **363**, 61 (1995); (44) G. Carlström, S.-M. Chen, S. Miick, and W. J. Chazin, *Methods Enzymol.* **261**, 163 (1995); (45) G. Carlström and W. J. Chazin, *Biochemistry* **35**, 3534 (1996); (46) R. Michalczyk, L. A. Silks III, and I. M. Russu, *Magn. Reson. Chem.* **34**, S97 (1996); (47) D. Leitner, W. Schröder, and K. Weisz, *J. Am. Chem. Soc.* **120**, 7123 (1998); (48) D. J. Patel, A. Pardi and K. Itakura, *Science* **216**, 581 (1982); (49) K. J. Walters and I. M. Russu, *Biopolymers* **33**, 943 (1993); (50) C. K. Brush, M. P. Stone, and T. M. Harris, *Biochemistry* **27**, 115, (1988); (51) C. K. Brush, M. P. Stone, and T. M. Harris, *J. Am. Chem. Soc.* **110**, 4405 (1988); (52) S.-I. Yamakage, T. V. Maltseva, F. P. R. Nilson, A. Földesi, and J. Chattopadhyaya, *Nucleic Acids Res.* **22**, 1404 (1994); (53) P. Agback, T. V. Maltseva, S.-I. Yamakage, F. P. R. Nilson, P. Agback, and J. Chattopadhyaya, *Tetrahedron* **51**, 10065 (1995); (55) A. Ono, T. Makita, S. Tate, E. Kawashima, Y. Ishido, and M. Kainosho, *Magn. Reson. Chem.* **34**, S40 (1996); (56) X. Huang, P. Yu, E. LeProust, and X. Gao, *Nucleic Acids Res.* **25**, 4758 (1997); (57) J. Yang, L. Silks, R. Wu, N. Isern, C. Unkerfer, and M. A. Kennedy, *J. Magn. Reson.* **129**, 212 (1997); (58) A. Földesi, T. V. Maltseva, Z. Dinya, and J. Chattopadhyaya, *Tetrahedron* **54**, 14487 (1998); (59) T. V. Maltseva, A. Földesi, and J. Chattopadhyaya, *Tetrahedron* **54**, 14515 (1998); (60) C. Kojima, E. Kawashima, K. Toyama, K. Ohshima, Y. Ishido, M. Kainosho, and Y. Kyogoku, *J. Biomol. NMR* **11**, 103 (1998); (61) B. B. Ulyanov, V. I. Ivanov, E. E. Minyat, E. B. Khomyakova, M. V. Petrova, K. Lesiak, and T. L. James, *Biochemistry* **37**, 12715 (1998); (62) A. Kintanar, W.-C. Huang, D. C. Schindele, D. E. Wemmer, and G. Drobny, *J. Chem. Phys.* **92**, 6840 (1990); (63) T. M. Alam and G. Drobny, *Biochemistry* **29**, 3421 (1990); (64) T. M. Alam and G. Drobny, *J. Chem. Phys.* **92**, 6840 (1990); (65) T. M. Alam, J. Orban, and G. Drobny, *Biochemistry* **29**, 9610 (1990); (66) W.-C. Huang, J. Orban, A. Kintanar, B. R. Reid, and G. P. Drobny, *J. Am. Chem. Soc.* **112**, 9059 (1990); (67) T. M. Alam, J. Orban, and G. P. Drobny, *Biochemistry* **30**, 9229 (1991); (68) A. C. Wang, M. A. Kennedy, B. R. Reid, and G. P. Drobny, *J. Am. Chem. Soc.* **114**, 6583 (1992); (69) M.

scalar couplings between ^{15}N atoms through hydrogen bonds.[44] In these studies, DNA oligomers labeled at specific residues are effectively used to obtain unambiguous NMR data. A similar approach has been used by other groups for the analyses of relatively large DNA oligomers, such as a DNA triplex and a Holliday junction.[45,46]

A ^{13}C relaxation study of a DNA oligomer requires atom-specific labeling, since the contribution of the directly bonded neighboring ^{13}C is not negligible, and the quantitative motional analysis of such a complicated system is quite difficult. In such applications, atom-specific ^{13}C labeling at the 6 position and the 1′ position of thymidine has been used.[47,48] As an alternative approach to eliminate the neighboring ^{13}C contribution, DNA oligomers with a lower fraction of isotope (15% ^{13}C-enrichment) have been chemically prepared and successfully applied to the studies of nucleic acid dynamics.[49]

Another recent quite exciting application is the determination of scalar couplings across a base-paired hydrogen bond.[44,50] Such $^{2h}J_{NN}$ scalar couplings are found between the N^1 of purine nucleotides and the N^3 of pyrimidine nucleotides. For uniformly ^{15}N-labeled DNA oligomers the overlapped ^{15}N signals and the presence of ^{15}N–^{15}N scalar couplings within the base ring interfere with the detection

[42b] T. Szyperski, C. Fernández, A. Ono, K. Wüthrich, and M. Kainosho, *J. Magn. Reson.* **140**, 491 (1999).

[42c] T. Szyperski, C. Fernández, A. Ono, M. Kainosho, and K. Wüthrich, *J. Am. Chem. Soc.* **120**, 821 (1998).

[43a] C. Fernández, T. Szyperski, H. Iwai, S. Tate, A. Ono, K. Wüthrich, and M. Kainosho, *J. Biomol. NMR* **12**, 25 (1998).

[43b] C. Fernández, T. Szyperski, M. Billter, H. Iwai, A. Ono, M. Kainosho, and K. Wüthrich, *J. Mol. Biol.* **292**, 609 (1999).

[44a] K. Pervushin, A. Ono, C. Fernández, T. Szyperski, M. Kainosho, and K. Wüthrich, *Proc. Natl. Acad. Sci. U.S.A.* **95**, 14147 (1998).

[44b] K. Pervushin, C. Fernández, R. Riek, A. Ono, M. Kainosho, and K. Wüthrich, *J. Biomol. NMR* **16**, 39 (2000).

[45a] G. Carlström, S.-M. Chen, S. Miick, and W. J. Chazin, *Methods Enzymol.* **261**, 163 (1995).

[45b] G. Carlström and W. J. Chazin, *Biochemistry* **35**, 3534 (1996).

[46a] O. Bornet, G. Lancelot, L. Chanteloup, N. T. Thuong, and J.-M. Beau, *J. Biomol. NMR* **4**, 575 (1994).

[46b] O. Bornet and G. Lancelot, *J. Biomol. Struct. Dynam.* **12**, 803 (1995).

[47a] J. R. Williamson and S. G. Boxer, *Nucleic Acids Res.* **16**, 1529 (1988).

[47b] J. R. Williamson and S. G. Boxer, *Biochemistry* **28**, 2819 (1989).

[48a] F. Gaudin, F. Paquet, L. Chanteloup, J.-M. Beau, N. T. Thuong, and G. Lancelot, *J. Biomol. NMR* **5**, 49 (1995).

[48b] F. Paquet, F. Gaudin, and G. Lancelot, *J. Biomol. NMR* **8**, 252 (1996).

[49a] C. Kojima, A. Ono, M. Kainosho, and T. L. James, *J. Magn. Reson.* **135**, 310 (1998).

[49b] C. Kojima, A. Ono, M. Kainosho, and T. L. James, *J. Magn. Reson.* **136**, 169 (1999).

[50a] A. J. Dingley and S. Grzesiek, *J. Am. Chem. Soc.* **120**, 8293 (1998).

[50b] S. Grzesiek, F. Cordier, and A. J. Dingley, *Methods Enzymol.* **338** [4] 2001 (this issue).

[50c] C. Kojima, A. Ono, and M. Kainosho, *J. Biomol. NMR* **18**, 269 (2000).

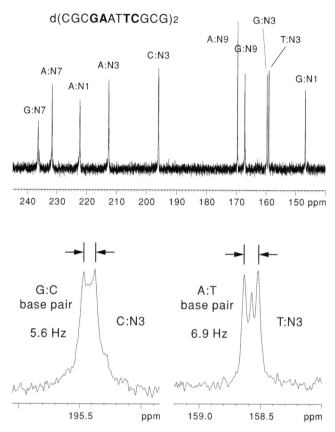

FIG. 5. Directly observed ^{15}N spectra of an atom- and site-specifically ^{15}N-labeled DNA dodecamer d(CGCGAATTCGCG)$_2$ whole spectrum (top) and the expanded spectra around cytidine N^3 (bottom left) and thymidine N^3 (bottom right). Oligomer sequence with labeled residues in bolface type, ^{15}N signal assignments, and $^{2h}J_{NN}$ values are shown. [Adapted from C. Kojima, A. Ono, and M. Kainosho, *J. Biomol. NMR* **18**, 269 (2000), with kind permission from Kluwer Academic Publishers.]

of the scalar couplings through hydrogen bonds. In Fig. 5, the directly observed ^{15}N signals of a DNA dodecamer duplex are shown.

^{15}N enriched 2′-deoxycytidine and thymidine specifically labeled in the N^3 position are incorporated into the sequence, d(CGC**GA**AT**TC**GCG)$_2$, where the bold and underlined letters are ^{15}N-enriched residues and the 2′-deoxyadenosine and 2′-deoxyguanosine are uniformly ^{15}N-enriched. The bottom panel in Fig. 5 depicts the expanded region of the 2′-deoxycytidine N^3 and thymidine N^3. The center peaks arise from the ^{14}N component of the purine nucleotides at the N^1 positions. Atom- and site-specific labeling facilitates easy determination of the scalar coupling constants across hydrogen bonds.[50c]

Stereoselectively ^{2}H/^{13}C Doubly Labeled DNA

The site-selectively and stereoselectively ^{2}H/^{13}C-labeled nucleosides are used for the accurate determination of the $^3J_{HH}$, $^3J_{HC}$, and $^3J_{HP}$ coupling constants.[51,52] The DNA oligomers containing (2'R)- and (2'S)-isotopomers of 2'-[1',2',3',4',5'-^{13}C$_5$; 2'-^{2}H]deoxyadenosine are used for the determination of $^3J_{H1'H2'}$, $^3J_{H1'H2''}$, $^3J_{H2'H3'}$, $^3J_{H2''H3'}$, and $^3J_{H3'H4'}$.[51] Usually, these coupling constants are difficult to determine with good precision since the H2' and H2'' signals are apparently broadened because of the scalar and dipolar couplings between methylene protons. Substituting one proton at the 2'-carbon with a deuterium eliminates the couplings between the methylene protons, and thus the signals for the remaining proton observed in HSQC and HCCH-E.COSY spectra become sharper, as shown in Fig. 6.

The absence of the passive couplings allows the precise and accurate determination of $^3J_{H3'H4'}$. Stereoselective deuteration at the 5'-position, even without ^{13}C-labeling, of DNA oligomers is useful for accurate signal assignments and structural studies of the H5' (pro-S) and H5'' (pro-R) proton signals.[52,53] When the methods of Kawashima et al. are used for 5' deuteration, the peak intensities of the H5' and H5'' signals are proportional to the residual proton concentrations (approximately 1 : 2), and the assignment of these signals is easily and reliably performed. The stereoselectively ^{2}H/^{13}C doubly labeled nucleotides at the 5'-position are used to determine the backbone β angle highly precisely, using the $^3J_{HP}$ coupling constants.[52] In this application, the 5'(R/S) signals disappears, as a result of relaxation without deuterium labeling. The preparation of these stereoselective ^{2}H/^{13}C doubly labeled samples becomes important for detailed and precise structural studies of relatively large DNA oligomers.

Summary

The solid-phase chemical synthesis method has a strong advantage over the enzymatic method for preparing selectively labeled DNA oligomers. Atom-specific and fully labeled 2'-deoxynucleosides are economically prepared with routinely available isotope precursors using this synthetic route. Special DNA oligomers prepared by advanced labeling techniques are needed for advanced NMR applications, and chemical synthesis is the method of choice to respond to such demands.

[51] J. Kurita, M. Kawaguchi, T. Shiina, S. Tate, A. M. Ono, A. Ono, and M. Kainosho, *Nucleic Acids Symp. Ser.* **32**, 49 (1995).

[52] S. Tate, Y. Kubo, A. Ono, and M. Kainosho, *J. Am. Chem. Soc.* **117**, 7277 (1995).

[53a] A. Ono, T. Makita, S. Tate, E. Kawashima, Y. Ishido, and M. Kainosho, *Magn. Reson. Chem.* **34**, S40 (1996).

[53b] C. Kojima, E. Kawashima, K. Toyama, K. Ohshima, Y. Ishido, M. Kainosho, and Y. Kyogoku, *J. Biomol. NMR* **11**, 103 (1998).

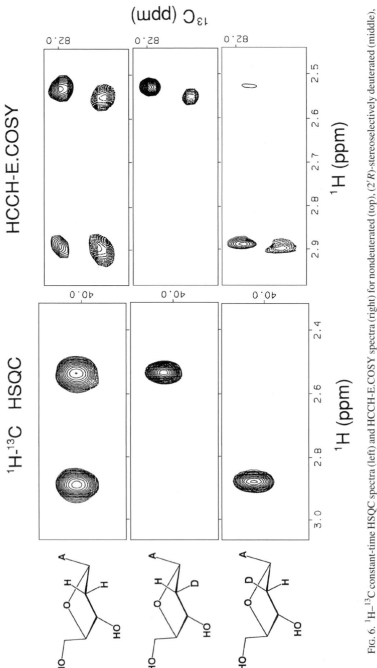

FIG. 6. ^{1}H–^{13}C constant-time HSQC spectra (left) and HCCH-E.COSY spectra (right) for nondeuterated (top), (2'R)-stereoselectively deuterated (middle), and (2'S)-stereoselectively deuterated (bottom) ^{13}C-labeled d(CGCGAATTCGCG)$_2$. [Adapted from J. Kurita, M. Kawaguchi, T. Shiina, S. Tate, A. M. Or, A. S. Ono, and M. Kainosho. *Nucleic Acids Symp. Ser.* **34**, 49 (1995).]

As a summary of this chapter, two tables are given. Table I lists the labeled nu-cleosides reported to be available by chemical syntheses. Table II lists the NMR studies using labeled DNA oligomers that were prepared by chemical syntheses.

Acknowledgments

This work was supported by CREST (Core Research for Evolutional Science and Technology) of the Japan Science and Technology Cooperation (JST) and by a Grant-in-Aid for Scientific Research on Priority Area from the Ministry of Education, Science and Culture (09235237).

[12] Uniform $^{13}C/^{15}N$-Labeling of DNA by Tandem Repeat Amplification

By MILTON H. WERNER, VINEET GUPTA, LESTER J. LAMBERT, and TAKASHI NAGATA

Introduction

Multidimensional heteronuclear nuclear magnetic resonance (NMR) has be-come a standard technique to determine the three-dimensional structure of proteins and RNA in solution.[1,2] One of the most important advances in the application of NMR spectroscopy to the study of biological systems has been the ease of in-corporation of ^{13}C and/or ^{15}N into proteins[3–6] and RNA.[7–9] The enrichment of macromolecules in these stable isotopes allows for the dispersion of ^1H, ^{13}C, and ^{15}N chemical shifts into multiple spectral dimensions in a manner that preserves the chemical and/or spatial relationship between atoms within a molecule of interest.[1,2]

[1] G. M. Clore and A. M. Gronenborn, in "Biological Magnetic Resonance" (N. R. Krishna and L. J. Berliner, eds.), Vol. 16, p. 3. 1998.

[2] K. Y. Chang and G. Varani, Nat. Struct. Biol. 4(Suppl.), 854 (1997).

[3] L. P. McIntosh and F. W. Dahlquist, Quart. Rev. Biophys. 23, 1 (1990).

[4] A. P. Hansen, A. M. Petros, A. P. Mazar, T. M. Pederson, A. Rueter, and S. W. Fesik, Biochemistry 31, 12713 (1992).

[5] D. LeMaster, Prog. NMR Spect. 26, 371 (1994).

[6] K. H. Gardner and L. E. Kay, in "Biological Magnetic Resonance" (N. R. Krishna and L. J. Berliner, eds.), Vol. 16, p. 27. 1998.

[7] R. T. Batey, M. Inada, E. Kujawinski, J. D. Puglisi, and J. R. Williamson, Nucleic Acids Res. 20, 4515 (1992).

[8] E. P. Nikonowicz, A. Sirr, P. Legault, F. M. Jucker, L. M. Baer, and A. Pardi, Nucleic Acids Res. 20, 4507 (1992).

[9] R. T. Batey, J. L. Battiste, and J. R. Williamson, Methods Enzymol. 261, 300 (1995).

Correlation of ^{1}H resonances with their attached ^{13}C and/or ^{15}N chemical shifts enhances the spectral resolution and facilitates the analysis of macromolecular structure from both angular (J coupling) and distance (nuclear Overhouser effect, NOE) restraints. Many of the experimental benefits of ^{13}C and/or ^{15}N enrichment realized for protein and RNA spectroscopy may also be applicable to analysis of DNA structure and its complexes with proteins. In pursuit of such a goal, three different approaches to the preparation of isotopically enriched DNA have been described in recent years.[10–18] Solid-phase chemical synthesis using either phosphonate[10] or phosphoramidite[11,12] chemistry permits the greatest flexibility in labeling schemes, but may be technically demanding depending on circumstance. For this reason, enzymatic synthesis has become a more popular alternative for the preparation of uniformly labeled deoxyribonucleic acids.[13–18]

Two fundamentally different enzymatic strategies have been described to date. The first approach, proposed by Zimmer and Crothers,[13] employed a duplex DNA with a single-stranded overhang that served as a template for the synthesis of a complementary strand of user-specified sequence. Since its introduction, three different polymerases have been used to synthesize labeled DNA using this template-directed approach, each with its particular advantages: DNA polymerase Klenow,[13,14] *Taq* polymerase,[15] and reverse transcriptase.[18] An alternative enzymatic strategy proposed by Louis et al.[16] employs a tandem repeat of user-defined sequence that can be amplified by thermal cycling. This second approach was originally developed to create long, tandemly repeated DNA sequences for the study of protein binding[19,20] and for the production of hybridization probes,[21] a technique known as the concatemer chain reaction.[21] Irrespective of the method employed, enzymatic syntheses rely on the availability of milligram quantities of uniformly labeled ^{13}C-and/or ^{15}N-deoxynucleotide triphosphates (dNTPs), which must be prepared from the genomic DNA of a suitable microorganism.

[10] R. A. Jones, *Methods Mol. Biol.* **26**, 207 (1994).

[11] A. Ono, S. Tate, and M. Kainosho, *Tanpakushitsu Kakusan Koso* **40**, 1509 (1995).

[12] S. Quant, R. W. Weschelberger, M. A. Wolter, K.-H. Wörner, P. Schell, J. W. Engels, C. Griesinger, and H. Schwalbe, *Tetrahedron Lett.* **35**, 6649 (1994).

[13] D. P. Zimmer and D. M. Crothers, *Proc. Natl. Acad. Sci. U.S.A.* **92**, 3091 (1995).

[14] D. E. Smith, J.-Y. Su, and F. M. Jucker, *J. Biomol. NMR* **10**, 245 (1997).

[15] J. E. Masse, P. Bortman, T. Dieckmann, and J. Feigon, *Nucleic Acids Res.* **26**, 2618 (1998).

[16] J. M. Louis, R. G. Martin, G. M. Clore, and A. M. Gronenborn, *J. Biol. Chem.* **273**, 2374 (1998).

[17] X. Chen, S. V. S. Marriappan, J. J. Kelly III, J. H. Bushweller, E. M. Bradbury, and G. Gupta, *FEBS Lett.* **436**, 372 (1998).

[18] A. Kettani, S. Bouaziz, E. Skripkin, A. Majumdar, W. Wang, R. A. Jones, and D. J. Patel, *Structure* **7**, 803 (1999).

[19] W. A. Rudert and M. Trucco, *Nucleic Acids Res.* **18**, 6460 (1990).

[20] F. Hemat and K. McEntee, *Biochem. Biophys. Res. Commun.* **205**, 475 (1994).

[21] M. J. White, B. W. Fristensky, and W. F. Thompson, *Anal. Biochem.* **199**, 184 (1991).

In this report, we describe detailed procedures for the large-scale synthesis of uniformly labeled duplex DNAs using polymerase chain reaction (PCR)-amplified, tandemly repeated templates of desired sequence. Because of the ready availability of labeled nucleic acids as a by-product of the production of labeled proteins expressed in *Escherichia coli,* our method focuses on the simultaneous preparation of labeled proteins and nucleic acids and their independent isolation from a single growth in this organism. We also describe the preparation of the kinasing enzymes and polymerase required for routine 10–50 mg scale synthesis of isotope-enriched DNA. The method is robust, providing a minimum amplification of 800-fold with essentially no synthetic by-products.

High Density Growth of *Escherichia coli* by Adaptive Control Fermentation

Reagents

> *Escherichia coli* BL21 (47092, ATCC, Rockville, MD)
> [^{13}C] Glucose
> [^{15}N] Ammonium chloride
> Minimal growth media

Equipment

> Bench scale fermentor equipped with a mass flow analyzer, dissolved oxygen, pH, and temperature probes.
> YSI 2000 glucose analyzer (optional)
> Glucose test strips

Procedure

1. *Minimal growth media.* Prepare minimal growth media by autoclaving 6.8 g anhydrous sodium phosphate dibasic, 3.0 g anhydrous potassium phosphate monobasic, 0.5 g NaCl dissolved in 1 liter of deionized water. Separately prepare stock solutions by sterile filtration of the following: 1 M MgSO$_4$, vitamin solution (5 mg/ml niacin, 5 mg/ml thiamin) (store at 4°, good for 1 year), and 1000 × trace elements[22,23] (40.8 mM CaCl$_2$ · H$_2$O, 21.6 mM FeSO$_4$ · 7H$_2$O, 5.8 mM MnCl$_2$ · 4H$_2$O, 3.4 mM CoCl$_2$ · 6H$_2$O, 2.4 mM ZnSO$_4$ · 7H$_2$O, 1.6 mM CuCl$_2$ · H$_2$O, 0.64 mM boric acid, 0.2 mM (NH$_4$)$_6$Mo$_7$O$_{24}$ · 4H$_2$O) (store in the dark at room temperature). Immediately prior to use, add to each liter of autoclaved

[22] S. Bauer and J. Shiloach, *Biotech. Bioengineer.* **16,** 933 (1974).

[23] M. Cai, Y. Huang, K. Sakaguchi, G. M. Clore, A. M. Gronenborn, and R. Craigie, *J. Biomol. NMR* **11,** 97 (1998).

salts: 3 ml 1 M MgSO$_4$, 5 ml vitamin solution, 1 ml trace elements. For ^{15}N labeling, add 2.0 g/liter of [^{15}N]ammonium chloride and 12 ml of 50% (w/v) anhydrous
glucose solution. For ^{13}C/^{15}N labeling, add 2.0 g/liter [^{15}N]ammonium chloride
and 5.0 g/liter [^{13}C]glucose. Add the appropriate antibiotic if protein labeling is to
be done simultaneously with nucleic acid labeling.

2. Take a small scratching of bacterial glycerol stock, inoculate a 1 ml culture
in Luria broth (LB), and grow to saturation. Transfer 10 μl of this culture to 1 ml
of minimal media and grow to saturation. The entire minimal saturated culture is
then used to inoculate a 25 ml minimal culture and growth is continued for 6 more
hours. The entire 25 ml minimal culture is then transferred to 1/5 volume of the
desired final culture and grown overnight.

Note for ^2H *labeling:* This protocol is also suitable for fractional or perdeuteration of proteins or nucleic acids. The level of deuteration using protonated glucose
is approximately 18% less than the percentage of D$_2$O used to prepare the media.
Therefore, for deuteration levels above 82%, [^{13}C,^{2}H$_7$]glucose must be used. For
^{2}H labeling, the anhydrous phosphate salts are dissolved in D$_2$O, sterile filtered,
and the remaining minimal media reagents added as dry solids directly to the
D$_2$O–phosphate solution just before use. Trace elements may be added as an H$_2$O
solution since they represent only 0.1% of the growth medium by volume. To adapt
the cells to D$_2$O medium, a 1 ml minimal H$_2$O culture is first diluted 10-fold into
a 1 ml minimal medium containing 50% D$_2$O (v/v) and grown to saturation. This
1 ml 50% D$_2$O culture is then diluted 50-fold into a 1 ml "100%" D$_2$O medium
and grown to saturation. The remaining procedures for large-scale growth are the
same as those described for minimal medium H$_2$O cultures.

3. *Adaptive control fermentation.* The objective of adaptive control fermentation is to maintain the bacterial culture in log phase growth for as long as possible to
maximize the production of biomass from which nucleic acids or an overexpressed
protein can be isolated (Fig. 1). To achieve this end, all of the nutrients are added to
the medium at the beginning and the growth is monitored for pH, temperature, and
dissolved oxygen continuously using a microprocessor-controlled measurement
system (Fig. 1). The adaptive controller receives real-time reports on the dissolved
oxygen level (pO_2) and attempts to hold the culture at 30% pO_2 by linear variation
of both stir rate and air flow into the medium. To accomplish this task, a recursive
least-squares method is employed based on the algorithm of Hsiao *et al.*[24] To reach
the target value for pO_2, the adaptive controller uses a regression calculation to
"learn" what percentage change in stir rate and air flow is necessary to maintain
the culture at the target level.[24] Since pO_2 is evaluated every 15 ms, the controller
responds effectively to the change in growth rate as the cell density increases.
We have used this control mechanism to grow bacterial cultures to 20 A_{600} units

[24] J. Hsiao, M. Ahluwalia, J. B. Kaufman, T. R. Clem, and J. Shiloach, *Ann. NY Acad. Sci.* **665,** 320 (1992).

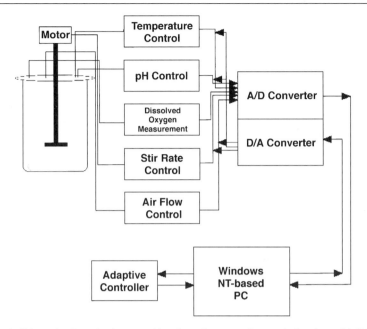

FIG. 1. *Schematic of an adaptive control bench-top fermentor.* A water-jacketed vessel is linked to a 4° water source with a computer-controlled heater. Individual measurement probes are inserted into the vessel to monitor temperature, stir rate, air flow, pH, and dissolved oxygen. The measured control parameters are digitized and fed into a Windows NT–based software package supplied by the fermentor manufacturer. The adaptive controller is a stand-alone module that receives the measured stir rate, air flow, and dissolved oxygen values from the NT computer and returns adjusted stir rate and air flow values necessary to maintain the target value for dissolved oxygen. This system has been implemented for both B. Braun Biotech DCU and New Brunswick BioFlow 2000 bench-top fermentation systems.

in minimal media and to >100 A_{600} units in LB in a single liter. The system is self-contained and requires minimal user input during the growth period. Because of the optimal growth condition under adaptive control, we routinely induce bacterial cultures at 3–5 A_{600} units for protein production, providing a 5- to 10-fold increase in biomass per unit volume. Off-line monitoring of carbon sources permits the quantitative consumption of labeling precursors to ensure maximal utilization (Fig. 2).

Preparation of Bulk Nucleic Acids

Reagents

Liquified phenol (Baker, Philipsburg, NJ)
Hydroxyquinoline (Fluka, Ronkonkoma, NY)

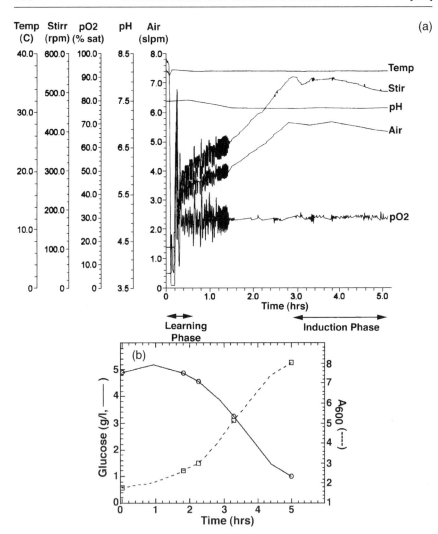

FIG. 2. *Time course of growth parameters during adaptive control fermentation.* (a) The various control parameters are monitored in real time and plotted to follow the time course of growth of the culture. The learning phase (0–0.5 hr) is accompanied by a large oscillation in the target function (pO_2) as the controller "learns" how to vary stir rate and air flow to hold the culture at 30% pO_2. Once the variation needed is learned, a linear increase in these parameters is observed. The oscillations in pO_2 dampen as the culture density increases. At the point of protein expression induction, the culture slows cellular replication, resulting in a plateau in the measured growth parameters, consistent with the switch from a growth to an expression phase for an inducible protein. When applied to the labeling of nucleic acids alone, the culture will continue in log-phase growth to very high densities ($\geq 20 \ A_{600}$) as long as nutrients (carbon and nitrogen sources) remain in the medium. (b) Variation in optical density and glucose consumption during adaptive control fermentation. The consumption of glucose in this example was approximately 5 g/liter in a 1.2 liter culture grown to 8 A_{600} units.

Chloroform
Isoamyl alcohol
100% Ethanol (USP grade)
3 M Sodium acetate, pH 5.3
Lysis buffer:

Nucleic acids only: 10 mM Tris-HCl (pH 8.0), 100 mM NaCl, 1 mM Na$_2$EDTA
Nucleic acids + soluble His-tagged protein: 50 mM Tris-HCl (pH 8.0), 1 M
    NaCl, 10 mM benzamidine hydrochloride
Nucleic acids + soluble untagged protein: 50 mM Tris-HCl (pH 8.0), 50 mM
    NaCl, 10 mM benzamidine hydrochloride, 1 mM dithiothreitol (DTT)
Lysozyme (Sigma, St. Louis, MO)
Pronase (Sigma)
Hydrated diethyl ether: Diethyl ether treated with an equal volume of 50 mM
    Tris-HCl, 1 mM Na$_2$EDTA. Aqueous layer (bottom) is removed before use.

Equipment

Homogenizer (Fisher Power Gen 700)
500 ml separatory funnel
French press
1000 Molecular weight cutoff (MWCO) dialysis membrane

Procedure

1. Prepare phenol. Equilibrate phenol according to established procedures.[25]
Mix equilibrated phenol, chloroform, and isoamyl alcohol in a 25 : 24 : 1 (v : v : v)
ratio. This may be stored at 4° in an amber bottle for 1–2 months.

2. Homogenize cell paste. Volumes indicated are suitable for the preparation
of nucleic acids from 35 g of wet cells and may be scaled linearly for different
amounts of cell paste. The cells are suspended with a homogenizer in 4 ml lysis
buffer/gram wet cell paste until a smooth slurry is created. The choice of lysis
buffer is dependent on whether proteins are to be simultaneously isolated and by
what method; the buffer composition for several common examples is described
under *Reagents*.

3. Process lysate. (a) *For nucleic acids only:* The homogenized cells are passed
through a French press twice at 500–1000 psi, 1 mg/ml lysozyme is added, and the
lysate is incubated with nutation at 37° for 1 hr. Pronase (1.5 mg/ml) is added and
the lysate incubated with nutation at 37° a further 8–12 hrs.[26,27] Sodium dodecyl

[25] J. Sambrook, E. F. Fritsch, and T. Maniatis, "Molecular Cloning." (Cold Spring Harbor Laboratory
    Press), Cold Spring Harbor, NY, 1989.
[26] C. A. Thomas, K. I. Berns, and T. J. Kelley, *Proced. Nucl. Acids Res.* **1**, 535 (1966).
[27] M. G. Smith, *Methods Enzymol.* **12**, 545 (1967).

sulfate (SDS) (10% w/v) is added with gentle stirring until the opaque, caramel-colored lysate becomes translucent. One volume phenol : chloroform : isoamyl alcohol is added and stirred at room temperature for 30 min in a 500 ml centrifuge bottle in a chemical hood. (Phenol is caustic and appropriate safety precautions should be observed.) Spin the phenolic solution at $5000g$ for 10 min and transfer the top (aqueous) layer with a glass pipette into a new centrifuge bottle. This aqueous layer will be a translucent, pinkish solution. Reextract the phenolic layer with a second equal volume of lysis buffer, shake vigorously, and centrifuge. Combine two aqueous layers and add an equal volume of hydrated diethyl ether, shake briefly, and centrifuge. Recover aqueous layer (bottom) with a separatory funnel. Add 1/10 volume 3 M sodium acetate (pH 5.3) and 3 volumes 100% ethanol to precipitate nucleic acids. Allow precipitation to continue for at least 12 hr at $-20°C$.

(b) *For nucleic acids + soluble His-tagged proteins:* Cells are homogenized and French pressed in the appropriate lysis buffer (see *Reagents*) and centrifuged at $5000g$. The pellet is kept frozen until recombined with the soluble nucleic acid fraction as described below. The supernatant is centrifuged at $100,000g$ in a preparative ultracentrifuge at 4° for 1 hr and subsequently passed over Ni^{2+}-chelate affinity resin after adding imidazole to the supernatant to 30 mM. The flow-through of the affinity column, which contains the nucleic acids, is combined with the membrane pellet, lyophilized, and resuspended with homogenization in the lysis buffer described in section (a), for *nucleic acids only,* and processed as described above.

(c) *For nucleic acids + soluble untagged proteins:* Cells are homogenized and French pressed in the appropriate lysis buffer (see *Reagents*) and centrifuged at $5000g$. The pellet is kept frozen until recombined with the soluble nucleic acid fraction as described below. NaCl is added to the supernatant to a final concentration of 500 mM and passed over anion-exchange resin (Q-sepharose) at 2 ml/min (2.6×40 cm). Elution of the nucleic acid fraction is achieved with a linear gradient of 5 column volumes from 500 mM to 1.2 M NaCl in 50 mM Tris-HCl (pH 8.0). The eluted nucleic acids are then lyophilized, combined with the membrane pellet, and treated identically as described above. The desired protein from the flow-through of the anion-exchange column can be subsequently purified with a user-defined protocol following removal of excess salts by dialysis.

Notes on preparation of lysate. Most of the chromosomal DNA is membrane-bound following French press.[26,27] The use of lysozyme and pronase solubilizes the DNA, permitting quantitative recovery of this nucleic acid fraction.[26,27] Pronase, a mixture of three aggressive endoproteinases from *Streptomyces griseus,* will completely digest all proteins present in the lysate. Failure to use lysozyme and/or pronase will hinder the ability to recover the DNA reproducibly. Heating the phenolic solution or blending the phenolic solution is less reliable for DNA recovery in our experience.[7–9,13–15] Enzyme treatment also has an added advantage when large scale ($\geq$20 g) extraction of cells is performed. Following phenolic extraction, a pronase-treated lysate will display only a thin white interface layer that is no more than a few millimeters thick. This improves the ease with which the aqueous layer can be recovered with a glass pipette. Stirring in SDS prior to phenolic extraction also helps in the recovery of the nucleic acid layer by shearing the chromosomal DNA, thereby reducing the viscosity of the aqueous layer following separation of the phenolic solution by centrifugation. It is imperative to add sufficient SDS that the lysate is completely clarified prior to phenolic extraction; otherwise, lipid and protein impurities will be carried along with the DNA that are not as easily removed in subsequent steps.

4. Prepare nucleic acids for digestion. The precipitated nucleic acids are centrifuged at 10,000g for 1 hr and the supernatant carefully decanted. The slightly yellowish pellet is air dried, resuspended in 75–125 ml 10 mM Tris-HCl, pH 8.0, and dialyzed in 1000 MWCO tubing against 4 liters of the same buffer for 8–12 hr at room temperature. At least two changes of this buffer should be done during dialysis to remove nonnucleic acid impurities and residual diethyl ether. The nucleic acids are subsequently centrifuged at 100,000g for 1 hr and the pH of the supernatant adjusted with NaOH to 7.5–8.0. Using the procedures outlined above, we routinely recover 28–30 mg nucleic acid/gram of wet cells following dialysis. (Because the nucleic acids are largely RNA, we assume a concentration of 40 μg/ml per A_{260} unit for the resuspended nucleic acids following dialysis.)

Preparation of Deoxynucleotide Monophosphates

Reagents

RNase-free DNase I (Boehringer-Mannheim)
Nuclease P1 (Sigma)
RNase A (Pharmacia, Piscataway, NJ)
MgSO$_4$
ZnSO$_4$
Affi-Gel 601 (Bio-Rad, Hercules, CA)
1 M Triethylammonium bicarbonate (TEABC)
1 M Triethylammonium acetate (TEAA)
1 M Ammonium bicarbonate (ABC, optional)

Equipment

5.0 × 20 cm medium pressure chromatography column
25 × 250 mm C$_{18}$ reversed-phase (RP) semipreparative HPLC column (Varian)

Procedures

1. *Digestion of duplex DNA.* The bulk nucleic acids are digested for 2 hr at 37°, pH 8.0, by the addition of 1 unit RNase-free DNase I per 660 μg total nucleic acid.

DNase I is shock sensitive, so very slow nutation of the nucleic acid solution should be used. DNA digestion is monitored by 0.7% (w/v) agarose gel electrophoresis. There should be no nucleic acid species binding ethidium bromide greater than 5000 base pairs prior to the next step, otherwise duplex DNA digestion could be incomplete. If necessary, add an additional aliquot of DNase I and incubate a second time.

2. *Digestion of RNA and single-strand DNA.* Following duplex DNA digestion, the nucleic acids are denatured by heating in a water bath at 95° for 10 min and rapidly chilled on ice. One unit of nuclease P1 is added per 24 mg total nucleic acid. Digestion is carried out at pH 8.0, 55°, for 2 hr, and the nucleic acids are again denatured and rapidly cooled on ice. A second, equal amount of Nuclease P1 is added without denaturation and digestion allowed to proceed for 2 more hr at 55°. At this point, 0.7% agarose gel electrophoresis should reveal the presence of only a minor band of ethidium bromide stainable nucleic acid of 100–200 base pairs in length. This material is resistant to mung bean nuclease, nuclease S1, exonucleases I and III, and *Bal*31 nuclease. It is easily digested with the addition of RNase A, suggesting that it is RNA (most likely tRNA given its size). One mg RNase A per 80 mg total nucleic acid is sufficient to completely digest this material within 2 hr at 37°.

Note on RNA digestion: We routinely observe a nuclease P1-resistant RNA species in *E. coli* nucleic acid preparations irrespective of the pH used, quantity of enzyme, or length of time allowed for digestion. This resistant RNA is present whether or not the cells are grown to saturation. If the nucleic acids are not denatured following the second P1 digestion, P1 will remove the 3′-phosphate from the mononucleotides released by RNase A. Although the product of RNase A/nuclease P1 digestion are ribonucleosides, they form only a minute fraction of the total RNA-derived products. It is essential that there be no ethidium bromide stainable nucleic acids remaining prior to separation of dNMPs from rNMPs by boronate affinity chromatography. Undigested nucleic acids, *whether DNA or RNA*, do not stick to the boronate affinity resin under the conditions described below, contaminating the dNMP fraction.

3. *Separation of dNMPs from rNMPs.* The dNMPs are separated from rNMPs by boronate affinity chromatography (Affi-Gel 601) as previously described.[7–9,28] To prepare the column, swell the resin in 20 volumes of cold, CO_2-acidified water (pH 4–4.5) and pack a 5.0 × 20 cm column at 1.0 ml/min, 4°. Once the bed height has stabilized, equilibrate the column with 150 ml 1 M TEABC at 4°. The column resin will shrink to one-half its original height. Do not readjust the column adapter since elution is conducted at acid pH which will reswell the gel. Fifteen g Affi-Gel 601 will efficiently separate the nucleotide monophosphates derived from 500 mg of bulk nucleic acid.

[28] M. Rosenberg, J. L. Wiebers, and P. T. Gilham, *Biochemistry* **11**, 3623 (1972).

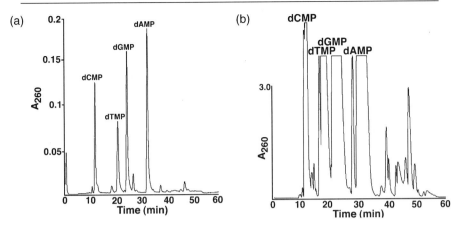

FIG. 3. *Digestion and purification of dNMPs.* (a) RP-HPLC chromatogram of dNMPs following boronate affinity chromatography. The dNMPs are relative pure, with a few minor products that represent small single-stranded nucleotide fragments resistant to digestion. (b) Preparative RP-HPLC of dNMP fraction following boronate affinity chromatography. The flow-through fractions from boronate affinity chromatography are lyophilized and resuspended in 20 ml 0.1 *M* TEAA. Five ml is then injected and purified by RP-HPLC as described in the text. The individual nucleotides are collected and lyophilized prior to phosphorylation. Reinjection of preparative RP-HPLC dNMPs on an analytical 4.6 × 25 cm RP-HPLC column reveals each nucleotide solution to be >98% pure.

The digested nucleic acids are dried by centrifugal lyophilization and resuspended in cold 1 *M* TEABC. Chill the nucleotide solution and pump onto the column at 4°, 0.5 ml/min. The column is washed with 1 *M* TEABC, 4°, 0.5 ml/min, until the A_{254} returns to baseline on the detector (250 ml for a column prepared from 15 g Affi-Gel 601). The flow-through fractions, containing the dNMPs, are combined and lyophilized. The rNMPs are eluted by washing the column with 250 ml CO_2-acidified water (pH 4–4.5). Purity of the nucleotide monophosphates can be assessed by C_{18} reversed-phase HPLC (RP-HPLC) in 100 m*M* TEAA pH 6.8 using a linear gradient of 2–18% methanol, 10 ml/min for 60 min (Fig. 3). Preparative purification of individual nucleotides can be accomplished with the same gradient (Fig. 3). Typical yields are 90–100 mg each dCMP, dAMP, dGMP, and 60–80 mg of dTMP after preparative RP-HPLC of processed nucleic acids derived from 35 g wet cell paste.

Notes: TEABC can be replaced by 1 *M* ammonium bicarbonate (ABC), pH 9.5. The advantage of ABC is its easier removal by lyophilization relative to TEABC. (One *M* TEABC is prepared by stirring 141 ml HPLC-grade triethylamine in 859 ml deionized water on ice in a chemical hood. CO_2 is bubbled into the mixture with vigorous stirring until most of the triethylamine is dissolved in the water. The pH is then monitored with further bubbling until it reaches 9–9.5. One *M* TEAA stock solution can be prepared by mixing 141 ml HPLC-grade triethylamine, 57 ml glacial acetic acid, and 750 ml deionized water at room temperature in a chemical hood. The pH is then adjusted with additional glacial acetic acid until it reaches pH 6.8. The stock solution is filtered and stable at room temperature for at least 1 year.)

Preparation of Deoxynucleotide Triphosphates

Reagents

Yeast thymidylate kinase (TMPK) (laboratory prepared)
E. coli Cytidylate monophosphate kinase (CMPK) (laboratory prepared)
Myokinase (MK) (Sigma)
Pyruvate kinase (PK) (Sigma)
Guanylate kinase (GK) (Sigma)
Phospho(enol)pyruvate (PEP) (Sigma)
$10\times$ Phosphorylation buffer (800 mM Tris-HCl, pH 7.5; 200 mM KCl, 200 mM MgCl$_2$)
Benzamidine hydrochloride
Glutathione agarose (Sigma)
Blue Sepharose (Pharmacia)

Equipment

1.6 × 20 cm medium pressure column
2.6 × 20 cm medium pressure column

Procedure

1. *Preparation of TMPK.* All procedures should be carried out at 4°. Yeast TMPK is expressed as a glutathione transferase (GST) fusion[14] and grown by conventional methods. Cells are disrupted by French press in 20 mM sodium phosphate, pH 7, 150 mM NaCl, 10mM benzamidine hydrochloride. Triton X-100 is added to 1% (v/v) and the lysate clarified by centrifugation at 5000g. The supernatant is ultracentrifuged at 100,000g for 1 hr and loaded onto a 1.6 × 15 cm glutathione–agarose column at 1 ml/min. The column is extensively washed with 20 mM sodium phosphate, pH 7, 150 mM sodium chloride until the absorbance at 280 nm returns to baseline. The protein is then eluted isocratically with 50 mM Tris-HCl, pH 8, 100 mM NaCl, 15 mM reduced glutathione, 1 mM DTT, 0.05% (v/v) nonidet P-40 (NP-40). The eluted protein is extensively dialyzed against 25 mM HEPES–KOH, pH 7.4, 100 mM NaCl, 2 mM DTT, 1 mM phenyl-methylsulfonyl fluoride (PMSF), 10% (v/v) glycerol, 0.05% NP-40. The protein is concentrated to 1.7 mg/ml and stored in aliquots at −80°.

2. *Preparation of CMPK.* All procedures should be carried out at 4°. *E. coli* CMPK is expressed by conventional methods in pET22b as described.[29] Cells are disrupted by French press in 50 m*M* Tris-HCl, pH 7.5, 10 m*M* benzamidine hydrochloride, and clarified by centrifugation at 5000*g*. The supernatant is subsequently centrifuged at 100,000*g* and loaded onto a 2.6 × 15 cm Blue Sepharose column equilibrated with 50 m*M* Tris-HCl, pH 7.5. The column is washed until the absorbance at 280 nm is returned to baseline, and the protein is eluted isocratically with 50 m*M* Tris-HCl, pH 7.5, 1.2 *M* NaCl. The eluted protein is extensively dialyzed against 50 m*M* Tris-HCl, pH 7.5, 1 m*M* DTT. The protein is concentrated by ultrafiltration to 2.5 mg/ml, glycerol is added to 50% (v/v), and the samples are stored in aliquots at −20°.

3. *Enzymatic phosphorylation.* The individually purified dNMPs are converted >95% to their corresponding dNTPs in four individual reactions by a modification of established procedures,[7-9,13-15,30,31] to include the utilization of recombinant TMPK and CMPK to phosphorylate dTMP and dCMP, respectively,[14,15,29] Myokinase (MK, final concentration = 1 unit/μl) and guanylate kinase (GK, final concentration = 0.1 unit/μl) are dissolved in degassed 50 m*M* Tris-HCl, pH 7.5, 1 m*M* DTT, 50% glycerol (v/v), and stored in a screw-cap vial at −20°. TMPK and CMPK are prepared under similar conditions as described above. The phosphorylation reaction for dAMP, dCMP, and dGMP contains 1× phosphorylation buffer, 10 m*M* PEP, 1 m*M* dNMP. TMPK efficiently catalyzes the conversion of dTMP to dTTP; therefore, PEP and PK are excluded from this reaction. Kinasing enzymes and rATP are added to a final concentration as follows: 69 units MK/mg dAMP, 28 units PK/mg dAMP, 0.1 m *M* rATP; 90 μg CMPK/mg dCMP, 28 units PK/mg dCMP, 0.1mM rATP; 0.46 units GK/mg dGMP, 28 units PK/mg dGMP, 0.5 m*M* rATP; 90 μg TMPK/mg dTMP, 5 m*M* rATP. The reaction is assembled in the following order: buffer, dNMP, PEP, rATP, PK, and finally the appropriate nucleotide monophosphate kinase. Prior to the addition of enzymes, the reaction solution is extensively sparged with argon. The reaction is incubated at 37° for 8 hr (dAMP, dCMP), 12 hr (dTMP), or 20 hr (dGMP) with the conversion monitored by analytical RP-HPLC using a 0→20% gradient of 98 m*M* TEAA, 2% CH₃OH vs 78 m*M* TEAA, 30% CH₃OH, pH 6.7, over 30 min at 0.5 ml/min. For 40 × 500 μl reactions, 3.3–3.9 mg of each dNTP is required (Fig. 4).

Following phosphorylation, the reaction is quenched by the addition of Na₂EDTA to 20 m*M* and the dNTPs purified by semipreparative C₁₈ RP-HPLC

[29] N. Bucurenci, H. Sakamoto, P. Briozzo, N. Palibroda, L. Serina, R. S. Sarfati, G. Labesse, G. Briand, A. Danchin, O. Barzu, and A-M. Gilles, *J. Biol. Chem.* **271,** 2856 (1996).
[30] E. S. Simon, S. Grabowski, and G. M. Whitesides, *J. Org. Chem.* **55,** 1834 (1990).
[31] J. V. Hines, S. M. Landry, G. Varani, and I. Tinoco, *J. Am. Chem. Soc.* **116,** 5823 (1994).

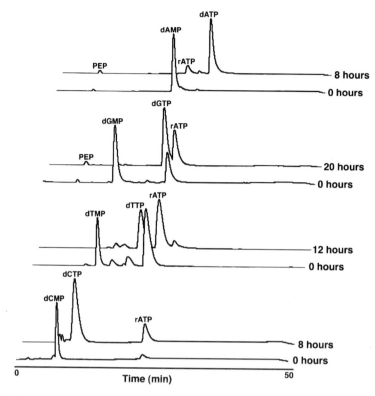

FIG. 4. *Conversion of dNMPs to dNTPs*. Analytical RP-HPLC chromatograms of the beginning and end-point composition of enzymatic phosphorylation reactions. For each pair of chromatograms, the zero time point appears in the foreground and the end point appears in the background, offset diagonally to demonstrate the nearly complete conversion of dNMP to dNTP. Note that for dTMP→dTTP, a 25-fold greater concentration of rATP is used. At this higher concentration, rAMP and rADP are observed in the dTMP zero time-point chromatogram because of spontaneous dephosphorylation of the rATP stock solution. For dCMP→dCTP, the PEP retention time is coincident with that of dCMP itself, accounting for the presence of a small peak in the dCTP chromatogram.

using the conditions described under *"Preparation of deoxynucleotide monophosphates."* The purified dNTPs are concentrated by lyophilization and stored at −80°.

PCR Amplification of Tandemly Repeated Oligonucleotides

Reagents

Pfu DNA polymerase (ATCC 87496)
Whatman P11 phosphocellulose
Anion-exchange resin

Polyethyleneimine (Fluka)
$(NH_4)_2SO_4$ (ultrapure enzyme grade)
$10 \times Pfu$ PCR buffer: 200 mM Tris-HCl, pH 8.8, 100 mM KCl, 100 mM
$(NH_4)SO_4$, 1.0% Triton X-100, 1.2 mg/ml bovine serum albumin (BSA)
100 mM MgSO$_4$
20 mM labeled dNTPs
Gel-purified oligonucleotide templates
EcoRV (Promega, Madison, WI)

Equipment

Thermal cycler (non-Peltier model preferred)
2.6 × 20 cm medium pressure chromatography column
2.5 × 10 cm Vydac 301 VHP DEAE HPLC column

Procedures

1. Preparation of Pfu polymerase. The expression vector for Pfu polymerase
is transformed into E. coli HMS174(DE3) pLysS using standard procedures.[25]
The cells can be grown by conventional methods and induced with 0.4 mM IPTG
for 4 hr. The cells are harvested by centrifugation at 5000g and lysed in 50 mM
Tris-HCl, 1 M NaCl, 10 mM Na$_2$EDTA, 10 mM benzamidine hydrochloride using the homogenization/French press approach described under "Preparation of
bulk nucleic acids." Unless indicated otherwise, all subsequent steps should be
performed at 4°. The lysate is clarified by centrifugation at 5000g and the nucleic
acids precipitated by the addition of 7.5 ml 10% (v/v) polyethyleneimine per 100 ml
lysate. The white precipitate is removed by centrifugation at 5000g. The supernatant is transferred to a 500 ml centrifuge bottle and heated in a water bath at 74°
for 20 min, then rapidly chilled on ice. Most E. coli proteins will be precipitated at
this stage and can be removed by centrifugation. Pfu polymerase is then precipitated by adding solid ammonium sulfate to 85% saturation. The protein precipitate
is recovered by centrifugation at 10,000g, resuspended in 50 mM Tris-HCl, pH
7.5, 1 mM Na$_2$EDTA, 10 mM benzamidine hydrochloride and dialyzed against
4 liters of the same at 4° for at least 12 hr. The protein solution is centrifuged
at 100,000g for 1 hr and loaded at 1.5 ml/min onto a 2.6 × 20 cm column of
Whatman P11 (phosphocellulose) equilibrated with 50 mM Tris-HCl, pH 7.5. Pfu
polymerase is eluted with a two column volume gradient from 0→1.2 M NaCl.
Pfu polymerase elutes at approximately 180 mM NaCl. The eluted fractions are
pooled, dialyzed against 10 mM Tris-HCl, pH 8.0, until the measured conductivity
of the protein solution is within twofold of the buffer. Pfu is further purified by
Mono Q (1.6 × 20 cm) (or any quaternary anion-exchange resin) using a 5 column
volume linear gradient from 0 to 18% 1 M KCl. Pfu polymerase elutes in three
chromatographically distinct peaks, with the most abundant fraction retained for

use as a thermal cycling enzyme. The main fraction is pooled and dialyzed against 10 mM Tris-HCl, pH 8.2, 0.2 mM EDTA extensively. The protein solution is then analyzed for purity by SDS–PAGE and concentrated by ultrafiltration to 1.3 mg/ml. Nuclease digestion assays should be performed to confirm that no nuclease has copurified with the enzyme.[25] The protein is then diluted $2\times$ with 2 mM DTT, 0.2% (v/v) NP-40, 0.2% (v/v) Tween 80, 100% (w/v) glycerol, and aliquoted for storage at $-20°$. Long-term storage can be at $-80°$, but the enzyme should not be repeatedly thawed and refrozen at this temperature. One μl of a fivefold dilution of enzyme is typically adequate for standard (cloning-type) PCR reactions. Undiluted enzyme is used for the thermal cycling procedure described below.

2. *PCR amplification of tandemly repeated oligonucleotide templates.* DNA synthesis is conducted in 40×500 μl thin-walled reaction vials in a gradient thermal cycler equipped with a "hot" top to avoid the use of mineral oil. A tandem repeat of the desired sequence is synthesized at the 1 μmol scale without trityl groups and gel purified on sequencing length gels.[25] The two step reaction procedure is similar in design to the concatemer chain reaction of Thompson and co-workers[21] as well as the adaptation of this approach to DNA-labeling by Louis *et al.,*[16] with the exception that the anealing temperature in each step is optimized for each template sequence using an optimized cycling protocol (Figs. 5 and 6). In step 1, the anealing temperature is optimized by systematic variation over 14° and examining the total nucleic acid content and length distribution of the product DNA by 0.7% agarose gel electrophoresis (Fig. 6a and Table I). For step 2, the annealing temperature is optimized over a similar temperature range by quantifying the digested DNA product in 15% polyacrylamide gels ($0.5\times$ TBE) (Fig. 6b and Table I).

A typical step 1 reaction contains 0.1 μM each of gel purified template, 2 mM labeled dNTPs (0.5 mM each nucleotide), 4 mM MgSO$_4$, $1\times$ *Pfu* reaction buffer, and 3.25 μg recombinant *Pfu* DNA polymerase. Thermal cycling procedure: 95° for 5 min, 25 cycles of 95° for 45 s, T_1 for 2 min (see Table I), and 72° for 4 min.

A typical step 2 reaction contains 50 μl step 1 product, 2 mM dNTPs, 4 mM MgSO$_4$, $1\times$ *Pfu* reaction buffer, 3.25 μg recombinant *Pfu* DNA polymerase. Single repeat DNA ($\approx$35 pmol) derived from restriction of step 1 reaction products or by restriction of the original templates may be added in step 2 to enhance overall product yield by as much as twofold (Figs. 5 and 6). Thermal cycling procedure: 95° for 5 min followed by 60 cycles of 95° for 45 s, T_2 for 2 min (see Table I), and 78° for 4 min. Thermal cycling was followed by incubation at 78° for 10 min.

Note on choice of thermal cycler: Although the PCR cycling scheme outlined above will work in any thermal cycler, Peltier-driven machines were found to be relatively inefficient at amplification of tandemly repeated templates. We recommend the use of a cycler that does not ramp temperature in a single block, such as the Stratagene (La Jolla, CA) Robocycler 40. For optimization of a Peltier-driven machine using 500 μl tubes, it will be necessary to actually optimize the ramping times and rates in addition to the temperature of each step in the cycle. The conditions

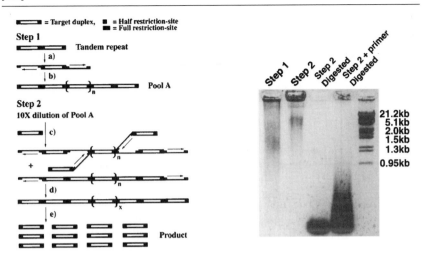

FIG. 5. *PCR amplification of tandemly repeated templates.* A two-step synthesis is employed,[16,21] the first of which prepares a self-priming/self-propagating template pool and the second of which synthesizes long, tandemly repeated DNA. The course of synthesis is followed by 0.7% agarose gel electrophoresis (right). (a) A tandem repeat of the desired sequence is added to the reaction mixture as a blunt-ended duplex. (b) Thermal cycling converts the blunt-ended duplex into a self-priming repeat, creating a pool of different length DNAs. (c) Step 1 products serve as templates for a second round of amplification. The step 1 products are diluted 10-fold into a series of step 2 reactions that create long tandem repeats. At the beginning of step 2, additional duplex DNA containing a single repeat of the desired sequence can be added to increase the overall yield by as much as twofold. Extensive thermal cycling (d) followed by restriction with *Eco*RV (e) results in milligram quantities of single-length DNA product of the desired sequence.

described are the results of extensive optimization on a non-temperature-ramping thermal cycler. Under the described conditions, product yields are dependent on neither sequence composition nor length.

3. *Endonuclease cleavage.* The product DNA from 40 reactions is combined, DTT added to 1 mM, NaCl added to 125 mM, and MgCl$_2$ added to a total final concentration of 10 mM. The pH is lowered to 7.9 with HCl and the DNA digested with 150 units of *Eco*RV per 500 μl step 2 reaction at 37° with continual mixing for 4 hr. Another 150 units of *Eco*RV is added for an additional incubation of no more than 4 hr. Digestion is monitored by 15% polyacrylamide gels (0.5× TBE) (Figs. 6 and 8).

Note: The choice of *Eco*RV here is essentially one of cost. Any restriction endonuclease may be used here depending on the properties of the ends that are desired.

4. *Purification of product DNA and recovery of unincorporated nucleotides.* Digested DNA is 0.2 μm filtered and purified by DEAE ion-exchange HPLC using a preparative Vydac 301 VHP column (Fig. 7). The column is equilibrated with 25 mM sodium phosphate, pH 7.4, 90 mM NaCl. The digested product DNA

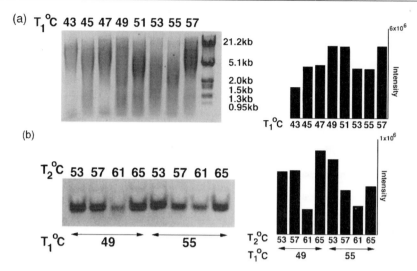

FIG. 6. *Optimization of synthetic yield by gradient thermal cycling.* An example of the annealing temperature optimization procedure for template **2** involves two different tests that can be performed with unlabeled dNTPs. (a) Optimization of step 1 annealing temperature. Ten μl of 500 μl step 1 reactions are analyzed with 0.7% agarose gels that display the length distribution and total quantity of nucleic acid at different annealing temperatures. The bar graph quantitates the total fluorescent intensity, expressed in fluorescent "counts" in a Molecular Dynamics Storm System, illustrating that the total nucleic acid content is roughly equal at $T_1 = 49°, 51°$, and $57°$. However, the length distribution varies widely at these three annealing temperatures as evidenced by the extent of smearing in the gel lanes. It is preferable to use the annealing temperature that displays a wide length distribution and maximal total nucleic acid content so that efficient priming in step 2 occurs. For this reason, the $T_1 = 49°$ nucleic acid pool was chosen for further amplification in step 2. (b) Optimization of step 2 annealing temperature. Ten μl of 500 μl reaction is analyzed with 15% polyacrylamide, 0.5× TBE gels following *Eco*RV digestion as described in the text. For $T_1 = 49°$, priming in step 2 is more efficient overall when compared to $T_1 = 55°$ as evidenced by the higher product yield in three of the four reactions shown. This reflects not only the higher total nucleic acid content of the $T_1 = 49°$C reaction relative to $T_1 = 55°$, but also the greater dispersion in product lengths at $T_1 = 49°$. The greatest product yield is observed for the $T_1 = 49°/T_2 = 65°$ combination in this example (see Table I).

is diluted with phosphate buffer to lower the initial [NaCl] to ≤ 90 mM, injected (5 ml aliquots), washed over the column to remove unincorporated nucleotides, then eluted using a gradient of 90 m$M \rightarrow 360$ mM NaCl over 15 min at 10 ml/min. Fractions containing the main DNA peak were collected and dialyzed against 1 mM sodium phosphate, pH 7.0, 1 mM Na$_2$EDTA and concentrated by lyophilization. The yield of product DNA was determined by measuring the absorbance at 260 nm assuming 50 μg/ml per A_{260} unit. A 500 μL reaction yields a minimum of 8.5 nmol of isotopically enriched single-length DNA from 5 pmol of unlabeled template, an 800 : 1 product : template yield. The pool of unincorporated nucleotides can be desalted by preparative C18 RP-HPLC column as described above, lyophilized, and stored at $-80°$ for rephosphorylation and reuse.

TABLE I

DNA TEMPLATES AND TARGET SEQUENCES FOR *ULTRA*[a]

DNA template	Product sequence
ATCAGGATGCGGTTACT**GATATC**AGGATGCGGTTACT**GAT**	**AT**CAGGATGCGGTTACT**GAT**
TAGTCCTACGCCAATGA**CTATAG**TCCTACGCCAATGA**CTA**	**TAG**TCCTACGCCAATGA**CTA**
1	2
ATCCAGAGGATGTGGCTTCT**GATATC**CAGAGGATGTGGCTTCT**GAT**	**AT**CCAGAGGATGTGGCTTCT**GAT**
TAGGTCTCCTACACCGAAGA**CTATAG**GTCTCCTACACCGAAGA**CTA**	**TAG**GTCTCCTACACCGAAGA**CTA**
3	4
ATCGTTTGTC**GATATC**GTTTGTCGAT	**AT**CGTTTGTCGAT
TAGCAAACAG**CTATAG**CAAACAGCTA	**TAG**CAAACAGCTA
5	6

[a]ULTRA: Uniform Labeling by Tandem Repeat Amplification. *Eco*RV restriction site is shown in boldface type, a half-site at each end and full site separating the tandem repeat of desired sequence.

5. *Fill-in of overdigested product.* Occasionally, incubation of product DNA with EcoRV longer than 12 hr results in overdigestion, creating multiple product lengths. Overdigestion results in recessed 3′ ends that can be filled-in with Klenow DNA polymerase and labeled dNTPs (Fig. 8). A typical fill-in reaction contains 0.1 mM DNA, 10 mM Tris-HCl pH 7.5, 5 mM MgCl$_2$, 7.5 mM DTT, 2 mM dNTP mixture, and 50 units/ml Klenow. Fill-in is complete within 2 hr at 37°. The reaction

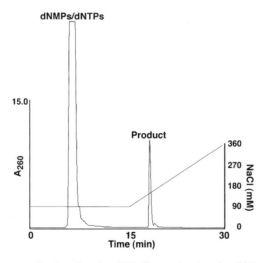

FIG. 7. *Single-step purification of product DNA.* The restricted product DNA can be purified in a single step by DEAE ion-exchange HPLC in 25 mM sodium phosphate employing a biphasic gradient. Isocratic elution of a 5 ml injection at 90 mM NaCl separates the unreacted nucleotides from the product, permitting their recovery and reuse. A linear gradient over 15 min from 90 to 360 mM NaCl elutes the product DNA as essentially a single peak.

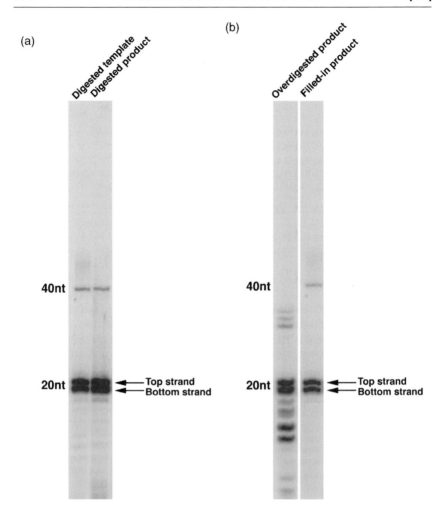

FIG. 8. *Purity of product DNA and fill-in of overdigested product.* (a) Purity of product DNA. 20% urea–polyacrylamide gel of product DNAs demonstrates that the product is identical in length to that derived from digested template DNA. There are two bands observed for this product, each of which represents one of the strands of the duplex DNA that are resolved in a 20% sequencing-length gel. We frequently observe a faint slower mobility band in these gels that represents product digested to only dimer length (<0.1%). The faint ladder seen below the product bands results from slight overdigestion and represents <0.3% of the sample. (b) Fill-in of intentionally overdigested product. Product DNA was digested for 14 hr, revealing substantial (≈20%) overdigestion to a length 1–2 nucleotides shorter than the main product. Fill-in with labeled dNTPs and Klenow results in only full-length product plus the small fraction of dimer as described in (a). The dimer can be removed by gel filtration, if desired.

is quenched by the addition of Na_2EDTA to 10 mM and the DNA recovered by ion-exchange chromatography as described above.

Sequence Fidelity of Product DNA

The sequence fidelity and extent of isotopic enrichment can be assessed by mass spectrometry. However, we find it to be easier to assess the spectroscopic homogeneity of the product by two simple experiments. The composition of the product DNA can be determined from a constant-time $^{13}C-^1H$ HSQC experiment (Fig. 9).[32] Simply counting the number of H_5-C_5 cross peaks of cytidine residues and the H_2-C_2 cross peaks of adenine residues will permit the direct assessment of the correct number of CG and AT base pairs, respectively. The digested product from template 2 (see Table I), for example, displays 11 adenine H_2-C_2 correlations (two cross peaks have two-proton intensities) and 9 cytidine H_5-C_5 correlations (one cross peak has two-proton intensity) as expected. The symmetry-related nucleotides in the EcoRV half-sites at each end of the product duplex are magnetically equivalent (Fig. 9). An additional verification of sequence composition can be accomplished by counting the number of cross peaks of a specific residue type in a constant-time multiple-quantum Hb(Cb)Nb experiment which correlates the base proton (Hb) to the N9 or N1 nitrogen (Nb) via the C8 or C6 base carbon (Cb) (Fig. 9b).[33] These spectroscopic analyses coupled with assessment of the purity of the digested DNA product by denaturing urea–PAGE are sufficient to be confident that the product is of high purity and homogeneity.

Summary

An optimized procedure has been described for the large-scale production of stable isotopeenriched duplex oligonucleotides of designed sequence. Large-scale production of labeled nucleotide triphosphates can be produced in this procedure simultaneously with labeled proteins, thereby providing synthetic dNMP precursors at no additional cost. The procedure is robust, with a minimum product : template yield of 800 : 1 overall, and produces >99% single-length product. Tandem repeat PCR amplification is a general approach to large scale synthesis of duplex oligonucleotides and may have applications to both NMR and X-ray methods, particularly for product lengths in excess of 25 base pairs where failed sequences from solid-phase synthesis can be difficult to remove chromatographically. A drawback of the present approach is that the product is a duplex of two equal-length strands, making single-stranded products more difficult to prepare. For this reason, it could be preferable to produce single-stranded products by the

---

[32] G. Vuister and A. Bax, *J. Magn. Reson.* **98,** 428 (1992).
[33] V. Sklenar, T. Dieckmann, S. E. Butcher, and J. Feigon, *J. Mag. Reson.* **130,** 119 (1998).

(a)   (b)

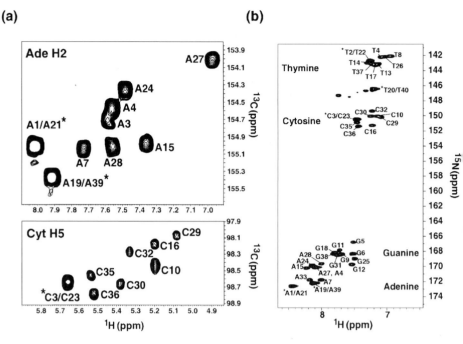

1-ATCAGGATGCGGTTACTGAT-20
40-TAGTCCTACGCCAATGACTA-21

FIG. 9. *Sequence fidelity of DNA products.* (a) The sequence fidelity can be assessed in a single constant-time $^{13}C-^1H$ HSQC spectrum.[32] Sequence 2 contains 9 GC and 11 AT basepairs (see Table I), which can be verified by counting the number of cytidine C_5-H_5 and adenine C_2-H_2 cross peaks. Symmetry-related nucleotides in the *Eco*RV half-sites at each end of the duplex are magnetically equivalent and are indicated with asterisks in the figure. (b) Residue-type assignment of the nucleotides of sequence 2 by a constant-time Hb(Cb)Nb experiment.[33] The nucleotides group by residue type according to their N 9 (purine) or N1 (pyrimidine) chemical shifts. Magnetically equivalent residues are indicated by asterisks. For T20/T40, there is an alternate conformer seen at H6 = 7.65 ppm/^{15}N = 146.5 ppm. This is most likely due to fraying at the ends of the sequence. The remainder of the spectrum is nearly fully resolved for every nucleotide. The spectra were collected at 600 MHz, 36°, in pH 7 phosphate buffer as described.[32,33]

method of Zimmer and Crothers.[13–15] Although a single base type can be selectively enriched in this approach, chemical synthesis will provide greater flexibility for labeled DNAs requiring site-specific labels at only one or a small number of nucleotide positions in the sequence.[10–12] Therefore, maximum flexibility in labeling patterns can be realized by judicious choice of labeling method appropriate to the type of DNA product and extent of isotopic enrichment desired.

[13] ^{13}C Isotopic Enrichment for Nuclear Magnetic Resonance Studies of Carbohydrates and Glycoconjugates

By DAVID LIVE, LOUIS A. "PETE" SILKS III, and JURGEN SCHMIDT

Carbohydrates and glycoconjugate molecules, while having wide occurrence, are disproportionally underrepresented in databases of biopolymer structures. This is in spite of the fact that in many instances the oligosaccharide moieties are involved in molecular recognition events where they are mediators of cell adhesion and act as antigenic determinants.[1] Our interest in this area focuses on tumor-associated carbohydrate antigens that can serve as markers for tumor cells and as targets in antitumor vaccine development.[2] The difficulty in obtaining adequate amounts of homogeneous material from natural sources has hindered conformational analysis of carbohydrates and glycoconjugates. Preparation through chemical synthesis offers an alternative that can provide a well-defined product, but it is not a trivial exercise because of the issues raised by the large number of functional groups on each residue and the stereochemical aspects of the glycosidic linkages. *In vitro* enzymatic synthesis is an appealing approach, but requires expression and isolation of appropriate enzymes, and preparation of the substrates.[3] This is in contrast to the preparation of peptides and oligonucleotides where practical synthetic methods are available, or of proteins, for which there are bacterial expression systems. In considering approaches to conformational analysis, the success of X-ray crystallography has been modest when applied to oligosaccharides, with mostly disaccharide structures reported.[4] Nuclear magnetic resonance (NMR) has thus emerged as the method of choice, since this circumvents the difficulties in crystallizing carbohydrate containing species and provides an approach to evaluating the degree of conformational averaging. Nonetheless, there have been challenges in applying ^{1}H NMR methods as well. The chemical diversity of sites and therefore spectral dispersion in sugars is limited, as are the number of interresidue nuclear Overhauser effects (NOE), with their interpretation sometimes complicated by conformational averaging about the glycosidic linkage.[4–6] The absence of suitably positioned protons about the glycosidic torsion angles ϕ and ψ prevents

[1] A. Varki, *Glycobiology* **3**, 97 (1993).

[2] D. H. Live, L. J. Williams, S. D. Kuduk, J. B. Schwarz, P. W. Glunz, X.-T. Chen, D. Sames, R. A. Kumar, and S. J. Danishefsky, *Proc. Nat. Acad. Sci. U.S.A.* **96**, 3489 (1999).

[3] G. M. Watt, P. A. S. Lowden, and S. Flitsch, *Curr. Opin. Struct. Biol.* **7**, 652 (1997).

[4] C. A. Bush, M. Martin-Pastor, and A. Imberty, *Annu. Rev. Biophys. Biomol Struct.* **28**, 269 (1999).

[5] H. van Halbeek, *Curr. Opin. Struct. Biol.* **4**, 697 (1994).

[6] T. Peters and B. M. Pinto, *Curr. Opin. Struct. Biol.* **6**, 710 (1996).

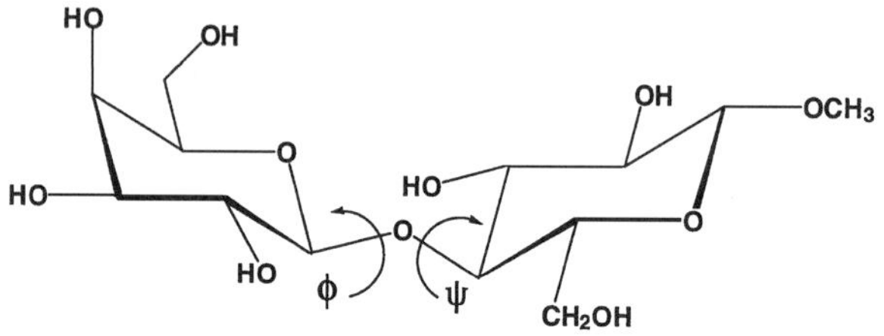

FIG. 1. Illustration of glycosidic torsion angles.

the use of proton J couplings in characterizing the glycosidic linkage (Fig. 1). ^{1}H NMR has therefore generally been used in concert with extensive computational approaches to arrive at a reasonable structural representation.[6] Having encountered the limitations of ^{1}H NMR for these molecules, investigators have looked to using information from ^{13}C NMR. With the many instrumental improvements and the maturation of pulse sequences for indirect detection experiments, particularly using pulsed field gradients,[5] it is possible to extract ^{13}C information from [1]H–[13]C two-dimensional (2D) correlation experiments for samples at concentrations on the order of 5 mM, or even less, at natural isotopic abundance. The enhanced dispersion for the carbon nuclei are of great help in resolving all the sites in more complex carbohydrates, and proton–carbon multiple through-bond correlation experiments have become a key aid in assigning primary structure, particularly in establishing the connections across the glycosidic linkages. In favorable instances, particularly at higher concentrations, vicinal [13]C–[1]H couplings can be measured at natural abundance, providing some information on the glycosidic torsions, but still not enough for an unequivocal interpretation.[7] To realize the full potential of NMR in the conformational analysis of carbohydrates, ^{13}C isotopic enrichment is needed, as this opens up or enhances a variety of additional options including determination of [13]C–[13]C spin couplings, dipolar interactions involving ^{13}C, ^{13}C dynamics measurements, ^{13}C isotopic filtering in complexes, and utilization of the greater ^{13}C shifts in 3D ^{13}C separated proton NOESY (nuclear Overhauser effect spectroscopy) or ROESY (rotating frame nuclear Overhauser effect spectroscopy) experiments to reveal cross peaks obscured by overlap in the homonuclear experiments.[4,5,8] From the structural perspective, the most interest has been in glycans associated with glycoproteins and glycolipids, including gangliosides and blood-group antigens. The constituent sugars in these are illustrated in Fig. 2. Our focus in this chapter is to bring together some illustrative examples of the preparation of ^{13}C-labeled building blocks and their assembly into oligosaccharides along with their application in NMR studies.

[7] T. J. Rutherford, J. Partridge, C. T. Weller, and S. W. Homans, *Biochemistry* **32**, 12715 (1993).
[8] Q. W. Xu and C. A. Bush, *Biochemistry* **35**, 14512 (1996).

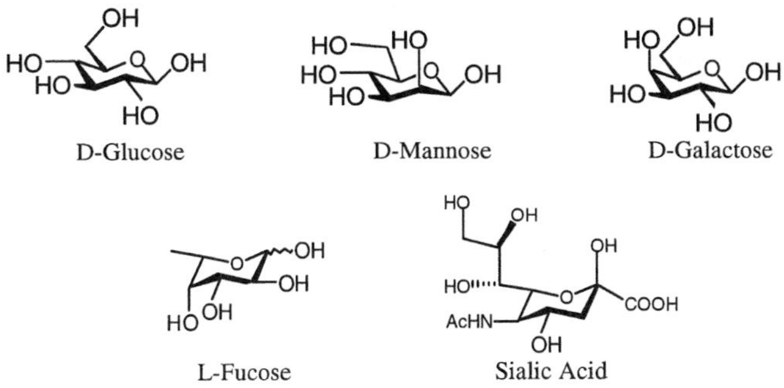

D-Glucose   D-Mannose   D-Galactose

L-Fucose   Sialic Acid

FIG. 2. Structures of some of the sugar residues discussed in the text that have been prepared with ^{13}C enrichment.

Synthesis of ^{13}C-Labeled Carbohydrates

Although there have been substantial improvements in the ability to prepare carbohydrates in recent years, no singly applicable general protocol has emerged, in part because of the number of possible sites that can participate in glycosidic linkages, and the variations in the residues. This variability contrasts with the comparative chemical consistency of peptide bond formation in peptide synthesis. Several techniques have been successfully applied in the preparation of isotopically labeled material, including production *in vivo,* enzymatic synthesis, and chemical synthesis.

Although limited in application, *in vivo* methods have been used in preparation of enriched yeast and bacterial cell wall materials for NMR, growing the organisms in regular or enriched media supplemented with ^{13}C-enriched glucose.[9,10] The labeled glycoprotein human chorionic gonadotropin was successfully prepared using transfected Chinese hamster ovary (CHO) cells in tissue culture medium employing [^{13}C] glucose in addition to isotopically enriched nutrients from algal hydrolyzates.[11] *In vivo* methods have the advantage of using relatively simple labeled species without modification, but cannot easily be generalized to efficiently produce a wide variety of interesting carbohydrate targets. *In vitro* biosynthesis methods exploit the power of enzymes to mediate glycosidic bond formation in the assembly of oligosaccharides.[12] The use of enzymes circumvents the need for the protection and deprotection steps of chemical synthesis; however, because

[9] L. Yu, R. Goldman, P. Sullivan, G. F. Walker, and S. W. Fesik, *J. Biomol. NMR* **3,** 429 (1993).
[10] R. Gitti, G. X. Long, and C. A. Bush, *Biopolymers* **34,** 1327 (1994).
[11] J. W. Lustbader, S. Birken, S. Pollak, A. Pound, B. T. Chait, U. A. Mirza, S. Ramnarain, R. E. Canfield, and J. M. Brown, *J. Biomol. NMR* **7,** 295 (1996).
[12] C.-H. Wong, R. L. Halcomb, Y. Ichikawa, and T. Kajimoto, *Angew. Chem., Int. Ed. Engl.* **34,** 521 (1995).

of the specificity of the enzymes for particular linkages and substrates, a variety of these enzymes may be needed to assemble a particular target. The appropriate ^{13}C-labeled substrate has to be prepared as well, using chemical or enzymatic methods. To this end elegant methods have been demonstrated employing several enzymes *in situ* to prepare the labeled nucleotide–sugar and mediate the formation of the glycosidic linkage to the growing oligosaccharide.[12] A number of ^{13}C-labeled oligosaccharides have been prepared for NMR studies by these methods, including the G_{M3} ganglioside, and the Lewis^x and sialyl Lewis^x antigens.[13,14]

Chemical synthesis can provide access to a wide range of oligosaccharide and glycoconjugate structures. The variety of strategies that are called into play for forming glycosidic linkages require preparation of more complex sugar derivatives incorporating protection and activation of appropriate functional groups. Overall, the chemical approaches typically involve a large number of steps, with a cost in terms of product yield. On the one hand, improved synthetic methodologies are mitigating the impact of this, and on the other, the molecular weights for structures of interest are such that even 1 or 2 mg of the desired product is adequate for NMR investigation with today's spectrometer sensitivity. The chemical synthesis approach has been used in the preparation of a number of glycolipids with ^{13}C-labeling of the carbohydrate component.[15–20] As part of our own work, a hexasaccharide containing the Lewis^y epitope has been prepared chemically incorporating labeled galactose and glucosamine components in the central lactosamine core of the epitope, which was then incorporated in the full construct following a published route.[21,22]

Because D-[U-^{13}C]glucose is now produced commercially on a kilogram scale, it is an economical starting point for a variety of labeled building blocks needed in polysaccharide synthesis, whether chemical or enzymatic. These are generally pyranose or glycal forms. Glycal (1,2-dideoxy sugar) building blocks have proven to be versatile elements in oligosaccharide synthesis, reducing the number of hydroxyl groups to be protected, and being able to function both as glycosyl donors and acceptors in the assembly of an oligosaccharide. Glucose can be converted readily into the 1,2-unsaturated protected glucal derivatives that may be

[13] M. A. Probert, M. J. Milton, R. Harris, S. Schenkman, J. M. Brown, S. W. Homans, and R. A. Field, *Tetrahedron Lett.* **38**, 5861 (1997).

[14] Y. Ichikawa, Y.-C. Lin, D. P. Dumas, G.-J. Shen, E. Garcia-Junceda, M. A. Williams, R. Bayer, C. Ketcham, L. E. Walker, J. C. Paulson, and C.-H. Wong, *J. Am. Chem. Soc.* **114**, 9283 (1992).

[15] B. A. Salvatore and J. H. Prestegard, *Tetrahedron Lett.* **39**, 9319 (1998).

[16] B. J. Hare, C. R. Sanders II, S. E. McIntyre, and J. H. Prestegard, *Chem. Phys. Lipids* **66**, 155 (1993).

[17] Y. Aubin, Y. Ito, J. C. Paulson, and J. H. Prestegard, *Biochemistry* **32**, 13405 (1993).

[18] Y. Aubin and J. H. Prestegard, *Biochemistry* **32**, 3422 (1993).

[19] B. J. Hare, F. Rise, Y. Aubin, and J. H. Prestegard, *Biochemistry* **33**, 10137 (1994).

[20] C. R. Sanders II and J. H. Prestegard, *J. Am. Chem. Soc.* **113**, 1987 (1991).

[21] W. G. Bornmann, S. Hintermann, and D. Live, unpublished results.

[22] S. J. Danishefsky, V. Behar, J. T. Randolph, and K. O. Lloyd, *J. Am. Chem. Soc.* **117**, 5701 (1995).

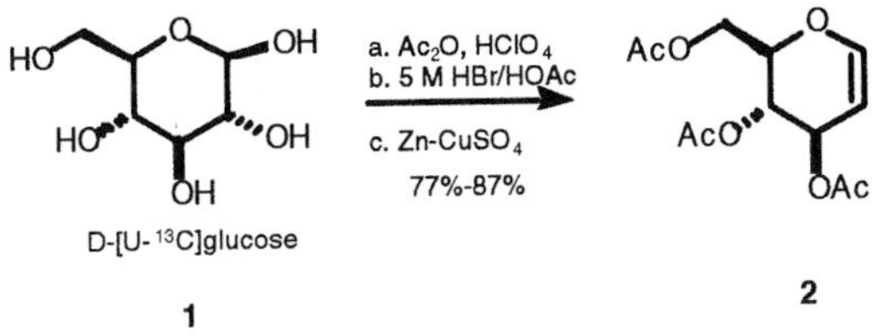

FIG. 3. Scheme for preparing tri-*O*-acetyl-D-glucal (**2**) from D-[U-^{13}C])glucose(**1**).

used as starting materials in the synthesis of labeled mannose and galactose building blocks. Gram quantities of the labeled glucal have been prepared from glucose at the Los Alamos Stable Isotope Resource by the following procedure.[23]

D-[U-^{13}C]Glucose is peracetylated with acetic anhydride in the presence of a catalytic amount of hydroperchloric acid, followed after 2 hr by treatment with glacial acetic acid/HBr. After an additional hour of reaction time the sole product is tetra-*O*-acetyl-D-glucosyl bromide. In the presence of zinc this is converted to tri-*O*-acetyl-D-[1,2,3,4,5,6-^{13}C$_6$]glucal in yields ranging from 77 to 87% (Fig. 3). The glucal has been used as a starting material for preparation of ^{13}C$_6$-labeled α-methyl-D-mannose with inversion of the C-2 center.[24] Treatment of the glucal with methanol and boron trifluoride diethyl etherate gives the α-methyl 2,3-dideoxy-4,6-di-*O*-acetyl-D-glucopyranoside. The 2,3 double bond is reacted with OsO$_4$, giving α-methyl-D-mannose. By substituting another alcohol for methanol, other glycosides such as glycolipids with ^{13}C-labeled mannose have been prepared for NMR studies. Yields for the preparation of glycolipids by this scheme are on the order of 20% relative to the starting D-glucose.[25] A variety of methods for producing mannose building blocks from the glucose-derived glucal are illustrated in a report on the preparation of a high-mannose core pentasaccharide.[26] A particularly intriguing example is the conversion of a glucosyl residue, already incorporated in the oligosaccharide, into a mannose residue.[26] This is accomplished by oxidizing the C2 carbon to a ketone and then producing the mannose skeleton on reduction.

In order to effect the conversion of the glucose skeleton to galactose, the C4 center needs to be inverted. Preparation of a D-galactose building block from D-[U^{-13}C]glucose as part of the preparation of a labeled ganglioside has been reported by Shimizu *et al.*[27] (Fig. 4). Free glucose is first converted to the pentaacetate to give **3**. Substitution of the 1-*O*-acetyl group is accomplished using

[23] R. Wu and L. A. Silks, *Carbohydrate Lett.* **2**, 363 (1997).

[24] E. W. Sayers, J. L. Weaver, and J. H. Prestegard, *J. Biomol. NMR* **12**, 209 (1998).

[25] B. J. Hare, C. R. Sanders II, S. E. McIntyre, and J. H. Prestegard, *Chem. Phys. Lipids* **66**, 155 (1993).

[26] S. J. Danishefsky, S. Hu, P. F. Cirillo, M. Eckhardt, and P. H. Seeberger, *Chem. Eur. J.* **3**, 1617 (1997).

[27] H. Shimizu, J. M. Brown, S. W. Homans, and R. A. Field, *Tetrahedron* **54**, 9489 (1998).

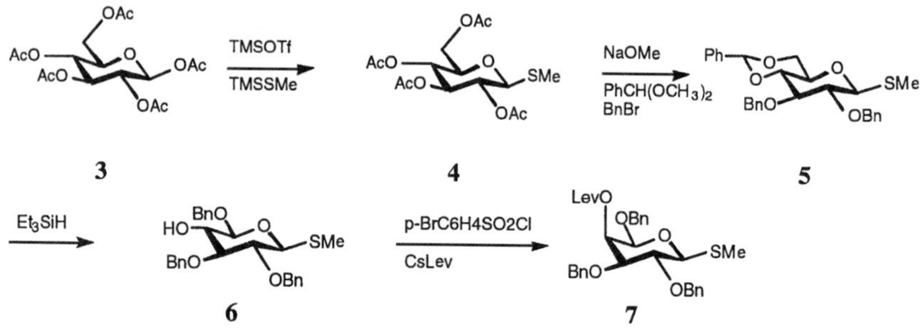

FIG. 4. Scheme for preparing the protected thiogalactosyl donor from peracetylated glucose.

trimethylsilyl triflate with trimethylsilylmethyl sulfide to give **4** in 93% yield. Hydrolysis of the remaining acetyl groups with sodium methoxide followed by selective protection of the 4,6-diol unit with benzaldehyde dimethyl acetal gives **5** in 98% yield. Acetal reduction gives rise to **6** in 92% yield. Reaction of a sulfonyl chloride with the 4-hydroxyl, followed by treatment with cesium levulinoate and 18-*crown*-6, gives the inverted galactose product ester **7** with about 90% yield. This is then carried forward in a series of steps in 72% overall yield to give the building block 1-chloro-2,3,4,6-tetra-*O*-benzyl[U-^{13}C]galactose for assembly into G_{b3} ganglioside. Although a number of steps are involved, yields are high for each of them. NMR studies of the Lewis^y antigen have been of interest in our own research, and as part of the efforts the antigen has been prepared with ^{13}C enrichment in the central lactosamine core. This is assembled in the initial step by coupling of glucal and galactal components[21] (Fig. 5). The efficient labeling of the galactal was developed at the Los Alamos Stable Isotope Resource starting from the tri-*O*-acetyl-D-glucal they had already prepared.[23] The acetyl groups are removed with sodium methoxide, with the primary alcohol functionality then protected with *tert*-butyldiphenylsilyl chloride (TBDPSiCl), and the 3-hydroxyl group with benzoyl chloride. An alternative one-pot route, deprotecting with ammonia, and after removal of solvents taking the material up in pyridine/CH$_2$Cl$_2$ with

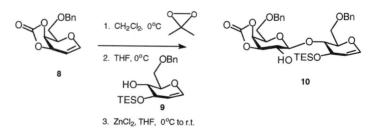

FIG. 5. Preparation of lactal derivative from glycal components.

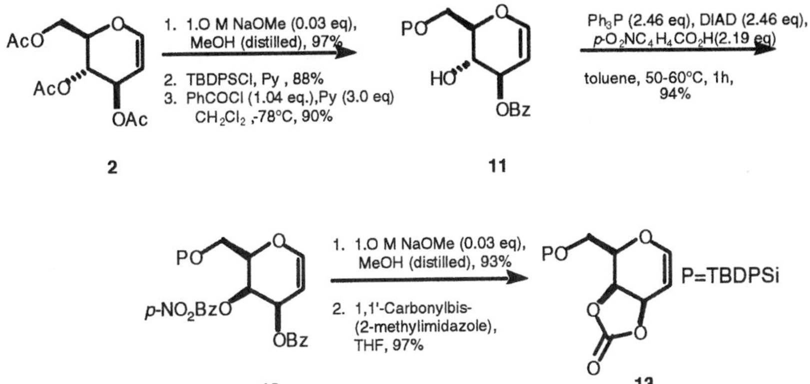

FIG. 6. Preparation of protected galactal from [U-^{13}C]tri-*O*-acetyl-D-glucal.

TBDPSiCl, followed finally by addition of the benzoyl chloride solution at $-78°$, has proved more efficient. The 4-hydroxyl group can be inserted using the Mitsunobu conditions with *p*-nitrobenzoic acid. The esters are then cleaved and the diol is converted into the carbonate (Fig. 6). The protecting group on C6 is converted to benzyl, providing the form used in preparing the lactosamine.

The preparation of ^{13}C-labeled L-fucose used in conjunction with the synthesis of several ^{13}C-labeled carbohydrates has been reported starting from L-[U-^{13}C] galactose.[13] The 6-hydroxyl of the L-galactose is converted to a bromide with the other OH groups protected, and then reduced to produce a methyl group at C6. This is an appealing approach, with the starting L-galactose obtained from an algal strain grown on ^{13}CO$_2$; however, the details of the L-galactose preparation remain unpublished. Wong and co-workers have published an enzymatic synthesis of L-fucose starting from dihydroxyacetone monophosphate (DHAP) and DL-lactaldehyde, using L-fucose-1-phosphate aldolase, followed by acid phosphatase and L-fucose isomerase.[28] The acid phosphatase is commercially available and the other enzymes can be prepared from appropriate strains of *Escherichia coli* available from the ATCC (Rockville, MD). This procedure can be carried out as a one-pot synthesis and has been done at a scale that produces 0.543g in 33% yield from DHAP. Fucose appears as a terminal residue at the nonreducing end of the carbohydrates of interest, so for many NMR experiments, particularly those using ^{13}C–^{13}C coupling across the glycosidic linkage, labeling even at only positions 1 and 2 can be valuable. The DHAP contributes carbons 1,2, and 3 to the product, so having just this component of the reaction labeled is useful. A

---

[28] C.-H. Wong, R. Alajarin, F. Morisvaras, O. Blanco, and E. Garcia-Junceda, *J. Org. Chem.* **60,** 7360 (1995).

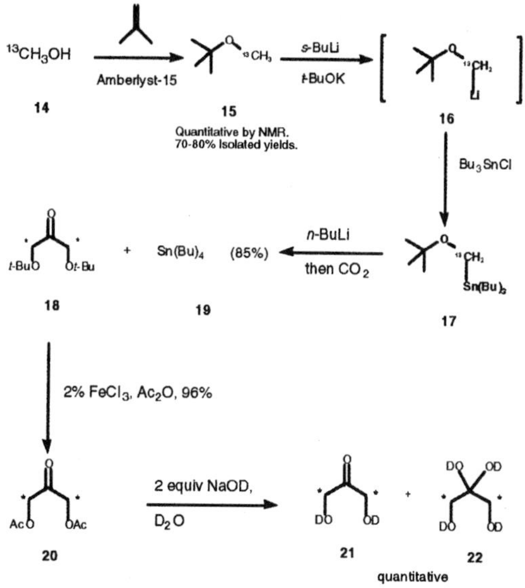

FIG. 7. Scheme for preparation of dihydroxyacetone from [^{13}C]methanol.

four-step synthesis of [U-^{13}C]DHAP has been reported from the same laboratory starting from [U-^{13}C]glycerol.[29] After protection of the primary alcohols, the secondary alcohol is oxidized, followed by deprotection that yields dihydroxyacetone (DHA), which can be converted into the phosphate using ATP and the commercially available glycerokinase. A newer method for preparing DHA from methanol prepared from ^{13}CO has been developed at the Los Alamos Stable Isotope Resource[30] (Fig. 7). [^{13}C]Methanol and isobutylene are combined to make tert-butyl methyl ether in 70–80% isolated yields, which is then converted to tri-n-butyl[(1,1-dimethylethoxy)[^{13}C]methyl]stannane in 60–80% isolated yield. The labeled solvent is recovered by capturing it in a double liquid nitrogen trap. The resulting viscous material is carefully quenched with water and the mixture is then extracted with ether. The reaction has been scaled to give 20–30 g per run, and it is believed that it can be scaled up further. Conversion of the labeled tri-n-butyl[(1,1-dimethylethoxy)methyl]stannane to the lithio-tert-butylmethyl anion, followed by trapping with 0.5 equivalents of CO_2 or the N,N-dimethylchlorocarbamate, affords the di-tert-butyl[1,3-^{13}C]DHA. The tert-butyl groups are removed using

[29] C.-H. Wong, E. Garcia-Junceda, L. R. Chen, O. Blanco, H. J. M. Gijsen, and D. H. Steenesma, J. Am. Chem. Soc. 117, 3333 (1995).

[30] R. Michalczyk, Z. Li, C. Orji, R. Wu, R. A. Martinez, C. Unkefer, and L. A. Silks, Trends Org. Chem. 7, 115 (1998).

acetic anhydride with 2% $FeCl_3$ to form the diacetate, followed by cleavage of the acetates with 2.0 equivalents of NaOH to give a neutral aqueous solution of the DHA. In addition, using labeled $^{13}CO_2$, we have constructed multigram quantities of [1,2,3-^{13}C]DHA with this process. The DHA can be phosphory-lated as indicated above. The value of this reagent goes beyond preparation of the molecules discussed here, as DHAP is a substrate for many useful aldolase-catalyzed reactions.[31]

The preparation of a building block for incorporation of N-acetyl-D-[U-^{13}C] neuraminic acid (sialic acid or NeuNAc) starting from enriched glucose has been reported along with its use in the preparation of several labeled oligosaccha-rides for NMR studies. The glucose is converted to 2-azido-2-deoxymannose and then N-acetyl-D-mannosamine, which, in the presence of [U-^{13}C]pyruvic acid and NeuNAc-aldolase provides the sialic acid skeleton that is activated as a p-nitrophenyl glycoside for enzymatic coupling to the oligosaccharide in question.[13] Likewise, labeling at just the C1, C2, and C3 positions of sialic acid has been achieved by the enzymatic condensation of N-acetylmannosamine with [U-^{13}C]pyruvate and NeuNAc-aldolase.[15]

An additional important modification of glucose and galactose that needs to be considered is the replacement of the 2-hydroxyl with an N-acetyl group, mak-ing glucosamine and galactosamine. Glucosamine and galactosamine precursors can be readily derived from the respective protected glycals using benzenesul-fonylamine. This has been done for a glucosamine residue in the synthesis of the labeled Lewis^y antigen,[22] and for galactosamine in the labeled G_{M4} lactam glycolipid.[15] In the latter case the ^{15}N is incorporated as well by preparing the benzenesulfonylamine with [^{15}N]ammonium chloride with the resulting ^{15}N la-beling of the N-acetyl group. Although we are not aware of chemical preparation of ^{13}C-labeled carbohydrate components of N- or O-linked glycopeptides, there are methods available. For coupling to an aspartic acid in forming the N-linked amino acid, anthracenesulfonylamine has been used in a manner similar to the benzen-sulfonyl chloride above to ultimately generate a 1-amino-2-N-acetyl residue that forms an amide linkage between the glucosamine and the side chain of aspar-tic acid.[26] In a procedure for preparing N-acetylgalactosamine O-linked serine or threonine, the galactal can be converted to the 2-azido-1-fluoro derivative, which is then coupled to the amino acid.[32]

The above discussion is not presented as a comprehensive review of the field; however, we have endeavored to emphasize approaches that have been applied to, or show promise in, the preparation of labeled oligosaccharides and glycoconjugates.

[31] S. Takayama, G. J. McGarvey, and C.-H. Wong, Ann. Rev. Microbiol. **51**, 285 (1997).
[32] S. D. Kuduk, J. B. Schwarz, X.-T. Chen, P. W. Glunz, D. Sames, G. Ragupathi, P. O. Livingston, and S. J. Danishefsky, J. Am. Chem. Soc. **120**, 12474 (1998).

NMR Applications

Coupling Constants

The glycosidic torsion angles are the most significant variables for characterizing the conformation of oligosaccharides, as they relate the orientation of essentially rigid sugar residues to each other. Vicinal J couplings provide the most direct measurement of these in the NMR spectrum. This coupling between any two particular vicinal substituents follows the characteristic Karplus curve dependency with several possible torsion angles being solutions for a typical coupling value.[33,34] In the event of conformational averaging, possible additional solutions are introduced as well. Thus, although the measurement of only one of the couplings limits the possibilities, it does not provide an unequivocal constraint.[7] The determination can be somewhat improved by measuring additional couplings involving the several substituents juxtaposed about a particular bond, since the magnitude of the couplings varies between different combinations of nuclei, and the phases of the curves are offset from each other by the respective angular relationships of different substituents about the central bond. With several vicinal couplings to define a torsion angle, the number of possible consistent solutions is reduced.[34] As pointed out already, there are not suitably disposed protons about the ϕ and ψ glycosidic torsions for ^{1}H homonuclear couplings to be useful. For each of the glycosidic torsions there is typically only one $^3J_{CH}$ that can be measured. (There are two of these for the ψ angle when the linkage involves a C6 carbon.) There are examples of these being measured at natural abundance by examining the cross-peak offsets or their fine structure in 2D ^{1}H–^{13}C correlation experiments,[35,36] or ^{1}H–^{13}C quantitative J type experiments.[37,38] However, the low natural abundance of ^{13}C limits the sensitivity of such experiments, making isotopic enrichment desirable. Short of using uniformly ^{13}C-labeled material, specific enrichment at the C1 position has been employed. For a number of residues, the ^{13}C-labeled C1 form is commercially available while, with the exception of glucose, uniformly labeled material is not, so it is tempting to use the C1-labeled material to avoid a larger synthetic effort, even if the information obtained is limited. As an example, the C1 of galactose was labeled in the sialyl Lewis^x epitope for NMR studies.[14] Additional coupling measurements relative to the glycosidic torsion angles can be obtained from ^{13}C–^{13}C couplings. In the absence of ^{13}C labeling, however,

[33] I. Tvaroska and F. R. Taravel, *Adv. Carbohdrate Chem. Biochem.* **51**, 15 (1995).
[34] B. Bose, S. Zhao, R. Stenutz, F. Cloran, P. B. Bondo, G. Bondo, B. Hertz, I. Carmichel, and A. S. Serianni, *J. Am. Chem. Soc.* **120**, 11158 (1998).
[35] W. Kozminski and D. Nanz, *J. Magn. Reson.* **124**, 383 (1997).
[36] L. Poppe, R. Stuike-Prill, B. Meyer, and H. van Halbeek, *J. Biomol. NMR* **2**, 109 (1992).
[37] G. Zhu, A. Renwick, and A. Bax, *J. Magn. Reson. Ser. A* **110**, 257 (1994).
[38] Q. W. Xu and C. A. Bush, *Carbohydrate Res.* **306**, 335 (1998).

there is very little prospect for their measurement. Selective labeling at multiple sites could be employed to extract such couplings, but it is more efficient and general to employ uniform labeling since this provides access to the maximum number of both J couplings and sites of linkage. Although it might seem that there would be a problem in unraveling the multiple couplings to a given ^{13}C-labeled site in a uniformly labeled oligosaccharide, this has already been elegantly addressed with experiments previously devised for labeled proteins.[39] Such experiments were used in the study of uniformly labeled sialyl Lewis[x] antigen and ganglioside G_{M3}, where the ^{13}C–^{13}C couplings were used with ROE constraints obtained from ^{13}C-separated ROESY-HSQC (ROESY-heteronuclear single quantum correlation spectroscopy) experiments to establish the solution conformation.[40,41] A similar approach was also applied to a streptococcal cell wall polysaccharide.[42]

The relationships between the $^3J_{HCOC}$ and $^3J_{CCOC}$ couplings and torsion angles have been examined with the preferred parameterization given in Eq. (1) and (2)[33,34]:

$$^3J_{COCC} = 3.49 \cos^2 \theta + 0.16 \tag{1}$$

$$^3J_{COCH} = 5.7 \cos^2 \theta - 0.6 \cos \theta + 0.5 \tag{2}$$

The potential for using $^2J_{COC}$ couplings across the glycosidic linkage as a measure of bond angle has been suggested as well.[43]

Although the desire to use spin–spin couplings in conformational analysis provided an early motivation for the isotopic enrichment of sugars, this also facilitates measurement of dipolar couplings between ^{13}C nuclei. These can be determined using high-resolution NMR techniques by reducing the degrees of motional freedom in partially oriented systems, and the couplings could be used effectively in structure determination, as was first done in the Prestegard laboratory in studies of glycolipids.[20,44] The partial orientation is achieved with liquid crystal bicelles that spontaneously align in the magnetic field, and into which glycolipids have been incorporated. The bicelle orientation in the magnetic field essentially fixes the axis of the bicelle normal, while leaving the carbohydrate headgroup freedom of motion in the other two spacial directions. The proton line widths for the glycolipid headgroups are quite large in this situation. The ^{1}H-decoupled ^{13}C lines are not

[39] A. Bax, D. Max, and D. Zax, *J. Am. Chem. Soc.* **114**, 6923 (1992).

[40] M. J. Milton, R. Harris, M. A. Probert, R. A. Field, and S. W. Homans, *Glycobiology* **8**, 147 (1998).

[41] R. Harris, G. R. Kiddle, R. A. Field, M. J. Milton, B. Ernst, J. L. Magnani, and S. W. Homans, *J. Am. Chem. Soc.* **121**, 2546 (1999).

[42] M. Martin-Pastor and C. A. Bush, *Biochemistry* **38**, 8045 (1999).

[43] F. Cloran, I. Carmichael, and A. S. Serianni, *J. Am. Chem. Soc.* **122**, 396 (2000).

[44] C. R. Sanders II, B. J. Hare, K. P. Howard, and J. H. Prestegard, *Prog. NMR Spectrosc.* **26**, 421 (1994).

as broad, so this and their better spectral dispersion make ^{13}C more useful. The dipolar couplings are extracted from ^{13}C–^{13}C COSY (correlation spectroscopy) or INADEQUATE (incredible natural abundance double quantum transfer experiment) experiments, and for such measurements ^{13}C isotopic enrichment is a necessity.[17–20,44–46] The dipolar couplings have a dependence on the angle made by the relevant internuclear vector and the external magnetic field, which in turn can be related to a molecular axis system.[44] In the case of oligosaccharides, the goal is to obtain adequate information to determine the orientation of each residue so their relationship to each other about the glycosidic linkage can be established. When the principal ordering axis is common to each of the residues, usually the bilayer normal for these glycolipid systems, the relative orientation of the molecular axis systems of each residue to the ordering axis system, can be used to establish the three-dimensional organization of the carbohydrate. These experiments provide a unique approach to studying the conformation of carbohydrates at the surface of a membrane. The preparation of some of these molecules was indicated above.

This work was an important prelude to the methods introduced by Bax and Tjandra that have exploited the effect of molecular asymmetry in collisional interactions between molecules and oriented bicelles in solution to impart a small bias to their molecular reorientation.[47,48] This makes the dipolar interactions evident. Because this mechanism for introducing anisotropy does not require integration into the bicelle, the latter strategy is applicable to a wider range of systems; however, the induced anisotropy is reduced relative to the glycolipids in bicelles discussed above. There have been several examples of applying this latter approach to smaller oligosaccharides.[49–51] In such experiments the dipolar contribution to the one-bond ^{13}C–^{1}H couplings are most easily determined. At least five independent couplings are needed to solve the expressions that describe the orientation of the coordinate frame of the molecular fragment.[44,49,52] A number of the C–H bonds in pyranoses are essentially parallel, so just using the direct dipolar interactions between these nuclei may compromise the accuracy of the result. With incorporation of some assumptions these can still be used effectively, as has been shown in some recent reports.[49–51] To augment the ^{13}C–^{1}H data, the use of ^{1}H–^{1}H

[45] K. P. Howard and J. H. Prestegard, *J. Am. Chem. Soc.* **117**, 5031 (1995).
[46] B. A. Salvatore, R. Ghose, and J. H. Prestegard, *J. Am. Chem. Soc.* **118**, 4001 (1996).
[47] N. Tjandra and A. Bax, *Science* **278**, 1111 (1997).
[48] A. Bax and N. Tjandra, *J. Biomol NMR* **10**, 289 (1997).
[49] M. Martin-Pastor and C. A. Bush, *Carbohyd. Res.* **323**, 147 (2000).
[50] T. Rundolf, C. Landersjo, K. Lycknert, A. Maliniak, and G. Widmalm, *Magn. Reson. Chem.* **36**, 773 (1998).
[51] G. R. Kiddle and S. W. Homans, *FEBS Lett.* **436**, 128 (1998).
[52] P. J. Bolon, H. M. Al-Hashimi, and J. H. Prestegard, *J. Mol. Biol.* **293**, 107 (1999).

dipolar interactions has been suggested.[53,54] Because the pyranose framework is effectively rigid, distances between protons separated by several atoms are known accurately, as is the direction of the relevant internuclear vector to the molecular frame. The $1/r^3$ dependence of the dipolar coupling interaction gives it a more favorable distance dependence than the NOE. With the combined carbon and proton data the orientation can be evaluated. Although ^{13}C isotopic enrichment is not always a requirement for the ^{13}C–^{1}H dipolar measurement, it has been used and clearly improves the sensitivity at which the measurement is made. Also, with labeled material one could use ^{13}C–^{13}C dipolar couplings for weakly ordered solution systems, as has been done for the glycolipids incorporated in the bicelles. The magnitude of the ^{13}C–^{13}C dipolar coupling would be about a factor of four smaller than in the ^{13}C–^{1}H case since the gyromagnetic ratio of ^{13}C is one-quarter that of a proton.

The ability to apply these methods in complex situations involving carbohydrates has also been demonstrated. In the binding of a trisaccharide ligand to an *Escherichia coli* toxin, ligand dipolar couplings were measured in the free state, and in the complex where the ligand binding constant was known and the ligand was in fast exchange.[55] From these data the dipolar couplings for the bound ligand were extracted and used as restraints in dynamical simulated annealing refinements. The orientation of α-methyl mannose bound to mannose binding protein was examined using dipolar measurements in the presence of both rapid exchange between free and bound forms and with multiple binding sites. In this instance the order tensor for the ligand was calculated from the ^{13}C–^{1}H dipolar couplings, which can then be related to the principal order direction of the protein.[52] ^{13}C enrichment was important here in view of the concentrations involved, and was realized using the synthesis from glucose discussed above.

Isotopic Filtering

Using ^{13}C labeling to discriminate the resonances of sugar residues in the presence of protein can be useful. Although the sugar ^{13}C resonances are found in the region generally downfield of the protein C_α shifts, there can be some overlap that might complicate interpretation.[56] Employing ^{13}C-labeled sialyl Lewis^x in a study of its binding to E-selectin was helpful in that it allowed the use of the resolution of ^{13}C separated 3D experiments to assign the transferred NOE peaks used in determining the bound conformation.[41] Through metabolic or enzymatic addition, specific ^{13}C-labeled sugar residues have been incorporated into the pendant glycans

[53] F. Tian, P. J. Bolon, and J. H. Prestegard, *J. Am. Chem. Soc.* **121**, 7712 (1999).

[54] P. J. Bolon and J. H. Prestegard, *J. Am. Chem. Soc.* **120**, 9366 (1998).

[55] H. Shimizu, A. Donohue-Rolfe, and S. W. Homans, *J. Am. Chem. Soc.* **121**, 5815 (1999).

[56] T. de Beer, C. E. M. van Zuylen, K. Hard, R. Boelens, R. Kaptein, J. P. Kamerling, and J. F. G. Vliegenhart, *FEBS Lett.* **348**, 1 (1994).

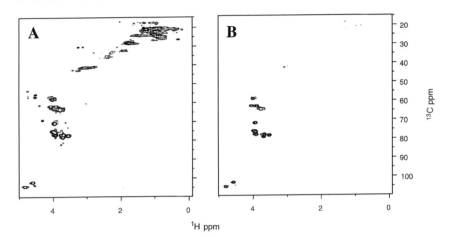

FIG. 8. (A) ^{1}H–^{13}C HSQC spectrum of a hexasaccharide with the tetrasaccharide Lewis^y epitope at the nonreducing end, and with the central lactosamine core of the Lewis y isotopically enriched with ^{13}C, in complex with the Fab fragment of the BR96 high-affinity antibody. (B) Spectrum from A with that of the BR96 alone subtracted, showing signals only from the 12 ^{13}C-labeled carbons.

on immunoglobulin G and its derived Fc fragment, making it possible to selectively probe the environment encountered by sites along the glycan chain.[57,58] Without enrichment, the complexity of the spectrum from all the residues in the biantennary glycan would preclude specific assignment even if they could be observed. When a carbohydrate component is fully labeled, however, specific assignments can be made using ^{13}C–^{13}C connectivities established through the sugar backbone with HCCH TOCSY (HCCH total correlated spectroscopy) experiments that overcome the overlap of the resonances.[59] This has been successfully done in the context of a fully labeled human chorionic gonadotropin glycoprotein, with additional heteronuclear 3D experiments allowing correlation through the amino acid side chain to the protein backbone as well. In our own work, we have used ^{13}C filtering to observe the Lewis^y carbohydrate antigen binding to the Fab fragment of a high-affinity antibody.[60] Here the ^{13}C labeling allows us to discriminate the signals arising from the ligand and those at natural abundance from such a large protein. Figure 8 shows the spectrum of the antibody–antigen complex, and, in addition,

[57] Y. Yamaguchi, K. Kato, M. Shindo, S. Aoki, K. Furusho, K. Koga, N. Takahashi, Y. Arata, and I. Shimada, J. Biomol. NMR 12, 385 (1998).
[58] A. M. Gilhespy-Muskett, J. Partridge, R. Jefferis, and S. W. Homans, Glycobiology 4, 485 (1994).
[59] C. T. Weller, J. Lustbader, K. Seshadri, J. M. Brown, C. A. Chadwick, C. E. Kolthof, S. Rammarian, S. Pollak, R. Canfield, and S. W. Homans, Biochemistry 35, 8815 (1996).
[60] D. E. Yelton, M. J. Rosok, G. Cruz, W. L. Cosand, J. Bajorath, I. Hellstrom, K. E. Hellstrom, W. D. Huse, and S. M. Glaser, J. Immunol. 155, 1994 (1995).

the spectrum of the complex with that of the free antibody subtracted. In the case of a protein of this size, the protein signals at natural abundance are comparable to that of the labeled ligand.

With the limited spectral range of the sugar protons, the dispersion in ^{13}C-separated spectra has proven valuable in extracting the maximum number of experimental constraints, as well as in enhancing water suppression.[24,61] Together these have allowed determination of NOE cross peaks between water and sugar hydroxyl protons that not only provide additional structural constraints, but also aid in describing interactions with solvent or binding site.

Future Prospects

With enhanced accessibility of labeled carbohydrates, there should be a significant increase in the number and quality of experimentally derived structures, with less reliance on computations to fill in the gaps in the data. At the same time, the increase in experimental data should enhance the parameterization of force fields for computational methods, improving their performance. The continuing experimental advances in the determination of dipolar couplings in weakly aligned systems will have a major impact in studies of carbohydrates, offering a much improved experimental basis for determining structures of carbohydrates in solution. ^{13}C-Labeled material will also be amenable to solid state multiple quantum techniques that use the dipolar coupling between spins to provide detailed structural information.[62] In more complex molecular situations where multiple or clustered sites of glycosylation are present, it will be possible to use labeling of a specific glycan in the system to investigate the interactions between labeled and unlabeled carbohydrate components that would otherwise be masked. Even if the amounts of labeled material available for a particular oligosaccharide and glycoconjugate are limited, the imminent appearance of cryoprobe technology on NMR spectrometers and its attendant severalfold increase in sensitivity will facilitate data collection.

Acknowledgments

This work was supported in part by NIH grant RR 02231 and XL2G (LAS) of the Los Alamos National Laboratory Directed Research and Development Fund to the National Stable Isotope Resource at Los Alamos.

[61] R. Harris, T. J. Rutherford, M. J. Milton, and S. W. Homans, *J. Biomol. NMR* **9**, 47 (1997).
[62] O. N. Antzutkin and R. Tycko, *J. Chem. Phys.* **110**, 2749 (1999).

[14] Preparation and Use of ^{2}H-Labeled RNA Oligonucleotides in Nuclear Magnetic Resonance Studies

By Edward P. Nikonowicz

Introduction

Multidimensional double- and triple-resonance nuclear magnetic resonance (NMR) techniques have extended the size and complexity of proteins and nucleic acids that can be studied in solution by increasing spectral resolution and providing the means to correlate resonances exclusively through scalar coupled pathways. The widespread application of these experiments has been fueled by the development and refinement of methods for facile preparation of uniformly ^{13}C,^{15}N-labeled molecules. However, the sensitivity of these heteronuclear experiments and the resolution of the resulting spectra degrade as molecular correlation time increases, leading to more efficient relaxation of the nuclear spins.[1,2]

Deuterium enrichment is a general strategy used to simplify ^{1}H NMR spectra and to lessen the problems associated with efficient relaxation. For more than 30 years, deuterium isotope enrichment has aided solution-state studies of biological macromolecules.[3] Prior to 1993, deuteration was employed in homonuclear NMR for resonance editing, linewidth narrowing, and spectral simplification through reduction of ^{1}H–^{1}H scalar coupling.[4] Since then, deuterium enrichment has been coupled with biosynthetic ^{13}C and ^{15}N labeling of large ($>$20 kDa) proteins and has led to the development of a multitude of heteronuclear experiments that utilize the favorable relaxation properties imparted by the ^{2}H nucleus.[1,2,5]

Deuteration of nucleic acids and NMR methodological applications for these molecules have been slower to evolve, in part due to the composition of nucleic acid spin systems relative to proteins. For example, there are no exchangeable sites along the phosphate backbone, analogous to the amide positions, that can be used for scalar sequential resonance assignment of perdeuterated molecules. This article describes the biosynthetic preparation of deuterated ^{13}C- and ^{15}N-labeled RNA oligonucleotides. The effects of deuterium enrichment on the relaxation of ^{13}C- and ^{15}N-bound protons and on the quality of scalar and nuclear overhauser effects (NOE) based spectra also are presented.

[1] S. Grzesiek, J. Anglister, H. Ren, and A. Bax, *J. Am. Chem. Soc.* **115,** 4369 (1993).

[2] T. Yamazaki, W. Lee, M. Revington, D. L. Mattiello, F. W. Dahlquist, C. H. Arrowsmith, and L. E. Kay, *J. Am. Chem. Soc.* **116,** 6464 (1994).

[3] J. L. Markley, I. Putter, and O. Jardetzky, *Science* **161,** 1249 (1968).

[4] D. M. LeMaster, *Q. Rev. Biophys.* **23,** 133 (1990).

[5] K. H. Gardner and L. E. Kay, *Ann. Rev. Biophys. Biomol. Struct.* **27,** 357 (1998).

The labeled 5'-nucleoside triphosphates (5'-NTPs) used in the T7 RNA polymerase directed synthesis of RNA oligonucleotides were prepared from ribosomal RNA isolated from *Escherichia coli* cultured in appropriately labeled minimal media.[6,7] Fractional deuteration of rRNA is achieved using minimal media composed of 50–90% ^{2}H$_2$O and containing protonated sodium acetate ([^{1}H$_3$]acetate), and perdeuteration is achieved using minimal media composed of 90% ^{2}H$_2$O and sodium deuterioacetate ([^{2}H$_3$]acetate). The purine C8 and pyrimidine C5 atoms of the deuterated 5'-nucleoside monophosphates (5'-NMPs) can be protonated using bisulfite modification before conversion to the 5'-NTPs to prepare deuterated oligonucleotides "labeled" with H5 and H8. Bisulfite modification also can be used to prepare ^{13}C- and ^{15}N-labeled RNA oligonucleotides that are specifically deuterated at C5 or C8. Deuteration of nucleic acids increases the signal-to-noise of NOE-based experiments and improves the resolution of NOE-based and scalar-correlated experiments.

Procedures

Protocol for Preparation of Labeled 5'-NMPs

Materials

Escherichia coli (BL21) (Novagen)
[^{15}N]Ammonium sulfate; sodium [^{13}C]acetate [1-^{13}C, 2-^{13}C, and 1,2-^{13}C$_2$]; sodium [^{2}H]acetate ([^{2}H$_3$]acetate); 99.8% ^{2}H$_2$O (Cambridge Isotope Laboratories, Andover, MA)
Minimal media (per liter): 10.2 g Na$_2$HPO$_4$, 4.5 g KH$_2$PO$_4$, pH 7.0, 0.5 g NaCl, 246 mg MgSO$_4$, 7.5 mg thiamin, 1.3 mg nicotinamide, 660 mg (NH$_4$)$_2$SO$_4$, and 3.0 g sodium acetate, 500 μl trace metals solution
Trace metals solution (per 50 ml): 1.50 g FeCl$_3$ · 6H$_2$O, 78 mg ZnCl$_2$, 146 mg CaCl$_2$ · 2H$_2$O, 198 mg MnCl$_2$ · 4H$_2$O, 4.5 ml 12.5 M HCl, and 45.5 ml H$_2$O

The 90% ^{2}H$_2$O culture medium is prepared as a 1 liter solution containing the phosphates and sodium chloride using 100 sterile ^{1}H$_2$O and 900 ml sterile ^{2}H$_2$O. The pH is adjusted to 7.0 using 5 M NaOD and the medium is divided into three sterile 2-liter baffled flasks. The cultures grown on sodium acetate reached higher density using a phosphate concentration 50% greater than that of M9 medium,[8]

[6] E. P. Nikonowicz, M. Michnicka, K. Kaluarachchi, and E. DeJong, *Nucleic Acids Res.* **25**, 390 (1997).

[7] E. P. Nikonowicz, K. Kalurachchi, and E. DeJong, *FEBS Lett.* **415**, 109 (1997).

[8] J. Sambrook, E. F. Fritsch, and T. Maniatis, "Molecular Cloning: A Laboratory Manual." Cold Spring Harbor Laboratory Press, Cold Spring Harbor, New York, 1989.

presumably due to the additional buffering capacity of the media. Stock solutions of $MgSO_4$ (1 M), thiamin (2.5 mg/ml), and nicotinamide (1.3 mg/ml) are prepared using 99.8% sterile ^2H_2O and added to each flask in the amounts 675 μl, 330 μl, and 110 μl, respectively. The $(NH_4)_2SO_4$ (220 mg) is added to each flask as a powder. The sodium acetate is dissolved in 15.0 ml 99.8% ^2H_2O and the pH* adjusted to 7.3 (uncorrected for deuterium) with ≈60 μl 12.5 M HCl, and 5.0 ml is added to each flask. The trace metals solution (167 μl) is added last.

The cells are adapted to ^2H_2O and to acetate as the sole carbon source by passage through a series of increasingly stringent culture conditions (Fig. 1). Five ml of LB medium (composed of 10% ^2H_2O and 90% ^1H_2O) is inoculated with a single colony of E. coli grown overnight on an agar plate (1× LB media with 1.5% agar).[8] The culture is incubated at 37° with shaking (200 rpm) until it reaches an optical density of 1.2 (A_{600}). The culture is diluted 1 : 25 into 10 ml of prewarmed minimal medium composed of 25% ^2H_2O, 0.2% glucose (w/v), and 0.1% sodium acetate (w/v). The cells are passed sequentially through minimal media (50%, 75%, and 92% ^2H_2O) with 0.3% (w/v) sodium acetate as the sole carbon source, each time diluting 1 : 25. The 92% ^2H_2O culture is used as the inoculum for the preparative scale labeled cultures. At an A_{600} of 0.8, the 92% ^2H_2O culture is divided among 2 liter baffled flasks at a 1 : 20 dilution. The 330 ml cultures are incubated at 37° with shaking (240 rpm) until the A_{600} is 1.6, and the cells are harvested by centrifugation and frozen at −20°. The doubling time of the cultures is 3–4 hr in ^1H_2O for both [$^1H_3$]acetate and [$^2H_3$]acetate. The doubling time increases to 6–7 hr when the cells are cultured in 90% ^2H_2O.

To reduce the cost of deuterated NMP production incurred by the ^2H_2O, the water from the preparative-scale cultures can be recycled. After the cells are harvested, each liter of medium is filtered through glass wool, stirred for 1 hr with 15 g of 50–200 mesh activated carbon, and filtered through a Whatman (Clifton, NJ) No. 1 filter disk. The pH of the filtered water is set to 12 using 40% NaOD and the water is refluxed for 2 hr. After cooling, the pH is adjusted to ≈8.2 with 12.5 M DCl and the water distilled under reduced pressure at 60° using rotary evaporation. The ^2H_2O content of the recycled water varies <2% (using the buoyant density method as detailed[9]) from the theoretical estimate based on the initial ^2H_2O content of the cell culture medium.

Isolation of Ribosomes

Materials

Deoxyribonuclease I Type II (Sigma)
RB [10 mM Tris-Cl (pH 7.6), 10 mM magnesium acetate, 60 mM NH$_4$Cl, 7 mM 2-mercaptoethanol]

[9] P. B. Moore, *Methods Enzymol.* **59**, 639 (1979).

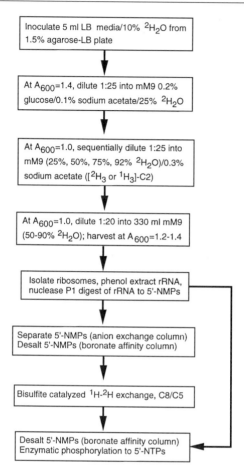

FIG. 1. Flow chart diagramming the production of deuterium-enriched 5'-NTPs from *Escherichia coli*. The modified M9 minimal medium is abreviated mM9. The doubling time for the cells cultured in 80–90% ^{2}H$_2$O was 6–7 hr.

All steps are performed on ice. The frozen cell paste ($\approx$5.8 g/liter of culture) is thawed and 5 ml of RB and 0.1 mg of DNase I are added. A Dounce tissue grinder (Kontes) is used to homogenize the cell suspension. An additional 5 ml of RB is used to rinse the tissue grinder, combined with the cell suspension, and the mixture incubated on ice for 15 min. The cells are then disrupted by passage through a prechilled (4°) French press five times at 1200 psi. The cell debris is removed by centrifugation for 20 min at 12,000 rpm using a Beckman JA-20 fixed angle rotor. The supernate is decanted and the cell debris is washed twice with RB (5 ml per wash). The supernates ($\approx$20 ml) are combined, and the ribosomes purified by ultracentrifugation for 16 hr at 43,000 rpm using a Beckman Ti-55.2 fixed angle rotor. The supernate is decanted and the ribosome pellet stored at $-20°$.

Extraction of rRNA

Materials

0.5 M EDTA (pH 8.0)
0.1 M Sodium acetate (pH 8.0)
10% Sodium dodecyl sulfate (pH 7.2 with 5 M HCl)
Phenol (with 0.05% 8-hydroxyquinoline and equilibrated against 100 mM sodium acetate)
Chloroform : isoamyl alcohol (24 : 1) solution

All steps are performed on ice and all glassware was baked at 200° for 4 hr. The ribosome pellet is thawed on ice and 5 ml of prechilled 0.1 M sodium acetate and 100 μl of 0.5 M EDTA are added. The ribosomes are dispersed by sonication using a cup attachment (Ultrasonics Incorporated, Plainview, NY). SDS (10%) is added to a final concentration of 0.5% and the suspension is mixed by vortexing. For the first extraction of the ribosomes, 4 ml of phenol is added, followed by 1 ml of chloroform : isoamyl alcohol solution, and the aqueous phase (top layer) is removed to a clean tube using a Pasteur pipette. For the second extraction, the volume ratio for water : phenol:chloroform is 2 : 1 : 1 and the aqueous phase is again transferred to a clean tube. Using the same volume ratio, two more extractions are performed, followed by a final chloroform : isoamyl alcohol extraction of the aqueous phase. The organic phases of the four phenol : chloroform extractions and the final chloroform extraction are themselves back-extracted three times each with 0.1 M sodium acetate ($\approx$5 ml each time). All aqueous phases are combined, the volume ($\approx$30 ml) reduced 50% by rotary evaporation, and the rRNA precipitated from ethanol for 16 hr at $-20°$. After centrifugation, the liquid is decanted and the rRNA pellet is dried under vacuum for 10 min.

Nuclease P1 Digestion of rRNA

Materials

Nuclease P1 (Roche Biochemicals, Indianpolis, IN)
5.0 mM Sodium acetate (pH 5.3)
10 mM ZnCl$_2$

The dried rRNA pellet is dissolved in 10 ml of distilled water and dialyzed against 5.0 mM sodium acetate (1 : 10,000) using dialysis tubing (3500 molecular weight cutoff). Next, the A_{260} of the dialyzed rRNA solution is measured and ZnCl$_2$

is added to a final concentration of 0.1 mM. Nuclease P1 is added (1 IU/300 A_{260} OD units) and the solution incubated at 45° for 2 hr. The digestion is monitored by thin layer chromatography (TLC) using polyethyleneimine TLC plates and 1.0 M LiCl as the developing solution.[10] After completion of the reaction, EDTA is added to 1.0 mM and the solution heated to 90° for 1 min and then frozen at −20°.

Chromatographic Separation of 5′-NMPs

Milligram quantities of each of the four 5′-NMPs can be resolved using the anion-exchange procedure of Cohn.[11] The 5′-NMPs are separated in order to prepare ribonucleotides that are uniformly deuterated except at the purine C8 and pyrimidine C5 or C6 positions. Fully protonated ^{13}C,^{15}N-labeled 5′-NMPs also are separated to prepare ribonucleotides that are specifically deuterated at C8, and C5 or C6.

Materials

Anion-exchange resin (AG 1-X2, 100-200 mesh) (Bio-Rad, Hercules, CA)
Affi-Gel 601 boronate-derivatized polyacrylamide gel (Bio-Rad)
Triethylamine (neat)
1 M Formic acid
1 M Triethylammonium bicarbonate (TEABC), pH 9.4
1 M NaOH
HNO$_3$
1% AgNO$_3$

Preparation of Anion-Exchange Column. The AG 1-X2 resin is prepared by exchange from the chloride form to the formate form. Approximately 50 ml of preswollen resin is mixed with 100 ml of distilled water to form a slurry and is poured into a 2.5 cm i.d. chromatography column forming a 13 cm bed (50 ml). The resin is then washed with 1.3 liter of 1.0 M NaOH. The column wash is tested for the presence of chloride ion by acidifying a 1 ml aliquot of effluent with HNO$_3$ and adding 100 μl of 1% AgNO$_3$. A clear solution indicates the exchange is complete and a white precipitate indicates that chloride ion is still present on the resin and that the resin requires additional NaOH wash. The resin is next washed with H$_2$O until the effluent pH is <9. Next, 1.0 M formic acid is applied to the column until the pH of the effluent is <2, and then the resin is rinsed with H$_2$O until the pH of the effluent is >5.

[10] E. P. Nikonowicz, A. Sirr, P. Legault, F. M. Jucker, L. M. Baer, and A. Pardi, *Nucleic Acids Res.* **20**, 4507 (1992).
[11] W. E. Cohn, *J. Am. Chem. Soc.* **72**, 1471 (1950).

Separation and Desalting of 5'-NMPs. Ethanol precipitated 5'-NMPs ($\approx$3100 A_{260} OD units or 100 mg calculated using an average $\varepsilon_{260} = 11,100$) are dissolved in 10 ml H_2O, the pH is adjusted to 8.0 with 3.0 M KOH, and the sample is loaded onto the column. The column is washed with distilled H_2O at a flow rate of 7 ml/min until the A_{260} absorbance of the effluent no longer changes and a steady A_{260} baseline is achieved. 5'-CMP is eluted from the column first using 0.1 M formic acid (pH 2.3–2.4). After the A_{260} absorbance peak corresponding to elution of the 5'-CMP begins to decrease, the column buffer is changed to 1.0 M formic acid. 5'-AMP elutes second and is followed by 5'-GMP. 5'-UMP is the last to elute from the column, and its removal is accelerated by bringing the column elution buffer to 0.1 M HCl. The typical elution volumes that we have generated for 100 mg of mixed 5'-NMPs are 160, 90, 860, and 190 ml for 5'-CMP, 5'-AMP, 5'-GMP, and 5'-UMP, respectively. The 5'-NMP fractions are neutralized using NH_4OH and rotary evaporated to reduce the volume of each 5'-NMP solution to $\approx$10 ml.

The 5'-NMPs are desalted using the Affi-Gel 601 resin, which is specific for *cis*-diols (i.e., 5'-rNMPs).[12] Triethylamine (1.7–3.0 ml) is added to each solution to raise the pH to 9.4 before loading the 5'-NMPs onto the affinity column. Once loaded, the column is washed with 150 ml of 1 M TEABC (prepared by bubbling CO_2 through a 1 M solution of triethylamine until pH 9.4 is reached). The 5'-NMPs are eluted using water acidified with bicarbonate to pH 4.5 (by bubbling CO_2 through distilled water) and lyophilized to remove triethylammonium bicarbonate salt. A total of 2950 A_{260} OD units is recovered.

Reverse Protonation and Deuteration of Base Positions C5, C6, and C8 in 5'-NMPs

The exchangeable NH and NH_2 protons are key to the determination of nucleic acid secondary structure and to identification of global folds through tertiary contacts. Additional details of nucleic acid structure such as base stacking and glycosidic torsion angle can be extracted using the 5, 6, and 8 base protons. However, since the biosynthetic methods for preparing deuterated nucleotides lead to extensive deuteration of the base carbons, it is necessary to selectively protonate these positions.

Materials

0.9 M $(NH_4)_2SO_3$ (pH 7.8 and 8.0 using sodium metabisulfite)
0.9 MNa$_2$S$_2$O$_5$

The C5 and C8 protium–deuterium exchange reactions are performed using bisulfite anion as a catalyst.[13] Two 0.9 M solutions of $(NH_4)_2SO_3$ are prepared

---

[12] R. T. Batey, J. L. Battiste, and J. R. Williamson, *Methods Enzymol.* **261**, 300 (1995).
[13] H. Hayatsu, *Prog. Nucl. Acid Res. Mol. Biol.* **16**, 75 (1976).

fresh and their pH is adjusted to 8.0 and 7.8 using freshly prepared 0.9 M Na$_2$S$_2$O$_5$ [≈7 ml and 15 ml, respectively, per 100 ml (NH$_4$)$_2$SO$_3$ solution]. The C8 and C5 exchange reactions are carried out at pH 8.0 and pH 7.8, respectively. The exchange reactions contain 30 mM 5′-NMPs and 300 mM (NH$_4$)$_2$SO$_3$ and are incubated at 65°. The reactions are monitored using ^{1}H NMR, and exchange at C8 and C5 is >95% complete after ≈60 hr and ≈72 hr, respectively. Exchange at the ribose carbons, adenine C2, or pyrimidine C6 has not been detected. After the reactions are complete, the 5′-NMPs are precipitated at −20° by addition of three volumes of ethanol, resuspended in 1M TEABC (pH 9.4), and desalted using the *cis*-diol affinity column (as above). Although exchange at the pyrimidine C6 positions is not catalyzed by this method, protonation of C6 can be accomplished using an alkaline solution of 40% ^{1}H$_2$O and 60% dimethyl sulfoxide (DMSO).[14] This method also leads to exchange at C5 and proceeds to ≈70% completion before the 5′-UMP and 5′-CMP begin to decompose.

Preparation of 5′-NTPs and Synthesis of RNA Oligonucleotides. The labeled 5′-NTPs are enzymatically phosphorylated at 37° by reacting the desalted 5′-NMPs with 2.5 equivalents of phosphoenolpyruvate (PEP) (Bachen, Torrance, CA) under N$_2$ as detailed.[10,12] Progress of the reaction is monitored by HPLC using an analytical anion-exchange column (Vydac) and usually is complete within 12 hr. The ^{1}H and ^{13}C NMR spectra of the 5′-NTPs indicate that phosphorylation and purification of the NTPs do not lead to exchange at the deuterated carbon positions. Published protocols for *in vitro* transcription using T7 RNA polymerase are used to produce labeled RNA molecules.[15] Yields from transcription reactions using ^{2}H/^{13}C/^{15}N-labeled 5′-NTPs and ^{13}C/^{15}N-labeled 5′-NTPs are comparable.

Preparation of Deuterated RNA

The biosynthetic approach to deuteration of RNA provides a simple route to RNAs that are uniformly ^{13}C,^{15}N-labeled and specifically protonated in the bases. This approach does not result in highly selective deuteration, but chemical methods can be used to selectively protonate pyrimidine C5 and C6 and purine C8 positions. Cells cultured in 90% ^{2}H$_2$O using either [^{2}H$_3$]acetate or [^{1}H$_3$]acetate yield comparable amounts of rRNA, 660 μmol/g and 600 μmol/g, respectively. (The number of moles is calculated using an average extinction coefficient of $\varepsilon_{260} = 11,100$ at 260 nm for the mixture of 5′-NMPs generated from the nuclease P1 digest of rRNA.) This yield is approximately 65% less per gram of carbon source than we have obtained from cells cultured using glucose and H$_2$O. However, the hydrogen atoms of glucose do not undergo significant exchange with the solvent during nucleoside biosynthesis, necessitating the use of [^{2}H$_7$]glucose to produce

[14] X. N. Huang, P. L. Yu, E. Leproust, and X. L. Gao, *Nucleic Acids Res.* **25**, 4758 (1997).
[15] J. F. Milligan, D. R. Groebe, G. W. Witherell, and O. C. Uhlenbeck, *Nucleic Acids Res.* **15**, 8783 (1987).

TABLE I
PATTERNS OF ^{2}H AND ^{13}C ENRICHMENT IN NUCLEOSIDES

Sodium acetate isotope label

Position	^{2}H enrichment (%)[a]			^{13}C enrichment (%)	
	^{2}H$_3$, 90% ^{2}H$_2$O	^{1}H$_3$, 90% ^{2}H$_2$O	^{13}C-2[b]	50% ^{13}C-1,2, 50% ^{13}C-2	75% ^{13}C-1,2, 25% ^{13}C-2
A,G 8	92	65	92	95	>95
A 2	88	80	95	>95	>95
C,U 5	92	64	90	94	>95
C,U 6	90	77	88	93	95
1'	≈95	75	80	90	95
2'	92	83	60	80	90
3'	90	78	20	60	80
4'	93	68	>95	>95	>95
5'	>95	56, 38	>95	>95	>95

[a] The percent isotopic enrichments were estimated from integrated peak areas in one-dimensional ^{2}H/^{13}C/^{15}N-decoupled ^{1}H and ^{1}H/^{2}H/^{15}N-decoupled ^{13}C spectra. The recovery delays for the ^{1}H and ^{13}C experiments were 10 s and 15 s, respectively.
[b] The distribution of ^{13}C isotope using ^{13}C-1 sodium acetate as the sole carbon source is consistent with the distribution that ^{13}C-2 sodium acetate yields.

high levels of deuteration. The fractional contributions of solvent and acetate to the ^{2}H enrichment of ribonucleotide carbons were estimated from one-dimensional ^{13}C NMR spectra and are listed in Table I. In medium that is 80% ^{2}H$_2$O, [^{2}H$_3$]acetate leads to 90–95% ^{2}H enrichment at all of the ribose carbon positions. Enrichment at the adenine C2 and purine C8 is 85% and 90%, respectively, and at the pyrimidine C6 and C5 is 90% and 88%, respectively. If [^{1}H$_3$]acetate is used as the carbon source in 90% ^{2}H$_2$O media, the enrichment of the ribose C1'–C4' is extensive with >68% deuteration. The C5' is approximately 56% enriched with one deuteron and has a slight stereospecific preference (12–15%) for deuterium at the *pro-S* position. Finally, 38% of the ribose C5' positions are enriched with two deuterons and 6% remain fully protonated. Figure 2 shows the ribose region of the ^{13}C spectrum of 5'-CMP.

The cost of preparing perdeuterated NMPs that are uniformly ^{13}C-labeled cannot be significantly reduced by using the less expensive ^{13}C-labeled C1 sodium acetate as the sole carbon source for the culture media. The distribution of ^{13}C label in the nucleosides resulting from either [^{13}C$_2$]C1, C2 or [^{13}C]C2 sodium acetate as the carbon source is shown in Table I. All proton-bound carbons except the ribose C2' and C3' are >80% derived from the acetate C2. Nonetheless, solvent exchange at the acetate C2 leads to substantial levels of ^{2}H enrichment in 90%

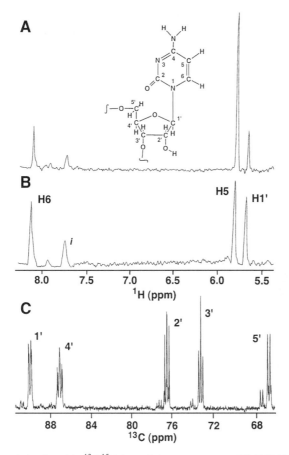

FIG. 2. Base and 1' region of the ^{13}C/^{15}N decoupled proton spectrum of 5'-CMP (A) after the proton exchange reaction and (B) before the proton exchange reaction. *E. coli* were cultured in 90% ^{2}H$_2$O and sodium [^{1}H$_3$, ^{13}C$_2$]acetate was used as the sole carbon source. An impurity that is not ^{13}C-coupled is marked *i*. There was no conversion of 5'-CMP to 5'-UMP as determined by HPLC. (C) The ribose region of the ^{1}H/^{2}H/^{15}N decoupled ^{13}C spectrum of perdeuterated ^{13}C/^{15}N-labeled 5'-GMP. *E. coli* were cultured in 90% ^{2}H$_2$O and sodium [^{2}H$_3$, ^{13}C$_2$]acetate was used as the sole carbon source. The C4' peak exhibits residual ^{31}P coupling.

^{2}H$_2$O, which can be increased by using 99% ^{2}H$_2$O or a mixture of [^{1}H$_3$]- and [^{2}H$_3$]acetates.

Since the degree of selectivity that can be achieved using biosynthetic methods for deuteration is limited, chemical methods are employed to site-specifically protonate C5, C6, and C8 atoms. Figure 2 compares the H6–H5 region of the ^{1}H spectrum of 5'-CMP before and after treatment with the ammonium bisulfite solution. The C8 and C5 atoms can be protonated site-specifically to levels >95%

and >92%, respectively. However, this method leads to >75% nonspecific cleavage of oligomeric RNA within 24 hr and therefore must be performed using the NMPs. A possible side reaction of the exchange is the proton-catalyzed deamination of cytidine to uridine.[13] Although we have not detected the presence of uridine in NMR spectra or HPLC traces of the cytidine exchange reaction, partial cytidine deamination (to form uridine) is not a serious problem since the exchange is performed on the NMPs, which can be easily separated, and may even be desirable since uridine is approximately 25% less abundant than cytidine in E. coli rRNA.

Alternative methods utilizing chemical or enzymatic protocols for the deuteration of nucleic acids have been developed.[16,17] At the present time, chemical methods for synthesis of multimilligram quantities of ^{2}H,^{13}C,^{15}N-enriched oligonucleotides only lend themselves to selectively labeled preparations. However, enzymatic methods have been used to prepare milligram quantities of RNA oligonucleotides with ^{13}C-labeled and specifically deuterated ribose carbons.[16] These methods utilize enzymes from the pentose phosphate pathway to prepare [3, 4, 5, 5-^{2}H$_4$]-phospho-D-ribosyl α-1-pyrophosphate (PRPP) from uniformly ^{13}C-enriched D-[1, 2, 3, 4, 5, 6, 6 ^{2}H$_7$]glucose. Protonation at the 1′ and 2′ carbon positions of ribose is an indirect process that results from the reactions catalyzed by glucose-6-phosphate isomerase and ribose-5-phosphate isomerase. The 5′-NMPs are prepared by enzymatic coupling of the bases (adenine, guanine, and uracil) to PRPP and then phosphorylated to the 5′-NTPs. 5′-CTP is prepared from 5′-UTP using CTP synthetase. The in vitro enzymatic reactions are an efficient route to specifically deuterated ([3′, 4′, 5′, 5″-^{2}H$_4$] and [2′, 3′, 4′, 5′, 5″-^{2}H$_5$]) nucleotides with ≈87% incorporation of glucose into 5′-NTPs. Preparation of deuterated nucleotides containing ^{13}C- and ^{15}N-enriched bases is necessary for sugar–base and intrabase scalar experiments but is more costly.

Effects of Deuteration on Relaxation of Exchangeable Base ^{1}H Magnetization

The recovery of NH proton longitudinal magnetization was measured for the perdeuterated (90% ^{2}H$_2$O, [^{2}H$_3$]acetate) and protonated forms of a 44 nucleotide RNA hairpin. Figure 3 compares the distribution of NH proton longitudinal recovery times for the two molecules at the same concentration and under the same buffer conditions. The recovery times for the NH protons of the perdeuterated hairpin are on average 33% longer than those of the protonated RNA molecule. This decrease of the recovery rate of the NH proton longitudinal magnetization in the perdeuterated RNA oligonucleotide is the result of fewer dipolar interactions

---

[16] T. J. Tolbert and J. R. Williamson, Jr., J. Am. Chem. Soc. 119, 12100 (1997).
[17] C. Glemarec, J. Kufel, A. Foldesi, T. Maltseva, A. Sandstrom, L. A. Kirsebom, and J. Chattopadhyaya, Nucleic Acids Res. 24, 2022 (1996).

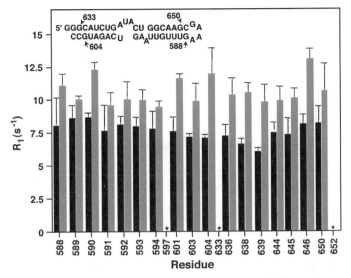

FIG. 3. Graphical presentation of the longitudinal recovery rates for a ^{2}H,^{15}N–labeled 45 nucleotide RNA hairpin corresponding to a variant of the binding site for *E. coli* ribosomal protein S8. The filled and speckled bars correspond to the deuterated and protonated, respectively, RNA hairpins. The deuterated RNA was synthesized from 5′-NTPs prepared from rRNA isolated from *E. coli* cultivated using 90% ^{2}H$_2$O and sodium [^{2}H$_3$]acetate as the deuterium sources. The recovery of magnetization from residues G$_{597}$, G$_{633}$, and G$_{652}$ is dominated by exchange with the solvent and cannot be estimated. The recovery rates contain contributions from both dipolar relaxation and from chemical exchange with the solvent. The nonselective inversion–recovery method was used and the solvent was suppressed using a 1$\overline{1}$ read pulse. The error bars indicate the standard deviations calculated using three experimental data sets.

being available to the NH protons. Only the ^{1}H dipolar interactions contribute significantly to the different relaxation rates since deuteration does not influence the chemical exchange processes between NH and solvent protons. Deuteration of the nonexchangeable sites in proteins also decreases the longitudinal relaxation rates of the amide protons by almost 35% relative to those of the protonated protein.[18]

In the absence of chemical exchange, the line width of a resonance is determined by its transverse relaxation rate. The line widths of the NH proton resonances of the perdeuterated RNA hairpin are 10–15% narrower than the corresponding resonances of the protonated RNA hairpin ($\approx$28 and $\approx$33 Hz, respectively). This indicates that the ^{1}H transverse relaxation rates decrease only a small amount in the perdeuterated RNA hairpin and suggests that chemical exchange significantly contributes to the NH proton resonance line width relative to the contribution of proton cross relaxation.

[18] K. A. Markus, K. T. Dayie, P. Matsudaira, and G. Wagner, *J. Mag. Reson., Series B* **105**, 192 (1994).

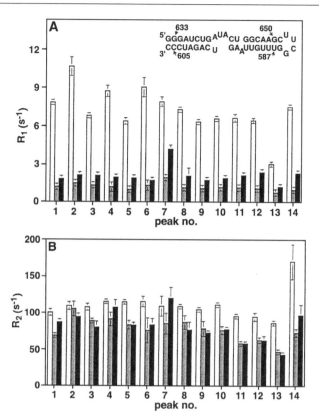

FIG. 4. Graphical presentation of the longitudinal (R_1) and transverse (R_2) relaxation rates of a [^{2}H, ^{13}C, ^{15}N]-labeled 44-nucleotide RNA hairpin corresponding to the binding site for *E. coli* ribosomal protein S8. (A) The H8 R_1 values for the deuterated (hatched) and protonated (open) hairpins in ^{2}H$_2$O and the deuterated hairpin in 90% ^{1}H$_2$O (filled). (B) The H8 R_2 values for the deuterated (hatched) and protonated (open) hairpins in ^{2}H$_2$O and the deuterated hairpin in 90% ^{1}H$_2$O (filled). The 5'-NTPs used to synthesize the RNA molecule were prepared from *E. coli* cultured on sodium [^{13}C$_2$]acetate as the sole carbon source and 90% ^{2}H$_2$O as the sole deuterium source. (A) The nonselective inversion–recovery method was used and rates were calculated by fitting cross-peak intensities from 2D ct-HSQC experiments to single exponential decays. (B) The spin–lock method was used and rates were determined as for (A). Error bars indicate standard deviations calculated using three experimental data sets.

Effects of Deuteration on Relaxation of Nonexchangeable Base ^{1}H Magnetization

The recovery of the H5 and H8 longitudinal and transverse magnetization for ^{13}C,^{15}N-labeled protonated and highly deuterated RNA hairpins generally parallels that of the exchangeable resonances. Figure 4 compares the distribution of H8 relaxation rates for ribose-protonated and ribose-deuterated molecules. Deuteration slows the longitudinal relaxation of H8 resonances 3.9-fold on average, and the H5 resonances are slowed 2.5-fold (not shown). The relative changes of the H8 and H5 relaxation rates are consistent with the predicted number of proximal dipole relaxation partners. In canonical A-form helices, three to four nonexchangeable protons are located within 3.0 Å of H8 while only one proton, H6, is within 3.0 Å of H5. Substitution of these relaxation partners with deuterium reduces the efficiency of the H5 and H8 longitudinal relaxation processes, but because of the presence of fewer proximal protons, H5 relaxation is reduced less.

The longitudinal and transverse relaxation rates of the H8 resonances of the ribose-deuterated RNA hairpin also have been measured in 90% ^{1}H$_2$O. Under this condition, the NH and NH$_2$ positions are protonated and therefore have the potential to contribute to the relaxation of the nonexchangeable base protons. The longitudinal relaxation rates increase an average of 25% in the presence of the NH and NH$_2$ protons, but the transverse relaxation rates are largely unaffected (Fig. 4). The small change of the H8 longitudinal relaxation is not unexpected since the H8–NH/NH$_2$ proton distances are ≈4.5 Å in regular A-form RNA helices. Although the H5 relaxation rates are only qualitatively determined, they generally are shorter because of the proximity of NH$_2$ protons in G·C and A·U base pairs.

Deuteration also reduces the transverse relaxation rates of the H8 and H5 protons, but the effect is modest compared to the effects on longitudinal relaxation. Figure 4 shows that the H8 transverse relaxation rates of the deuterated RNA molecule are 35% slower on average. The slower rates lead to a 10–25% line-width reduction of the H8 (and H5) resonances of the RNA hairpin. This improvement is not as dramatic as the 50% linewidth reduction that has been measured for the amide resonances of perdeuterated (≈80% ^{2}H enrichment) proteins.[18] Although the density of proximal ^{1}H neighbors is different for amide protons and nucleotide base protons, it is also important to note that H5 and H8 are ^{13}C-bound. The ^{13}C–^{1}H dipolar contribution to ^{1}H transverse relaxation is about twice the ^{15}N–^{1}H dipolar contribution.[19] Although the heteronuclear dipolar interaction limits the effectiveness of deuteration for base ^{1}H linewidth improvement, the ^{13}C nucleus also makes possible the coupling of deuteration and TROSY based experiments.

Scalar Correlation Spectra using Deuterated RNAs

^{15}N–^{1}H Spectra

The relative cross-peak heights for NH and NH$_2$ resonances in HSQC spectra of the two RNA hairpins are consistent with resonances that are dominated by

---

[19] R. R. Ernst, G. Bodenhausen, and A. Wokaun, "Principles of Nuclear Magnetic Resonance in One and Two Dimensions," p. 504. Oxford University Press, New York, 1992.

chemical exchange. An increase in signal-to-noise after deuteration is expected if proton cross relaxation significantly contributes to the line width of the exchangeable protons. In the case of the NH proton resonances, there is only a small increase of the signal-to-noise on deuteration. However, the cytidine NH_2 resonances of the deuterated hairpin exhibit a range of sensitivity enhancement, 10–60% signal-to-noise increase, over the corresponding resonances of the protonated hairpin. Nonetheless, this sensitivity gain is modest compared to the 250% increase observed for the amide resonances of proteins.[20] Relaxation measurements of the amide proton resonances indicate that transverse relaxation rates can be reduced by 50% when the side-chain positions are deuterated.[18] Thus, the sensitivity gains that are expected from decreased transverse relaxation rates are lost to chemical exchange even in regions of regular RNA secondary structure.

$^{13}C-^{1}H$ Spectra

The linewidths of the H8 resonances of the deuterated RNA hairpin are 20–25% narrower than the corresponding resonances of the protonated RNA hairpin ($\approx$27 and $\approx$34 Hz, respectively). Similarly, the linewidths of the H5 resonances of the deuterated RNA molecule are about 10–20% narrower than the corresponding resonances of the protonated RNA molecule, $\approx$26 Hz and 32 Hz, respectively. These narrower resonances result in a modest (30–40%) signal-to-noise increase of H8–C8 and H5–C5 cross peaks in the HSQC spectrum of the deuterated RNA molecule.

In addition to improving the spectral quality of the base ^{1}H resonances, deuteration also provides benefits for the resonances of the sugar ring. Assignment of the ribose spin systems of slowly tumbling RNAs is complicated by rapid transverse relaxation of the ribose ^{13}C nuclei. For these molecules, the HCCH-TOCSY experiment that is employed in the conventional heteronuclear assignment strategy becomes inefficient and leads to incomplete spin system correlations. Selective deuteration of C3'–C5' increases the transverse relaxation times of those nuclei and improves the efficiency of magnetization transfer through the ribose ring, simplifying cross-peak pattern analysis.[21] Figure 5 compares the 2D CT-C(CC)H TOCSY spectra of protonated and 3', 4', 5', 5'' deuterated TAR RNA hairpins. In the deuterated molecule, the H1' and H2' resonances of each ribose group yield a set of cross peaks for that entire ^{13}C ribose spin system. These correlations are incomplete in the spectrum of the protonated molecule.

$^{31}P-^{1}H$ Spectra

The phosphate backbone is a fundamental structural element of nucleic acids and is a key mediator of protein–nucleic acid contacts. The conformation of the

[20] R. A. Venters, W. J. Metzler, L. D. Spicer, L. Mueller, and B. T. Farmer II, *J. Am. Chem. Soc.* **117,** 9592 (1995).
[21] K. T. Dayie, T. J. Tolbert, and J. R. Williamson, *J. Mag. Reson.* **130,** 97 (1998).

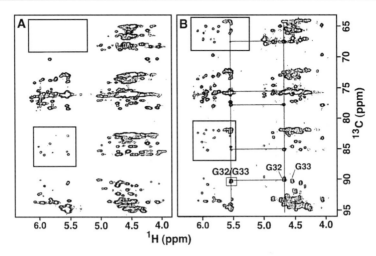

FIG. 5. Comparison of 2D CT-C(CC)H TOCSY spectra acquired at 25° of TAR RNA oligonu-cleotides that are ^{13}C-enriched and either (A) fully protonated or (B) selectively deuterated at the 3′, 4′, 5′ carbon atoms. The TAR RNA oligonucleotide, (5′ GGCCAGAUUGACCUGGGAGCUCU-CUGGCC 3′) forms a hairpin and with a stem containing an internal loop. The vertical lines (B) are drawn at the H1′ and H2′ frequencies of G32. Pairs of H1′ to C1′–C5′ and H2′ to C1′–C5′ correla-tions are connected by horizontal lines, producing a ladder of intraresidue connectivities that simplifies the resonance assignment. The fully protonated oligonucleotide does not yield the analogous set of correlations. The selectively deuterated 5′-NTPs used for the T7 RNA polymerase reaction were pre-pared beginning with [^{13}C]glucose and [^{2}H,^{13}C]-glucose using enzymes from the pentose phosphate pathway.[16] [Reproduced with permission from Dayie *et al., J. Mag. Reson.* **130,** 97 (1998). Copyright © 1998 Academic Press].

backbone is defined by six torsional angles. Two of these angles, β and ε, are accessible through measurement of ^{31}P–^{1}H coupling constants. Figure 6 compares the ^{31}P–^{1}H HetCor spectra of ≈75% deuterated and fully protonated RNA hairpin molecules. The passive couplings have been decreased below 10%, thus enabling observation of many of the often weak P–H5′, H5″, and P–H4′ correlations and permitting measurement of the $^3J_{HP}$ values directly from the spectrum. Although the effective concentration of the deuterated sample is less than one-third that of the fully protonated sample, the deuterium-assisted narrowing of the ribose ^{1}H line widths increases the experimental sensitivity of the deuterated RNA relative to the fully protonated RNA. Thus, fractional deuteration can be used to improve the quality of ^{31}P–^{1}H (and ^{1}H–^{1}H) correlated spectra and facilitate the extraction of structurally important coupling constant information.

NOE-Based Experiments using Deuterated RNA Oligonucleotides

The exchangeable NH and NH$_2$ protons participate in hydrogen bonding net-works that are central to base–base interactions and are the source of NOEs that are

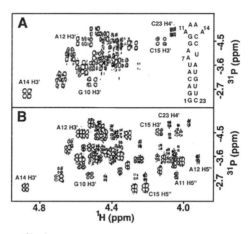

FIG. 6. Comparison of ^{31}P–^{1}H HetCor spectra of (A) fully protonated RNA hairpin and (B) ≈75% random ^{2}H-enriched RNA hairpin. The 5′-NTPs used to synthesize the ^{2}H-enriched sample were prepared from *E. coli* cultured in 80% ^{2}H$_{2}$O with sodium [^{2}H$_{3}$]acetate as the sole carbon source. The samples were ≈1.3 m*M* and spectra A and B were acquired in 13 and 16 hr, respectively. The nearly complete elimination of passive couplings in the proton dimension of B permits observation of the weakly coupled P–H5′, H5″ resonances and direct measurement of $^{3}J_{HP}$ values.

used to establish secondary structure. In addition, NOEs involving these protons are the source of more global structural information and can provide conformational details on a variety of nucleic acid structure elements including nonstandard base pairs, base triples, and coaxial stacked helices. However, NOE cross peaks between NH and NH$_2$ protons can be weak or may overlap with carbon-bound protons, making them difficult to identify and interpret. Figure 7 compares the NH regions of 260 ms mixing time NOESY spectra of the perdeuterated and protonated 45 nucleotide RNA hairpins. NOE cross peaks U^{644}H3–G^{597}H1 and U^{644}H3–A^{596}NH$_2$ that are present in the spectrum of the perdeuterated RNA molecule define the secondary structure in the central portion of the hairpin stem. These cross peaks are too weak to be observed in the spectrum of the protonated molecule.

Figure 8 compares one-dimensional vectors along ₁ of the 260 ms NOESY spectra for the U$_{603}$ NH resonance, demonstrating the increased sensitivity of the experiment for the deuterated RNA molecule. Both cross peaks and diagonal peaks in the spectrum of the perdeuterated molecule have 2 to 5-fold higher signal-to-noise ratios than the corresponding peaks of the fully protonated RNA molecule. The increased sensitivity results from the reduced number of cross relaxation pathways for the NH protons in the perdeuterated RNA molecule rather than longer lived transverse magnetization of the participating resonances in the deuterated molecule or more rapid NOE cross-peak decay in the protonated molecule.

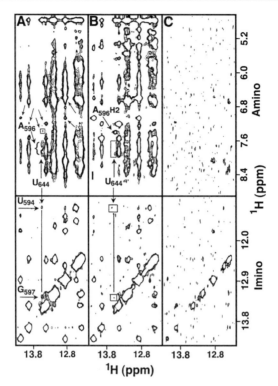

FIG. 7. Comparison of NH regions of the 45 nucleotide RNA hairpin (Fig. 8) for the NOESY spectra of (A) the deuterated and (B) the protonated RNA hairpin molecules and (C) the ω_1 ^{15}N-selected NOESY spectrum of the protonated RNA molecule. The NH–NH and NH–NH$_2$ regions are divided. The NOE mixing time was 260 ms and the concentration of the RNA molecule was the same for each experiment. The small arrows in A indicate the H2–H3 cross peaks corresponding to the six A·U base pairs. The corresponding cross peaks are considerably more intense in the spectrum of the protonated RNA hairpin. Positions of cross peaks that are absent from the spectrum of the protonated RNA hairpin are indicated by boxes in B. Cross peaks between both NH$_2$ protons of A596 and U644H3 are present in A, but only one of the two cross peaks is present in B. The deuterated 5'-NTPs were prepared from *E. coli* cultured in 90% ^{2}H$_2$O with sodium [^{2}H$_3$]acetate as the sole carbon source.

The NOESY spectrum of the deuterated RNA hairpin contains cross peaks primarily between the exchangeable protons, although weak cross peaks involving the adenine H2 resonances also can be observed (Fig. 7A). To observe cross peaks only between exchangeable resonances using the fully protonated RNA molecule, an ω_1-selected NOESY experiment must be employed. However, chemical exchange and relaxation during the filter delay periods dramatically reduce the sensitivity of this experiment relative to the NOESY and results in the absence of most cross peaks (Fig. 7C). The ω_1-selected NOESY spectrum of the deuterated molecule

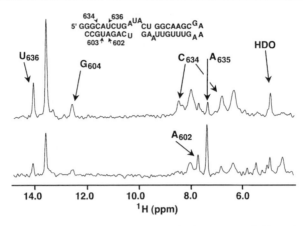

FIG. 8. ω_1 vectors from the 2D NOESY spectra shown in Fig. 7 at $\omega_2 = 13.34$ ppm for the deuterated (top) and protonated (bottom) RNA hairpins. The spectra are plotted to show the same level of noise.

retains several cross peaks, but still has fewer than the corresponding NOESY spectrum. Assignment of the NOE cross peaks using only the 2D $^1H-^1H$ spectrum is difficult, but many of the correlations can be assigned using 2D $^{15}N-^1H$ and 3D ^{15}N-edited NOESY spectra.[6] Although chemical exchange processes involving solvent protons remain active, deuteration of RNA oligonucleotides increases the signal-to-noise of NOE–based spectra and diminishes the effects of spin diffusion. Thus, deuteration permits extraction of additional and more accurate structural constraints.

Deuterium enrichment of [^{13}C, ^{15}N]-labeled RNA oligonucleotides can enhance the sensitivity of NOE–based 2D and 3D experiments and simplify cross-peak assignments in the resulting spectra. The conventional sequence specific assignment strategy for the H5 resonances using the H6/H8–H1' region of the NOESY spectrum is complicated by the broad lines resulting from slow molecular tumbling. An alternative strategy employing a 3D ^{13}C-separated HMQC-NOESY spectrum has been used to assign H5 resonances of the 44 nucleotide RNA hairpin through the NH-H5 cross peaks.[7] The intrabase UH5-UH3 and intrabase pair CH5-GH1 cross peaks primarily arise from spin diffusion through the adenine and cytidine NH_2 protons. The signal-to-noise of the same experiment acquired using the protonated RNA hairpin was very poor and did not yield identifiable cross peaks. Because the H5 protons contribute structural information in the form of interbase NOEs, assignment of the H5 resonances is important, but their assignment also is one step toward the sequence-specific assignment of the ribose groups in deuterated molecules. Using triple-resonance experiments, it should be possible to use the H5 (and H8) to link the base and deuterated ribose groups while taking advantage of the extended transverse relaxation offered by deuteration of the 1′ carbon.

The proton longitudinal relaxation rates of deuterium enriched molecules can degrade their experimental utility. Deuteration increases the time required for the protons to return to thermal equilibrium, but long recovery delays are impractical in multidimensional NOE–based experiments. However, the exchange of NH and NH_2 protons with solvent is sufficiently rapid so that the steady-state ^{1}H magnetization is largely dependent upon the recovery of the solvent protons and not the longitudinal relaxation of the NH and NH_2 protons. Thus, the majority of NH and NH_2 proton magnetization is replenished during the recovery periods as long as excitation of the solvent is limited. The carbon–bound H5 and H8 within the deuterated environment have longitudinal relaxation times around 1.0 s (Fig. 4). The directly attached ^{13}C nuclei facilitate the relaxation of longitudinal magnetization and permit the use of recovery delays that lead to reasonable experimental acquisition times.

Site-Specific Deuteration of ^{13}C-Labeled RNA Oligonucleotides

The resonance assignment of nucleic acids is primarily accomplished using sequential NOE cross peaks between adjacent base and 1′ protons and is facilitated by ^{13}C and ^{15}N isotopic enrichment that simplifies the NOESY spectrum and permits the scalar correlation of intraresidue base-1′ resonances.[22,23] Selective deuteration at the pyrimidine C5 using the bisulfite-catalyzed exchange method can be combined with uniform ^{13}C enrichment to yield additional spectral simplification. Figure 9 compares the NOE-based resonance assignment procedure applied to a C5 protonated (5-^{1}H) RNA molecule and the corresponding C5 deuterated (5-^{2}H) molecule. The cross peaks involving H6 are separated according to the chemical shifts of the directly attached C6 nuclei. In the 5-^{2}H molecule, the intensities of the residual H6-H5 cross peaks are significantly weaker than the H6-H1′ correlations (Fig. 9). The near absence of an H6-H5 cross peak in the spectrum of the 5-^{2}H molecule permits unambiguous identification of both Cyt-10 H6-H1′ correlations (Fig. 9). The interresidue crosspeak between Uri-12 H6 and Cyt-13 is not observed because of the conformation of the base. The absence of this interaction is immediately clear in the 5-^{2}H molecule spectrum. In the spectrum of the 5-^{1}H molecule, it is not possible to distinguish whether this cross peak is absent or present but concealed beneath the H6-H5 cross peak. The sensitivity of HMQC-NOESY-HMQC experiments was insufficient to provide comparable information through 1′C6/H6 correlations. Thus, ^{2}H substitution of C5 leads to ready identification of the H6-H1′

[22] A. Pardi, *Methods Enzymol.* **261**, 350 (1995).
[23] T. Dieckmann, E. Suzuki, G. K. Nakamura, and J. Feigon, *RNA* **2**, 628 (1996).

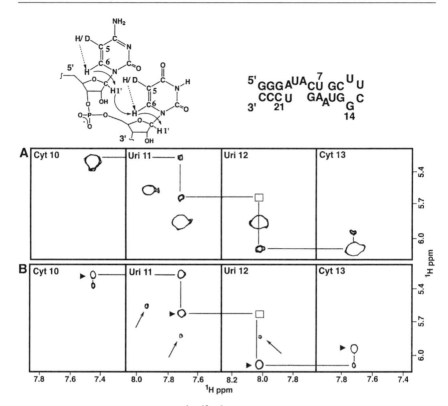

FIG. 9. $\omega_1-\omega_3$ planes extracted from {^{1}H, ^{13}C, ^{1}H} 3D NOESY-CT-HMQC spectra showing sequential base-1' proton correlations from Cyt-10 to Cyt-13 in (A) C5 protonated and (B) C5 deuterated RNA molecules. The secondary structure of the RNA sequence is shown at the top. In (B), the H5-H6 cross peaks are indicated by arrows and the intraresidue H6-H1' correlations are marked (c). The RNA molecules were prepared as reported using uniformly ^{13}C enriched and uniformly ^{13}C enriched/C5 deuterium labeled 5' NTPs. Sample concentrations were 2.1 mM and 1.8 mM for the deuterated and protonated molecules, respectively. [Adapted with permission from Nikonowicz et al., J. Am. Chem. Soc. 120, 3813 (1998). Copyright © 1998 American Chemical Society.]

correlations and allows the sequential resonance assignment to proceed alternately between C6/8 and C1' NOE planes without interruption.

Conclusion

^{13}C and ^{15}N isotopic enrichment strategies have increased the variety and complexity of RNA molecules amenable to solution-state structure analysis, and the inclusion of ^{2}H provides an additional tool to study these systems. The biosynthetic methods for deuteration of RNA are extensions of established protocols for ^{13}C and

^{15}N enrichment of RNA and so are straightforward to implement. Deuteration simplifies and improves the quality of NOE-and scalar-based spectra. It may also provide an avenue toward the development of triple resonance techniques that exploit the improved relaxation properties of deuterated ^{13}C nuclei or yield novel forms of structural information for nucleic acids and nucleic acid–protein complexes.

Acknowledgments

I thank the students and postdoctoral fellows in the laboratory who have contributed to this work, especially Dr. Margaret Michnicka. I also thank Drs. Theresa Koehler, Sean Moran, and Yousif Shamoo for critical reading of the manuscript. This work was supported by grants from the Welch Foundation (C-1277) and the National Institutes of Health (GM52115).

[15] Investigation of Unusual DNA Motifs

By Anh-Tuân Phan, Maurice Guéron, and Jean-Louis Leroy

1. Introduction

As new nucleic acid structures are found, nuclear magnetic resonance (NMR) continues to be a major player in their study. In a new structure, stoichiometry is unknown, and so is base pairing. Structural symmetries or quasi-symmetries, either static or involving motional averaging, may affect spectral resolution and give rise to ambiguities in the interpretation. Under such conditions, model-independent methods are required or strongly preferred. The results of model-dependent procedures should be checked by independent means.

The situation is much the same as with proteins. Multidimensional heteronuclear NMR and isotopic labeling, coupled to molecular refinement by methods such as distance geometry, are the basic tools of the trade.

There are also differences such as the restriction to four bases (mainly) compared to 20 amino acids. This enhances the difficulty of spectral assignments. But it also limits the chemistry and the variety of the interactions so that representative structures may be formed even by short oligonucleotides, which are easily synthesized by the phosphoramidite method for DNA and even RNA. This makes modulation of sequence a convenient and useful tool for structural investigations.

Another difference is the flexibility of the backbone and its often weak or poorly visible involvement in tertiary interactions: as a result, the NMR study of the backbone is not structurally rewarding. In contrast, the general occurrence of hydrogen-bonded base pairs provides powerful structure indicators, enhancing the value of exchangeable proton spectroscopy. The moderate pK of the imino protons is also of value, making possible the determination of the rate of internal motions using proton exchange catalyzed by proton acceptors such as ammonia. This is not possible with peptide protons.

In this review we describe procedures used in our laboratory for the investigation of different aspects of unusual DNA structures: stoichiometry, spectral assignments, structure characterization, symmetry and symmetry-related motions, associated water. The examples are taken mostly from studies of structures involving the i-motif, a four-stranded intercalated structure, but also DNA loops. The procedures are applicable to other cases, for instance, G-quartets,[1-4] the arrowhead motif,[5] and others.[6] Much relies on exchangeable protons whose investigation was discussed in an earlier volume.[7]

Three important and recent developments make use of line narrowing by cross-correlation (transverse relaxation-optimized spectroscopy or TROSY), J-coupling across H-bonds, and residual dipolar couplings in anisotropic media, of which the first two at least may require isotopic enrichment. We do not discuss them further since we have not used them in our laboratory. They are treated elsewhere in these volumes.

2. Determination of Stoichiometry

Determination of stoichiometry is fraught with difficulties. One can distinguish two types of methods. One type uses mass-related properties such as translational diffusion (centrifugation methods), rotational diffusion (light depolarization, NMR relaxation methods), size (gel filtration methods), number of electrons (X-ray scattering), and mass itself (mass spectrometry); light-scattering methods also make use of these properties. The other type uses the concentration-dependent equilibrium between structures with different stoichiometries.

2.1. NMR Titration of Monomer–Multimer Equilibrium

The reaction between an n-mer and an m-mer, according to:

$$n(m\text{-mer}) \rightleftharpoons m(n\text{-mer}) \tag{1}$$

leads to equilibrium concentrations C_n, C_m which are given by the mass action law:

$$(C_m)^n = K_{n,m}(C_n)^m \tag{2}$$

[1] F. W. Smith and J. Feigon, *Biochemistry* **32**, 8682 (1993).

[2] A. Kettani, S. Bouaziz, A. Gorin, H. Zhao, R. A. Jones, and D. J. Patel, *J. Mol. Biol.* **282**, 619 (1998).

[3] Y. Wang and D. J. Patel, *J. Mol. Biol.* **242**, 508 (1994).

[4] Y. Wang and D. J. Patel, *Biochemistry* **31**, 8112 (1992).

[5] A. Kettani, S. Bouaziz, E. Skripkin, A. Majumdar, W. Wang, R. A. Jones, and D. J. Patel, *Structure* **7**, 803 (1999).

[6] A. Kettani, S. Bouaziz, W. Wang, R. A. Jones, and D. J. Patel, *Nat. Struct. Biol.* **4**, 382 (1997).

[7] M. Guéron and J. L. Leroy, *Methods Enzymol.* **261**, 383 (1995).

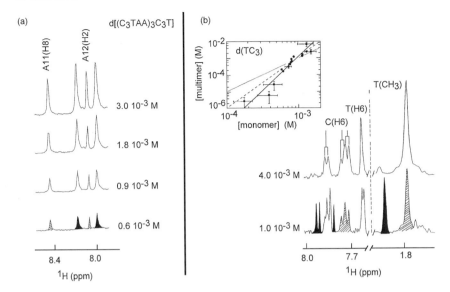

FIG. 1. NMR determination of stoichiometry. (a) NMR titration of the human telomeric DNA fragment d[(C₃TAA)₃C₃T]: part of the A(H8), A(H2) spectral region, showing the A11(H8) and A12(H2) peaks of the structured fragment (hatched) and the two peaks of the adenine aromatic protons of the unstructured, presumably monomeric fragment (in black). At equilibrium, the *relative* intensities of the two sets of peaks are independent of DNA concentration, demonstrating that the structured species is monomeric. Temperature, 30°; pH 6; proton frequency, 500 MHz. (b) The equilibrium concentrations of the d(TCCC) *i*-motif structure and of the unstructured monomer. The measurement uses NMR peaks that have been assigned to the spectra of the structured (hatched) and unstructured (black) species. The data fit a straight line of slope 4 (full line) on the log–log plot. Lines of slope 3 (dashed line) and 2 (dotted line) are also traced, showing the difficulty of a precise determination of stoichiometry, and the importance of measurements at the lowest and highest concentrations, for which error bars are presented. In contrast to (a), the determination of a stoichiometry different from one requires the determination of absolute concentrations. This is provided for instance by comparison of the methyl proton signals to an internal reference (DSS). Temperature, 6°; pH 4.2; 100 mM NaCl; proton frequency, 360 MHz. (b) is adapted from Ref. 8.

where $K_{n,m}$ is the equilibrium constant, in (mol/liter)$^{n-m}$. This relation holds even if other multimers are present. A double logarithmic plot of concentrations of the two species for different dilutions is a straight line whose slope, n/m, reveals the ratio of stoichiometries, and whose position gives the equilibrium constant. In a typical case, one seeks to derive the stoichiometry n of a multimer from the equilibrium with a monomer ($m = 1$).

The NMR titration is based on the intensity of peaks originating from the spectra of the n-mer and of the monomer in slow exchange. The intensities are proportional to the concentrations, except for errors such as those discussed below. A first example is that of Fig. 1a, a spectrum with peaks assigned to two forms of

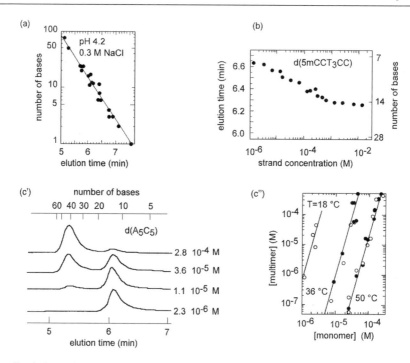

FIG. 2. Determination of stoichiometry by gel filtration chromatography, monitored by UV absorption. (a) Calibration of the Synchropack GPC 100 chromatographic column. The log of the number of residues of several nucleic acid reference samples (including thymidine, single-stranded oligonucleotides and duplexes, and yeast transfer RNA) is plotted as a function of the elution time. Temperature, 18°. (b) Elution time of d(5mCCTTTCC) as a function of the strand concentration of the injected sample, showing the transition between monomer and dimer. The monomer–dimer equilibrium is fast relative to the rate of separation in the migration process, so that the material elutes as a single peak, whatever the concentration. Temperature, 0°. (c') The elution profile of d(A₅C₅) consists of two well-separated components, indicating that conversion between them is negligible during the migration process. The elution times correspond to approximately 10 and 40 nucleotides, indicating a monomer–tetramer equilibrium. Temperature, 18°. (c'') The stoichiometry of the multimer of d(A₅C₅) is confirmed by the plot of (absolute) concentrations of the two species, as derived from the integrated UV absorbance. The data carrespond to solutions of the multimeric (d) or monomeric (s) forms, the latter being obtained by quenching from 100°. They were incubated at 18°, 36°, and 50° for 1 week prior to the elution measurement. Both symbols fall in the same pattern, indicating that equilibrium has been reached. The elution solution was buffered at pH 4.2 with sodium acetate (20 mM). It contained NaCl (300 mM) so as to reduce the DNA–matrix interaction. The volume of the chromatography sample was in the range of 1 to 20 μl. The flow rate for optimal resolution was 0.4 ml/min. The elution profiles were measured by the absorbance at 280 nm. (a), (c'), and (c'') are adapted from Ref. 57.

the human telomeric DNA fragment d[C$_3$TAA)$_3$C$_3$T]. The relative intensities are independent of dilution, showing that the two species have the same stoichiometry. One species is a single unstructured strand, as shown separately by the temperature and pH dependence, by gel filtration chromatography, etc. Hence, the structured form is monomeric.

A different case is that of d(TCCC) (Fig. 1b). The ratio of the two spectral components (black and hatched) is strongly concentration-dependent. The double logarithmic plot has a slope of 4, corresponding to a tetramer/monomer equilibrium. The plot shows the importance of measurements at low concentrations, and the difficulty of deriving a good value of the stoichiometry from the slope. In particular, the low-concentration measurements are affected by weak sensitivity, leading to random errors. Systematic errors arise if equilibrium is not reached, and one should be aware that the interconversion times may well depend on oligonucleotide concentration.[8] Interconversion may be detected and measured by structural exchange cross peaks for interconversion times as long as 1 min.

2.2. Gel Filtration Chromatography

The molecular size and hence the stoichiometry may be determined by gel filtration chromatography,[8] using a column calibrated with oligonucleotides of known molecular weight (Fig. 2a). Figure 2b shows the elution time of a sample whose equilibration between different forms is so fast that it migrates as a single peak. The migration rate indicates a monomer at low concentrations, and a dimer at high concentrations.

Figure 2c' shows a case of slow equilibration. The sample elutes as two separate peaks, which correspond to a monomer and a tetramer. The slope of the double logarithmic plot (Fig. 2c") provides independent confirmation of the stoichiometry.

3. Assignment Procedures

Assignment usually starts with the classification of resonances according to the type of nucleotide. Resonances with J couplings between them are said to be part of a "spin system." The nucleic bases may be recognized by characteristic short proton–proton distances and their corresponding NOESY cross peaks (Fig. 3), and by TOCSY cross peaks as well. After grouping the resonances by nucleotide, one must match each nucleotide, as characterized by its NMR resonances, to a position on the oligonucleotide sequence.

---

[8] J. L. Leroy, K. Gehring, A. Kettani, and M. Guéron, *Biochemistry* **32**, 6019 (1993).

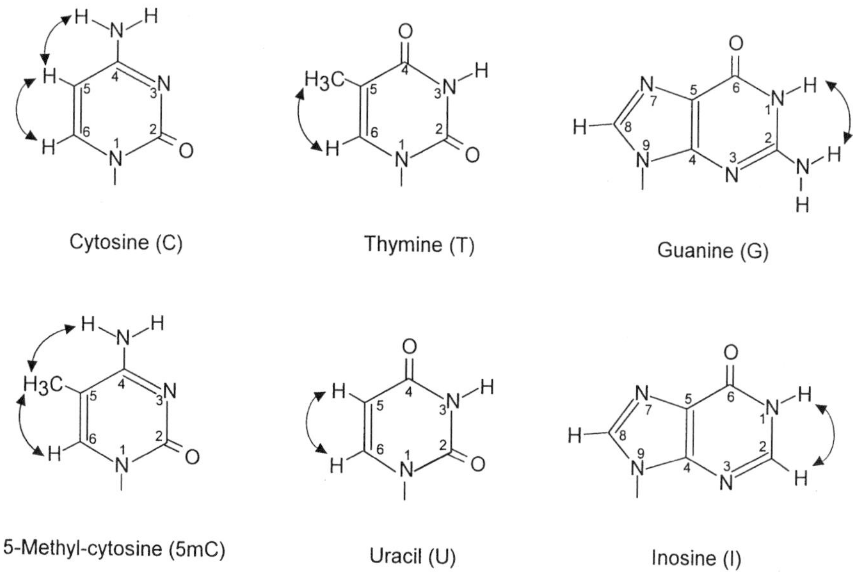

Cytosine (C)   Thymine (T)   Guanine (G)

5-Methyl-cytosine (5mC)   Uracil (U)   Inosine (I)

FIG. 3. Natural and modified bases, and their specific patterns of short interproton distances.

Except for the smallest oligonucleotides and the simplest cases, the assignments are limited by insufficient spectral resolution. The "royal road" to assignment is by selective isotope labeling, which may be used in many ways and provides unambiguous characterization by nuclear spin, without chemical alteration. A powerful spectroscopic method that is presently dependent on isotopic labeling is the correlation of resonances by *J*-couplings across a hydrogen bond.[9] This provides for the first time a direct assignment of paired bases.

Still, isotopic enrichment may be difficult and costly. This generates a niche for careful chemical modifications such as methylation, and for the exploitation of heteronuclei in natural abundance.

3.1. Chemical Modifications

Phosphoramidite derivatives are commercially available for modified deoxyribonucleotides such as deoxy-5-methylcytidine, deoxyuridine, deoxyinosine, and others, but apparently not for ribonucleotides. Once incorporated in an oligodeoxyribonucleotide, they are useful for spectral assignments in 1D and 2D spectra.[10,11]

---

[9] A. J. Dingley and S. Grzesiek, *J. Am. Chem. Soc.* **120**, 8293 (1998).
[10] F. Jiang, D. J. Patel, X. Zhang, H. Zhao, and R. A. Jones, *J. Biomol. NMR* **9**, 55 (1997).
[11] J. L. Leroy and M. Guéron, *Structure* **3**, 101 (1995).

One must of course watch for possible structural alterations. Consider for instance the intercalation topology of the *i*-motif tetramer of d(TCC) (only one strand of each duplex is represented):

$$5'\text{-T}\quad\text{C}\quad\text{C}$$
$$\text{C}\quad\text{C}\quad\text{T}$$

The same topology is observed for the tetramer of d(TCCC), and for the methylated derivative d(Tm5CCC):

$$5'\text{-T}\quad5\text{mC}\quad\text{C}\quad\text{C}$$
$$\text{C}\quad\text{C}\quad5\text{mC}\quad\text{T}$$

This is not surprising, since the methylated pairs of the two duplexes are not in contact.

The situation for d(Tm5CC) is more complex. The same topology occurs, even though it brings into contact the methylated pairs:

$$5'\text{-T}\quad5\text{mC}\quad\text{C}$$
$$\text{C}\quad5\text{mC}\quad\text{T}$$

However, this topology is in equilibrium with another one:

$$5'\text{-T}\quad5\text{mC}\quad\text{C}$$
$$\text{C}\quad5\text{mC}\quad\text{T}$$

which avoids contact between the methylated pairs while maintaining full stacking of the C · CH$^+$ pairs.[11]

Similar phenomena occur in the case of *i*-motif structures formed by intramolecular folding, for instance with the human telomeric fragment d[(CCCTAA)₃CCCT]. Methylation of C1 usually does not affect the structure, and it provides a good starting point for assignments, as previously observed.[12,13] However, methylation of C7 and/or demethylation of T16 within the sequence have interesting structural effects,[14] as discussed below.

3.2. Natural Abundance Heteronuclear Correlations

The through-bond correlations required for model-independent assignments may often be obtained without isotopic enrichment, using coherence transfer pathways that involve only one rare heteronucleus. These techniques apply both to nonexchangeable and to exchangeable protons. (a) Figure 4 shows how

[12] X. Han, J. L. Leroy, and M. Guéron, *J. Mol. Biol.* **278,** 949 (1998).
[13] A. T. Phan and J. L. Leroy, *J. Biomol. Struct. Dynam.* **Conversation 11,** 2, 245 (2000).
[14] A. T. Phan, M. Guéron, and J. L. Leroy, *J. Mol. Biol.* **299,** 123 (2000).

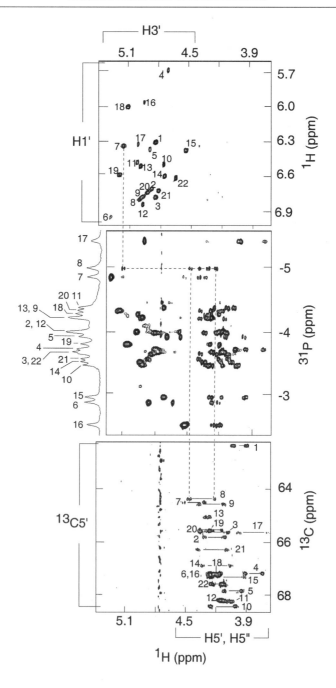

successive nucleotides may be recognized[14] by coordinating an H1′–H3′ intra-residue TOCSY[15] with a phosphorus–proton hetero-TOCSY[16] and a natural abundance ^{13}C5′–H5′/H5″ HSQC.[17] (b) An HMBC experiment provides long-range linkage of adenosine H2 and H8 protons through ^{13}C.[18,14] An example of this is shown in the spectrum of Fig. 5 (top). (c) A correlation of an imino proton to its geminal ^{15}N can be used to distinguish imino protons of guanosine and uridine.[19] (d) A JRHMBC experiment links imino and nonexchangeable protons through ^{13}C in guanosine and thymidine (Fig. 5, bottom) and uridine (unpublished). (e) Sugar and base protons can be identified by correlation of H1′ to H8/H6 through a combination of multibond and one-bond heteronuclear J-couplings (in preparation).

Further through-bond heteronuclear correlations in natural abundance can be considered in the base, in the sugar, between the base and sugar, and in the backbone.[20–24] The reader is referred to the quoted literature for a description of the pulse sequences used in these experiments. They make use of time delays comparable to $1/J$, during which relaxation causes signal loss. In practice, good sensitivity requires J values of a few hertz, which limits the correlations available for assignment purposes. Representative J values are shown in Fig. 5 (Refs. 20, 21, 24, and A. T. Phan, unpublished). They are of a few hertz, much larger than any homonuclear multibond proton–proton couplings, and this is of course the major value of heteronuclear correlations.

[15] A. Bax and D. G. Davis, *J. Magn. Reson.* **65**, 355 (1985).

[16] G. W. Kellogg, *J. Magn. Reson.* **98**, 176 (1992).

[17] G. Bodenhausen and D. J. Ruben, *Chem. Phys. Lett.* **69**, 185 (1980).

[18] M. J. P. van Dongen, S. S. Wijmenga, R. Eritja, F. Azorin, and C. W. Hilbers, *J. Biomol. NMR* **8**, 207 (1996).

[19] A. A. Szewczak, G. W. Kellogg, and P. B. Moore, *FEBS Lett.* **327**, 261 (1993).

[20] P. Schmieder, J. H. Ippel, H. van den Elst, G. A. van der Marel, J. H. van Boom, C. Altona, and H. Kessler, *Nucl. Acids Res.* **20**, 4747 (1992).

[21] G. Zhu, D. Live, and A. Bax, *J. Am. Chem. Soc.* **116**, 8370 (1994).

[22] V. V. Krishnamurthy, *J. Magn. Res. Ser. B* **109**, 117 (1995).

[23] A. T. Phan, Ph.D. thesis, Ecole Polytechnique, Palaiseau, France (1999).

[24] A. T. Phan, *J. Biomol. NMR* **16**, 175 (2000).

FIG. 4. Sequential assignment in the telomere-like sequence d(CCCTA$_2$5mCCCTA$_2$CCCUA$_2$ CCCT). The two substitutions (underlined) are used to break the quasi-symmetry between the TA$_2$ linkers 1 and 3 (see Fig. 9c) and the resulting internal motion. *Upper:* H1′/H3′ region of a TOCSY spectrum (mixing time 30 ms). *Middle:* ^{1}H–^{31}P hetero-TOCSY spectrum (mixing time, 50 ms) displaying the cross peaks between the (N-1, N) phosphate, H3′ of residue N-1, and H4′ and H5′/H5″ of residue N. *Lower:* HSQC connectivities between ^{13}C5′ and H5′/H5″. Note the high-field shift of the non-phosphorylated ^{13}C5′ of Cyt1 (i.e., cytidine 1). Some homonuclear splittings are resolved. Temperature, 30°; pH 5; proton frequency, 500 MHz. The dotted lines connect the TOCSY cross-peak 5mCyt7(H1′-H3′) to the hetero-TOCSY cross peaks 5mCyt7(H3′)-Cyt8(^{31}P) and Cyt8(^{31}P-H5′/H5″), then to the HSQC cross peaks (H5′/H5″-^{13}C5′). (From Ref. 14.)

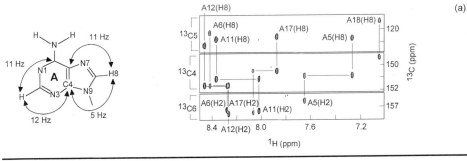

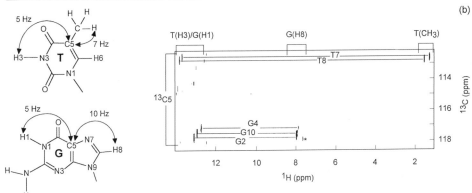

FIG. 5. Natural abundance, through-bond correlations of aromatic and imino protons in adenosine (H2-H8), guanosine (H1-H8), and thymidine (H3-CH$_3$), using the heteronuclear coherences indicated by arrows. (a) Correlation of adenosine H8 to H2 in d(CCCTAA5mCCCTAACCCUAACCCT) by an HMBC measurement. The spectrum shows the correlations between H8 and ^{13}C5 (*upper*), between H8/H2 and ^{13}C4 (*central*), and between H2 and ^{13}C6 (*lower*). Because of motional broadening, the expected correlations of A18(H2) are absent. Temperature, 30°; pH 5; strand concentration, 1.7 mM; measurement time, 52 hr. (b) Correlation between thymine imino and methyl protons, and between guanine imino and aromatic protons in [d(CGCGAATTCGCG)]$_2$ by a JRHMBC measurement (see text). Strand concentration, 1.9 mM; measurement time, 24 hr. (a) adapted from Ref. 14, supplementary material; (b) adapted from Ref. 24. Proton frequency, 500 MHz.

One might think that the low natural abundance of ^{13}C and ^{15}N (respectively 1.1% and 0.37%) makes these measurements very insensitive. However, apart from relaxation effects, the intensity of a ^{1}H–^{13}C–^{1}H correlation cross-peak in natural abundance should be 1.1% of that of a full proton peak, which is comparable to the intensity of a typical ^{1}H–^{1}H NOESY cross peak. Furthermore, a low isotopic abundance avoids the complication of dipolar and scalar interactions between vicinal magnetic heteronuclei. The spectrum of Fig. 5 (bottom), recorded in 24 hr, is representative of the sensitivity of natural abundance heteronuclear methods, in the case of ^{13}C.

3.3. Enrichment: How Much?

The commercial availability of isotopically labeled DNA phosphoramidites would make it easy to synthesize oligonucleotides in which an arbitrary choice of bases is labeled. Assignments are easily achieved with a few such syntheses. Based on the sensitivity of the natural abundance measurements described above, it appears that the required enrichment for assignment purposes is quite low: for instance, base resonances of a ^{15}N-labeled residue within an oligonucleotide of length 40 or less may be distinguished from unlabeled ones at 1% enrichment. With enrichments as low as this, DNA assignment strategies by isotope enrichment will become not only easy but affordable.

4. Structure Characterization

NMR structure determination is based primarily on short proton–proton distances detected by NOE, a second-order effect of the dipolar interaction, and on angles between covalent bonds derived from J couplings. An emerging method relies on the residual first-order dipolar coupling in anisotropic media.[25] Another is the characterization of hydrogen bonds by J-couplings across the bond,[9] as mentioned earlier. Other indicators, including kinetic ones, may be used in specific conditions.

4.1. Specific Cross Peaks

Most nucleic acid structures or motifs display short and characteristic internucleotide proton–proton distances that give rise to revealing NOESY cross peaks. Such distances have long been recognized for B-DNA, A-DNA, and Z-DNA,[26] for B'-DNA,[27] and for G-tetrads and triple helices.[28] In the i-motif, characteristic cross peaks arise from short distances between sugar protons from different strands[29] across narrow grooves (e.g., H1'–H1'), and also between H2'/H2'' and amino protons across the wide grooves[11] (Fig. 6). The H1'–H1' and amino-H2'/H2'' cross peaks occur in regions that are empty in B-DNA spectra. The intercalation topology is revealed by intersugar proton cross peaks (J. L. Leroy, in preparation), and independently by the amino–sugar proton cross peaks. The imino–amino and imino–imino proton cross peaks provide a third determination.

[25] N. Tjandra, J. G. Omichinski, A. M. Gronenborn, G. M. Clore, and A. Bax, *Nat. Struct. Biol.* **4**, 732 (1997).

[26] K. Wüthrich, "NMR of Proteins and Nucleic Acids." Wiley, New York (1986).

[27] V. P. Chuprina, A. A. Lipanov, O. Y. Fedoroff, S. G. Kim, A. Kintanar, and B. R. Reid, *Proc. Natl. Acad. Sci. U.S.A.* **88**, 9087 (1991).

[28] J. Feigon, K. M. Koshlap, and F. W. Smith, *Methods Enzymol.* **261**, 225 (1995).

[29] K. Gehring, J. L. Leroy, and M. Guéron, *Nature* **363**, 561 (1993).

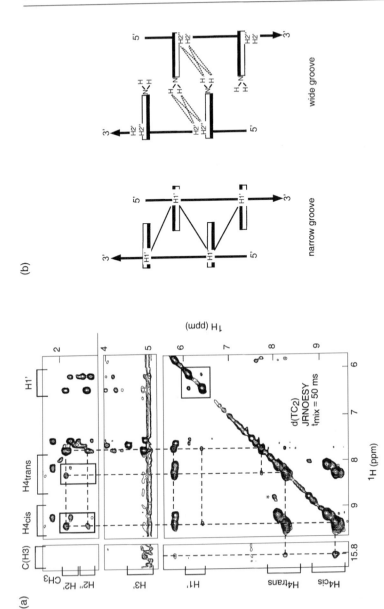

FIG. 6. Spectral characterization of the i-motif structure. (a) 360 MHz NOESY spectrum of [d(TCC)]$_4$. The specific amino-H2'/H2'' and H1'–H1' cross peaks are boxed in. Adapted from Ref. 57. (b) Some short cross-groove proton–proton distances that give rise to specific cross peaks. The amino-H2/H2'' cross peaks occur for "white–white" stacking (dashed lines) and the H1'–H1' cross-peaks for both "white–white" and "black–black" stacking (solid lines). The black and white faces of the bases have been defined by Lavery.[58] A nucleotide in *anti* conformation has the black face on its 5' side.

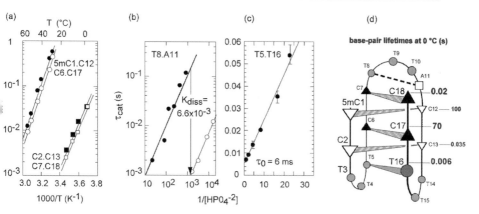

FIG. 7. Imino proton exchange and base-pair kinetics for structure exploration. The example of an intramolecular *i*-motif d(5mCCTTTCCTTTACCTTTCC): (a–c) proton exchange, (d) scheme of folding, with base-pair lifetimes. (a) The exchange times of cytidine imino protons (equal to the base-pair lifetime) vs temperature. The two internal pairs of the intercalated core have nearly equal lifetimes. Those of the external pairs, also nearly equal, are shorter. (b) The exchange time of thymidine imino protons vs the inverse of the concentration of the proton acceptor [HPO_4^{2-}] at $-5°$. The dissociation constant of the T8 · A11 pair is derived directly from the reduced efficiency of the exchange catalyst for the imino proton of T8 (d) relative to that of isolated thymidine (s). (c) Extrapolation of the imino proton exchange time to infinite phosphate concentration yields the lifetime of the T5 · T16 base pair. (d) A model of the structure, displaying the intercalation and loop topologies. The base-pair lifetimes are indicated. The difference between the lifetimes of internal and outer pairs provide strong independent evidence for the intercalation topology. (Adapted from Ref. 12).

4.2. Imino Proton Exchange and Base-Pair Kinetics

Imino proton exchange and base-pair lifetime depend strongly on nucleic acid structure.[7] In B-DNA, the base-pair lifetimes are in the range of 10 ms. Longer lifetimes are found in the A-tracts of B'-DNA duplexes, in Z-DNA, etc. Because of efficient intrinsic catalysis, the exchange time of the imino proton of $C \cdot CH^+$ pairs is equal to the base-pair lifetime.[8] In *i*-motif structures, a marked reduction of the lifetime of the outer pairs provides an independent confirmation of the intercalation topology, as displayed in Fig. 7 for an intramolecular *i*-motif. The structure computed from the NOESY spectra indicates the possibility of a T8 · A11 Hoogsteen pair in the second loop (on top) and of a T5 · T16 pair connecting loops 1 and 3 (bottom). This is well supported by protection of the imino protons of thymines 5, 8, and 16 against exchange with water, in contrast to those of the six other thymine residues, and by the imino proton chemical shifts. At high concentrations of exchange catalyst (phosphate), the (identical) exchange times of the imino protons of T5 and T16 extrapolate to 6 ms, a further demonstration of the formation of the T · T pair. Protection from exchange as compared to free thymidine corresponds to a dissociation factor of 5.6×10^{-4}. In the case of T8, the dissociation constant is 6.6×10^{-3}, but the base-pair lifetime is too short for measurement.

Another example is the T_3A loop formed in the i-motif structure of the dimer [d(5mCCTTTACC)]$_2$. In the computed structure, T3 is too far from A6 for base-pair formation, but the relative orientation of the bases is compatible with indirect hydrogen bonding through a water molecule. The chemical shift of the imino proton supports this model, and further support is provided by the exchange time, which is shorter than in free thymidine, as expected if exchange is enhanced by A6(N1) acting as an exchange catalyst.[30]

These are but two examples of the extensive and detailed contributions of exchange measurements to structural analysis.

5. Symmetry and Internal Motions

Many biological objects involve a multimeric assembly of chemically identical subunits, e.g., the two α or the two β subunits of hemoglobin, the duplex of the self-complementary DNA dodecamer [d(CGCGAATTCGCG)]$_2$, or the protein coat of tobacco mosaic virus. These objects usually display a large degree of symmetry, with close structural similarity of the chemically identical subunits and of the interfaces between them.

However, there is no fundamental requirement for structural identity, and this is amply illustrated by crystallographic structures. For instance, symmetry is relaxed in the shell of the tomato bushy stunt virus, which consists of 180 identical subunits.[31] In the insulin dimer, subunits display widely distributed differences.[32] The two strands of self-complementary DNA duplexes often show slight differences.[33] This is also the case for the four strands of i-motif tetramers, e.g., d(C$_3$T)[34] and others. One reason these differences are visible is that structural change within one multimeric assembly may be restricted by crystallographic constraints, while concerted change in many multimeric assemblies would be extremely slow.

In solution, structural identity of subunits is not expected any more than in the crystal. However, the time constant for structural averaging in solution may be much shorter. Measurements lacking sufficient time resolution will usually reflect only the same time-averaged structure for all subunits. For instance, the NMR

[30] S. Nonin, A. T. Phan, and J. L. Leroy, *Structure* **5**, 1231 (1997).

[31] L. Stryer, "Biochemistry," p. 731. Freeman, New York, 1981.

[32] T. L. Blundell and L. N. Johnson, "Protein Crystallography," p. 51. Academic Press, New York, 1976.

[33] H. R. Drew, R. M. Wing, T. Takano, C. Broka, S. Tanaka, K. Itakura, and R. E. Dickerson, *Proc. Natl. Acad. Sci. U.S.A.* **78**, 2179 (1981).

[34] C. H. Kang, I. Berger, C. Lockshin, R. Ratliff, R. Moyzis, and A. Rich, *Proc. Natl. Acad. Sci. U.S.A.* **91**, 11636 (1994).

spectrum of the *i*-motif tetramer [d(C$_3$T)]$_4$ corresponds to identical strands (J. L. Leroy, unpublished), in contrast to the crystal structure. The situation is the same with dimeric *i*-motifs,[30] and with tetrameric or dimeric G-quartets as well.[1-3] The case of multimeric proteins can be the same.[35]

Although it may seem paradoxical, one should consider that the apparent identity of the subunits of a multimeric object in solution is strong evidence for internal motions, masked by instrumental time averaging. Unseen motions may be of small amplitude and importance, but it is not always so.

In some cases, motion (symmetry-related or not) may be demonstrated even if it is too fast for direct observation. Thus, the set of *J*-couplings between protons of the sugar cycle of B-DNA[36,37] is found to be incompatible with a rigid conformation: there is "pseudo-rotation." NOEs also may be incompatible with a rigid structure. One early example is provided by the NOEs between H8 and numerous sugar protons of 5'-GMP,[38] which imply a modulation of the (*syn/anti*) glycosidic angle. In another example, described below, a proton is strongly NOE-coupled to both H8 and H2 of a single adenine, a feat impossible in a single structure.

A commonly observed motion is that between two different conformations, giving rise to spectral features designated as chemical or conformational exchange. When exchange is slow (compared to the difference in chemical shift), one sees two separate resonances; in fast exchange, there is only one, at the average chemical shift; in both cases, the motion gives rises to spectral broadening. In the intermediate case, the spectrum is strongly altered or even vanishes. One may change conditions (field, temperature) in order to switch between these different cases.

In the particular case when the motion is related to symmetry, one may attempt to alter or eliminate it by destroying the symmetry, for instance by an asymmetrical chemical modification.

We shall now present three examples of symmetry-related motions. They are taken from studies of *i*-motif structures, but similar situations are bound to occur in other multimeric systems.

5.1. $C \cdot CH^+$ Base Pair

In this hemiprotonated base pair, the covalent N–H linkage of the imino proton shifts quickly from one base to the other.[8] A proton position midway between the two nitrogens would probably be signaled by a marked change in chemical shift, as compared for instance to the imino proton in $G \cdot CH^+$ pairs of a triple helix,[39] and this is not observed.

[35] E. T. Baldwin, I. T. Weber, R. St. Charles, J. C. Xuan, E. Appella, M. Yamada, K. Matsushima, B. F. Edwards, G. M. Clore, and A. M. Gronenborn, *Proc. Natl. Acad. Sci. U.S.A.* **88**, 502 (1991).
[36] L. J. Rinkel and C. Altona, *J. Biomol. Struct. Dynom.* **4**, 621 (1987).
[37] F. J. M. Van de Ven and C. W. Hilbers, *Eur. J. Biochem.* **178**, 1 (1988).
[38] S. Tran-Dinh, W. Guschlbauer, and M. Guéron, *J. Am. Chem. Soc.* **94**, 7903 (1972).
[39] R. Macaya, E. Wang, P. Schultze, V. Sklenar, and J. Feigon, *J. Mol. Biol.* **225**, 755 (1992).

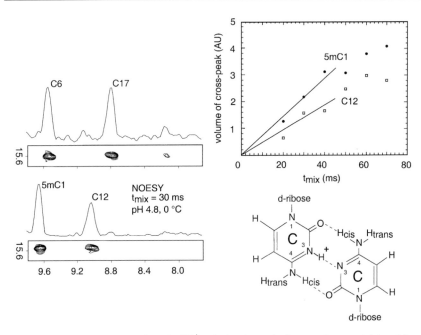

FIG. 8. A study of symmetry in a $C \cdot CH^+$ pair (a scheme is shown at bottom right) and in a $5mC \cdot CH^+$ pair of the intramolecular i-motif structure of $d(5mCCT_3C_2T_3AC_2T_3C_2)$. *Left:* 600 MHz NOESY cross peaks between imino and amino protons at 30 ms mixing time. The intensities of cross peaks of the imino proton with the *cis*-amino protons of C6 and of C17 are equal, as expected for equal distribution of the imino proton between the two cytidines of the $C6 \cdot C17H^+$ pair. The cross peaks with the *trans*-amino protons are within the noise (not shown). The weak cross peak at 8.2 ppm is a direct NOE to the *trans*-amino proton of C12. In the $5mC1 \cdot C12H^+$ pair, the ratio of cross peak intensities (areas) is 1.2 to 1.3. *Top (right):* Buildup of the cross peaks of the imino proton with the *cis*-amino protons of 5mC1(d) and of C12(h). It is linear up to 40 ms and more, showing that diffusion toward the *trans*-amino proton may be ignored at 30 ms. (Adapted from ref. 12, supplementary materials).

The investigation of the base pair requires a setting that breaks the (motion-averaged) spectral symmetry between its two bases. Consider for instance (Fig. 8) the $C6 \cdot C17H^+$ pair formed by the intramolecular i-motif structure of $d(5mCCT_3C_2T_3AC_2T_3C_2)$, schematized in Fig. 7. The amino protons of the two bases are inequivalent, so that cross peaks between the shared imino proton and the two *cis*-amino protons can be resolved, contrary to pairs of equivalent bases in a symmetrical tetramer such as $d(TCC)_4$. But the inequivalence is essentially magnetic (aromatic currents of the surroundings, etc.), and the effect on the structure should be negligible. Indeed, the equal cross-peak intensities indicate that the imino proton is equally shared between two structurally and chemically identical bases.

In the case of the $5mC1 \cdot C12H^+$ pair, the bases are chemically inequivalent, so it is not surprising that the cross-peak intensities are unequal (Fig. 8). One interpretation is that the distribution of the imino proton between the two bases is unequal, because of a pK difference. In fact, the pK of the 5mC nucleoside is higher than that of cytidine ($\approx$4.2) by only +0.05.[12] This is less than the log of the ratio of cross-peak intensities, +0.12, suggesting another interpretation: that the distance of the imino proton to the *cis*-amino proton is different in the methylated and in the nonmethylated base. A log(distance ratio) of −0.12/6, i.e., a ratio of 0.955, would explain the effect.

5.2. d(5mCCTCC) Tetramer: Spontaneously Broken Symmetry

This sequence forms an *i*-motif tetramer in which the strands are not identical.[40] Instead of the "one-strand" spectrum expected for a symmetrical tetramer, the spectrum consists of two such subspectra. For example, the 1D spectrum displays two distinct thymidine imino protons. In the NOESY spectrum, each cytidine imino proton resonance is connected to a single amino proton cross peak, showing that the C · CH$^+$ pairs are symmetrical. Therefore the *i*-motif is formed of symmetrical duplexes, and each type of duplex has its own subspectrum.

Further analysis requires that one distinguish between dipolar and conformational exchange cross-peaks. This is done with combined ROESY–NOESY excitation, adjusted so that the magnetization spends twice as much time along the main field as in the perpendicular direction (ROESY: rotating frame Overhauser enhancement spectroscopy; NOESY: nuclear Overhauser enhancement spectroscopy). The intensity of dipolar cross peaks is then zero, as may be checked by the cancellation of the strong intraresidue cytidine H5–H6 cross peaks, and one obtains a pure conformational exchange spectrum.[41,40]

The NOESY and ROESY-NOESY data are consistent with two "one-strand" subspectra of equal intensities: protons in the slow-exchange regime give rise to two resonances, connected by an exchange cross peak. The interconversion time is measured by magnetization transfer. It is 1.4 s at 0° and the temperature dependence corresponds to an activation energy of 94 kJ/mol.

Such data could originate from an equimolar mixture of two fully symmetrical tetrameric forms, or from a single tetramer composed of two different duplexes. The second interpretation is suggested by the equal intensities of the subspectra. It is proven by the observation of dipolar cross peaks between the two duplex forms: for instance, the C2 amino proton peak from one form has a cross peak with each of the two T3 imino proton resonances, that from the same duplex form, and that from the other form.

These observations show that the proton spectrum corresponds to a tetramer consisting of two different, interconverting duplexes as the sole structure. Interconversion leaves the global structure unchanged. In other words, the tetramer switches between two identical conformations.

[40] S. Nonin and J. L. Leroy, *J. Mol. Biol.* **261**, 399 (1996).
[41] J. Fejzo, W. M. Westler, S. Macura, and J. L. Markley, *J. Magn. Reson.* **92**, 20 (1991).

One may then proceed to the study of the subspectra by the usual spectral and computational methods. The main difference between the two duplexes comes from the inability of the intercalated structure to accommodate the two T · T pairs. The tetramer may be symbolized as:

$$C5^*\text{-}5mC1\text{-}C4^*\text{-}C2\text{-}(T3^*)\text{-}T3\text{-}C2^*\text{-}C4\text{-}5mC1^*\text{-}C5$$

where each pair is represented by a single base, and pairs of one duplex are labeled with an asterisk. Note, however, that the thymidines (T3*) of one duplex are unpaired and looped out in the wide grooves, so that the T3 · T3 pair is stacked between the sequentially adjacent C2 · C2H$^+$ pair and the C2* · C2*H$^+$ pair of the other duplex.

Exchange of the imino protons with water provides a detailed picture of duplex interconversion and suggests that it is driven by the concerted opening and closing of the thymidine pairs. The reader is referred to the original article for further analysis.[40]

The formation of an asymmetrical tetramer from four identical strands is an example of *spontaneously broken symmetry,* a phenomenon encountered in many branches of physical science. In the case of the d(5mCCTCC) tetramer, NMR provides us with a detailed description of the structure and motion that express the broken symmetry.

5.3. *Quasi-Symmetrical Structure of Fragment of C-Rich Strand of Human Telomeres*

The question of the biological relevance of the *i*-motif led us to investigate its formation by intramolecular folding of fragments of the C-rich strand of the human telomere containing four d(C$_3$) stretches. A scheme for the NMR structure of the natural sequence d(C$_3$TAAC$_3$TAAC$_3$TAAC$_3$T) is shown in Fig. 9. If one ignores the three loops and the terminal T, one has a tetramer of chemically identical d(C$_3$) strands. Adding the two bottom loops (loop 1 and loop 3) turns this into a dimer of chemically identical d(C$_3$TAAC$_3$) strands. Lastly, with incorporation of the top loop (loop 2) and of the terminal T, all symmetry is lost. However, there remains structural and spectroscopic traces of the lost symmetry, and this quasi-symmetry creates essential opportunities and pitfalls in the structural investigation.[13,14]

Quasi-symmetry is particularly marked in the lower part of the *i*-motif core and in the bottom loops. Thus, in Fig. 9a, nearly equal chemical shifts in both sugar and base give rise to couples of barely resolved intraresidue cross-peaks for couples of residues such as 4/16, 5/17, 6/18, and 7/19 in the 30° spectrum. Furthermore, many cross-peak intensities are rather close, indicating similar geometries of the bottom loops. Comparable features had been found previously in the spectrum of d(5mCCTTTCCTTTACCTTTCC), whose structure is schematized in Fig. 7.

The spectrum in Fig. 9a gives no indication of possible motions. However, drastic line broadening of all bottom loop resonances occurs at lower temperatures,

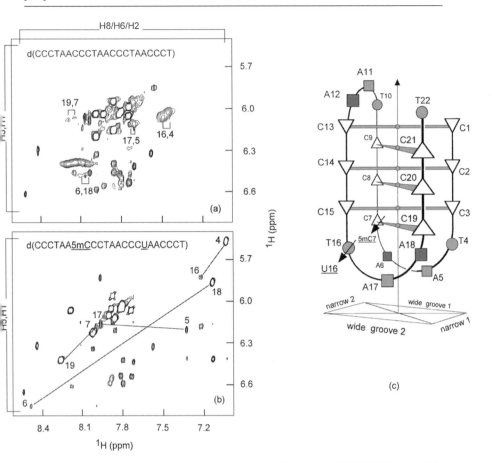

FIG. 9. Methylation and demethylation for symmetry breaking. (a) The H1′–H6/H8 region of the 500 MHz NOESY spectrum of the natural telomeric, sequence d(C_3TAAC$_3$TAAC$_3$TAAC$_3$T). Because of the quasi-symmetry which is apparent in the structural model (c), the chemical shifts of nucleotides 4–7 are nearly degenerate with those of nucleotides 16–19, as shown by the near-overlap of the NOESY intraresidue cross peaks (e.g., residues 4 and 16, 5 and 17, . . .). This is a fast-exchange spectrum resulting from mutual exchange of the configurations of the two bottom loops. Temperature, 30°. (b) Same as in (a), except that the sequence has been modified by methylation of C7 and demethylation of T16. As a result, the switch is frozen, with each loop stabilized in its own configuration. The chemical shifts of the two configurations differ sharply. The midpoint between cross-peaks of corresponding residues (e.g., 6, 18) is not too far from the quasi-degenerate position in the fast-exchange spectrum of the natural sequence. (c) Structural scheme displaying the intercalation and loop topologies of the natural and modified sequences. The arrows indicate the two base substitutions. *Note:* Further motion within each loop configuration, in the fast exchange regime, is ignored here; see the legend of Fig. 10.

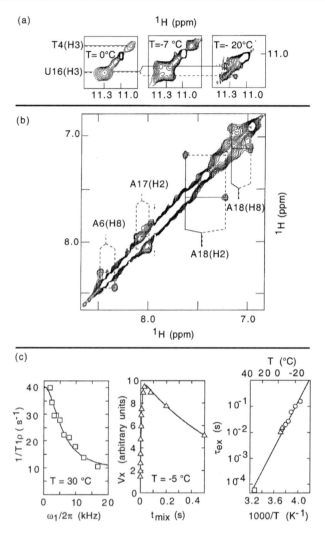

FIG. 10. Spectral characterization of internal motions in the intramolecular folded structure of the modified human telomere sequence d(CCCTAA5mCCCTAACCCUAACCCT). (a) From fast to slow conformational exchange. The diagonal U/T imino proton region of the 500 MHz NOESY (H$_2$O) spectra at different temperatures. At 0°, the conformational exchange rate of U16(H3) is fast, resulting in a single diagonal peak. At −7°, the diagonal peaks are distinct and the off-diagonal cross peaks between them correspond to intermediate exchange. At −20°, exchange is even slower and the cross peaks are weaker. The diagonal peak of T4(H3) does not split in this temperature range. Mixing time, 50 ms. (b) The diagonal aromatic region. Splitting of the adenine resonances indicates two conformations, in intermediate or slow exchange. The exchange rate is obtained from the NOESY (D$_2$O) cross-peak intensities. The complete investigation[14] shows that the splittings are all due to a *syn/anti*

a sign of intermediate conformational exchange in the intermediate regime. This implies that the 30° spectrum is in the fast exchange regime and should be used with caution for structural determination. Unfortunately, in most cases the slow exchange regime was not reached even at the lowest temperatures.

In order to go beyond the intercalation and looping topologies, which are easily derived from the high-temperature spectrum,[14] we proceeded to break the quasi-symmetry of the bottom part of the structure, so as to freeze the presumed motion in which the quasi-symmetrical bottom loops exchange their structures. The sequence was modified by cautious methylation and/or demethylation, using as first criterion the conservation of intercalation and looping topologies. Various sequences were tried, and a single methylation at position C7 was enough for structural studies.[13,14] The most detailed study involved the doubly modified sequence d(C₃TAA5mCC₂TAAC₃UAAC₃T). When compared to the natural sequence, the modified sequence reveals striking chemical shift differences between the bottom loops, e.g., between the cross peaks of adenines 6 and 18 (Fig. 9b): symmetry has been destroyed.

Nevertheless, some motion remains. For instance, at 40°, C19(H5) has strong dipolar cross peaks with both H2 and H8 of A18, an impossibility in a rigid structure. In this case, slow exchange conditions were reached at low temperatures. It was then possible to determine the kinetics and the structure of the mobile regions. The motion is less extensive than the symmetry-related one, being restricted to a single adenine. Figure 10 shows various spectral manifestations of motion, and the conditions required for elucidating them: in particular, a wide range of temperatures, short mixing times, $T1\rho$ measurements.

In the structure of the modified sequence, the bottom loops are well defined, by numerous NOE cross peaks, and they are indeed different. Loop 1 is rigid, with A6 *anti* and T4 *syn*, whereas loop 3 is flexible with A18 flipping between *syn* and *anti* (this is the only residual motion), and with U16 *anti*.

As a result of these measurements, we have a detailed description of the structure of the modified sequence, including that of the three loops, the interaction between the two bottom loops and the residual motion in loop 3. The application of this information to the structural analysis of the natural sequence is based on the notion that the chemical modifications have been chosen for minimal effect.

---

flip of A18. Mixing time, 5 ms; temperature, −5°. (c) Conformational kinetics, as reflected by NMR spectroscopy of A18(H2) and U16(H3). *Left:* The relaxation rate of A18(H2) in the rotating frame vs the spin-lock field (30°, fast exchange conditions). The fit is to a conformational exchange time of 50 μs. *Middle:* The volume of the exchange cross-peak between the conformers vs mixing time, fitted to an exchange time of 15 ms (−5°, slow exchange conditions). *Right:* The time for interconversion between the two conformations vs temperature. The high temperature point (h) is from the measurement in the fast exchange regime (left panel). The points at lower temperatures are derived from the volume of the slow-exchange cross-peaks of A18(H2) (△) and U16(H3) (s). (Adapted from Ref. 14).

We therefore made the tentative assumption that the structural elements of the natural sequence are the same as in the modified sequence, and that the main difference lies in the quasi-symmetry of the bottom loops and in the associated motion.

We started with the top region. Here, the spectral characteristics are similar for the two sequences, in terms of both structure and lack of motions. It was assumed that the structures in this region are identical, and also, therefore, the global topologies. Turning to the bottom loops of the natural sequence, we searched for similarity with the modified sequence. Assignments were obtained with the help of exchange cross peaks between the structured and melted forms at temperatures close to the melting temperature.

Quasi-symmetry implies that each loop of the natural sequence should assume successively the structures of loop 1 and loop 3 of the modified sequence. A first consequence is that the chemical shifts of the natural sequence in the fast exchange regime should be rather close to the average of those of the two loops of the modified sequence, and this is indeed observed. For instance, the quasi-superposed peaks of residues 6 and 18, or 17 and 5, in Fig. 9a are rather close (both vertically and horizontally) to the average of the well-spread-out ones in Fig. 9b. (The case of residue 16 is affected by the U/T substitution.) A second prediction is that the thymine of *each* loop should be part-time *syn* (like T4 of the modified sequence), part-time *anti* (like U16), and this is indeed supported by H1'-H6 cross peaks that act as indicators of the *syn* conformation, with the expected intensity. The reader is referred to the original article[14] for more details, including many indicators of the *syn/anti* motion of both A6 and A18 in the natural sequence, as expected on the basis of the motion of A18 (only) in the modified sequence.

In summary, this study showed how the quasi-symmetric structure and motions of the natural telomeric sequence could be solved. This is done with the help of an ancillary sequence, chemically modified in order to eliminate quasi-symmetry, and whose structure (and remaining internal motion: the flip of a single adenine) is solved in detail. The natural sequence exhibits quasi-symmetry and motion related to it. It switches between two structures, of which one is the same as that of the modified sequence, the other being derived from it by exchange of the structures of the two bottom loops.

6. Association of Water to Nucleic Acids

The interaction of macromolecules with water has generated much discussion. It has been stated that water is an integral part of nucleic acid structure.[42] A "spine of hydration" in the minor groove of B'-DNA has been observed in crystals, and

---

[42] E. Westhof, *Annu. Rev. Biophys. Chem.* **17**, 125 (1988).

its structural and energetic significance has been considered.[43,44] X-ray crystallography can locate water molecules and evaluate site occupancy, but not residence time. In contrast, NMR can provide both location and residence time of water molecules associated to macromolecules in solution.[45,46]

6.1. Determination of Water Kinetics by Zero-ROE and Zero-NOE

An Overhauser effect between a DNA proton and water reveals the presence of water molecules nearby. The sign of the NOE depends on the correlation time of the dipolar interaction (Fig. 11a), which is modulated by the rotation of the macromolecule, possibly by that of the water molecule, and by the association/dissociation of water. When the latter is the fastest motion, as is often the case, the correlation time is equal to the residence time. For a certain, field-dependent value of the correlation time (e.g., 0.297 ns at 600 MHz), the NOE is zero. This is the zero-NOE condition, which may in many cases be achieved by varying the proton NMR frequency and/or the temperature. For this measurement, spurious contributions to the NOE must be avoided, in particular those from neighboring exchangeable protons, and those from H3′ protons whose chemical shift is close to that of water.[47,48]

The zero-(off-resonance ROE) and zero-(mixed ROE-NOE) methods are variants that eliminate the need for a change in NMR frequency or temperature (Fig. 11c). They are implemented by continuous off-resonance irradiation[49] and by pulsed on-resonance irradiation,[41] respectively, during the mixing period.

The off-resonance measurement is characterized by θ, the angle between the static and effective fields. The angle θ_0 which corresponds to zero-enhancement is given by:

$$\tan^2 \theta_0 = [-J(0) + 6J(2\omega)]/[2J(0) + 3J(\omega)] \qquad (3)$$

where $J(\omega)$ is the spectral density.

The correspondence between (off-resonance ROE) and (mixed ROE-NOE) is given by:

$$\tan^2 \theta = t_{\text{ROE}}/t_{\text{NOE}} \qquad (4)$$

where t_{ROE} and t_{NOE} are the times during which the ROE and NOE conditions are satisfied in the ROE–NOE experiment (A. T. Phan, in preparation).

[43] H. R. Drew and R. E. Dickerson, *J. Mol. Biol.* **151,** 535 (1981).
[44] L. McFail-Isom, C. C. Sines, and L. D. Williams, *Curr. Opin. Struct. Biol.* **9,** 298 (1999).
[45] G. Otting and K. Wüthrich, *J. Am. Chem. Soc.* **111,** 1871 (1989).
[46] V. P. Denisov, G. Carlstrom, K. Venu, and B. Halle, *J. Mol. Biol.* **268,** 118 (1997).
[47] A. T. Phan, J. L. Leroy, and M. Guéron, *J. Mol. Biol.* **286,** 505 (1999).
[48] F. J. M. Van de Ven, H. G. J. M. Jansen, A. Graslund, and C. W. Hilbers, *J. Magn. Reson.* **79,** 221 (1988).
[49] H. Desvaux, P. Berthault, N. Birlirakis, and M. Goldman, *J. Magn. Reson. Ser. A* **108,** 219 (1994).

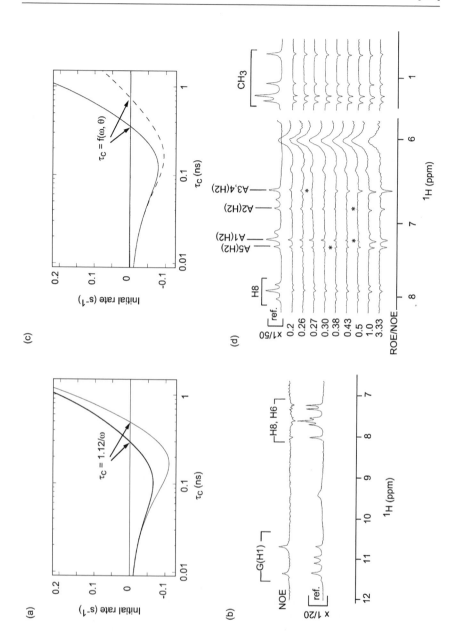

6.2. Examples

These methods have been applied to the comparison of residence times of water next to B-DNA and to B'-DNA. The latter is a duplex structure that generates DNA curvature.[50] It is formed by stretches of A · T base pairs and involves well-stacked, propeller-twisted pairs together with a narrowed minor groove. In the minor groove of the B'-DNA duplex [d(CGCGA<u>A</u>TTCGCG)]$_2$, the residence time of the water molecule next to H2 of A6 (underlined), is 0.6 ns at 10°. For water next to A5, the residence time is slightly but distinctly shorter, suggesting noncooperative departure of these two molecules, which are presumed to be part of the hydration spine. In [d(AAAAATTTTT)]$_2$, another B'-DNA duplex, residence times are up to twice as long. The activation enthalpies are about the same, ca 38 kJ/mol.

---

[50] H.-M. Wu and D. M. Crothers, *Nature* **308**, 509 (1984).

FIG. 11. Measurement of the residence time of water molecules at well-defined locations on nucleic acids, using the NOE from water to nucleic acid protons. (a) The "zero-NOE" method.[47] Plot of the initial NOE cross-relaxation rate as a function of the correlation time τ_c of the dipole–dipole interaction, for an interproton distance of 0.25 nm. Bold and thin lines are for proton frequencies of 600 and 360 MHz, respectively. In conditions where the initial rate is zero, the correlation time is determined without any calibration. (b) 500 MHz proton NMR spectrum of the G-tetrad [d(TG$_5$T)]$_4$, measured 100 ms after irradiation of water. The negative peaks of the H6/H8 protons are due to NOE from water molecules in the grooves, with residence times in the range of 0.1–0.35 ns. By decreasing the temperature and/or increasing the magnetic field, one could increase the NOE of a given resonance up to zero, thus providing the value of the correlation time at the corresponding temperature. Water molecules have been observed in G-tetrad grooves by X-ray crystallography.[51] Temperature, 10°; pH 7; strand concentration, 4 mM. (c) The zero ROE-NOE or off-resonance method (A. T. Phan, in preparation). Plot of the initial cross-relaxation rate in an ROE-NOE (or, equivalently in an off-resonance ROE) experiment vs the correlation time τ_c of the dipole–dipole interaction, for an interproton distance of 0.25 nm and a proton frequency of 500 MHz. Continuous and broken lines are for $\tan^2\theta = 0$ (NOE) and $\tan^2\theta = 0.3$. In conditions where the initial rate is zero, the correlation time is determined without any calibration. See text for the value, $\tan\theta_0$, which corresponds to zero-enhancement [Eq. (3)], and for the relations between the ratio of ROE and NOE periods and $\tan\theta$ [Eq. (4)]. (d) 500 MHz water ROE-NOE spectra of the d(A$_5$T$_5$) duplex. The excitation sequence is identical to that for NOE in high fields [Fig. 1A(a) in Ref. 47] except that the mixing period (70 ms) contains a series of phase-alternating 90° pulses along the x axis, i.e., orthogonal to the water proton magnetization, separated by rf-free intervals.[41] Water molecules with a long residence time in the minor groove are responsible for cross peaks to adenosine H2 that are negative for large values of (ROE period)/(NOE period) and positive for small ones. Zero enhancement is indicated by a star, and the correlation time is derived from the corresponding ratio of ROE and NOE periods according to Eqs. (3) and (4). The correlation times are 0.8, 0.8, and 1.2 ns for H2 of adenine residues 3, 4, and 5, respectively. Water molecules in the major groove give negative ROE-NOE cross peaks for H8 and CH$_3$ protons at all ratios, indicating shorter residence times. In the case of the terminal bases A1 and A2, the ROE-NOE cross peak to H2 is due to dipolar interaction with the thymine imino protons, whose magnetization is affected by fast exchange with water. Temperature, 10°; pH 7; strand concentration, 4 mM.

For comparison, the residence time in the minor groove of the regular B-DNA sequence d(CGCGATCGCG) is 0.3 ns at $10°$, shorter than in the case of the B'-DNA sequences, but not by a large factor. As for the major groove residence times, they are comparable for the three sequences, and a few times shorter than those of minor groove water. At $-8°$ the value is 0.36 ns, or even more if there is rotation of water, a phenomenon that may be considered more likely in the major than in the minor groove.

Water ROE-NOE spectra of the B'-DNA duplex $[d(A_5T_5)]_2$ are shown in Fig. 11d. Long-lived water molecules in the minor groove give a negative cross peak to H2 protons at large ROE/NOE ratios and a positive one at small ratios. From the zero crossing one can derive the correlation time according to Eqs. (3) and (4). Water molecules in the major groove give negative cross peaks to H8 and methyl groups at all ROE/NOE ratios, indicating shorter residence times.

We have also studied the water residence times in i-motif and in G-tetrad structures. Residence times around 0.5 ns were found in the TTTA loop of i-motif dimeric and monomeric structures.[30,12] In the G-tetrad tetramer $[d(TG_5T)]_4$, negative NOE cross peaks from water (Fig. 11b) to H8 and H6 protons (at $10°$, 500 MHz) point to water molecules with residence times of 0.1 to 0.35 ns, in groove locations where water has been detected in the crystal.[51]

7. NMR of Exchangeable Protons in High Fields

7.1. Radiation Damping

When working in H_2O at high fields, one is faced with so-called "radiation damping," the reorientation of the magnetic moment of water and of species whose chemical shift is close to that of water. The cause is the magnetic field of the current resulting from the electromotive force induced in the coil by the large precessing magnetic moment of water protons.[52] The transverse water proton magnetization (WPM) may be destroyed by a field gradient (thus avoiding radiation damping) and restored as needed by a gradient of the opposite sign ("gradient echo"), or by a $180°$ radio frequency (rf) pulse followed by a gradient of unchanged sign ("spin echo"). In combination with selective rf pulses, this type of procedure makes possible the manipulation of the water proton spins without radiation damping. For instance, the water proton spins may be selectively and efficiently inverted at the beginning of an excitation pulse sequence for NOE from water, or for magnetization transfer from water as in Fig. 12f. In this case, the first hard $90°$ pulse brings all magnetizations in the transverse plane where they are spread out by the first G_0 gradient pulses. For all spins except those isochronous with water proton spins, the second G_0 gradient

[51] G. Laughlan, A. I. H. Murchie, D. G. Norman, M. H. Moore, P. C. E. Moody, D. M. J. Lilley, and B. Luisi, Science 265, 520 (1994).

[52] M. Guéron, P. Plateau, and M. Decorps, in "Progress in NMR Spectroscopy" (J. W. Emsley, J. Feeney, and J. H. Sutcliffe, eds.), Vol. 23, p. 135. Pergamon Press, Oxford, 1991.

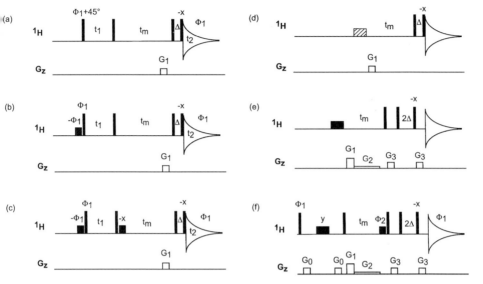

FIG. 12. Pulse sequences for NOESY spectra of exchangeable protons at short mixing time (a–c) and for proton exchange measurements (d–f) in high field. The minimum phase cycle is described. The pulse phase is $0°$ (x) unless indicated otherwise. Phase Φ_1 is cycled as ($x, -x$). In sequence (f), phase Φ_2 is such that the WPM is along $+Oz$ at the end of the mixing time, i.e., it is equal to Φ_1 in the common case where t_m is much shorter than the longitudinal relaxation time of the water protons.

pulse acts similarly. In contrast, the latter spins, including those of water protons, are affected by the long and weak $180°$ pulse, whose selectivity is good because it is applied in the absence of gradients, and which operates without radiation damping since the transverse WPM averaged over the coil region is zero. Because of the inversion, the second gradient pulse generates an echo for the water protons. For the isochronous macromolecular protons, one takes advantage of the shorter T_2 so as to weaken the echo, using a sufficiently long interval between the pulses. The result of these manipulations is that after the second G_0 gradient pulse, the WPM is refocused in the transverse plane, whereas the other protons have no transverse magnetization. The second hard pulse then brings the magnetization of water protons along $(+/-)Oz$ and that of the other protons in the transverse plane where they are spread out by G_1, resulting in the appropriate configuration for the "mixing" period.[47]

7.2. Jump-and-Return

For sensitive detection of quickly exchanging protons one needs a solvent suppression method that operates rapidly so as to avoid radiation damping, which avoids saturation of the WPM, and which is compatible with a short mixing time. The Jump-and-return (JR) method is a good candidate for such requirements. Its best features are simplicity (a common implementation is by two 90° hard pulses of opposite phases, at the water frequency), the short duration of the sequence, and the absence of phase shift (and therefore baseline roll).

On our home-built spectrometer, the JR sequence is tuned by adjusting the length and phase of a weak pulse placed immediately before the first strong pulse, whose flip angle is slightly less than 90°. On the Varian spectrometer, we use two weak orthogonal pulses to the same effect. On Bruker spectrometers, we have used length and phase adjustment of the first strong pulse. Each of the two adjustments affects exclusively either the in-phase or the out-of-phase component of the free precession. The JR sequence is easily tuned if one takes care to display on the screen these in-phase and out-of-phase components of the free precession of WPM. They can be zeroed independently, almost without cross-talk, as described in Fig. 13.

7.3. Applications

We distinguish two types of measurements, one in which the WPM should be perturbed as little as possible, and the other where its variation is required. A NOESY experiment for the detection of dipolar cross peaks between macro-molecular protons, or of cross peaks reflecting conformational exchange[14] belongs to the first type. An NOE experiment for measuring dipolar cross-peaks between water and macromolecular protons,[47] or a magnetization transfer experiment for measuring proton exchange with water[7] belong to the second type.

Figure 12(a–c) shows different variants of NOESY sequences with JR detection (JRNOESY). These are 2D measurements of the first type. WPM may be modified during the "preparation" subsequence, and the goal is to ensure that WPM is back in the $+Oz$ direction at the time of the JR detection subsequence, whatever the phase of the preparation pulses, whatever the labeling time t_1, and however short the mixing time t_m. The problem is specially acute when the phases of the two preparation pulses are identical so that the WPM after the second pulse is along $-Oz$. The recovery of the longitudinal WPM may then be slow, with an erratic effect of radiation damping.

In sequence (a), this situation is avoided by a 45° shift of the phase of the first pulse,[53] which ensures that the two-pulse subsequence brings the WPM to an orientation at least 45° away from the $-Oz$ direction at the beginning of the mixing period, so that reorientation of the WPM toward $+Oz$ by radiation damping proceeds in a stable and strong manner. Just before JR detection, the G_1 gradient pulse destroys the residual transverse WPM. Here and below, excitation is at the water proton frequency, unless otherwise noted.

In sequence (b), an initial selective pulse cancels the effect of the first hard pulse on the WPM, so that the WPM is transverse at the beginning of the mixing time.

---

[53] S. H. Smallcombe, *J. Am. Chem. Soc.* **115**, 4776 (1993).

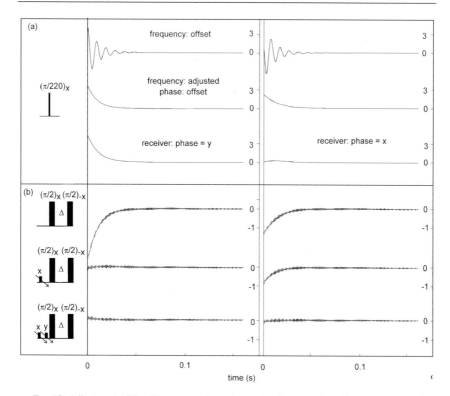

FIG. 13. Adjustment of the JR sequence for solvent signal suppression. The sequence consists principally of two pulses of the same amplitude and of opposite phase $(x, -x)$ at the solvent frequency. Since the solvent magnetization does not precess during the interval Δ, it is brought back along the $+z$ axis by the second pulse. If the pulse amplitude is $\pi/2$, the magnetizations of all spins are in the xOz plane immediately after the second pulse, and there is no first-order phase shift (whose correction would cause baseline distortion). Two weak pulses provide for orthogonal fine tuning of the excitation sequence, in order to correct for radiation damping and instrumental artifacts. For the tuning procedure, the spectrometer is set to display the free precession, as seen in the two orthogonal receiver channels, if possible in real time. **(a) Step 0** ("frequency: offset"): In a first stage, excitation is by a single weak pulse $(\pi/220)$. The receiver gain is set to its normal value, chosen so that the noise at the analog-to-digital converter is at least three times that which corresponds to the lowest significant bit. *The receiver gain remains unchanged during the entire procedure.* The oscillation of the water proton FID shows that the excitation frequency is misadjusted (offset). **Step 1** ("frequency: adjusted, phase: offset"): The excitation frequency is tuned so that the FID is monotonous. **Step 2** ("frequency: adjusted, phase $= x$"): The phase of the receiver is adjusted for zero initial intensity in the quadrature channel. **(b) Step 3:** In the second stage, excitation is changed to the JR sequence. Because of the strong excitation, the solute signals (thymidine, 10 mM) are visible as small oscillations of the FID trace. Without fine tuning, solvent suppression is imperfect. **Step 4:** The fine tuning x-phased pulse is used to cancel the residual initial in-phase solvent signal. **Step 5:** The fine tuning y-phased pulse is used to cancel the residual initial quadrature solvent signal. The in-phase adjustment is not, or nearly not, affected. This procedure applies also to the JR variations: second-order JR or SJR[59] and power-adapted JR or PJR.[60] Sample, thymidine, 10 mM; pH 5; temperature, 20°; proton frequency, 500 MHz.

In sequence (c), two selective pulses keep the WPM close to $+Oz$ from beginning to end of the preparation subsequence and during the subsequent mixing period.[54][56] Note that the limited selectivity of the selective pulses may affect the spectrum in the vicinity of the water proton frequency.

The sequences presented in Fig. 12(d–f) are used in proton exchange measurements,[7] or in other studies involving exchangeable protons.

Sequence (d), again of the first type, is used for a 1D measurement of longitudinal relaxation (e.g., of exchangeable protons) in protonated water.[7] The protons are inverted by a selective pulse at their resonance frequency, or by a DANTE sequence[55] at the water proton frequency, and JR detection follows after a variable delay t_m.

Sequences (e) and (f) generate spectra of magnetization transfer from water due to proton exchange or NOE. They are of the second type. In sequence (e), the WPM is inverted by a selective pulse at the water proton frequency. Then comes a variable transfer time t_m, followed by the detection subsequence, in which a spin echo is used to filter out the water protons: the first hard 90° pulse brings all magnetizations into the transverse plane where they are spread out by G_3; they are next refocused by a 180° JR sequence and the second gradient pulse G_3, except for the water protons, which are not refocused.[56]

Sequence (e) may not be adequate in high field. First, the selective inversion of the WPM is imperfect because of radiation damping. Second, the sequence generates an NOE from macromolecular protons that are isochronous with the water protons. Another problem is that the WPM is destroyed by the detection sequence, leading to a long waiting period between measurements.

These defects are remedied in sequence (f). First, the inversion subsequence is protected against radiation damping, as explained above in the section on radiation damping. Second, the spin echo detection subsequence begins with a selective pulse that, together with the contiguous 90° pulse, avoids generating any transverse WPM. Furthermore, the phase of this pulse is chosen so that the WPM ends up in the $+Oz$ direction, thus minimizing the required waiting period between measurements. Third, the initial magnetization of all protons including those isochronous with water is zero and therefore invariant during the phase cycle

[54] V. Sklenar, R. Tschudin, and A. Bax, *J. Magn. Res.* **96**, 541 (1992).

[55] G. A. Morris and R. Freeman, *J. Magn. Res.* **29**, 433 (1978).

[56] V. Sklenar and A. Bax, *J. Magn. Reson.* **74**, 469 (1987).

[57] J. L. Leroy, S. Nonin, X. Han, A. T. Phan, and M. Guéron, *in* "Structure, Motion, Interactions and Expression of Biological Macromolecules" (R. H. Sarma and M. H. Sarma, eds.), Vol. 1, p. 49. Adenine Press, New York, 1998.

[58] R. Lavery, K. Zakrzewska, J. S. Sun, and C. Harvey, *Nucl. Acids Res.* **20**, 5011 (1992).

[59] P. Plateau and M. Guéron, *J. Am. Chem. Soc.* **104**, 7310 (1982).

[60] M. Guéron, P. Plateau, A. Kettani, and M. Decorps, *J. Magn. Reson.* **96**, 541 (1992).

(during which the initial WPM alternates), so that any NOE from the isochronous protons cancels out.

Another advantage of sequence (f) is that the zero initial magnetization of the other protons helps in avoiding artifacts, arising for instance from imperfect signal, cancellation over the phase cycle. This is particularly useful for measurements with short mixing times, e.g., 5 ms in the case of proton exchange (J. L. Leroy, in preparation). In fact, a useful test of any sequence involving a mixing time is the examination of artifacts in the case of zero mixing time, for which the ideal result is strictly null.

[16] Resonance Assignment and Structure Determination for RNA

By JENNY CROMSIGT, BERND VAN BUUREN, JÜRGEN SCHLEUCHER, and SYBREN WIJMENGA

Introduction

RNA has many important biological functions, ranging from the delivery of genetic information to catalysis in RNA processing. These diverse functions require an equally diverse collection of structures. Nuclear magnetic resonance (NMR) has the unique capability to determine these structures in solution under physiological conditions. Advances in methods for synthesis of isotopically labeled RNA oligomers and developments in heteronuclear NMR techniques have made structure determination possible for RNA molecules up to a size of ca 50 nucleotides (ca 15,000 Da). Structure elucidation of larger oligonucleotides is still hampered by resonance overlap and line broadening. In this review we concentrate on two aspects of the structure determination of [13]C- and ^{15}N-labeled RNA molecules. First assignment strategies are described, and subsequently structure calculation procedures are discussed. We conclude with a section in which we consider new methods for improvement of structure determination and discuss prospects for extension of NMR studies to RNAs above the 50 nucleotide limit.

Assignment Strategies

Assignment of resonances is the first essential step in an NMR study of biomolecules aimed at determining their three-dimensional structure. The development of [13]C- and ^{15}N-labeling of RNAs[1,2] made it possible to devise RNA-specific heteronuclear through-bond assignment methods similar to the field of

0076-6879/00 $35.00

protein NMR. However, experience with these methods has now shown that for nucleic acids through-bond assignment complements rather than replaces the nuclear overhauser effect (NOE)-based assignment. We will therefore briefly outline the assignment strategy for unlabeled compounds (for reviews[3,4]), before describing the two isotope-aided strategies, namely assignment based on using [13]C- and ^{15}N-edited NOESY spectra and assignment via through-bond correlations (see for reviews[4,5]). Finally, we show how the two isotope-aided assignment strategies complement each other using a practical example.

Assignment without Isotopic Labeling

For unlabeled oligonucleotides, the sequential resonance assignment is based on a combination of through-bond and through-space ^{1}H to ^{1}H contacts (Fig. 1; for more extensive reviews[3,4,6]). The first step (step I) is the sequential assignment of the imino resonances via through-space (NOE) contacts. This establishes which residues are base paired and thus the secondary structure. In step II the nonexchangeable protons are partially assigned via NOEs from imino and amino protons. In step III the base H6/8 and sugar H1'/H2'/H3' resonances are sequentially assigned, mostly via a sequential walk in the H6/8–H1' region of the NOESY spectrum. This is often combined with identification via (^{1}H,[1]H) TOCSY/COSY of the H6–H5 spin systems in uridine and cytosine and the H1', H2', and H3' spin systems in the ribose. Finally, the ^{31}P resonances are assigned and the ribose spin systems assignment extended (to H3', H4', H5', H5'') via (^{1}H,[31]P) HETCOR.

The two main problems associated with this NOE-based sequential assignment of proton resonances are resonance overlap (most severe for the ribose ring) and the ambiguity of NOE contacts. NOE-based sequential assignment requires assumptions with regard to the conformation of structural elements of the sequence studied. Therefore, proof of correctness of the assignments has to come from the internal consistency of the NOE contacts. Because of these problems, resonance assignment in unlabeled compounds can be quite difficult. Nevertheless, the NOE-based assignment strategy is extremely powerful and detailed structures have been derived with this method for smaller RNAs.[7–12]

[1] E. P. Nikonowicz, A. Sirr, P. Legault, F. M. Jucker, L. M. Baer, and A. Pardi, *Nucleic Acids Res.* **20,** 4507 (1992).
[2] R. T. Batey, M. Inada, E. Kujawinski, J. D. Puglisi, and J. R. Williamson, *Nucleic Acids Res.* **20,** 4515 (1992).
[3] S. S. Wijmenga, M. M. Mooren, and C. W. Hilbers, *in* "NMR of Macromolecules, A Practical Approach" (G. C. K. Roberts, ed.), p. 217. Oxford University Press, New York, 1993.
[4] S. S. Wijmenga and B. N. M. van Buuren, *Prog. NMR Spec.* **32,** 287 (1998).
[5] G. Varani, F. Aboul-Ela, and F. H. Allain, *Prog. NMR Spec.* **29,** 51 (1996).
[6] M. M. Mooren, C. W. Hilbers, and S. S. Wijmenga, *J. Magn. Reson.* **94,** 101 (1991).

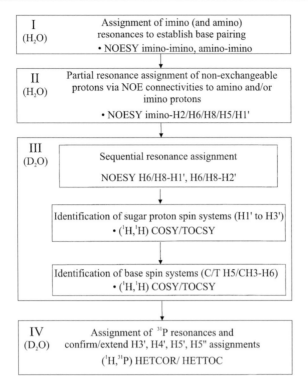

FIG. 1. Flow chart for resonance assignments in unlabeled RNA molecules; NOE-based assignment.

Assignment with Isotope Labeling

Isotope labeling with ^{13}C and ^{15}N has signaled the introduction of alternative and improved assignment schemes (for reviews, see Refs. 4, 5, and 13–15). Two

[7] C. W. Hilbers, S. S. Wijmenga, H. Hoppe, and H. A. Heus, *in* "Dynamics and the Problem of Recognition in Biological Macromolecules" (O. Jardetzky and J. F. Lefevre, eds.), p. 193. Plenum Press, New York, 1996.

[8] H. A. Heus, S. S. Wijmenga, H. Hoppe, and C. W. Hilbers, *J. Mol. Biol.* **271**, 147 (1997).

[9] M. J. P. van Dongen, S. S. Wijmenga, G. A. van der Marel, J. H. van Boom, and C. W. Hilbers, *J. Mol. Biol.* **263**, 715 (1996).

[10] B. A. Luxon and D. G. Gorenstein, *Methods Enzymol.* **261**, 45 (1995).

[11] N. B. Ulyanov and T. L. James, *Methods Enzymol.* **261**, 90 (1995).

[12] U. Schmitz and T. L. James, *Methods Enzymol.* **261**, 3 (1995).

[13] J. D. Puglisi and J. R. Wyatt, *Methods Enzymol.* **261**, 323 (1995).

[14] A. Pardi, *Methods Enzymol.* **261**, 350 (1995).

[15] K. B. Hall, *Methods Enzymol.* **261**, 542 (1995).

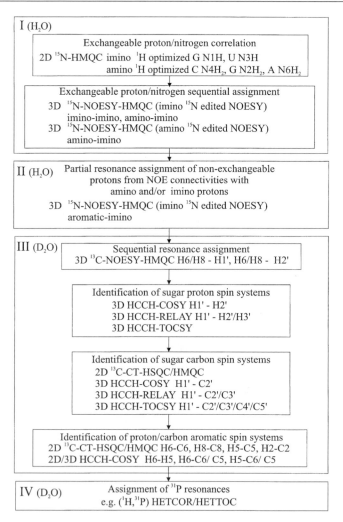

I (H₂O)

Exchangeable proton/nitrogen correlation
2D ¹⁵N-HMQC imino ¹H optimized G N1H, U N3H
amino ¹H optimized C N4H₂, G N2H₂, A N6H₂

Exchangeable proton/nitrogen sequential assignment
3D ¹⁵N-NOESY-HMQC (imino ¹⁵N edited NOESY)
imino-imino, amino-imino
3D ¹⁵N-NOESY-HMQC (amino ¹⁵N edited NOESY)
amino-imino

II (H₂O) Partial resonance assignment of non-exchangeable
protons from NOE connectivities with
amino and/or imino protons
3D ¹⁵N-NOESY-HMQC (imino ¹⁵N edited NOESY)
aromatic-imino

III (D₂O) Sequential resonance assignment
3D ¹³C-NOESY-HMQC H6/H8 - H1', H6/H8 - H2'

Identification of sugar proton spin systems
3D HCCH-COSY H1' - H2'
3D HCCH-RELAY H1' - H2'/H3'
3D HCCH-TOCSY

Identification of sugar carbon spin systems
2D ¹³C-CT-HSQC/HMQC
3D HCCH-COSY H1' - C2'
3D HCCH-RELAY H1' - C2'/C3'
3D HCCH-TOCSY H1' - C2'/C3'/C4'/C5'

Identification of proton/carbon aromatic spin systems
2D ¹³C-CT-HSQC/HMQC H6-C6, H8-C8, H5-C5, H2-C2
2D/3D HCCH-COSY H6-H5, H6-C6/ C5, H5-C6/ C5

IV (D₂O) Assignment of ³¹P resonances
e.g. (¹H,³¹P) HETCOR/HETTOC

FIG. 2. Flow chart for resonance assignments in labeled RNA molecules; NOE-based assignment.

main strategies can be recognized. The first, shown in Fig. 2, is still mainly based on NOE contacts and aims at improvement of the assignment process via reduction of resonance overlap using ¹³C/¹⁵N-edited (¹H,¹H) NOESY and TOCSY spectra. The second strategy is completely based on through-bond correlation (Fig. 3). Through-bond correlation resolves the ambiguity problem of NOE contacts. Resonance overlap is also reduced in this scheme via the ¹³C/¹⁵N editing achieved in the through-bond correlation experiments.

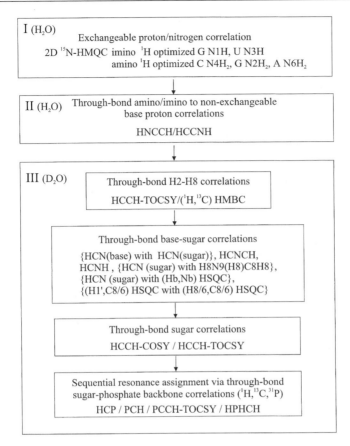

FIG. 3. Flow chart for resonance assignments in labeled RNA molecules; assignment via through-bond coherence transfer.

NOE-Based Correlation. The NOE-based strategy shown in Fig. 2 is a modified version of the one originally proposed by Nikonowicz and Pardi.[16] We have adapted the original scheme to emphasize the similarities with the assignment in unlabeled compounds, and to indicate the usual flow in the assignment process, which goes from imino/amino assignment to assignment of nonexchangeable protons. Briefly, the imino/amino protons are sequentially assigned via two-dimensional (2D) (^1H,^{15}N) HMQC and 3D (^1H,^{15}N,^1H) NOESY-HMQC experiments in H_2O (step I). These experiments also provide connectivities from exchangeable protons to nonexchangeable protons (step II). The most time-consuming step is the

[16] E. P. Nikonowicz and A. Pardi, *J. Mol. Biol.* **232**, 1141 (1993).

sequential assignment of H6/8 and H1' resonances that are obtained via NOE contacts in ^{13}C-edited NOESY spectra [^{2}3D(^{1}H,^{13}C,^{1}H) NOESY-HMQC; first block in step III]. The sugar spin systems are assigned using a set of 2D and 3D heteronuclear HCCH-COSY and HCCH-TOCSY experiments[16,17] (third and fourth block in step III; see also next section). Finally, ^{31}P resonances are assigned from (^{1}H,^{31}P) HETCOR experiments (step IV).

Through-Bond Correlations. The assignment scheme that aims at using through-bond correlations instead of NOE-based correlations is shown in Fig. 3. The first step is the assignment of imino resonances. This step is still essentially NOE-based. However, as shown recently by a number of authors,[18-21] a *J* coupling is present across the hydrogen bond between the bases, which can be used to establish the base pairing via a through-bond correlation. Rüdisser *et al.*[22] have proposed a complementary through-bond experiment that correlates the N3 of cytosine with its *trans*-amino proton.

Through-bond amino/imino to nonexchangeable proton correlation is step II. So far three sets of pulse sequences have been proposed.[4,23-26] These sequences have large overall similarities and differ only in the parameter settings used (see for a detailed review[4]). As for other through-bond correlation experiments, the pulse sequences are not straightforward, because of the complex network of *J*-couplings in the bases.[4,27]

For obtaining through-bond H2–H8 correlations in ^{13}C/^{15}N-labeled RNA two pulse sequences have been published.[28,29] Both sequences are closely related to the HCCH-TOCSY experiment originally proposed for proteins[30,31] and work well

[17] A. Pardi and E. P. Nikonowicz, *J. Am. Chem. Soc.* **114**, 9202 (1992).
[18] K. Pervushin, A. Ono, C. Fernández, T. Szyperski, M. Kainosho, and K. Wüthrich, *Proc. Natl. Acad. Sci. U.S.A.* **95**, 14147 (1998).
[19] A. J. Dingley and S. Grzesiek, *J. Am. Chem. Soc.* **120**, 8293 (1998).
[20] J. Wöhnert, A. J. Dingley, M. Stoldt, M. Görlach, S. Grzesiek, and L. R. Brown, *Nucleic Acids Res.* **27**, 3104 (1999).
[21] A. J. Dingley, J. E. Masse, R. D. Peterson, M. Barfield, J. Feigon, and S. Grzesiek, *J. Am. Chem. Soc.* **121**, 6019 (1999).
[22] S. Rüdisser, J. G. Pelton, and I. Tinoco, Jr., *J. Biomol. NMR* **15**, 173 (1999).
[23] J. P. Simorre, G. R. Zimmermann, A. Pardi, B. T. Farmer II, and L. Mueller, *J. Biomol. NMR* **6**, 427 (1995).
[24] J. P. Simorre, G. R. Zimmermann, L. Mueller, and A. Pardi, *J. Biomol. NMR* **7**, 153 (1996).
[25] V. Sklenar, T. Dieckmann, S. E. Butcher, and J. Feigon, *J. Biomol. NMR* **7**, 83 (1996).
[26] R. Fiala, F. Jiang, and D. J. Patel, *J. Am. Chem. Soc.* **118**, 689 (1996).
[27] J. H. Ippel, S. S. Wijmenga, R. de Jong, H. A. Heus, C. W. Hilbers, E. de Vroom, G. A. van der Marel, and J. H. van Boom, *Magn. Reson. Chem.* **34**, S156–S176 (1996).
[28] J. P. Marino, J. H. Prestegard, and D. M. Crothers, *J. Am. Chem. Soc.* **116**, 2205 (1994).
[29] P. Legault, B. T. Farmer II, L. Mueller, and A. Pardi, *J. Am. Chem. Soc.* **116**, 2203 (1994).
[30] E. R. Zuiderweg, *J. Magn. Reson.* **89**, 533 (1990).
[31] S. W. Fesik, H. L. Eaton, E. T. Olejniczak, E. R. Zuiderweg, L. P. McIntosh, and F. W. Dahlquist, *J. Am. Chem. Soc.* **112**, 886 (1990).

for larger RNAs.[32] This is an important experiment because the H2 resonance is readily identified from the strong imino(U)-H2 cross peak in a (^{1}H,^{1}H) NOESY spectrum in H_2O. Alternatively, one can obtain the H2 to H8 correlation from an HMBC experiment, via long-range H to C correlations. This experiment has not been used in labeled RNA, as far as we are aware, but has been used in DNA (unlabeled) to obtain H2 to H8 correlations.[33]

The NMR experiments that have been proposed for correlating base and sugar protons are usually given a name that indicates the magnetization transfer route. With these naming conventions the NMR experiments proposed for through-bond base–sugar correlation can be classified as HCN experiments[34,35] (correlating H1′, C1′, and N9 or N1, or alternatively H6/8, C6/8, and N1/9), HCNCH,[35–37] H8N9H8,[35] and HCNH experiments.[38] To improve the sensitivity of the HCN and HCNCH experiments, more recent versions employ multiple-quantum narrowing for the long period required for coherence transfer from C1′ to N9/N1.[39,40] The multiple-quantum coherences (e.g., H1′xC1′x) are not sensitive to the strong dipolar interaction between C1′ and H1′, and therefore relax more slowly than in-phase or antiphase C1′ coherences. They are, however, affected by the $J_{H1'H2}$ coupling during this period. Consequently, these experiments show improved sensitivity when the ribose is in an N-puckered state, when $J_{H1'H2}$ is small. We finally consider the possibility for using the $J_{H1'C6/8}$ coupling and the $J_{H1'C2/4}$ coupling to obtain base–sugar correlation via long-range HSQC or HMQC experiments. The $J_{H1'C6/8}$ and $J_{H1'C2/4}$ coupling depend on the χ torsion angle and have values of 4–5 Hz and 2 Hz, respectively, when the χ torsion angle is in the usual *anti* range.[27]

For the correlation through the ribose sugar ring (step III, Fig. 3) HCCH-TOCSY or HCCH-COSY experiments can be employed as well as a hybrid form, HCCH-TOCSY/COSY.[41,42] Compared to the (^{1}H,^{1}H) COSY or (^{1}H,^{1}H) TOCSY experiments used with unlabeled samples, the HCCH-type experiments have the advantage that resonance overlap is reduced and most importantly, that they use the large conformation-independent (42 Hz) J_{CC} couplings, instead of the smaller (2 to

[32] M. H. Kolk, S. S. Wijmenga, H. A. Heus, and C. W. Hilbers, *J. Biomol. NMR* **12**, 423 (1998).

[33] M. J. P. van Dongen, S. S. Wijmenga, R. Eritja, F. Azorín, and C. W. Hilbers, *J. Biomol. NMR* **8**, 207 (1996).

[34] V. Sklenar, R. D. Peterson, M. R. Rejante, and J. Feigon, *J. Biomol. NMR* **3**, 721 (1993).

[35] B. T. Farmer II, L. Mueller, E. P. Nikonowicz, and A. Pardi, *J. Biomol. NMR* **4**, 129 (1994).

[36] V. Sklenar, R. D. Peterson, M. R. Rejante, E. Wang, and J. Feigon, *J. Am. Chem. Soc.* **115**, 12181 (1993).

[37] B. T. Farmer II, L. Mueller, E. P. Nikonowicz, and A. Pardi, *J. Am. Chem. Soc.* **115**, 11040 (1993).

[38] S. Tate, A. Ono, and M. Kainosho, *J. Am. Chem. Soc.* **116**, 5977 (1994).

[39] J. P. Marino, J. L. Diener, P. B. Moore, and C. Griesinger, *J. Am. Chem. Soc.* **119**, 7361 (1997).

[40] V. Sklenar, T. Dieckmann, S. E. Butcher, and J. Feigon, *J. Magn. Reson.* **130**, 119 (1998).

[41] W. D. Hu, L. T. Kakalis, L. C. Jiang, F. Jiang, X. M. Ye, and A. Majumdar, *J. Biomol. NMR* **12**, 559 (1998).

[42] W. D. Hu and L. C. Jiang, *J. Biomol. NMR* **15**, 289 (1999).

8 Hz) conformation-dependent J_{HH} couplings. In all experiments the transfer from ^1H to ^{13}C is achieved either via a refocused INEPT[31,43] or via a cross-polarization sequence.[30,44,45] Many improved version of the HCCH experiments are available, e.g., with gradient coherence selection and sensitivity enhancement.[46–50] To remove the splittings of the cross peaks stemming from the large 42 Hz J_{CC} couplings in the ribose ring, the ^{13}C evolution is incorporated into a constant time period.[51–53] Incorporation of a CT period in a 3D HCCH-TOCSY provides maximal resolution, but may lead to insufficient sensitivity for larger RNAs.[32] In other words, one has to resort to a combination of CT 2D HCCH-TOCSY, 3D HCCH-TOCSY, and HCCH-COSY. The last has only one single (C,C) transfer step and therefore higher sensitivity.[16,32] Improved resolution can also be obtained via the 3D HCCH-COSY-TOCSY proposed by Hu *et al.*,[41] in which the COSY transfers and TOCSY steps are combined in one experiment. A third alternative is to make use of dispersion of the N9/N1 resonances and create a N9/N1-edited CT-HCCH-TOCSY (HCN-CCH-TOCSY).[42] The transfer out and back from C1′ to N1/N9 is done via multiple-quantum line narrowing.

Backbone assignment by through-bond coherence transfer (last box in step III, Fig. 3) can be achieved via the generally rather large $J_{C4'iPi}$ and $J_{C4'i+1Pi}$ in a 3D HCP experiment.[28,54] The HCP pulse sequence generates H4′ coherence, which is transferred to C4′, then to P, and back. The alternative PCH experiment starts on P and transfers the P coherence via C4′ to H4′ and shows better sensitivity for larger systems.[32] Outside helical regions the C4′ resonances are remarkably well resolved.[32] The H4′, C4′, and P resonances of residues located in helical regions tend to overlap. It is then better to transfer the magnetization to more dispersed resonances such as the H1′ resonances as done in the PCCH-TOCSY sequence.[55] For larger RNAs this may not be feasible because of reduced sensitivity.[32] Varani *et al.*[56] have proposed to perform an additional triple resonance experiment (HPHCH).

[43] A. Bax, G. M. Clore, and A. Gronenborn, *J. Magn. Reson.* **88,** 425 (1990).

[44] H. Wang and E. R. Zuiderweg, *J. Biomol. NMR* **5,** 207 (1995).

[45] A. Majumdar, H. Wang, R. C. Morshauer, and E. R. Zuiderweg, *J. Biomol. NMR* **3,** 387 (1993).

[46] L. E. Kay, P. Keifer, and T. Saarinen, *J. Am. Chem. Soc.* **114,** 10663 (1992).

[47] J. Schleucher, M. Schwendinger, M. Sattler, P. Schmidt, O. Schedletzky, S. J. Glaser, O. W. Sorensen, and C. Griesinger, *J. Biomol. NMR* **4,** 301 (1994).

[48] S. S. Wijmenga, E. Steensma, and C. P. M. van Mierlo, *J. Magn. Reson.* **124,** 459 (1997).

[49] M. Sattler, M. Schwendinger, J. Schleucher, and C. Griesinger, *J. Biomol. NMR* **6,** 11 (1995).

[50] M. Sattler, P. Schmidt, J. Schleucher, O. Schedletzky, S. J. Glaser, and C. Griesinger, *J. Magn. Reson.* **B108,** 235 (1995).

[51] F. J. M. van de Ven and M. E. P. Philippens, *J. Magn. Reson.* **97,** 202 (1992).

[52] L. E. Kay, M. Ikura, and A. Bax, *J. Magn. Reson.* **91,** 84 (1991).

[53] L. E. Kay, M. Ikura, R. Tschudin, and A. Bax, *J. Magn. Reson.* **89,** 496 (1990).

[54] H. A. Heus, S. S. Wijmenga, F. J. van de Ven, and C. W. Hilbers, *J. Am. Chem. Soc.* **116,** 4983 (1994).

[55] S. S. Wijmenga, H. A. Heus, H. A. Leeuw, H. Hoppe, M. van der Graaf, and C. W. Hilbers, *J. Biomol. NMR* **5,** 82 (1995).

[56] G. Varani, F. Aboulela, F. Allain, and C. C. Gubser, *J. Biomol. NMR* **5,** 315 (1995).

This is a ^{13}C-edited version of regular (^{1}H,^{31}P)-HETCOR. The advantage of employing such an additional experiment is that certain cross peaks absent in the HCP/PCH experiment (due to small $J_{C4'P}$ couplings when the torsion angles ε and β have unusual values) are likely to show up in the HPHCH experiment (for an analysis see Wijmenga and van Buuren[4]). Finally, Brown and co-workers[57] have introduced a simultaneous HCP/HCN experiment that in principle combines the sequential backbone assignment via through-bond coherence transfer and sugar to base correlation via through-bond coherence transfer. As for the PCCH-TOCSY experiment, the long transfer routes may make this experiment too insensitive to be useful for larger RNAs such as the 44 nucleotide TYMV pseudoknot.

The optimal transfer efficiencies for most of the different through-bond experiments discussed above have been compared,[4] assuming a transverse relaxation time (T_2) of the relevant coherences of 3 s and 30 ms. The main conclusions that can be drawn are as follows. (1) No significant differences in sensitivity between the different versions proposed for a certain through-bond correlation; for the sugar–base through-bond correlation new versions have been published since the original comparison was made. These experiments employ multiple-quantum line narrowing and have improved sensitivity for residues with an N-puckered ribose and are thus optimal for helical regions. (2) Sufficient transfer is still obtained for most experiments even for $T_2 = 30$ ms, which corresponds to a RNA of ca 50 nucleotides.[4] Consequently, whether sequential backbone assignments is possible for systems up to 50 nucleotides depends mainly on the degree of resonance overlap. A caveat is that broadening due to conformational exchange on the millisecond time scale may adversely affect the through-bond experiments.

A Practical Case

For proteins, a more or less standard set of through-bond experiments is used to yield backbone resonance assignments.[58] For nucleic acids, several experiments have been proposed but a standard set has not yet emerged, which is largely due to resonance overlap and the fundamentally different nature of nucleic acids. Consequently, the two assignment schemes described in Figs. 2 and 3 are practically never carried through in full, but rather a combination of experiments is taken from both schemes. In addition, residue-specific labeling has turned out to be necessary in many cases.[32,59–61] It seems therefore valuable to briefly describe how the actual assignment was carried out for a practical case.

[57] R. Ramachandran, C. Sich, M. Grune, V. Soskic, and L. R. Brown, *J. Biomol. NMR* **7**, 251 (1996).
[58] M. Sattler, J. Schleucher, and C. Griesinger, *Prog. NMR Spec.* **34**, 93 (1999).
[59] D. Fourmy, M. I. Recht, S. C. Blanchard, and J. D. Puglisi, *Science* **274**, 1367 (1996).
[60] T. Dieckmann and J. Feigon, *J. Biomol. NMR* **9**, 259 (1997).
[61] M. H. Kolk, M. van der Graaf, S. S. Wijmenga, C. W. A. Pleij, H. A. Heus, and C. W. Hilbers, *Science* **280**, 434 (1998).

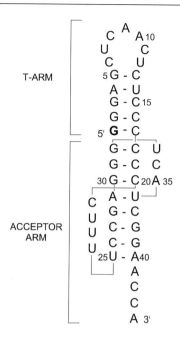

FIG. 4. Secondary structure of the T-arm and pseudoknotted acceptor stem of the tRNA-like structure of TYMV RNA. The 5'-terminal residue is indicated in bold for clarity.

TYMV Pseudoknot. For the 44nt TYMV pseudoknot (pk44) two main obstacles had to be overcome (Fig. 4). The first problem was the resonance overlap due to the relatively large number of residues. The second problem was the strong line broadening observed for many residues present in the pseudoknotted region (residues 18–40). In the T-arm (residues 1–17) and the 3' tail only the problem of resonance overlap was encountered. Here, assignment could be carried out essentially via NOE-based methods (Fig. 2). For the other parts several through-bond experiments had to be adjusted to achieve full resonance assignment.

T-arm and 3'-tail. Nearly complete assignments were achieved for the T-arm and 3'-tail via homonuclear 2D NOESY and ^{13}C-edited NOESY-HMQC experiments performed at 750 MHz. The sequential contacts were obtained using the standard H1'-H6/8 sequential NOEs as indicated in Fig. 2III. Confirmation of part of the assignments was obtained by the identification of A- and U-related resonances in A- and U-selectively labeled samples. Finally, for one residue the imino resonance was identified using the HNCCCH experiment (U14, imino to H6). All adenine H2 and H8 resonances were correlated via a 3D HCCH-TOCSY experiment.[29]

Pseudoknotted region. The sequential walk (NOE) into this region from the T-arm and 3'-tail was interrupted because of severe overlap and the absence of sequential NOEs. Also, broadening due to conformational exchange on the millisecond time scale was observed for many residues. As a result NOE contacts were absent in the 750 MHz ^{13}C-edited NOESY or 2D NOESY spectra. Therefore the assignment was based on spectra recorded at 400 MHz, in which the resonances narrow to acceptable line widths. Because of the overlap, the use of selectively A- and U-labeled samples proved indispensable. Residues U22 through U25 in loop1 could only be assigned using a U-labeled sample via doubly X-filtered NOESY, ^{13}C-NOESY-HMQC, and 3D PCH experiments. A sensitivity-optimized PCH experiment provided P-C4'-H4' connectivities. The H4' resonances were connected to the H1' resonances via a separate CT (H)CC(H)-TOCSY in which the ^{1}H–^{13}C transfer was achieved via hetero-TOCSY steps. The assignment of the G and C residues was achieved after all other resonances had been identified. This was done first via careful comparison of ^{13}C-edited NOESY with 2D NOESY and TOCSY. This yielded complete sequential assignments of the aromatic and H1' resonances and part of the H2' and H3' resonances. Subsequently, part of the remaining sugar spin systems could be assigned via 3D and 2D CT HCCH-TOCSY experiments. Because of the CT period the 3D experiment was in many cases too insensitive to show sufficient cross peaks while the 2D version contained too much overlap. Therefore, the high transfer efficiency in the 3D relayed HCCH-COSY was used to identify the H3' resonances from the well dispersed H1'-C1' cross peaks, after which the C3' resonances were identified from a CT 2D (H)CC(H)-TOCSY. The H2' resonances were identified from a 3D HCCH-COSY experiment that lacked the relay step. In a similar way the H5' and H5'' were assigned from the H4'-C4' cross peaks. The latter are remarkably well dispersed and could largely be identified in ^{13}C NOESY-HMQC from intraresidue NOEs to H6/H8/H1'. The assignments were verified via the above-mentioned 2D and 3D HCCH-TOCSY experiments.

Conclusions

Assignment of resonances in RNAs is despite the use of isotope enrichment with ^{13}C and ^{15}N still not a straightforward task. This is not because through-bond experiments have insufficient sensitivity. They perform well for RNAs with a size up to 50 nucleotides,[4] although broadening due to conformational exchange on the ms-time scale may further shorten T_2 values. Therefore, for RNAs up to 50 nucleotides the main problem is resonance overlap and one has to resort to residue-specific labeling to accomplish complete assignment. Above 50 nucleotides through-bond experiments start to fail because of too short T_2 values.[4,62] Introduction of ^{2}H labeling can easily expand the range of through-bond experiments to over 100 nucleotides.[4,62] The conclusion is again that the main assignment problem is resonance overlap. To resolve this overlap problem improved labeling

[62] J. A. M. T. C. Cromsigt, J. Schleucher, K. Kidd-Ljunggren, and S. S. Wijmenga, *J. Biomol. Struct. Dyn.* **Conversation 11,** 211 (2000).

schemes are needed.[4,62] Further improvements in assignment can come from better use of chemical shifts as well as from the inclusion of building blocks with a known structure.[4,63,64] We will discuss some possibilities in the final section.

Structure Calculation

Structure calculation from NMR data generally progresses as follows.[4,5] A set of random/semirandom starting structures is generated. Subsequently, one obtains a global fold, which in the next steps may be further refined. The final structures are then analyzed to establish their quality. The most widely used approach is based on restrained molecular dynamics either in Cartesian coordinate space or in torsion angle space. We recommend for nucleic acids to calculate the set of structures using torsion angle dynamics (TAD),[65] incorporated into, e.g., XPLOR[66] or DYANA.[67] TAD is computationally less demanding than molecular dynamics in Cartesian coordinate space, because the chemical structure is maintained during the dynamics. In addition, the TAD protocol is robust and has a high convergence rate[9,61,68–70] (up to 80%; see below). Furthermore, parts of the molecule with a known structure can easily be kept fixed in their conformation during TAD.

Alternative calculation methods exist. For example, the restrained Monte Carlo method (rMC) by Ulyanov et al.,[71] which employs Metropolis Monte Carlo methods to simulate an annealing process and uses the much reduced helix parameter space rather than Cartesian coordinates space. This rMC method can be used to refine DNA helix structures[71] from a B-DNA starting structure or from a set of structures generated by DYANA.[72]

Structure Calculation Protocol Based on Use of TAD

For convenience the parameters that go into the protocol are listed in Table I.

I. Holonomic Restraints. Nonexperimental constraints are the well-established holonomic constraints, i.e., bond lengths, bond angles, dihedrals, and impropers.

[63] F. J. J. Overmars, J. A. Pikkemaat, H. van den Elst, J. H. van Boom, and C. Altona, J. Mol. Biol. 255, 702 (1996).
[64] C. Altona, J. A. Pikkemaat, and F. J. J. Overmars, Curr. Opin. Struct. Biol. 6, 305 (1996).
[65] E. G. Stein, L. M. Rice, and A. T. Brunger, J. Magn. Reson. 124, 154 (1997).
[66] A. T. Brunger, in "XPLOR manual," Yale University Press, New Haven, CT, 1993.
[67] P. Güntert, C. Mumenthaler, and K. Wüthrich, J. Mol. Biol. 273, 283 (1997).
[68] M. J. P. van Dongen, M. M. W. Mooren, E. F. A. Willems, G. A. van der Marel, J. H. van Boom, S. S. Wijmenga, and C. W. Hilbers, Nucleic Acids Res. 25, 1537 (1997).
[69] M. J. P. van Dongen, J. F. Doreleijers, G. A. van der Marel, J. H. van Boom, C. W. Hilbers, and S. S. Wijmenga, Nature Struct. Biol. 6, 854 (1999).
[70] A. Dallas and P. B. Moore, Structure 5, 1639 (1997).
[71] N. B. Ulyanov, U. Schmitz, and T. L. James, J. Biomol. NMR 3, 547 (1993).
[72] N. B. Ulyanov, V. I. Ivanov, E. E. Minyat, E. B. Khomyakova, M. V. Petrova, K. Lesiak, and T. L. James, Biochemistry 37, 12715 (1998).

TABLE I
TORSION ANGLE DYNAMICS PROTOCOL[a]

I. Holonomic restraints	II. Experimental restraints	III. Starting structures
Parameter file	Distance (NOE)	Random extended
psf file	Dihedral	Locally random
	Hydrogen bond restraints for base pairs	
	Planarity restraints for base pairs	
	Chemical shifts	
	Dipolar coupings	

[a] Based on the original protocol by Stein et al.[65] and modified in a number of instances.

The most recent set of holonomic constraints can be found in Ref. 73. Also, the parameters for the force field are read in. Generally, a simplified force field is used in the structure calculation protocol. First, van der Waals interactions are introduced simply as soft-sphere repulsive constraints with reduced radius. Electrostatics are turned off. Dihedral force field terms that confine the torsion angles to the different staggered regions are generally not employed. The scaling of van der Waals radii, turning off of electrostatics, and omission of the dihedral terms give larger conformational freedom and aim at avoiding the use of less well-defined force field terms (e.g., electrostatics), which might bias the set of final structures.[4,5,74] The impact of the use of these less well-defined force field terms on the final structures is not clear, but probably depends on the level of precision to which the final structures are determined by experimental restraints.

II. Experimental Restraints. The experimental constraints are generally divided into subsets: NOE (distance) restraints, dihedral restraints, hydrogen bond restraints for base pairs, and planarity restraints for base pairs. Experimental restraints (distances and torsion angles) are usually given as upper and lower bounds, with a quadratic increase in penalty outside the allowed range or, in the case of distance restraints, a weaker penalty for longer distance errors. The distance between upper and lower bound in the estimated uncertainty in the restraint value.

Of the experimental restraints the distance restraints are the most important.[4,5,74] The derivation of distances from the cross-peak intensities in NOESY spectra has been the subject of intense research to devise methods that correct for spin diffusion (for detailed reviews, see Refs. 4, 5, 10, 12, 75, and 76). One can use NOESY spectra with short mixing times and calibrate against known distances

[73] G. Parkinson, J. Vojtechovsky, L. Clowney, A. T. Brunger, and H. M. Berman, *Act. Crystallogr. D* **52**, 57 (1996).

[74] F. H. T. Allain and G. Varani, *J. Mol. Biol.* **267**, 338 (1997).

[75] U. Schmitz, D. A. Pearlman, and T. L. James, *J. Mol. Biol.* **221**, 271 (1991).

[76] H. Liu, H. P. Spielmann, N. B. Ulyanov, D. E. Wemmer, and T. L. James, *J. Biomol. NMR* **6**, 390 (1995).

using an r^{-6} dependence (isolated spin pair approach), or correct for spin diffusion using a modified isolated spin approach,[4,8] or use full relaxation matrix methods to derive distances (using, e.g., MARDIGRAS[77]). Instead of aiming at deriving precise distances via the relaxation matrix methods, the most commonly used approach is to take the former simpler methods and conservative error bounds (± 20 to 50%),[4,5,8,74] which are large enough to account for all errors (see comments). Allain and Varani[74] (and others[12,78]) have systematically tested different restraint sets and their effect on the ultimate structures. The former study confirmed that the use of a large number of conservative distance restraints leads to high-precision structures [root mean square deviation (rmsd) < 1 Å]. It also showed that the number of restraints is more important than the error bounds. The number of restraints should at least be 10 per residue and evenly spread throughout the chemical structure.[4,5,12] Consequently, one should include not only sequential base–sugar and base–base distances, but also sequential sugar–sugar distances and distances derived from exchangeable protons (see also Refs. 4 and 8). To improve the convergence and the final structure's quality, non-NOEs are often included.[4,9,12,32,70] This should only be done after carefully checking the experimental NOESY spectra for the absence of an NOE. A very valuable approach to check the restraints against the experimental data is back-calculation of the NOE data during each of the various rounds of structure refinement to ensure that all the cross peaks have been correctly interpreted.

The dihedral restraints are important to improve the precision of the structure mainly in the final stages of the refinement.[74] Here again the general approach is to derive as many dihedral restraints as possible but with conservative error bounds, so that the torsion angle is only confined to one staggered region.[4,5,32]

A variety of methods have been developed to determine the J couplings that determine the various torsion angles.[4,5,79] We briefly illustrate the determination of torsion angles for a large RNA (the 44 nucleotide TYMV pseudoknot). For TYMV the J couplings could not determined from cross peak splittings or E.COSY type patterns.[32] Instead, a nearly complete set of torsion angles was obtained using intensity-based methods.[4,32,80,81] Most sugar puckers and γ torsion angles could be determined from a 3D (^{1}H,^{1}H) TOCSY-CT-HSQC experiment. For the sugar pucker one obtains then $J_{H1'H2'}$ couplings at the C1' position and $J_{H3'H4'}$ couplings at C4' position. This same slice also gives $J_{H4'H5'}$ and $J_{H4'H5''}$ for the γ torsion angles. Note that the effect of transverse relaxation can be eliminated in these spectra by normalizing the transfer cross peaks against the diagonal cross peaks (both the diagonal and transfer cross peaks are well separated). The torsion

[77] B. A. Borgias and T. L. James, *J. Magn. Reson.* **87**, 475 (1990).
[78] K. Kaluarachchi, R. P. Meadows, and D. G. Gorenstein, *Biochemistry* **30**, 8785 (1991).
[79] J. P. Marino, H. Schwalbe, and C. Griesinger, *Acc. Chem. Res.* **32**, 614 (1999).
[80] G. W. Vuister, A. C. Wang, and A. Bax, *J. Am. Chem. Soc.* **115**, 5334 (1993).
[81] P. Legault, F. M. Jucker, and A. Pardi, *FEBS Lett.* **362**, 156 (1995).

angles ε and β could be determined from a combination of two ^{31}P spin–echo difference experiments. To determine $J_{C2'P}$, a new spin–echo difference experiment was developed, the spin–echo difference CT-HCCH. In this experiment, the overlap in the C2'–H2' region is resolved by transferring the C2' coherence via C1' to H1'. The $J_{C2'P}$ couplings can then be measured from the less crowded H1'-C2' cross peaks. Because the torsion angle ε is geometrically restrained to the range of 150° to 300°, a one-to-one relationship exists between the $J_{C2'P}$ coupling and the torsion angle ε. The evaluation of ε circumvents the ambiguity problem associated with $J_{C4'P}$, which can accurately be determined from a spin–echo difference CT-HSQC (also $J_{C2'P}$ when overlap is less severe). If ε is known, the value of $J_{C4'P3'}$ can be calculated from the Karplus equation[3] and $J_{C4'P5'}$ determined from the cross-peak intensity ratio at H4–C4' and thus the torsion angle β. Finally, the torsion angles α and ζ can be derived from the ^{31}P chemical shifts, because the (α, ζ) pair is $(g\text{-}, g\text{-})$ when ^{31}P falls in the normal range. The glycosidic angle χ can best be determined from H6/8–H2' or H6/8–H1' NOEs. We note that to correctly define the puckering of the ribose ring, the sugar is best restrained by using all 5 ribose torsion angles (υ_0, υ_1, υ_2, υ_3, and υ_4).

Hydrogen bond restraints are invoked to maintain base pairing, often together with weak planarity restraints to prevent buckling of the base pairs.[8,61,69] As pointed out by Butcher et al.,[82] planarity restraints should only be used in a helical environment.

In addition to the more traditional restraints, novel experimental parameters start to become available as restraints, namely chemical shifts and dipolar couplings, including the related information-derived T_1 and T_2 relaxation for an anisotropic tumbling molecule. They have not yet extensively been employed in nucleic acids, but more so in the field of proteins.[83–85] A careful analysis of chemical shifts of DNA and RNA has shown that with proper reference values ^{1}H chemical shifts can be calculated with good accuracy.[86,87] We have incorporated a module into XPLOR that employs these chemical shift parameters so that ^{1}H shifts are correctly (back) calculated given a structure and that allows the ^{1}H chemical shifts to be used as restraints in structure calculations.[88,89] We have seen that with

[82] S. E. Butcher, F. H. T. Allain, and J. Feigon, *Nature Struct. Biol.* **6**, 212 (1999).

[83] J. L. Baber, D. Libutti, D. Levens, and N. Tjandra, *J. Mol. Biol.* **289**, 949 (1999).

[84] G. M. Clore, A. M. Gronenborn, and N. Tjandra, *J. Magn. Reson.* **131**, 159 (1998).

[85] N. Tjandra, J. G. Omichinski, A. M. Gronenborn, G. M. Clore, and A. Bax, *Nature Struct. Biol.* **4**, 732 (1997).

[86] S. S. Wijmenga, M. Kruithof, and C. W. Hilbers, *J. Biomol. NMR* **10**, 337 (1997).

[87] J. A. M. T. C. Cromsigt, C. W. Hilbers, and S. S. Wijmenga, *J. Biomol. NMR*, in press (2001).

[88] J. A. M. T. C. Cromsigt, B. N. M. van Buuren, J. Zdunek, J. Schleucher, C. W. Hilbers, and S. S. Wijmenga, *in* "Magnetic Resonance and Related Phenomena, 13th ISMAR" (D. Ziessow, W. Lubitz, and F. Lendzian, eds.), p. 132. Berlin University Press, 1998.

[89] J. Zdunek, J. A. M. T. C. Cromsigt, J. H. Ippel, B. N. M. van Buuren, and S. S. Wijmenga, to be published (2001).

the chemical shift restraints, the precision and number of distance and dihedral restraints required to obtain precise structures is considerably reduced.[88,89] Alternatively, ^{1}H chemical shifts provide an ideal independent check on the quality of the derived structures.[90] Finally, dipolar couplings, which give global structure information, can be introduced as restraints to increase the quality and quantity of NMR structure information.[83,91] Tjandra et al.[91] have derived an exceptionally high-quality structure of the Dickerson dodecamer using a large number of dipolar couplings in combination with traditional restraints. These issues will be briefly described in the section on new methods.

III. Starting Structures. Extended random structures are generated to scan conformational space as much as possible. Alternatively, it is possible to randomize only the essential part of the structure, ensuring in this way that conformational space is still sufficiently scanned, while the computational demands are smaller. In their study of the Dickerson dodecamer using dipolar couplings, Tjandra et al.[91] have used A-DNA and B-DNA helices as starting models. These starting structures are relatively close to the final refined structure. In principle, this holds the danger that certain conformations are missed. In this case, this may not be relevant since it involves refining a duplex structure where the changes are relatively minor. The example does illustrate that the type of starting structure that needs to be used depends on the context.

The protocol. Figure 5 shows a flow chart of a typical TAD protocol. First, all relevant parameters are read in (points 1a and b). The outer loop (Loop Fam) ensures that a whole family of structures is calculated, one or more for each starting structure (read in at point 1c). The inner loop (Loop main) controls how many structures are generated from each starting structure. This loop also contains the actual structure calculation protocol. First a high-temperature TAD is carried out that gives the global fold (DYNA TORS); 6000 steps of TAD are carried out at a temperature of 20,000, although we found that a temperature of 2000 was sufficient to obtain convergent folding in the case of a DNA three-way junction. Prior to the actual high-temperature TAD, the parameters are set. A simple soft-sphere repulsive van der Waals potential is used with the van der Waals radii reduced to 0.1 of their original value (vdw_tad = 0.1). The NOE restraints are given a relatively strong force constant (150) while torsion angles restraints are kept relatively weak (force constant 5).[8,61,69] The topology statement describes which parts of the structure are kept with a fixed geometry using the fix command during TAD with a certain force constant (kdihmax = 300; note that this does not mean that the positions are fixed in Cartesian space). In the protocol shown, not only the bases but also larger parts of the molecule (helices) are kept with a fixed

[90] B. N. M. van Buuren, F. J. J. Overmars, J. H. Ippel, C. Altona, and S. S. Wijmenga, J. Mol. Biol. **304**, 371 (2000).

[91] N. Tjandra, S. I. Tate, A. Ono, M. Kainosho, and A. Bax, J. Am. Chem. Soc. **122**, 6190 (2000).

Start

1a. read holonomic restraints and force-field parameter (parameter file, psf file)

1b. read experimental restraints

Loop Fam (loop through a set of starting structures)

1c read starting structure coordinates

    minimize powell step 300, drop 10 end

       Loop Main (loop until one or more structures have been generated from each starting structure)

          (set parameters, topology and do high temperature dynamics)

          vdw_tad = 0.1; noe 150; dihedral 5;

          topology

            kdihmax = 300; maxchn = 1000; cut C3'-C4' bond; fix bases

            fix helix arms fix group

          end topology

          DYNA TORS; n-steps = 6000; timestep = 0.015; tcoupling = true; tbath=20000

          Loop cool TAD (decrease temp from 20000 to 300, while increasing vdw_tad to 0.8)

            noe scale = 150; dihedral = 5; temp = 20000;

            while temp > 300

            DYNA TORS nsteps= (20000-300)/50 = ca. 400; timestep = 0.015

            end while

            change temp to current temp (= temp-25)

          end Loop cool TAD

          DYNA TORS; topology reset end; end;

          Loop cool Verlet

            start temp = 1000; noe 150; dihedral 100; vdw_verlet 0.8

            while temp > 300

             DYNA Verlet; nsteps=(1000-300)/25=ca. 30; timestep=0.003

            end while

            change temp to current temp (= temp-25)

          end Loop cool Verlet

          noe 50; dihedral 200

          Minimize Powell; 1000; drop 10;

          DYNA TORS; topology reset end; end;

          Evaluate violations; rms energy etc.

       end Loop Main

end Loop Fam

End

FIG. 5. Torsion angle dynamics protocol: how it actually works. Parameters are given in dynamics units, i.e., temperature in K, length in Å, energy in kcal/mol, time in fs, angles in °, force constant in kcal/(mol Å^2) for distance and in kcal/[mol (°)^2] for angles; see XPLOR manual for further details.[66]

geometry. Note also that TAD cannot handle rings that can change conformation. Therefore, during TAD the C3′–C4′ bond is cut to allow for torsion motion in the ribose. In the next step the system is cooled to 300 K (Loop cool TAD) in a number of steps while the van der Waals radii are slowly increased to 80% of their real values. The protocol ends with a final minimization, which consists of either a Powell minimization (Minimize Powell) or a short SA run followed by a Powell minimization (Loop cool Verlet and Minimize Powell). Most importantly, the dihedral restraints are given a higher force constant during this stage (100). The converged structures are often subjected to rounds of further refinement using this Verlet and Powell refinement protocol.

A good way to identify the converged structures is to order the structures according to total energy (E_{tot}). A plot of the ordered E_{tot} against the structure number shows a more or less flat plateau region at lower energies, indicating the converged structures, while at higher energies a considerable increase is found. One obtains approximately 50% to 80% convergence starting from either random extended structures[61,69,70,92] or locally randomized structures as in the case of a DNA three-way junction.[90]

Sometimes a final minimization is carried out using the "full" force field (see, e.g.,[12,93]) that is using the full van der Waals forces and possibly electrostatics. Whether one should include less well-defined terms such as electrostatics in the final refinement is a matter of debate as it may bias the final result (for discussions see, e.g., Refs. 5 and 11). In the protocol we show here, the structure calculation is based on experimental restraints supplemented only with the well-defined force field terms.

Validation and Analysis

In the final stage the derived structures have be to be analyzed and validated. The usual approach is to present an overview of structure calculation statistics[8,61,69,70] and check the set of structures for correct geometry.

Table II shows the structural characteristics of the 10 lowest energy structures of the 44 nucleotide TYMV RNA pseudoknot[61] and, for comparison, the statistics of the 12 lowest energy structures of the 30 nucleotide Hy-5 DNA triple helix.[69] First, the number of experimental restraints is listed. As can be seen in both cases, the number of distance restraints is well over 10 per residue, while essentially all torsion angles are restrained. Note that weak planarity restraints are included for base pairs/triplets in a helical environment (force constant 2 to 5).[61,69] Subsequently, the mean residual violations are given of the experimental restraints. The mean residual violations describe how much the restraints deviate above the upper

[92] S. C. Stallings and P. B. Moore, *Structure* 5, 1173 (1997).
[93] U. Schmitz, T. L. James, P. Lukavsky, and P. Walter, *Nature Struct. Biol.* 6, 634 (1999).

TABLE II
STRUCTURAL CHARACTERISTICS[a]

Parameter	TYMV (40 nucleotides)	Hy-5 triple helix (30 nucleotides)
Restraints		
Internucleotide	258	323
Intranucleotide[b]	193	186
Dihedral restraints	257	268
Repulse[c]		83
Base pair restraints[d]	13	54
Rmsd of restraints		
Distance restraints (Å)	0.085 ± 0.003	0.0472 ± 0.0012
Dihedral restraints (°)	2.9 ± 1.0	1.77 ± 0.11
Rmsd from idealized geometry		
Bonds (Å)	0.0094 ± 0.0004	0.0083 ± 0.0002
Angles (°)	1.96 ± 0.11	1.58 ± 0.02
Improper (°)	1.06 ± 0.03	1.58 ± 0.01
Restraint violations		
Number of distance violations[e]	4 ± 1 (>0.4 Å)	12 ± 1 (>0.2 Å)
Number of dihedral restraint violations	5 ± 1 (>5°)	8 ± 2 (>5°)[f]
Atomic rmsd[g] (Å)		
All	2.10 ± 0.23^{h}	
Helix region		1.03 ± 0.25^{i}
Loop regions	0.99 ± 0.28^{j}	1.56 ± 0.45^{k}

[a] Statistics are for 10 and 12 lowest-energy structures of TYMV and the Hy-5 triple helix respectively.

[b] Conformationally relevant[4].

[c] Non-NOE (see text).

[d] The base pair restraints were hydrogen bonding and weak planarity restraints imposed for residues in helix region on the basis of the imino spectra and observed NOE and chemical shifts.

[e] No violations >0.5Å.

[f] No violations >8.7°.

[g] Average pairwise rmsd to heavy atoms only.

[h] rmsd of all atoms.

[i] Average rmsd of triple helix.

[j] Average rmsd for residues 6 through 12, 21 through 24, and 33 through 35.

[k] Average rmsd of loop-tail region.

or below the lower bound. These values should be small compared to the error bounds, and the values shown are typical. Then the rmsd from idealized geometry is given. Again these values should be small compared to error bounds (given in Parkinson et al.[73]); otherwise, the structure is evidently distorted, possibly because of errors in the distance restraints. The number of violations in the experimental restraints is given next. The numbers given here are typical. It is important that the number is small and that no large violations occur. To further test the structure set one can provide the R-factor (see Gonzalez et al.[94] for definition), which contains the correspondence between experimental NOEs and back-calculated NOEs. For nucleic acids typical good values are ca 10% or lower. Furthermore, back calculation of experimental parameters that are not used as structural restraints (e.g., chemical shifts[86,87] or dipolar couplings[83–85] is an excellent way to validate the derived structures.

The above aspects all address the issue of accuracy of the derived structures. The rmsd of the set of final structures reflects the precision in the derived structures. Because of the extended nature of most nucleic acid structures and short-range nature of NMR restraints, it is good to indicate the rmsd per region (local rmsd) containing no more than six base pairs. In this way, one can make a fair comparison between different structures. As can be seen from Table II, both structures are locally well defined with local rmsd values around 1 Å. Higher precision requires a larger number of NOEs per residue. Two examples are the solution structure of the RNA duplex that contains a tandem $G \cdot A$ base pair[8] and the P1 helix from the group I intron.[95] With more than 30 distance restraints per residue, covering the whole range from sugar–sugar distances to distances involving exchangeable protons, very precise structures are obtained with a local rmsd of ca 0.6 Å.

To further investigate whether the set of structures is physically reasonable, one can perform an analysis of the geometry. Two aspects are important. The first concerns checks on holonomic restraints, i.e., bond lengths, bond angles, etc. This can be done with WhatCheck (standard when depositing in the PDB), which also directly provides checks on van der Waals clashes, unusual hydrogen bond donor/acceptor approaches, etc. The second type of check concerns unusual torsion angles. For proteins various programs exist for this task. No such programs are in use for nucleic acids. This is unfortunate, but understandable, since the structure of nucleic acids is determined by a much larger number of torsion angles; the nucleic acid backbone is determined by six variable torsion angles, while that of proteins is determined by only two. Nevertheless, the possible ranges of these torsion angles are restricted: in particular the torsion angle ε, which is confined to the range 150° to 300° (see, e.g., Mooren et al.[96]).

In conclusion, although the table of structural characteristics is not perfect (criteria are not strictly defined), it does provide together with additional geometry

[94] C. Gonzalez, J. A. C. Rullmann, A. M. J. J. Bonvin, R. Boelens, and R. Kaptein, J. Magn. Reson. **91,** 659 (1991).

[95] F. H. Allain and G. Varani, J. Mol. Biol. **250,** 333 (1995).

[96] M. M. Mooren, S. S. Wijmenga, G. A. van der Marel, J. H. van Boom, and C. W. Hilbers, Nucleic Acids Res. **22,** 2658 (1994).

checks a reasonable indication of the quality (precision/accuracy) of the set of NMR structures.

Comments on Validity of Structure Calculation Procedure

There are a number of potential problems with the use of such a protocol and with using NMR data for deriving structures of nucleic acids in general: incompleteness of the restraint set, criteria for assessment of the quality of NMR-derived structures, search of conformational space, precision/accuracy of NMR-derived distances and torsion angles, dynamics, and short range of distances and torsion angles. This list may not be complete, but it contains the most important issues. Some of them are discussed below in more detail; others have already been addressed at least in part above.

Incompleteness of the Restraint Set. Distance restraints from NOE data and torsion angle restraints from *J* couplings are the most often used experimental parameters for structure determination. Generally, the number of *J* couplings measured is fairly complete (see Table II). In addition, tests[74] have shown that torsion angle information makes a relatively minor contribution to improved the precision for less precise structures (local rmsd > 1 Å)[74]; it does further refine structures that already have small local rmsd(<1 Å). Consequently, there is little room for improvement.

Increasing the number of distance restraints improves the precision and accuracy of the derived structures.[4,12,74] As a rule of thumb, 10 distance restraints per residue evenly spread throughout the structure is the lower limit for obtaining good precision (local rmsd of ca 1 Å or lower). Approximately 60^4 to 45^{74} distance restraints per residue can be measured by NMR, of which 37^4 to 26^{74} are relatively easily obtained. In other words, the lower limit (10 per residue) corresponds to only 27% to 38% of the latter distances. For TYMV and Hy-5 (Table II) the completeness is ca 30–46% and 46–65%, respectively. Obviously, there is still quite some room for improvement, when overlap problems can be resolved, for example, via improved labeling schemes (see Novel Methods).

A problem with presently used distance restraint sets is that important structural features often depend on just a few NOE contacts. In various examples, a few wrong assignments have led to wrong structural features.[97–100] Increasing the number of distance restraints can resolve this problem, although only to some degree because

[97] G. M. Clore, J. G. Omichinski, K. Sakaguchi, N. Zambrano, H. Sakamoto, E. Appella, and A. M. Gronenborn, *Science* **265**, 386 (1994).

[98] G. M. Clore, J. G. Omichinski, K. Sakaguchi, N. Zambrano, H. Sakamoto, E. Appella, and A. M. Gronenborn, *Science* **267**, 1515 (1995).

[99] G. Varani, C. Cheong, and I. Tinoco, Jr., *Biochemistry* **30**, 3280 (1991).

[100] K. Gehring, J. L. Leroy, and M. Gueron, *Nature* **363**, 561 (1993).

the proton density is relatively small and unevenly spread throughout the structure. Chemical shifts and dipolar couplings are more likely to provide the much-needed additional structural information. The global information from dipolar couplings resolves in addition the problem of the short-range nature of the traditional NMR data.

Dynamics. In contrast to the situation considered above where the molecule was assumed to be rigid, NOEs, *J* couplings, chemical shifts, and dipolar couplings all represent average data in the presence of dynamics. Strictly speaking, restrained molecular dynamics cannot be applied when internal dynamics affects the restraints, except when they are only affected by small-scale motions around a single average such as libration motions. Experimental restraints can then be corrected for the effect of dynamics, e.g., the NOE rates are scaled by the angular and radial order parameters that describe the extent of the dynamics.[101] For situations such as the sugar ring flips between *N* and *S* states, and torsion angle flips between different rotamer states, distances can be small in one case and large in the other, and the above approach to structure calculation does not apply. To tackle this problem, timeaveraging[102,103] or ensemble-averaging methods can be employed.

Ensemble averaging or multicopy refinement seems the most promising approach, and three different algorithms have been published for ensemble-averaged (or multicopy) molecular dynamics.[104-107] Bonvin and Brunger[105] showed that different loop conformers could be identified. In a subsequent study they attempted to determine relative populations of loop conformers in a protein using synthetic data and cross-validation.[104] They came to the unfortunate conclusion that with the qualitative distance restraints most commonly used in NMR structure determination, cross-validation is unsuccessful in providing the correct answer. The new method proposed by Görler *et al.*[108] aims at performing multicopy refinement of nucleic acids structures using DNAminiCarlo, which performs the calculations in helical parameter space.[71] Görler *et al.* essentially determine conformer populations from NOE data with internally inconsistent NOE data as the driving force. The method was tested using different synthetic data sets. From a mixture of

[101] R. Bruschweiler, B. Roux, M. Blackledge, C. Griesinger, M. Karplus, and R. R. Ernst, *J. Am. Chem. Soc.* **114,** 2289 (1992).

[102] A. E. Torda, R. Scheek, and W. F. van Gunsteren, *J. Mol. Biol.* **214,** 223 (1990).

[103] A. E. Torda, R. M. Brunne, T. Huber, H. Kessler, and W. F. van Gunsteren, *J. Biomol. NMR* **3,** 55 (1993).

[104] A. M. Bonvin and A. T. Brunger, *J. Biomol. NMR* **7,** 72 (1996).

[105] A. M. J. J. Bonvin and A. T. Brunger, *J. Mol. Biol.* **250,** 80 (1995).

[106] J. Kemmink and R. Scheek, *J. Biomol. NMR* **6,** 33 (1995).

[107] J. Fennen, A. E. Torda, and W. F. van Gunsteren, *J. Biomol. NMR* **6,** 163 (1995).

[108] A. Görler, N. B. Ulyanov, and T. L. James, *J. Biomol. NMR* **16,** 147 (2000).

A- and B-DNA minor fractions as low as 10% could be determined using ca 11 NOEs per residue. For NOEs rates with an error smaller than 20%, about 80% of the refinements were successful, but the success rate dropped to around 50% or lower for qualitative NOEs with error rates of around 50%. For B-DNA with a local mixture of three instead of two different sugar puckers, the minor component could not be determined reliably. For helical DNA the influence NOEs on conformation has been thoroughly analyzed by various groups and the effects are quite well known. As the authors point out, prior to application of this algorithm to an experimental system, the presence of internal inconsistencies in the experimental data should be ascertained. The results are encouraging, and with even more complete distance sets and the inclusion of scalar couplings as well as possibly chemical shifts and dipolar couplings, the ensemble averaging method may prove to work well for nucleic acids.

Novel Methods

Nucleic acid structures are now routinely determined with NMR. However, as discussed above, there is still room for improvement both with regard to assignment of resonances and with regard to calculating structures. These improvements might come from some of the new developments described below.

Chemical Shifts as Tool for Assignment

Chemical shifts are already used to some degree in the assignment process, as each type of resonance has a certain range[4]; a detailed list is given in Wijmenga and van Buuren.[4] For example, the spectral region between 88 and 97 ppm contains only the $C1'$ resonances, which directly identifies them as such. Similarly, for other heteronuclei distinct resonance regions can be recognized. Unfortunately, for some nuclei the resonance regions overlap, e.g., $H3'$ and $H2'$ cover more or less the same shift region of 4.2–5.0 ppm.

A detailed analysis of ^1H resonances in DNA has shown that in a helical environment certain shifts directly depend on the local triplet sequence.[86,109] Consequently, in a double helix the $H1'$, $H2'$, $H2'$, H5, and H6/8 of the center nucleotide of a triplet have characteristic shifts that determine its 5'- and 3''-neighbor.[109] This knowledge can be used to assign the ^1H resonances from the sequence alone in a double helical environment.[109] This suggests that for RNA a similar chemical shift based assignment might also be possible. Until a few years ago the number of well-determined RNA NMR structures was limited and therefore such an analysis had not been done. Cromsigt et al.[87] have now performed a detailed ^1H

[109] C. Altona, D. H. Faber, and A. J. A. Westra Hoekzema, *Magn. Reson. Chem.* **38**, 95 (2000).

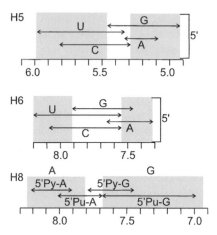

Fig. 6. Chemical shift distribution (in ppm) of H5, H6, and H8 in RNA. The values given are for residues involved in Watson–Crick base pairs in a double helix environment, i.e., the residues have a canonical pair on both the 5′- and the 3′-sites. The range given corresponds to the 95% confidence limit [95% chance that an observed shift falls within this range assuming a normal distribution (2 times standard deviation)].

shift analysis of a database with 30 RNA structures. It was found that in a helical environment the chemical shifts of certain nonexchangeable protons show distinct patterns depending on the type of sequence.[87] This implies that also for RNA, ^{1}H chemical shift information can be very useful in assignment procedures. We next describe an example involving the base protons.

Figure 6 shows the experimental chemical shift values of H5, H6, and H8 protons in a helical environment, sorted by their 5′-neighbor (no effect of the 3′ neighbor on the shifts could be detected). No distinction could be made between the chemical shifts of H5 of U and C; the same applies for H6 of U and C. However, the H5 and H6 protons with a purine or pyrimidine as a 5′-neighbor resonate in nearly distinct chemical shift ranges (shaded regions, top and middle panel, Fig. 6). This implies that for these protons, on the basis of chemical shift alone, a 5′-neighbor can be assigned. For H8 protons, one can establish whether the proton belongs to an adenine or a guanine (shaded regions, bottom panel, Fig. 6). The chemical shifts of H8 protons are also influenced by their 5′-neighbors, although not to the same extent as for H5 and H6. In the extreme regions one can identify the nature of the 5′-neighbor: a pyrimidine 5′-neighbor for adenine (5′-Py-A) or a purine 5′-neighbor for guanine (5′-Pu-G). In summary, when a proton is identified as being H5 or H6 it is straightforward to assign the 5′-neighbor. For a resonance identified as H8 one can assign it to 5′-N-A or 5′-N-G (shaded regions, Fig. 6), and in the extreme cases to 5′-Py-A or 5′-Pu-G.

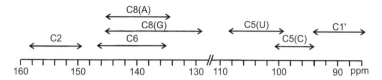

FIG. 7. Chemical shift distribution (in ppm) of carbon nuclei in RNA. The values given are for residues involved in Watson–Crick base pairs in a double helix environment, i.e., the residues have a canonical pair on both their 5′ and 3′ sites. The range given corresponds to the 98% confidence limit [98% chance that an observed shift falls within this range assuming a normal distribution (2.2 standard deviation)].

On the basis of the proton shifts alone it is not straightforward to identify a resonance as being H5, H6, and H8. For example, the H5 protons have their chemical shift in the same region as the H1′ protons and may thus be confused with H1′ protons. To resolve this ambiguity one usually correlates the H5 with H6 via a (^{1}H,^{1}H) TOCSY/COSY experiment or more directly one uses the C5 carbon shifts, because C5 and C1′ resonate in separate spectral regions (Fig. 7). Moreover, on the basis of carbon shifts a C5 resonance can be assigned to either a U or a C residue. Thus, from the C5 and H5 resonance positions it can directly be determined that they belong to a doublet of the type 5′-Py-U, 5′-Pu-U, 5′-Py-C, or 5′-Pu-C. The H6, H8, and H2 all resonate between 7.1 ppm and 8.2 ppm. They therefore cannot be distinguished on the basis of ^{1}H chemical shifts. However, as discussed the H6 can be identified via correlation to H5 in a (^{1}H,^{1}H) TOCSY/COSY experiment and thereby distinguished from H8 and H2 resonances. Consequently, the H6 resonances provide confirmation of the same doublets already determined via H5. This leaves the overlapping H8 and H2 resonances. The C2 resonate in a separate region from C8, which means that one can distinguish the H2 resonances from H8. Consequently, the H8 resonance can be identified and assigned to a doublet as described above.

As the above example illustrates, in RNAs the H6/8 and H5 proton shifts allow identification of doublets. Such knowledge is extremely useful in assignment, as it limits the number of possible placements in the sequence. We finally note that the difference in chemical shift patterns found in RNA and DNA can be understood from the difference in helical environment (A-type helix for RNA versus B-type helix for DNA).[87]

Deuteration and Residue-Specific Labeling

Although the assignment procedures for ^{15}N/^{13}C-labeled RNA described above work well for small (<50 nucleotide) RNA molecules, unambiguous assignments of larger RNAs is quite difficult if not impossible because of extensive spectral overlap and short carbon relaxation times. These problems can be overcome by the use of specifically deuterated RNAs[62,110,111] and/or base-type selective ^{13}C, ^{15}N, and ^{2}H labeling.[62]

The resonances of the ribose exhibit extreme overlap, most severely in the proton dimension but also in the carbon dimension. Uniform ^{13}C-labeling resolves only some of the overlap. In addition, the introduction of isotopic labels results in broader line widths for protons, which increases the overlap problem. Selective deuteration of ribose protons reduces overlap of important resonances by removing nonessential resonances. Furthermore, strong dipolar interactions are reduced, resulting in smaller line widths. Two different selective deuteration schemes have been proposed: (1) deuterium labels on the 3', 4', and 5' carbons, leaving the H1' and H2' resonances present[110,111]; (2) deuterium labels on the 1', 3', 4', 5' carbons (^{12}C or ^{13}C), leaving only H2' resonances present.[62]

Scheme (2) is achieved by a modification of the labeling procedure used for Scheme (1). For both schemes the torsion angles, sugar pucker, base pairing, and sequential information can be obtained from either NOEs or J couplings or both, while the spectral assignment is remarkably simplified. For example, deuteration of the H3' removes the H2' and H3' overlap and makes C2' and C3' distinguishable. The first scheme has the advantage that the well-dispersed H1' resonances are still present. The second scheme has the advantage that the dipolar coupling between the H1' and C1' is removed, so that magnetization transfer via C1' to N9/N1 is more efficient,[4] which should benefit through-bond experiments (see assignment section).

Assignment procedures rely heavily on through-bond NMR experiments. Figure 8 shows estimates of transverse relaxation rates as a function of the number of nucleotides for different labeling patterns. Figure 8A shows the transverse relaxation rate of hydrogens (H2') when the RNA is uniformly labeled with ^{13}C and no deuteration is applied. Figure 8C shows the corresponding ^{13}C (C2') relaxation. The similarity in relaxation behavior demonstrates the dominating role of the dipolar interaction between ^{1}H and ^{13}C. For RNA molecules, uniformly labeled with ^{13}C and fully protonated, through-bond NMR experiments begin to fail at rather short helix lengths of 25 bp (50 nucleotides). In comparison, the relaxation behavior of RNA molecules is largely improved when they are deuterated in the ribose according to Scheme (2) (Figs. 8B and 8D). Figure 8B shows the strongly reduced relaxation of a ^{1}H nucleus, when the directly bound ^{13}C is replaced by a ^{12}C (and neighboring protons are deuterated). Figure 8D shows the strongly reduced relaxation of a ^{13}C nucleus, when the directly bound ^{1}H is replaced by a ^{2}H (shown for a C1'). This immediately demonstrates that using deuteration Scheme (2), HCN

[110] T. J. Tolbert and J. R. Williamson, *J. Am. Chem. Soc.* **118**, 7929 (1996).
[111] T. J. Tolbert and J. R. Williamson, *J. Am. Chem. Soc.* **119**, 12100 (1997).

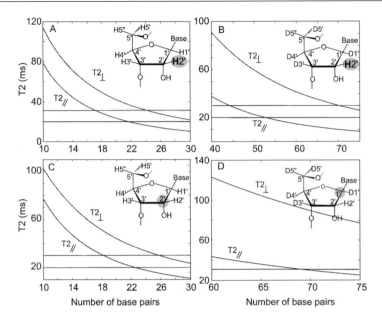

FIG. 8. Transverse relaxation rates (T_2) of different nuclei versus the number of base pairs for different labeling patterns. Two situations are considered: when the relaxation vector lies parallel ($T_{2\parallel}$) to the helix axis, and when it lies perpendicular to it ($T_{2\perp}$). The length of the molecule is calculated as 2.7 Å times the number of base pairs and the width is taken to be 22 Å. Cylindrical symmetry is assumed and the relaxation rates are calculated taking anisotropic tumbling into account via the equations given in Wijmenga and van Buuren.[4] (A) Transverse relaxation rates for an H2' proton in ^{13}C-labeled ribose. (B) The improved relaxation rates of an H2' proton in a nonlabeled and deuterated ribose. (C) Transverse relaxation rates of a ^{13}C nucleus in a ^{13}C-labeled ribose. (D) The improved relaxation rates of a ^{13}C nucleus in a ^{13}C-labeled and deuterated ribose. The nucleus considered is shaded in gray. The horizontally drawn lines indicate the limit of relaxation times for breakdown of through-bond NMR experiments, 20–30 Hz.[4,62]

type experiments are still efficient for RNAs up to 70 bp (140 nucleotides).[62] Note that the strong dipolar interaction between the single remaining proton (H2') and its directly bound ^{13}C can be removed via multiple-quantum coherences in various NMR experiments.

For large RNA molecules uniform ^{13}C/^{15}N-labeling is not sufficient to achieve complete resonance assignment. The examples of the ATP-binding RNA aptamer,[60] the antibiotic binding 16S RNA,[59] and the TYMV pseudoknot[32] show that residue-type specific labeling is necessary to obtain unambiguous assignment in crowded regions for RNAs up to a size of 50 nucleotides. A combination of residue-specific labeling and deuteration will open the way for structure elucidation by means of NMR of larger RNA molecules (>50 nucleotides).[32] Different labeling patterns can be generated, specific for each molecule under investigation to obtain sufficient distance data and angle information for structure calculation.[62]

Dipolar Couplings

All parameters traditionally used in NMR structures depend only on the local spatial relation of the interacting nuclei. Global structural features such as domain–domain orientation in proteins or the angle between arms of branched nucleic acid molecules relate to biological function. To determine global structural features, NMR parameters must depend on properties of the molecule as a whole. Dipolar interactions fulfill this requirement; if the spherical symmetry is broken by an alignment of the molecule with respect to the B_0 field, then residual dipolar couplings (which hold the global structural information) can be measured. The alignment can be induced by the magnetic field[86,89,112] or by external factors.[113–118] For a comprehensive review on the use of dipolar couplings in biomolecular NMR and their theoretical background we refer to Ref. 118a.

The abundant aromatic rings in a nucleic acid molecule have a magnetic anisotropy, which causes the molecule to align in the magnetic field. This fractional alignment increases with the squared of the magnetic field strength (B_0^2),[85,89,112] and gives small but measurable residual dipolar couplings for NMR-size nucleic acids. We will demonstrate their usefulness for the example of a DNA four-way junction.[119] This molecule can be viewed as two stacked helices that make a certain interhelix angle with each other. This interhelix angle is a global structural feature of the molecule and is therefore difficult to determine by traditional (short-range) NMR restraints. Since the magnetic anisotropy of the individual bases is known, the magnetic anisotropy of a helix can easily be calculated. The same holds for the two helices in a four-way junction. The magnetic anisotropy of the whole molecule then depends on the interhelix angle and, hence, the alignment tensor can be calculated for different interhelix angles, and dipolar couplings can be predicted. The experimental interhelix angle can then be determined by taking the best fit of theoretically calculated dipolar couplings with the experimentally observed

[112] H. C. Kung, K. Y. Wang, I. Goljer, and P. H. Bolton, *J. Magn. Reson. Series B* **109**, 323 (1995).

[113] R. D. Beger, V. M. Marathias, B. F. Volkman, and P. H. Bolton, *J. Magn. Reson.* **135**, 256 (1998).

[114] N. Tjandra and A. Bax, *Science* **278**, 1111 (1997).

[115] P. Bayer, L. Varani, and G. Varani, *J. Biomol. NMR* **14**, 149 (1999).

[116] M. R. Hansen, L. Mueller, and A. Pardi, *Nature Struct. Biol.* **5**, 1065 (1998).

[117] J. Sass, F. Cordier, A. Hoffmann, A. Cousin, J. G. Omichinski, H. Lowen, and S. Grzesiek, *J. Am. Chem. Soc.* **121**, 2047 (1999).

[118] B. W. Koenig, J. S. Hu, M. Ottiger, S. Bose, R. W. Hendler, and A. Bax, *J. Am. Chem. Soc.* **121**, 1385 (1999).

[118a] A. Bax, G. Kontaxis, and N. Tjandra, *Methods Enzymol.* **339**, [8] (2001).

[119] B. N. M. van Buuren, J. Schleucher, C. Griesinger, H. Schwalbe, and S. S. Wijmenga, to be published (2001).

dipolar couplings. From such an approach we deduced the interhelical angle of the four-way junction to be approximately 70°.[119] For future larger systems and higher magnetic fields, the magnetic field alignment increases and dipolar couplings will become more easy to measure. Alternatively, one can align molecules using interaction with a paramagnetic ligand.[113]

A second way to align molecules exploits long-range order of liquid crystalline media, such as liquid crystal detergents,[114,115] phages,[116] or purple membranes.[117,118] Because the alignment-inducing interaction between the sample and the alignment medium cannot be predicted, the size and orientation of the alignment tensor have to be determined, and the liquid crystalline medium to use has to be empirically determined. The drawback is that the assumption to be made to derive the alignment tensor only holds if the vectors for which dipolar couplings are measured sample all orientations in the molecular frame. Whether this is the case for a double-helical molecule with repetitive units such as RNA (and DNA) molecules remains to be seen. The advantage with this approach is that the degree of alignment is tunable, so that one still maintains high-resolution conditions, but the residual dipolar couplings will be higher—and therefore easier to measure—than with alignment by the magnetic field. That also this approach works nicely is demonstrated by the results of Tjandra, Bax, and co-workers on the Dickerson dodecamer.[91] Another example is the study by Bayer et al.[115] in which they refined a protein–RNA complex.

Conclusions

Strategies for structure determination of RNA molecules by means of NMR have emerged in the past decade and are now used routinely. Structures up to 50 nucleotides have been determined. However, NMR studies of RNA molecules beyond this size are still hindered by resonance overlap and line broadening. Deuteration and specific labeling approaches are likely to reduce these problems. New NMR parameters such as global structural information and dynamic studies will provide additional structural information that was not attainable before. The extended use of proton chemical shifts will simplify the assignment procedures and provides a new means of structure evaluation.

[17] Solution Nuclear Magnetic Resonance Probing of Cation Binding Sites on Nucleic Acids

By JULI FEIGON, SAMUEL E. BUTCHER, L. DAVID FINGER, and
NICHOLAS V. HUD

Introduction

The structure of DNA and RNA in the cell is modulated by monovalent and divalent cations, polyamines, and proteins.[1] Binding of cations can affect the intrinsic flexibility of the DNA and can drive the conversion of DNA from one conformation to another, e.g., B-DNA to Z-DNA,[2,3] duplex to triple-helical H-DNA,[4,5] and folding into DNA quadruplexes.[6] Both monovalent and divalent cations have been shown to play essential roles in the secondary and tertiary folding of biologically important RNAs.[7]

In this paper we review NMR methods developed and/or being used in our laboratory for localizing monovalent and multivalent cation binding sites on nucleic acids in solution. We have found that ammonium ion is an excellent probe for alkali metal ion-binding sites on nucleic acids. The use of this method is reviewed for ions in slow exchange and in fast exchange with the nucleic acid. We also present results of experiments using thallium ion as the monovalent cation. For multivalent cation binding sites, we have found that the combined use of several different cations, i.e., Mg^{2+}, Mn^{2+}, and $Co(NH_3)_3^{3+}$, is particularly useful. Applications to three specific nucleic acid structures are presented: (1) monovalent cation binding to DNA quadruplexes,[8,9] (2) monovalent cation binding sites on DNA duplexes containing A-tracts,[10] and (3) multivalent cation binding sites on the hairpin ribozyme.[11] The possible structural implications of these binding sites are also discussed.

[1] V. A. Bloomfield, D. M. Crothers, and I. Tinoco, eds., "Nucleic Acids: Structures, Properties, and Functions." University Science Books, Sausalito, CA, 2000.

[2] A. Rich, A. Nordheim, and A. H.-J. Wang, *Annu. Rev. Biochem.* **53,** 791 (1984).

[3] M. Guéron, J. P. Demaret, and M. Filoche, *Biophys. J.* **78,** 1070 (2000).

[4] S. M. Mirkin and M. D. Frank-Kamenetskii, *Ann. Rev. Biophys. Biomol. Struct.* **23,** 541 (1994).

[5] V. N. Soyfer and V. N. Potaman, "Triple-Helical Nucleic Acids." Springer-Verlag, New York, 1996.

[6] J. R. Williamson, *Ann. Rev. Biophys. Biomol. Str.* **23,** 703 (1994).

[7] D. E. Draper and V. K. Misra, *Nature Struct. Biol.* **5,** 927 (1998).

[8] N. V. Hud, P. Schultze, and J. Feigon, *J. Am. Chem. Soc.* **120,** 6403 (1998).

[9] N. V. Hud, P. Schultze, V. Sklenar, and J. Feigon, *J. Mol. Biol.* **285,** 233 (1999).

[10] N. V. Hud, V. Sklenar, and J. Feigon, *J. Mol. Biol.* **286,** 651 (1999).

[11] S. E. Butcher, F. H.-T. Allain, and J. Feigon, *Biochemistry* **39,** 2174 (2000).

Localization of Monovalent Cations on DNA and RNA

In spite of their essential role in nucleic acid folding and tertiary structure, relatively few specific cation binding sites have been identified in crystal and solution structures of nucleic acids. Monovalent cations are notoriously difficult to distinguish from water in crystallographic structures. Na^+ and H_2O have the same number of electrons; only in very high resolution crystal structures of nucleic acids can they be distinguished, usually based on valency arguments, and these have been controversial.[12–15] The same problem holds for other alkali cations and often even for divalent cations. Thallium ion is much more easily detected by crystallography,[16] and, since it is a relatively sensitive spin 1/2 nucleus, it has also been proposed to be potentially useful for studying monovalent cation binding to nucleic acids in solution.[17] In solution, binding sites of alkali ions can only be studied indirectly. Most of the NMR investigations of nucleic acid–monovalent cation interactions have until recently been restricted to relaxation measurements of the quadrupolar $^{23}Na^+$ resonance. These studies revealed that bulk monovalent cations are in fast exchange with duplex DNA, with line broadening demonstrating that there is some sequence dependence that is likely associated with secondary structure (e.g., B-form versus A-form).[18,19] The unique cation coordination associated with G-quartet formation has also been studied rather extensively by ^{23}Na NMR. Such studies provided the first solid evidence of the coordination of dehydrated cations on G-quartet formation[20,21] and provided a lower limit on the bound lifetime for Na^+ within a DNA quadruplex.[22] However, the first direct localization of coordinated Na^+ within a DNA quadruplex was provided by X-ray crystallography.[23] The proton NMR spectra of nucleic acids with specific monovalent cation binding sites can be sensitive to the species of associated cation, and this can also be used to monitor the binding and displacement of

[12] X. Q. Shui, L. McFail-Isom, G. G. Hu, and L. D. Williams, *Biochemistry* **37**, 8341 (1998).

[13] X. Q. Shui, C. C. Sines, L. McFail-Isom, D. VanDerveer, and L. D. Williams, *Biochemistry* **37**, 16877 (1998).

[14] T. K. Chiu, M. Kaczor-Grzeskowiak, and R. E. Dickerson, *J. Mol. Biol.* **292**, 589 (1999).

[15] V. Tereshko, G. Minasov, and M. Egli, *J. Am. Chem. Soc.* **121**, 3590 (1999).

[16] S. Basu, R. P. Rambo, J. Strauss-Soukup, J. H. Cate, A. R. Ferré-D'Amaré, S. A. Strobel, and J. A. Doudna, *Nature Struct. Biol.* **5**, 986 (1998).

[17] S. Basu, A. A. Szewczak, M. Cocco, and S. A. Strobel, *J. Am. Chem. Soc.* **122**, 3240 (2000).

[18] L. Nordenskiöld, D. K. Chang, C. F. Anderson, and M. T. Record, Jr., *Biochemistry* **23**, 4309 (1984).

[19] W. H. Braunlin, in "Adv. Biophys. Chem." (C. A. Bush, ed.), Vol. 5, p. 89. JAI Press Inc., Greenwich, CT, 1995.

[20] C. Detellier and P. Laszlo, *J. Am. Chem. Soc.* **102**, 1135 (1980).

[21] M. Borzo, C. Detellier, P. Laszlo, and A. Paris, *J. Am. Chem. Soc.* **102**, 1124 (1980).

[22] H. Deng and W. H. Braunlin, *J. Mol. Biol.* **255**, 476 (1996).

[23] G. Laughlan, A. I. H. Murchie, D. G. Norman, M. H. Moore, P. C. E. Moody, D. M. J. Lilley, and B. Luisi, *Science* **265**, 520 (1994).

cations. In the case of a DNA quadruplex, where K^+ was titrated into a quadruplex formed in the presence of Na^+, analysis of proton chemical shifts as a function of K^+ concentration provided thermodynamic parameters on cation binding.[24] This study was particularly significant because it provided the first direct measurement of the free energy associated with cation selectivity for a DNA quadruplex. Binding of monovalent cations as well as water to nucleic acids can also be studied by magnetic relaxation dispersion, or MRD.[24a] Nevertheless, even in the most favorable cases, none of the methods discussed above allow one to directly determine the binding site(s) of the monovalent cation on the DNA or RNA.

Ammonium Ion as Probe of Monovalent Cation Sites on Nucleic Acids

The first direct localization of a monovalent cation binding site on a nucleic acid in the solution state resulted from our titration of $^{15}NH_4Cl$ into a DNA quadruplex with Na^+ as the principle counterion (Fig. 1). This experiment was motivated by our desire for a technique capable of localizing monovalent cation binding sites in the solution state. The DNA quadruplex $[d(G_4T_4G_4)]_2$ (Oxy-1.5)[25,26] was chosen for these experiments since our previous studies had shown that the lifetime of coordinated potassium ions within this quadruplex could be long on the NMR time scale.[27,28] Based on its similar ionic radius, we hypothesized that the ammonium ion would behave in a similar fashion, allowing cation localization through the nuclear overhauser effect (NOE) between the ammonium protons and the protons of the DNA.

We have found that ammonium ion is an excellent NMR probe of monovalent cation binding sites on nucleic acids in solution. The most obvious advantage over alkali cations is that NOEs between the ammonium ion and the macromolecule can be detected, and therefore the ion binding sites can be directly determined. The ammonium ion has nearly the same van der Waals radius as K^{+29} and can substitute for K^+ in a multitude of K^+-dependent enzymes.[30,31] We have shown that both K^+ and NH_4^+ displace Na^+ coordinated in DNA quadruplexes in solution.[8,24] We have also found that NH_4^+ occupies coordination sites similar to those of Na^+, K^+, and

[24] N. V. Hud, F. W. Smith, F. A. L. Anet, and J. Feigon, *Biochemistry* **35**, 15383 (1996).
[24a] B. Halle and V. P. Denisov, *Methods Enzymol.* **338**, [7] 2001 (this volume).
[25] P. Schultze, F. W. Smith, and J. Feigon, *Structure* **2**, 221 (1994).
[26] F. W. Smith and J. Feigon, *Nature* **356**, 164 (1992).
[27] P. Schultze, N. V. Hud, F. W. Smith, and J. Feigon, *Nucleic Acids Res.* **27**, 3018 (1999).
[28] F. W. Smith and J. Feigon, *Biochemistry* **32**, 8682 (1993).
[29] A. A. Rashin and B. Honig, *J. Phys. Chem.* **89**, 5588 (1985).
[30] H. J. Evans and G. J. Sorger, *Annu. Rev. Plant Physiol.* **17**, 47 (1966).
[31] C. H. Suelter, *Science* **168**, 789 (1970).

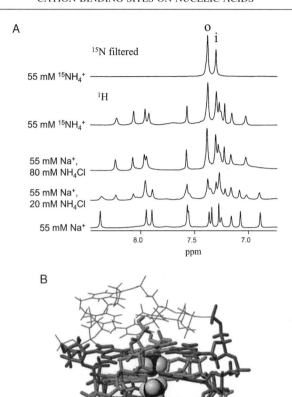

FIG. 1. (A) ^{1}H NMR spectra of 5.0 mM d(G$_4$T$_4$G$_4$) at 10° and pH 7.0 in the presence of: (bottom to top) 55 mM Na$^+$; 55 mM Na$^+$, 20 mM ^{15}NH$_4$Cl; 55 mM Na$^+$, 80 mM ^{15}NH$_4$Cl; 55 mM ^{15}NH$_4$$^+$; and 1D ^{15}N-filtered spectrum of 55 mM ^{15}NH$_4$$^+$ sample. The ammonium resonances are designated "i" (inner) and "o" (outer). (B) Structure of Oxy-1.5 with three NH$_4$$^+$ ions, one between the central two quartets (i) and two in symmetry related sites between the outer two quartets (o).

Rb$^+$ in high-resolution crystal structures of DNA quadruplexes,[32] dinucleotides,[33] and duplexes.[12,13,15] Although the tetrahedral geometry and ability to hydrogen

[32] K. Phillips, Z. Dauter, A. I. H. Murchie, D. M. J. Lilley, and B. Luisi, *J. Mol. Biol.* **273**, 171 (1997).
[33] N. C. Seeman, J. M. Rosenberg, F. L. Suddath, J. J. Kim, and A. Rich, *J. Mol. Biol.* **104**, 109 (1976).

bond make NH_4^+ arguably different from the alkali ions, it has been shown that selectivity between monovalent cations for binding is largely based on the geometry of the coordination site and the size of the cation.[34,35] In the cases we have studied, it does appear, however, that NH_4^+ is preferentially coordinated with respect to Na^+ and K^+.

Depending on the K_D of the bound ammonium ion, the ions may be in fast exchange or slow exchange with the DNA or RNA on the NMR time scale. In the case of slow exchange, the number of ions bound and their precise binding sites can be determined via the observed NOEs between the distinct bound ammonium ion proton resonances and the nucleic acid proton resonances. In the more usual case of fast exchange, ions can be localized via NOEs to the bulk ammonium ion resonance, but the occupancy of the site(s) cannot be determined. These two cases are discussed in detail below.

Sample Preparation

The DNA is synthesized chemically on a DNA synthesizer using standard protocols for 1 μM synthesis and purified by column chromatography on a Sephadex G-25 column as previously described[36,37] using only distilled H_2O for loading and elution. The DNA can also be purified by high-performance liquid chromatography (HPLC), but the simple sizing column has the advantage that the sample is eluted in H_2O only, avoiding the need to remove protonated buffers or solvents in preparing the NMR sample. Fractions containing the full-length DNA are lyophilized and stored until use. All of the experiments described below are also applicable to RNA. Sample preparation for RNA or DNA that is purified by polyacrylamide gel electrophoresis is described in the section on divalent cation binding.

For studies of DNA with Na^+ as the principal initial counterion, purified DNA samples are passed over a 1 m × 2.5 cm Sephadex G-15 column that has been rinsed extensively with a 1 mM NaCl solution followed by distilled H_2O. The DNA is eluted from the column with distilled H_2O and lyophilized. NMR samples are prepared by dissolving the DNA in H_2O (plus 5% D_2O) and adjusting the pH with NaOH. In most cases, a pH of 5.0 is optimal, as discussed below. The concentration of Na^+ in each sample can be determined by a comparison of the ^{23}Na integrated peak intensity with standard NaCl samples. In some cases the Na^+ concentration was adjusted by the addition of NaCl to a predetermined valued, e.g., 50 mM. The $^{15}NH_4^+$ is then added to the desired concentration. Samples with Li^+ as the principal counterion are prepared in a similar manner, except using LiCl and LiOH.

[34] G. Eisenman, *Biophys. J.* **2**, 259 (1962).
[35] G. Eisenman and R. Horn, *J. Memb. Biol.* **76**, 197 (1983).
[36] J. Feigon, V. Sklenar, E. Wang, D. E. Gilbert, R. F. Macaya, and P. Schultze, *Methods Enzymol.* **211**, 235 (1992).
[37] J. Feigon, K. M. Koshlap, and F. W. Smith, *Methods Enzymol.* **261**, 225 (1995).

To prepare samples with only $^{15}NH_4^+$ as counterion, the purified DNA is run on a 1 m $\times$ 2.5 cm Sephadex G-25 size-exclusion column equilibrated and eluted with 1 mM $^{15}NH_4Cl$. Peak fractions are lyophilized and then dissolved to the desired concentration for NMR studies (1–5 mM in oligonucleotide) in 5% D_2O/95% H_2O. Samples of around 2.5 μmol or greater were found to elute from the column with an associated concentration of $^{15}NH_4^+$ that was virtually an equimolar equivalent with DNA phosphates. Samples of around 1 μmol or less should be eluted with a lower concentration of $^{15}NH_4Cl$ to avoid a greater than molar equivalent of $^{15}NH_4^+$ to DNA phosphates. An excess of $^{15}NH_4^+$ can be troublesome, as indicated below, but can easily be reduced by dialysis with 500 molecular weight cutoff tubing against distilled H_2O. The pH of the sample is adjusted with LiOH. LiOH was chosen since Li^+ is not coordinated by G-quartets[38,39] and its hydrated radius is too big to bind in alkali metal ion binding sites of duplex DNA. The $^{15}NH_4^+$ concentration of a sample can be verified by comparing the integrated intensity of the bulk $^{15}NH_4^+$ resonance in an ^{15}N-filtered ^1H spectrum (pH 5.0, 273 K) with that of the same sample following the addition of a known amount of $^{15}NH_4Cl$. The concentration of Na^+ contamination in each sample can be determined by the comparison of the integrated ^{23}Na resonance with a set of NaCl standards. Na^+ contamination was found to be less than 100 μM in all samples prepared as described above.

The minimum amount of D_2O necessary to achieve a stable lock signal should be used (we used 5% D_2O). This is because the $^{15}NH_3D^+$ peak shows up as a shoulder on the $^{15}NH_4^+$ peak in both the ^{15}N and ^1H spectra (Figs. 1 and 2).

For cation localization with NMR, $^{15}NH_4^+$ has several advantages over the more common $^{14}NH_4^+$. First of all, the quadrupolar ^{14}N nucleus results in the splitting of the ammonium ^1H resonance into a triplet. Most NMR spectrometers do not have the capability to decouple the ^{14}N nucleus. The proton resonance of $^{15}NH_4^+$ is split into a doublet ($\sim$75 Hz) by the spin 1/2 ^{15}N nucleus, but this can be eliminated by decoupling on most spectrometers with a broadband amplifier. Furthermore, the spin 1/2 ^{15}N nucleus also allows the implementation of heteronuclear experiments, described below, which are very useful but impossible with the [^{14}N] ammonium.

Effect of pH

The width of the $^{15}NH_4^+$ proton resonance is very sensitive to pH, and this can be used to monitor fine adjustments in the pH of the samples. To ensure that the pH of all samples in a particular study is identical, the line widths of the $^{15}NH_4^+$ proton

[38] O. L. Acevedo, L. A. Dickinson, T. J. Macke, and C. A. Thomas, Jr., *Nucleic Acids Res.* **19**, 3409 (1991).

[39] J.-F. Chantot and W. Guschlbauer, *FEBS Lett.* **4**, 173 (1969).

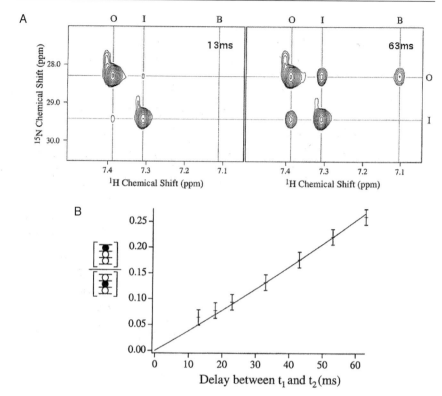

FIG. 2. (A) N_z-Ex HSQC spectra of the $^{15}NH_4^+$ form of Oxy-1.5 for the ^{15}N region containing the inner and outer NH_4^+ resonances acquired with a total mixing time of (left) 13 ms and (right) 63 ms. Spectra were acquired at 283 K for a 2.5 mM Oxy-1.5 quadruplex, 55 mM in NH_4^+ at pH 5.0 The ^1H and ^{15}N chemical shifts corresponding to the inner (I), outer (O), and bulk (B) NH_4^+ autocorrelation and cross peaks are labeled. The large autocorrelation peak from the bulk $^{15}NH_4^+$ (^{15}N chemical shift 19.8 ppm) is outside of the region shown. Note the "ears" on the cross peaks, which arise from the $^{15}NH_3D^+$ because of the 5% D_2O added for the lock signal. (B) Plot of the ratio of cross peak: autocorrelation peak volumes for the NH_4^+ that was initially in the inner coordination site but moved to an outer coordination site (7.31 and 7.38 ppm for protons, respectively; along "I" for nitrogen) as a function of mixing time (delay between t_1 and t_2) from a series of N_z-EX HSQC experiments. The line is the best fit of the data to a first-order rate equation for the movement of the inner ion to the outer coordination site.

resonances are compared in ^{15}N-filtered 1D experiments at 283 K and adjusted to the same line width by the addition of microliter quantities of LiOH or HCl (50–500 mM) in the NMR tube. At 283 K and pH 5.0, the bulk ammonium ion protons are in sufficiently slow exchange with H_2O to exhibit a resonance that is minimally affected by exchange. At pH 7.0, on the other hand, the bulk ammonium resonance is reduced to baseline.

For studies in which the ammonium ion is in fast exchange with the nucleic acid, it is essential that the pH be kept relatively low, or exchange of the ammonium ion protons with water will obliterate the signal. Low pH is less important for cations bound in sites with slow exchange, since then the bound ions are protected from exchange and can still be observed (discussed below). However, for detection of cation movement between slow exchange sites and solution to be possible, the sample will also need to be at lower pH.

NH_4^+ Ions in Slow Exchange: Application to DNA Quadruplex

We were interested in developing a general method to directly determine the number of bound monovalent cations and their precise locations on a DNA molecule in solution. With this objective, we studied the binding of ammonium ion to the quadruplex formed by d($G_4T_4G_4$) [Oxy-1.5]. This molecule dimerizes to form a symmetrical four-quartet quadruplex with T4 loops spanning the diagonal of each end quartet (Fig. 1B).

Titration Experiments

We had previously observed that the Na^+ form and the K^+ form of Oxy-1.5 exhibited distinct ^1H NMR resonances and were in slow exchange on the NMR time scale.[27,28] To assess the effect of $^{15}NH_4^+$ on the DNA quadruplex, $^{15}NH_4^+$ was added in the form of $^{15}NH_4Cl$ (98% isotopically pure, Sigma) to the Na^+ form of Oxy-1.5. One to 5 μl of $^{15}NH_4Cl$ from concentrated solutions was added to 500 μl samples of 5 mM DNA (in strand) in the NMR tube. Over the course of this titration the DNA exhibited a double set of proton resonances in 1D spectra, indicative of slow exchange between the Na^+ and $^{15}NH_4^+$ forms of the quadruplex (Fig. 1A). In addition, instead of seeing one resonance peak containing both the bulk and bound ammonium ion resonances, we found two shifted resonances with a 2 : 1 intensity ratio. These correspond to three bound ammonium ions, one between the inner two quartets and the other two in the symmetrical outer sites, as discussed below. These resonances can be easily resolved from the proton resonances of the DNA by acquiring a 1D ^{15}N-filtered spectrum. The observation of ammonium ion proton resonances that are chemically shifted from the bulk ammonium ion indicated that the ions are bound tightly to the DNA in slow exchange on the NMR time scale.

As a technical note, decoupler artifacts in the experiments described here can obscure the resonances of the bound ammonium ions and NOEs from ammonium ions. These artifacts are proportional to the intensity of the bulk ammonium ^1H resonance and, in our experience, are approximately 1/100 to 1/50 the intensity of the bulk resonance. Thus, to observe bound ions with ^{15}N decoupling, it is desirable to keep the bulk ammonium ion concentration to no more than approximately 20-fold the concentration of a bound ion. Also, the 2% $^{14}NH_4^+$ in commercially available $^{15}NH_4^+$ can result in the appearance of small $^{14}NH_4^+$ satellites around the much larger $^{15}NH_4^+$ resonance.

Effect of pH and Temperature

Both pH and temperature have a tremendous effect on the width of the bulk ammonium ion ^1H resonance. For our sample of Oxy-1.5 at pH 5.0 the bulk peak is observed to go from very sharp to baseline over the range of 267 to 333 K. The bound ions of the DNA quadruplex, on the other hand, show comparatively little change in line width over this temperature range. The line width of the bulk ammonium ion ^1H resonance at 273 K also goes from very sharp to baseline as the pH is raised from pH 5.0 to pH 7.0. Again, over this same pH range, the resonances from the bound ions show no significant change in line width. This illustrates that tightly coordinated ions can be observed over a considerable range of pH and temperature.

Localization of Bound Ammonium Ions

In the case of slow exchange of ammonium ions with their binding sites on the DNA or RNA, the ions can be precisely localized based on observed NOEs between the ammonium ion protons and protons on the nucleic acids. This was the case with Oxy-1.5, where the two shifted $^{15}NH_4$ resonances gave rise to NOE cross peaks to the guanine imino protons.[10] The peak labeled (i) in Fig. 1A arises from the ion located in the center of the quadruplex, whereas (o) arises from the two symmetrically located ions in the outer two quartets. This explains the two-to-one ratio of their intensities.

A ROESY rather than a NOESY experiment is used in order to unambiguously distinguish between cross peaks arising from NOE transfers and those from pure chemical exchange. The positive cross peaks in a ROESY spectrum arise from chemical exchange, whereas the negative cross peaks are from NOE transfers. Strong exchange cross peaks are observed between the amino protons of the outer G-quartets and water. Less intense exchange cross peaks are also observed between these amino protons and the bulk ammonium ions. All other cross peaks are negative and thus indicative of NOE transfers. However, outer G-quartet imino resonance (negative) cross peaks with water and the bulk ammonium ions are apparently exchange-relayed. That is, these cross peaks actually result from NOE transfers between amino and imino protons on the same guanine, where the amino proton had exchanged with a proton that had been a water or a bulk ammonium ion proton during the chemical shift evolution step (t_1) of the pulse sequence. In the case of Oxy-1.5 it is possible to assign these cross peaks as exchange-relayed NOEs based on the amino exchange cross peaks and because direct NOEs between G-imino protons and bulk ions would be inconsistent with other NOE data.[27] However, in many cases such cross peaks may not be obvious as being exchange-relayed and could lead to erroneous conclusions concerning ammonium ion coordination. Thus, additional precautions must be taken whenever possible to distinguish between true NOEs and exchange-relayed NOEs (see below).

Monitoring Exchange of Ions between Different Sites

In the cases of more than one ammonium ion bound in slow exchange with the DNA or RNA, exchange of these ions between sites and/or to the bulk solution can be observed in ROESY or NOESY spectra, where cross peaks between these sites will be present if such exchange occurs during the mixing time of the experiment. The movement of these ions from one site to another or to the bulk solution can be monitored using N_z-exchange HSQC experiments acquired with several different mixing times.[10] In the case of Oxy-1.5, three autocorrelation peaks are observed in the standard $^1H-^{15}N$ HSQC experiment, corresponding to the inner ammonium ion, the outer ammonium ion, and the bulk ammonium ions. Cross peaks between the inner and outer and between the outer and bulk autocorrelation peaks appear with increasing intensities as the mixing time in the N_z-exchange HSQC experiment is increased (Fig. 2A). These cross peaks can be quantitatively analyzed to determine the bound lifetime of the ion. The ratio of ions that had moved from the inner site to the outer site was calculated for each mixing time (delay between t_1 and t_2 in milliseconds) by calculating the relative integrated intensities of the cross peaks and autocorrelation peaks. These values were plotted vs the mixing time and the data were fitted with a first-order rate equation. This gave an estimated bound lifetime of 250 ms for the inner NH_4^+ (Fig. 2B).

NH_4^+ as Probe of Monovalent Cations in Fast Exchange with Nucleic Acids: Application to DNA A-Tracts

Experimental Methods and Observations

In the case of ammonium ions in fast exchange with binding sites on DNA or RNA, only a single ammonium resonance will be observed and it will have a chemical shift essentially identical to that of the bulk ammonium ions. In this situation, for exchangeable protons such as iminos it is essential to compare NOE and ROE (or NOESY and ROESY) spectra to distinguish between exchange and NOEs. In some cases, both events may be occurring. This can be determined by comparing the relative intensities of the NOE and ROE.

As noted above, the ROE and NOE experiments do not distinguish between exchange-relayed NOEs and direct NOEs, as these pathways produce cross peaks with the same sign with respect to the diagonal. In the case of ammonium ion localization, exchange-relayed NOEs with water as the intermediate present a distinct potential pathway because ammonium ion protons exchange with water under most conditions and the surface of nucleic acids is well hydrated in solution. However, the potential problem of water-relayed ammonium ion-DNA NOEs is easily circumvented by simply irradiating the H_2O resonance during the mixing time of the NOESY experiment,[10] a variation called the MINSY (mixing irradiation during a NOESY) experiment, first proposed by Massefski and Redfield.[40]

Specific NOEs can often most easily be observed using 1D ^{15}N-filtered NOESY experiments rather than two-dimensional NOESY experiments (Fig. 3A). This is because there will usually be only a few protons on the DNA or RNA that give rise to NOEs with the ammonium ion. Therefore, if it is possible to assign the resonance (s) in the 1D spectrum, this is the most efficient way to collect the data since a large number of acquisitions can be acquired for good signal-to-noise. NOEs from ammonium ions in fast exchange are typically very small in comparison to the bulk ammonium ion resonance, and we have found that decoupling the ^{15}N can give rise to artifacts that obscure these NOE peaks. Thus, best results are usually obtained by acquiring the spectra without decoupling. For the A-tract DNAs discussed below, we typically acquired 12K scans with a 60 ms mixing time on samples of 2 mM DNA.

Ammonium Ion Binding to DNA A-Tracts

We investigated the binding of NH_4^+ to the DNA duplexes [d(GCA$_4$T$_4$GC)]$_2$ (A4T4) and [d(CGT$_4$A$_4$CG)]$_2$ (T4A4) as well as d(GCA$_5$CG)·d(CGT$_5$GC).[10] In contrast to the case with Oxy-1.5, only one ^{1}H resonance is observed for the [15]NH$_4^+$ at the chemical shift and line width of [15]NH$_4^+$ in the absence of DNA, indicating that any bound ions are in fast exchange with the DNA. ^{15}N-Decoupled NOESY spectra and 1D ^{15}N-filtered NOESY revealed the presence of specific NOEs between the [15]NH$_4^+$ and the DNA.[10] These NOEs indicated that there were specific [15]NH$_4^+$ binding sites in the minor groove of the A-tracts of these DNAs. Furthermore,

[40] W. Massefski and A. G. Redfield, *J. Magn. Reson.* **78**, 150 (1988).

FIG. 3. (A) One-dimensional (a,c) ^{1}H and (b,d) ^{15}N-filtered NOESY spectra of (a,b) [d(GCA$_4$T$_4$GC)]$_2$ and (c,d) [d(GCT$_4$A$_4$GC)]$_2$. DNA samples are 2.0 mM in duplex, 60 mM LiCl, 4.0 mM [15]NH$_4$Cl, pH 5.5 at 273 K in 95% H_2O/5% D_2O. Spectra were collected with 4K complex points, spectral width of 10,000 Hz, and 128 and 12K scans for the ^{1}H and 1D ^{15}N-filtered NOESY ($\tau_m = 60$ ms) spectra, respectively. The H_2O peak was suppressed using a 1$\bar{1}$ echo pulse sequence.[70] The 1D ^{15}N-filtered NOESY spectra were baseline corrected using a Lorenzian functions to fit the [15]NH$_4^+$ doublet. The * resonance near 8.2 ppm is the result of a proton exchange with the amino group of the terminal C residues. (B) Schematic showing the location of NH_4^+ binding sites in the minor groove of [d(GCA$_4$T$_4$GC)]$_2$ (*top*) and [d(GCT$_4$A$_4$GC)]$_2$ (*bottom*) deduced from the observed NOEs to the ammonium ion proton resonance. These sites correlate with the relative minor groove width deduced from other studies (see Hud *et al.*[10] for references).

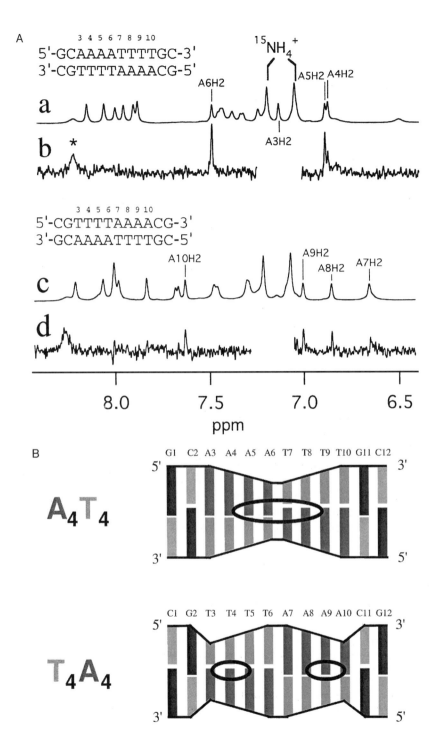

ammonium ion NOEs to the different sequences revealed the sequence-specific localization of the monovalent cations in the minor groove. We found that the binding of NH_4^+ to the sequence A4T4 is dramatically different from that of T4A4, as illustrated in Fig. 3B. For $[d(CGA_4T_4CG)]_2$, NH_4^+ binding is observed in the minor groove between A4 and T9 while for $[d(CGT_4A_4CG)]_2$ binding is observed at the two sites between A8 and A10. No NOEs were observed to the major groove or at any other sites, including to the molecule $[d(GCA_3CGT_3GC)]_2$, thus supporting the notion that binding of monovalent cations is specific to the minor groove of A-tract DNA. The cation distribution correlates very well with the minor groove width, with the greatest concentration of cations at the narrowest parts of the grooves (Fig. 3B). Based on the observed localization of NH_4^+ for different DNA sequences, we proposed that specific localization of monovalent cations may play an integral role in the observed sequence specific bending of A-tract DNAs.[10]

Thallium-205 NMR

The thallous(I) ion, like ammonium, is known to substitute for potassium in potassium dependent enzymes,[41,42] ribosome stabilization,[43] muscle contraction,[44] and, more recently, in G-quartet formation.[17] These observations are not surprising since Tl^+ is chemically similar to the alkali metals and is similar to K^+ in size and coordination geometry characteristics.[16] However, unlike the alkali metals, [205]Tl is a NMR active nuclide (spin $1/2$) with a high natural abundance of 70% and relative sensitivity to protons of 0.13. [205]Tl also exhibits a large chemical shift range of 3000 ppm due to the extreme sensitivity of the nuclide to the chemical nature of its ligands and coordination geometry.[45] Therefore, [205]Tl NMR of thallous(I) ion can be used to obtain relative binding constants of monovalent cations and information regarding the ion's chemical environment.[46] With these desirable properties, [205]Tl NMR has been successfully used to probe potassium ion binding sites in proteins such as pyruvate kinase,[47] S-adenosylmethionine synthetase,[48] protein C,[49] and gramicidin A.[50]

[41] F. J. Kayne, *Arch. Biochem. Biophys.* **143**, 232 (1971).

[42] J. S. Britten and M. Blank, *Biochim. Biophys. Acta* **159**, 160 (1968).

[43] T. Hultin and P. H. Näslund, *Chem.–Biol. Interact.* **8**, 315 (1974).

[44] I. Rusznyak, L. György, S. Ormai, and T. Millner, *Experientia* **24**, 809 (1968).

[45] J. J. Dechter and J. I. Zink, *Inorg. Chem.* **15**, 1690 (1976).

[46] J. F. Hinton, K. R. Metz, and R. W. Briggs, *in* "Annual Reports on NMR Spectroscopy" (G. A. Webb, ed.), Vol. 13A, p. 211. Academic Press, London, 1983.

[47] J. Reuben and F. J. Kayne, *J. Biol. Chem.* **246**, 6227 (1971).

[48] G. D. Markham, *J. Biol. Chem.* **261**, 1507 (1986).

[49] K. A. Hill, S. A. Steiner, and F. J. Castellino, *J. Biol. Chem.* **262**, 7098 (1987).

[50] J. F. Hinton, W. L. Whaley, D. Shungu, R. E. d. Koeppe, and F. S. Millett, *Biophys. J.* **50**, 539 (1986).

It is important to be aware that TlNO$_3$ is very toxic (Tl is commonly the active ingredient in rat poison) and therefore care must be taken to avoid exposure. TlCl has low solubility, so samples must be prepared without Cl$^-$ as an anion. Since TlNO$_3$ is quite soluble, this salt was used as the source of Tl$^+$ in our studies, and the NO$_3^-$ forms of other salts (e.g., Na$^+$) were also used over the course of sample preparation.

Tl$^+$ Binding to Oxy-1.5

The Strobel laboratory has proposed the use of ^{205}Tl NMR to study monovalent cation–nucleic acid interactions.[17] This initial study showed that Tl$^+$ could support G-quadruplex formation. In addition, using ^{205}Tl NMR, four Tl$^+$ signals were detected that originate from bulk ions and three ions in chemically distinct binding sites in the ion channel of d(TTGGGGTT). We have also conducted preliminary studies of Tl$^+$–DNA quadruplex interaction using a combination of ^{1}H, ^{23}Na, and ^{205}Tl NMR using Oxy-1.5. For these studies an Oxy-1.5 sample was originally prepared in 50 mM NaCl and then dialyzed against cesium nitrate followed by dialysis against distilled water. Dialysis against cesium nitrate allows for the removal of excess sodium ions, leaving only those that are necessary to maintain the structure of the quartets. A titration of TlNO$_3$ into this sample of Oxy-1.5 is shown in Fig. 4A. On three additions of TlNO$_3$ to a final concentration of 6 mM, we observed changes in the 1D ^{1}H similar to changes observed when potassium ion is titrated into the sodium form of Oxy-1.5.[27,28] The differences observed between the K$^+$ and Na$^+$ forms are known to be due to changes in the structure of the T4 loop.[27] To confirm that the changes observed in the 1D spectra were due to changes in the T4 loop, NOESY spectra (75 and 200 ms) were taken and showed that the NOE patterns for the Tl$^+$ form were similar to those obtained for the K$^+$ form of Oxy-1.5 (data not shown). To confirm that the changes observed in the 1D ^{1}H spectrum were due to displacement of the sodium ions in the ion channel and not due to binding of grooves and phosphates externally, ^{23}Na NMR spectra were taken concomitant with the 1D ^{1}H spectra. The ^{23}Na spectra show that the sodium ion resonance sharpens when the Tl$^+$ is added (Fig. 4B). This is presumably due to the increased population of free Na$^+$ as it is displaced from the ion channel of Oxy-1.5 by Tl$^+$. These results are consistent with similar studies of Na$^+$ displacement by K$^+$ from Oxy-1.5.[22] Taken together, we conclude that the Tl$^+$ does enter the ion channel of Oxy-1.5 and displaces sodium ions. In addition, since the changes observed in the T4 loop structure are similar to those observed with the addition of potassium ion, we conclude that the Tl$^+$, like the ammonium ion,[9] behaves more like potassium ions than sodium ions in Oxy-1.5.

We were also able to observe the ^{205}Tl directly using the NMR equipment already in the laboratory. Most commercially available broadband probes are incapable of detecting the ^{205}Tl signal (288 MHz on a 500 MHz spectrometer) since

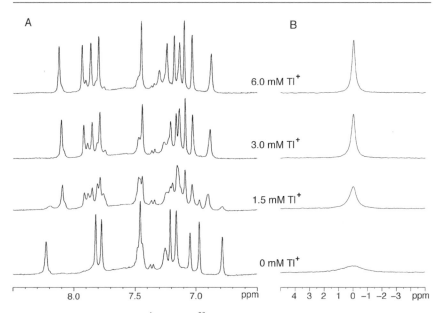

FIG. 4. One-dimensional (A) ^{1}H and (B) ^{23}Na spectra of Oxy-1.5 as a function of added TlNO$_3$. The aromatic region of the proton spectrum is shown. Sample conditions are 3.0 mM DNA strand, pH 6.0 at 25° with, 0, 1.5, 3.0, or 6.0 mM Tl$^+$. The sample was originally prepared in 50 mM NaCl, but was exhaustively dialyzed versus CsNO$_3$ followed by distilled H$_2$O in order to remove excess Na$^+$ and Cl$^-$. ^{1}H and ^{23}Na spectra were acquired at 500 MHz and 132.3 MHz, respectively, on a Bruker broadband probe. ^{23}Na spectra were acquired with 256 scans, 8K complex points, and a 50 ppm spectral width.

it lies well above the upper limit of the detectable frequency range. In addition, commercial amplifiers are incapable of generating radio frequencies corresponding to the ^{205}Tl resonance. To overcome these limitations, we inserted a 500 MHz broadband probe into a 360 MHz magnet. At the field strength of a 360 MHz magnet, the ^{205}Tl signal resonates at approximately the frequency of ^{31}P in a 500 MHz magnet (202 MHz). Thus, using a 500 MHz broadband probe in a 360 MHz magnet and a 500 MHz RF amplifier pulsing near the frequency of ^{31}P, we could detect the ^{205}Tl signal at varying concentrations of TlNO$_3$. Once we confirmed that we were able to detect ^{205}TlNO$_3$, we obtained an NMR spectrum on a sample of Oxy-1.5 with Tl$^+$ as counterion. Our results were similar to those reported by Basu et al.[17]

Tl$^+$ Binding to Duplex DNA

To investigate the utility of ^{205}Tl NMR in nucleic acid systems other than G-quartets, we monitored the interaction of the Tl$^+$ with A-tract containing DNA. ^{205}Tl NMR spectra of the DNA duplex d(GCAAAICTTTGC) at varying

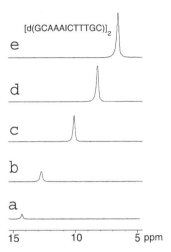

[d(GCAAAICTTTGC)]₂

e

d

c

b

a

15   10   5 ppm

FIG. 5. One-dimensional ^{205}Tl spectra of [d(GCAAAICTTTGC)]₂ as a function of added TlNO₃: (a) 5 mM, (b) 25 mM, (c) 50 mM, (d) 75 mM, and (e) 100 mM TlNO₃. Sample conditions are 2 mM strand, 25 mM NaNO₃, 10 mM Bis–Tris–propane–NO₃, pH 6.7. The spectra were acquired with 256 scans, 8 K complex points, and a 50 ppm spectral width, and processed with a line broadening of 1 Hz. The spectra were referenced to a 100 mM TlNO₃ sample.

concentrations of TlNO₃ (Fig. 5) show only one signal, suggesting that the ions are in fast exchange with the DNA. However, the Tl^+ signal is shifted by 14 ppm due to the presence of the duplex DNA. As the Tl^+ concentration increases, the signal shifts farther upfield toward the frequency of free Tl^+ because of a decreasing mole fraction of Tl^+ interacting with DNA. This sensitivity of Tl^+ to the presence of DNA should allow studies of cation interaction with DNA and RNA molecules similar to those that have been achieved by $^{23}Na^+$ NMR, but with greater sensitivity. For example, detecting the effect of changing a single base pair in a DNA duplex on monovalent cation binding may be possible with ^{205}Tl NMR. In addition, line shape analysis similar to studies conducted on protein–Tl^+ interaction[47] and ^{23}Na relaxation studies on A-tract DNA[51] could be used to estimate the occupancy and binding constants of monovalent cations to different DNAs.

Localizing Multivalent Metal Ions on Nucleic Acids

The interaction of divalent metal ions with nucleic acids can have profound effects on both structure and thermodynamic stability. Many RNAs and DNAs recruit divalent metal ions for electrostatic shielding and folding, particularly those

[51] V. P. Denisov and B. Halle, *Proc. Natl. Acad. Sci. U.S.A.* **97,** 629 (2000).

that must adopt compact tertiary structures to function.[37,52] Additionally, it has been elegantly demonstrated that divalent ions participate directly in the active site chemistry of some ribozymes.[53-55] Given the significant role that divalent metal ions play in nucleic acid structure and function, an important early step for most nucleic acid NMR investigations should be to ascertain whether the observed conformation is influenced by millimolar amounts of magnesium.

Sample Purification

It is critical for the NMR sample to be free of EDTA when analyzing divalent metal ion binding. If standard polyacryalamide gel purification is used, the sample will contain significant amounts of EDTA after elution from the gel. There are two methods that are commonly used to further purify nucleic acids after gel purification. The first method involves extensive dialysis. Although this method does remove EDTA, it does not remove acrylamide polymers. All traces of EDTA and acrylamide can be removed by purifying the sample on a small (4 × 1 cm) DEAE-Sepharose column. The sample is loaded in a buffer containing 20 mM Tris pH 7.4 and 150 mM NaCl, washed with five column volumes of the same buffer, and eluted with 20 mM Tris pH 7.4 and 1.5 M NaCl. The sample is then ethanol precipitated by adding 2 volumes of ethanol followed by centrifugation. A final desalting step is then performed with a 50 × 2 cm Sephadex G-15 desalting column. The sample is then lyophilized to dryness and can be resuspended in water or the buffer of choice and brought to the desired pH.

Chemical Shift Mapping

Divalent ion dependent structure can be examined rather quickly by titrating magnesium into an NMR sample and monitoring the 1D imino proton spectra as a function of magnesium concentration. Once purified, a 0.5 mM RNA (or DNA) sample is titrated with magnesium chloride in 1–2 mM increments, and 1D spectra of the imino proton resonances are acquired using the 1$\bar{1}$ echo pulse sequence[56] for water suppression. The magnesium is added in microliter amounts so as to minimize sample dilution. It has been reported that the ratio of Mg^{2+} : RNA is more important than the final metal ion concentration, and in some cases a ratio as high as 80 : 1 may be required to observe maximal folding of magnesium-dependent conformations.[57] One-dimensional imino proton spectra may fail to

[52] A. L. Feig and O. C. Uhlenbeck, *in* "The RNA World" (R. F. Gesteland, T. R. Cech, and J. F. Atkins, eds.), p. 287. Cold Spring Harbor Laboratory Press, Cold Spring Harbor, NY, 1999.

[53] J. A. Piccirilli, J. S. Vyle, M. H. Caruthers, and T. R. Cech, *Nature* **361,** 85 (1993).

[54] D. E. Ruffner and O. C. Uhlenbeck, *Nucleic Acids Res.* **18,** 6025 (1990).

[55] H. van Tol, J. M. Buzayan, P. A. Feldstein, F. Eckstein, and G. Bruening, *Nucleic Acids Res.* **18,** 1971 (1990).

[56] A. Bax, V. Sklenar, G. M. Clore, and A. M. Gronenborn, *J. Am. Chem. Soc.* **109,** 6511 (1987).

[57] M. Wu and I. Tinoco, *Proc. Natl. Acad. Sci. U.S.A.* **95,** 11555 (1998).

indicate magnesium dependent structure formation or metal ion binding, if the region of interest is adenine- and cytosine-rich. Therefore, nonexchangeable proton resonances must also be monitored by 2D TOCSY, NOESY, or ^1H–^{13}C HSQC spectra if the sample is labeled. Magnesium is not always required for nucleic acid structure formation, and its presence will decrease the half-life of RNA NMR samples because of nonspecific magnesium-catalyzed hydrolysis. However, we find that many RNA samples containing magnesium are stable for months if stored at 4° at pH 5–7.

The location of a bound metal ion and/or the regions of the RNA or DNA that are conformationally affected by the ion interaction can be determined by chemical shift mapping experiments. Changes in chemical shift (Δ ppm) monitored by ^1H–^{13}C HSQC spectra are calculated from the equation Δ ppm $= [(\Delta^1H$ ppm$)^2+(\Delta^{13}C$ppm$^*\alpha C)^2]^{0.5}$,[58] where Δ ppm is the difference in ppm between the chemical shifts of the RNA in the absence and presence of metal ions and αC is a scaling factor to normalize the magnitude of the carbon chemical shift changes relative to the proton scale. αC is calculated by dividing the proton chemical shift range by the carbon chemical shift range. Observed Δ ppm values of 0.1–0.2 have thus far been associated with metal binding events that do not change the RNA conformation,[11,59] whereas greater Δ ppm values may indicate RNA conformational change. RNA conformational change (or absence thereof) is confirmed by monitoring the NOE cross-peak patterns in 2D NOESY spectra. No change in NOE cross-peak patterns and small Δ ppm values are consistent with divalent ion binding to preformed sites.

Paramagnetic Effect

Manganese is a very useful probe for localizing divalent metal ion binding sites by NMR, because it has an ionic radius that is very similar to that of magnesium and is paramagnetic, causing the rapid relaxation of nuclear spins leading to line broadening with an r^{-6} distance dependence.[60] Thus far, all manganese binding has been observed within the fast exchange time regime, with off-rates of 1000 s^{-1} or greater.[11,61–64] Therefore, only micromolar amounts of manganese are required to completely and specifically broaden resonances when the RNA concentration is millimolar. We find that 30 μM MnCl$_2$ is sufficient to broaden resonances

[58] B. T. Farmer, K. Constantine, V. Goldfarb, M. S. Friedrichs, M. Wittekind, J. Yanchunas, J. G. Robertson, and L. Mueller, *Nat. Struct. Biol.* **3**, 995 (1996).

[59] P. Legault, C. G. Hoogstraten, E. Metlitzky, and A. Pardi, *J. Mol. Biol.* **284**, 325 (1998).

[60] I. Bertini and C. Luchinat, "NMR of Paramagnetic Molecules in Biological Systems." Benjamin/Cummings, Menlo Park, CA, 1986.

[61] G. Ott, L. Arnold, and S. Limmer, *Nucleic Acids Res.* **21**, 5859 (1993).

[62] F. H.-T. Allain and G. Varani, *Nucleic Acids Res.* **23**, 341 (1995).

[63] M. R. Hansen, J. P. Simorre, P. Hanson, V. Mokler, L. Bellon, L. Beigelman, and A. Pardi, *RNA* **5**, 1099 (1999).

[64] G. R. Zimmermann, C. L. Wick, T. P. Shields, R. D. Jenison, and A. Pardi, *RNA* **6**, 659 (2000).

surrounding a metal binding site to baseline when the sample concentration is 0.5–1 mM. Broadening of exchangeable resonances is monitored by 1D spectra using 11 echo water suppression, while ^{1}H–^{13}C HSQC spectra are used to monitor the line broadening of the aromatic and sugar resonances.

Cobalt Hexammine

Cobalt hexammine is similar in geometry to hexahydrated magnesium and manganese, but has an outer ligand shell of amino groups that preclude inner sphere coordination to nucleic acids. Because cobalt hexammine is trivalent, it binds approximately 10 times more tightly than magnesium or manganese to nucleic acid metal binding sites.[65] Cobalt hexammine binding can be detected directly via the NOE or by chemical shift mapping or both. These methods are reviewed in the chapter by Tinoco,[65a] and will therefore not be reviewed in this chapter.

Applications to B Domain of Hairpin Ribozyme

The hairpin ribozyme is a small catalytic RNA composed of two domains, A and B, which interact in the transition state.[66] A 38-nucleotide construct of the B domain of the hairpin ribozyme[67] interacts with at least four divalent metal ions.[11] Analysis of 2D NOESY spectra in the presence and absence of magnesium and cobalt hexammine indicated that the RNA conformation is not changed by the presence of the metal ions. Chemical shift mapping indicated that cobalt hexammine and magnesium interact with the same residues. On addition of manganese, paramagnetic line broadening was observed at the same residues that show chemical shift changes upon the addition of magnesium or cobalt hexammine (Fig. 6). Magnesium causes chemical shift changes at G2, G7, G13, G34, and G25 (Fig. 6B). Manganese specifically broadens the same residues to baseline, except for G13, which is only partially broadened (Fig. 6C). Neither ion affects G4, G15, or G23 (Fig. 6B,C), which are residues that do not participate in metal ion interactions. Taken together, these data indicate that the metal ions localize to the same preformed binding sites on the RNA, probably through outer sphere water contacts (or amino groups in the case of cobalt hexammine). Analysis of the paramagnetic line broadening data using 2D homonuclear NOESY and ^{1}H–^{13}C HSQC spectra indicated that there were two major binding sites, each of which showed 10–20 atoms that were close to the manganese ion.

The location of the ions was modeled by using the r^{-6} distance dependence of the paramagnetic effect. Cobalt hexammine also showed NOEs to these sites;

[65] R. L. Gonzalez and I. J. Tinoco, *J. Mol. Biol.* **289**, 1267 (1999).
[65a] R. L. Gonzalez, Jr. and I. J. Tinoco, *Methods Enzymol.* **338**, [18] (2001) (this volume).
[66] A. Hampel and R. Tritz, *Biochemistry* **28**, 4929 (1989).
[67] S. E. Butcher, F. H.-T. Allain, and J. Feigon, *Nature Struct. Biol.* **6**, 212 (1999).

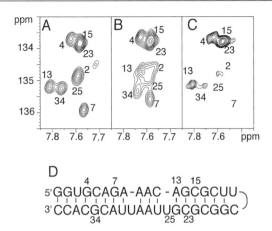

FIG. 6. (A) Portion of a ^{1}H–^{13}C HSQC spectrum of the B domain RNA of the hairpin ribozyme. Sample conditions are 0.5 mM RNA, 50 mM NaCl, pH 6.0. (B) Same as in (A), except after the addition of 6 mM MgCl$_2$. (C) Same as in (A), except after the addition of 30 μM MnCl$_2$. (D) Sequence of the 38 nucleotide B domain RNA construct.

however, we found that more restraints could be obtained using the paramagnetic line broadening data, and sample aggregation in the presence of millimolar concentrations of cobalt hexammine was a problem for the B domain RNA. Here we will review the methodology for localizing hexahydrated manganese ions in structure calculations using X-PLOR (3.1). The ion binding sites were calculated after the nucleic acid structure had been determined. Hexahydrated manganese ions are included in the calculations, from coordinates obtained in high-resolution crystal structures (1D49).[68]

Paramagnetic line broadening effects due to manganese interactions are included as weak distance restraints (<7 Å) for atoms that display complete line broadening at 20–30 μM MnCl$_2$. We find that 10 or more of these distances are sufficient to precisely position the ion. An equivalent number of repulsive distances (>7 Å) for nonbroadened atoms can also be included in the calculations. Simulated annealing is done with X-PLOR (3.1).[69] A molecular dynamics step is initially done at 1000 K for 105 ps (15,000 steps of 0.0007 ps) in the absence of van der Waals forces, to allow the ion to pass freely through the RNA. This is followed by molecular dynamics and simulated annealing in the presence of van der Waals forces. For this round of calculations, an initial temperature of 2000 K is used with cooling to 100 K in 28 ps (40,000 steps with a 0.0007 ps time

[68] J. R. Quintana, K. Grzeskowiak, K. Yanagi, and R. E. Dickerson, *J. Mol. Biol.* **225,** 379 (1992).
[69] A. T. Brünger, "X-PLOR (Version 3.1) Manual." Yale University Press, New Haven, CT, 1992.
[70] V. Sklenar and A. Bax, *J. Magn. Reson.* **75,** 378 (1987).

FIG. 7. Superposition of the five lowest energy structures of the B domain, including two bound hexahydrated manganese ions. Only nucleotides 1–7 and 31–38 are shown. Overall heavy atom pairwise rmsd for this region of the ensemble is 1.2 Å. The hexahydrated manganese ions are shown as thick black lines.

step), followed by 500 steps of energy minimization using the Powell algorithm. The results from a typical calculation, including two hexahydrated manganese ions and the 38 nucleotide loop B domain from the hairpin ribozyme, are shown in Fig. 7.

Acknowledgments

The authors thank Dr. Jane Strouse for help in setting up the thallium NMR experiments. The work described here was supported by NIH grants GM48123 and GM37254 and NSF grant MCB-980872 to J.F.

[18] Identification and Characterization of Metal Ion Binding Sites in RNA

By RUBEN L. GONZALEZ, JR. and IGNACIO TINOCO, JR.

Introduction

Metal ions are an integral component of RNA molecules. The highly complex tertiary structures adopted by large, biologically active RNA molecules necessarily involve the close packing of negatively charged backbone phosphate groups. This leads to formation of negatively charged, geometrically unique pockets ideal for the binding of positively charged metal ion complexes. Nuclear magnetic resonance (NMR) methods for studying the interaction of metal ions with RNA in solution have been available for some time. However, until recently, the structural characterization of metal ion binding sites in RNA has been restricted to crystallographic studies of RNA.[1-4] More recently, structural and thermodynamic studies of metal ion binding sites in RNA using solution state NMR have appeared. Here we review currently utilized NMR methods to study metal ion binding to RNA; we emphasize $Mg(H_2O)_6^{2+}$ metal ion binding sites characterized by intermolecular NOEs from $Co(NH_3)_6^{3+}$ protons to RNA protons. These studies include an RNA molecule with tandem imino hydrogen bonded G · A mismatches as well as a GAAA tetraloop[5] derived from the P5 junction region of the *Tetrahymena thermophila* group I intron (Fig. 1A); two RNA molecules containing different arrangements of tandem G · U wobble base pairs[6,7] derived from the P5b and P5 helices of the *T. thermophila* group I intron (Figs. 1B and 1C); and the VPK viral RNA pseudoknot involved in −1 frameshifting in mouse mammary tumor virus (Fig. 2).[8] NMR methods that have been successfully applied to metal ion binding in proteins and could be effectively applied to RNA–metal ion studies are also discussed.

Sample Preparation

One of the most important steps in a successful NMR study is sample preparation. RNA samples for metal ion studies are prepared as previously described,[5-8]

---

[1] S. R. Holbrook, J. L. Sussman, R. W. Warrant, G. M. Church, and S. H. Kim, *Nucl. Acids Res.* **8,** 2811 (1971).
[2] W. G. Scott, J. T. Finch, and A. Klug, *Cell* **81,** 991 (1995).
[3] J. H. Cate and J. A. Doudna, *Structure* **4,** 1221 (1996).
[4] C. C. Correl, B. Freeborn, P. B. Moore, and T. A. Steitz, *Cell* **91,** 705 (1997).
[5] S. Rüdisser and I. Tinoco, Jr., *J. Mol. Biol.* **295,** 1211 (2000).
[6] J. S. Kieft and I. Tinoco, Jr., *Structure* **5,** 713 (1997).

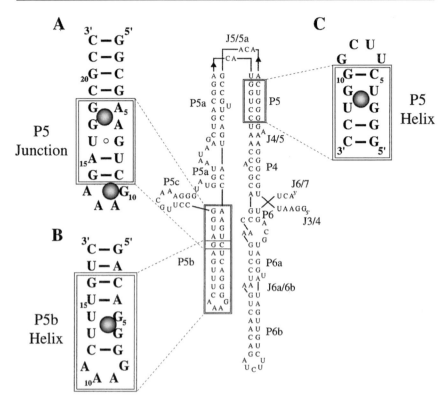

FIG. 1. Divalent metal ion binding sites characterized by NMR studies and derived from the P5 junction (A), P5b helix (B), and P5 helix (C) regions of the *Tetrahymena thermophila* group I intron P4-P6 domain. These sites include tandem imino–hydrogen bonded G · A mismatched base pairs (A), a GAAA tetraloop (A), and two sets of tandem G · U wobble base pairs (B) and (C). The gray spheres indicate the approximate positions of the metal ion in each binding site. The detailed structures of the binding sites at the two sets of tandem G · U base pairs and at the GAAA tetraloop were solved using intermolecular NOEs from the protons of $Co(NH_3)_6^{3+}$ to RNA protons. In the P5b helix binding site (B), the metal ion binds in the major groove of the helix adjacent to the guanines in the G · U base pairs (G5 and G6) and forms hydrogen bonds with the N7 and O6 of these guanines. The metal ion in the P5 helix (C) also binds in the major groove at the center of the two G · U base pairs and can form hydrogen bonds with the O6 and N7 of the guanines in the G · U base pairs (G3 and G11) and the guanines in the G · C base pairs above and below the G · U base pairs (G2 and G10), as well as the O4 of the uracils in the G · U base pairs (U4 and U12). Both structures are in excellent agreement with metal ion binding sites observed in the P4-P6 crystal structure (see Ref. 3). The metal ion at the GAAA tetraloop had not been previously observed in crystal structures and binds in the major groove with electrostatic and hydrogen bonding interactions with the nonbridging phosphate oxygens of A12 and the N7 of G10. The binding site at the tandem G · A base pairs was identified using chemical shift changes and paramagnetic line broadening, but no intermolecular NOEs were observed and therefore the detailed structure of the binding site could not be solved.

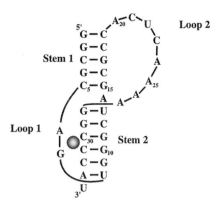

FIG. 2. Divalent metal ion binding site in the VPK mRNA pseudoknot derived from the -1 frameshifter pseudoknot in mouse mammary tumor virus. The gray sphere indicates the approximate position of the metal ion in the binding site. The detailed structure of the metal ion binding site was solved using intermolecular NOEs from the $Co(NH_3)_6^{3+}$ amino protons to RNA protons. The metal ion binds in the major groove of stem 2 stabilizing the tight turn between G7 and U8 and mediating the close packing of the two nucleotides of loop 1 against the major groove of stem 2. Within the family of structures, hydrogen bonds are possible to backbone nonbridging phosphate oxygens, sugar $2'$-hydroxyl oxygens, and base nitrogens and oxygens in loop 1 and stem 2.

with the following special considerations. RNA concentrations[7] are generally 500 μM for one-dimensional titration experiments and 2 mM for structural studies by two-dimensional NMR experiments. Volumes for NMR samples are typically 400 μl in 5 mm Aldrich NMR tubes or 250 μl in 5 mm Shigemi tubes. After in vitro transcription and purification,[9,10] RNA pellets are dissolved in a 5 mM EDTA solution. In order to remove trace metal ions, the RNA solution is dialyzed for 24 hr at 4° against 5 mM EDTA followed by dialysis at 4° against a 10 mM sodium phosphate (pH 6.4), 200 mM NaCl buffer solution. Using relatively high concentrations of sodium ion for the NMR experiments saturates the negatively charged phosphate backbone and minimizes nonspecific electrostatic interaction between the phosphate backbone and the metal ion complex under investigation. In some cases 100 μM EDTA will be added to the final dialysis buffer to chelate any trace divalent metal ion contaminants not removed by dialysis. Because of the generally low-affinity binding of divalent metal ions to RNA, large concentrations of divalent metal ions are used in these studies and this trace amount of EDTA contributes insignificantly to the free metal ion concentration.

[7] G. Colmenarejo and I. Tinoco, Jr., *J. Mol. Biol.* **290,** 119 (1999).

[8] R. L. Gonzalez, Jr. and I. Tinoco, Jr., *J. Mol. Biol.* **289,** 1267 (1999).

[9] J. F. Milligan, D. R. Groebe, G. M. Witherall, and O. C. Uhlenbeck, *Nucl. Acids Res.* **15,** 8783 (1987).

[10] J. R. Wyatt, M. Chastain, and J. D. Puglisi, *Biotechniques* **11,** 764 (1991).

A final consideration is that of buffer system choice. Organic buffer systems that do not interact significantly with the metal ions can be used. However, if not deuterated, these buffers will produce very large ^{1}H NMR resonances. A phosphate buffer system will not produce ^{1}H NMR signals, but has the disadvantage that phosphate anions interact with metal ions and affect the free metal ion concentration. In the pH range used in many metal ion binding studies (pH 6–7), the dominant phosphate species present in phosphate buffer is $H_2PO_4^-$.[11–13] Binding of divalent metal ions to $H_2PO_4^-$ has not been studied in detail, but equilibrium dissociation constants for Mg^{2+}, Ca^{2+}, Mn^{2+}, or Sr^{2+} bound to HPO_4^{2-} have been reported and are in the range of 2.5–30 mM.[14] The binding of metal ions to $H_2PO_4^-$ is expected to be even weaker because of its reduced negative charge. However, the binding of metal ions to components of the buffer will affect binding constants to RNA. If the concentration of metal ion is established by equilibrium dialysis, the free ion concentration (the ion concentration in equilibrium with the RNA solution) includes that bound to buffer components. The apparent binding constant to the RNA will thus depend on the buffer used. Even if there is no metal ion binding to the buffer anions, the binding constants to the RNA will depend on the concentrations and identity of the added cations, because of competition for the metal ion binding sites on the RNA molecule.

Diamagnetic Metal Ion Probes

Addition of a diamagnetic metal ion complex such as Mg^{2+} or $Co(NH_3)_6^{3+}$ to an RNA molecule can lead to changes in NMR spectral features of RNA nuclei near the metal ion binding site and can therefore serve as probes for binding site identification (Fig. 3).[5–8,15–19] Typically the RNA sample is titrated with the diamagnetic metal ion and one-dimensional experiments are recorded in either 90% H_2O/10% D_2O or 99.9% D_2O solution. The chemical shifts and/or resonance line widths for well-resolved RNA resonances can then be recorded as a function of metal ion concentration. The imino proton region of the RNA NMR spectrum is usually well resolved and provides information about metal ions binding near base pairs regardless of whether binding takes place in the major or minor groove.

[11] R. G. Bates, *J. Res. Natl. Bur. Standards* **47**, 127 (1951).
[12] R. G. Bates and S. F. Acree, *J. Res. Natl. Bur. Standards* **30**, 129 (1943).
[13] J. J. Christensen, R. M. Izatt, L. D. Hansen, and J. A. Partridge, *J. Phys. Chem.* **70**, 2003 (1966).
[14] R. M. Smith and R. A. Alberty, *J. Am. Chem. Soc.* **78**, 2376 (1956).
[15] S. Limmer, H.-P. Hoffman, G. Ott, and M. Sprinzl, *PNAS* **90**, 6199 (1993).
[16] K. Kalurachchi and E. P. Nikonowicz, *J. Mol. Biol.* **280**, 639 (1999).
[17] P. L. Nixon, C. A. Theimer, and D. P. Giedroc, *Biopolymers* **50**, 443 (1999).
[18] M. R. Hansen, J. P. Simorre, P. Hanson, V. Mokler, L. Bellont, L. Beigelman, and A. Pardi, *RNA* **5**, 1099 (1999).
[19] S. E. Butcher, F. H.-T. Allain, and J. Feigon, *Biochemistry* **39**, 2174 (2000).

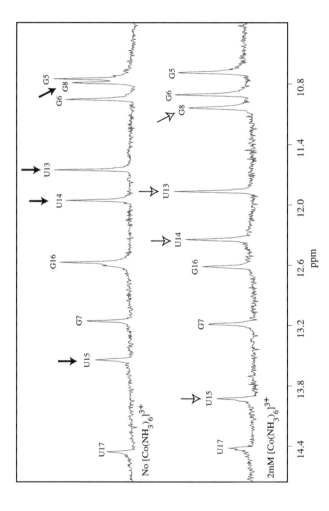

Fig. 3. Imino proton portion of a one-dimensional NMR spectrum of the P5b stem loop from the *T. thermophila* group I intron. Resonances corresponding to the U13, U14, U15, and G8 imino protons, denoted by arrows, shift downfield with no significant change in intensity and/or line width on addition of cobalt(III) hexammine. This is consistent with a fast exchange process on the chemical shift time scale. Similar trends are observed on the addition of Mg²⁺ ion. [Reprinted with permission from J. S. Kieft and I. Tinoco, Jr., *Structure* **5**, 713 (1997)].

For these reasons this region of the spectrum provides a very good place to begin spectral analysis.

Generally, chemical shift and resonance line-width changes observed during diamagnetic metal ion titrations will be small. Spectral changes will depend on (i) direct changes in the magnetic environment of the observed nucleus caused by the bound ion, (ii) indirect effects caused by RNA conformational changes upon metal ion binding, and (iii) the exchange regime of the observed resonance on the time scale of the appropriate NMR parameter (e.g., chemical shift, scalar coupling, relaxation rate). In order to identify a binding site one would ideally prefer to separate the direct effect of the metal ion from effects due to conformational changes. Detection of intermolecular nuclear Overhauser effect (NOE) cross peaks between protons from a metal ion complex and protons from an RNA molecule or the use of paramagnetic metal ion probes provide methods of separating these two effects.

Intermolecular Nuclear Overhauser Effects

Intermolecular nuclear Overhauser effect (NOE) cross peaks between protons from a metal ion complex and protons from an RNA molecule provide distance constraints that can be used to determine the structure of the metal ion binding site. Generally the native metal ion complexes that interact with RNA possess H_2O ligands, which are unfavorable for the detection of intermolecular NOEs due to fast exchange with the bulk solvent water. Metal ion complexes with ligands other than water that are in slow exchange with the solvent can be used to mimic the native metal ion complex. $Co(NH_3)_6^{3+}$ has been successfully used as a model for $Mg(H_2O)_6^{2+}$ to solve the structures of several $Mg(H_2O)_6^{2+}$ binding sites (Fig. 4).[5–8] $Co(NH_3)_6^{3+}$ has been proposed as a substitute for $Mg(H_2O)_6^{2+}$ based on geometric similarities.[20]

Intermolecular NOEs from $Co(NH_3)_6^{3+}$ protons to RNA protons can be observed in a conventional H_2O NOESY experiment (Figs. 4C and 4D). In a typical experiment,[6] a 2 mM RNA sample in 10 mM sodium phosphate (pH 6.4), 200 mM NaCl, 100 μM EDTA, and 2 mM $Co(NH_3)_6^{3+}$ in 90% H_2O/10% D_2O is utilized. An H_2O NOESY experiment is performed using the jump-return[21] water suppression scheme. Usually, the t_2 dimension will contain a very strong resonance for the $Co(NH_3)_6^{3+}$ protons, and it will be difficult to detect the intermolecular NOEs. Intermolecular NOEs, however, are observable in the t_1 dimension at the $Co(NH_3)_6^{3+}$ proton chemical shift of 3.65 ppm. Imino, amino, and aromatic proton to $Co(NH_3)_6^{3+}$ proton cross peaks will be easily detected. Intermolecular cross peaks to the sugar protons and the H5 pyrimidine protons will be more difficult

---

[20] J. A. Cowan, *J. Inorg. Biochem.* **49,** 171 (1993).
[21] P. Plateau and M. Guéron, *J. Am. Chem. Soc.* **104,** 7310 (1982).

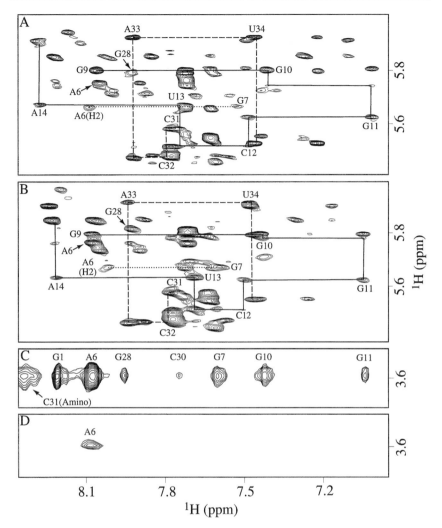

FIG. 4. Aromatic to H1' region of a 400 ms D_2O NOESY of the VPK pseudoknot in the presence of (A) 5 mM Mg^{2+} or (B) 2 mM $Co(NH_3)_6^{3+}$. The continuous and broken lines indicate standard A-form walks for the two strands of stem 2. The dotted line indicates the A6H2 to G7H8 connectivity that is observed in loop 1. Arrows denote the A6 and G7 intranucleotide H8 to H1' NOE. (C) The $Co(NH_3)_6^{3+}$ proton to H6/H8 region of a 300 ms H_2O NOESY experiment. Intermolecular NOE cross peaks are observed between $Co(NH_3)_6^{3+}$ protons and the H6/H8 proton of the denoted nucleotide. (D) The $Co(NH_3)_6^{3+}$ proton to H6/H8 region of a ^{13}C-resolved HSQC-NOESY on a VPK sample selectively ^{13}C-labeled at the adenine C8 and the uracil C6. Only one cross peak, to the A6H8, is observed. Sample conditions in (A) are 2 mM VPK, 5 mM Mg^{2+}, 10 mM sodium phosphate (pH 6.4), 200 mM NaCl, and 100 μM EDTA at a temperature of 30°. Sample conditions in (B), (C), and (D) are identical, with the exception of 2 mM $Co(NH_3)_6^{3+}$ instead of 5 mM Mg^{2+}, and a temperature of 35°. [Reprinted with permission from R. L. Gonzalez, Jr. and I. Tinoco, Jr., *J. Mol. Biol.* **289,** 1267 (1999).]

to detect because of severe overlap of the sugar region and close proximity to the suppressed water resonance.

One important difference between $Mg(H_2O)_6^{2+}$ and $Co(NH_3)_6^{3+}$ is that $Mg(H_2O)_6^{2+}$ can give up H_2O ligands and form direct inner-sphere coordinate bonds to RNA. Clearly, when direct coordinate bonds between the metal ion center and RNA functional groups are present, $Co(NH_3)_6^{3+}$ will not be a good mimic for the interaction.

There are several ways to test if $Co(NH_3)_6^{3+}$ is a valid substitute for a metal ion binding site. Two transitions in the ultraviolet absorbance melting curves of the VPK pseudoknot in sodium ion displayed very similar stabilization effects as a function of $Co(NH_3)_6^{3+}$ and of $Mg(H_2O)_6^{2+}$ concentration. Additionally, both $Co(NH_3)_6^{3+}$ and $Mg(H_2O)_6^{2+}$ had similar effects on the VPK imino proton chemical shifts. Furthermore, D_2O NOESYs recorded in the presence of $Mg(H_2O)_6^{2+}$ or $Co(NH_3)_6^{3+}$ exhibit analogous NOE connectivities and intensities, thus indicating that the RNA structure in Mg^{2+} is analogous to the structure in $Co(NH_3)_6^{3+}$ (Figs. 4A and 4B). These results corroborate the role of $Co(NH_3)_6^{3+}$ as a probe of the Mg^{2+} binding site and indicate that the hexahydrated magnesium ion complex is the form of magnesium ion interacting with the VPK pseudoknot. In cases of direct coordination of the metal ion, a more appropriate probe can be used. For example, in order to mimic a metal ion binding site involving a pentahydrated magnesium ion with one direct coordinate bond to RNA, $Co(NH_3)_5^{3+}$ could be used as a mimic of $Mg(H_2O)_5^{2+}$ where both metal ion complexes can give up one H_2O ligand and form a direct coordinate bond. Similarly, various other complexes of cobalt, or other transition metals, can be used as analogs of directly coordinated magnesium ion complexes.

Paramagnetic Metal Ion Probes

Unlike titrations utilizing diamagnetic metal ions, titration of an RNA molecule with paramagnetic metal ions can cause relatively large and dramatic NMR spectral changes. As with diamagnetic probes, the RNA sample is titrated with metal ion and one-dimensional experiments are recorded in either 90% H_2O/10% D_2O or 99.9% D_2O solution. The chemical shifts and/or line widths of well-resolved RNA resonances are then recorded as a function of metal ion concentration. Again, the favorable properties of the imino proton region of the spectra provide a good place to begin spectral analysis. The large changes observed in the chemical shifts and/or line widths of the RNA resonances on addition of paramagnetic metal ions are easier to interpret than changes observed during titration with diamagnetic metal ions. This is because the spectral changes observed in the paramagnetic case are dominated by the strong interactions between the unpaired electron spin and nearby ($\sim$10 Å) nuclear spins in a distance-dependent manner.[19,22,23] In most cases, small chemical shift or line-width changes associated with minor conformational changes on metal ion binding will be easily separated from the relatively larger effect of an unpaired electron with a very large magnetic moment in the vicinity of an NMR-active nucleus.[15]

Relaxation Rate Enhancement by Paramagnetic Metal Ions

Perhaps the most widely employed technique for identifying metal ion binding sites in RNA is paramagnetic line broadening. When a paramagnetic ion, such as Mn^{2+}, is located within ~ 10 Å of an NMR-active nucleus, the nucleus will experience an increased rate of relaxation. Enhanced rate of spin–spin (T_2) relaxation, in particular, leads to the observation of line broadening in titrations with paramagnetic metal ions.[23] In most cases, the dipolar interaction between the unpaired electron and the nucleus dominates the observed relaxation rate enhancement because of the large electronic magnetic moment. Relaxation rate enhancement due to this dipolar mechanism has an r^{-6} distance dependence where r is the distance between the paramagnetic metal ion center and the observed nucleus.[18,19,23-30]

The metal ion in an RNA–metal ion complex is typically weakly bound and has a correspondingly short residence lifetime. Therefore, a single paramagnetic ion can cause enhanced relaxation rates of NMR-active nuclei on multiple sites and multiple RNA molecules. As a result, substoichiometric amounts of Mn^{2+} will lead to complete broadening of nuclei in the vicinity of the binding site and at higher concentrations will begin to affect all RNA resonances in a nonspecific manner.

Paramagnetic line broadening of imino proton resonances caused by Mn^{2+} has been used to identify metal ion binding sites in tandem $G \cdot U$ wobble base pairs[7] and in the GAAA tetraloop.[5] In both cases the paramagnetic broadening data agree well with the binding site identified through chemical shift changes observed with Mg^{2+} or $Co(NH_3)_6^{3+}$ titrations, and with intermolecular NOEs observed from the $Co(NH_3)_6^{3+}$ protons to RNA protons at the binding site. In addition, paramagnetic line broadening by Mn^{2+} has been used to identify and localize Mg^{2+} binding sites in numerous RNA or RNA–protein complexes,[16,18,19,27,31,32] including the P1 helix from the *T. thermophila* group I intron (Fig. 5).[26] In addition,

[22] R. E. Hurd, E. Azhderian, and B. R. Reid, *Biochemistry* **18**, 4012 (1979).
[23] D. J. Craik and K. A. Higgins, *Ann. Rep. NMR Spectr.* **22**, 61 (1989).
[24] I. Solomon, *Phys. Rev.* **99**, 559 (1959).
[25] N. Bloembergen, *J. Chem. Phys.* **27**, 572 (1957).
[26] F. H.-T. Allain and G. Varani, *Nucl. Acids Res.* **23**, 341 (1995).
[27] B. L. Bean, R. Koren, and A. S. Mildvan, *Biochemistry* **16**, 3322 (1977).
[28] D. Bentrop, I. Bertini, M. A. Cremonini, S. Forsâen, C. Luchinat, and A. Malmendal, *Biochemistry* **36**, 11605 (1997).
[29] K. Tu and M. Gochin, *J. Am. Chem. Soc.* **121**, 9276 (1999).
[30] M. Gochin, *Structure* **8**, 441 (2000).
[31] P. F. Agris and S. C. Brown, *Methods Enzymol.* **261**, 270 (1995).
[32] D. G. Gorenstein, E. M. Goldfield, R. Chen, K. Kovarand, and B. A. Luxon, *Biochemistry* **20**, 2141 (1981).

A

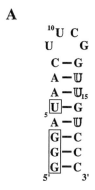

B

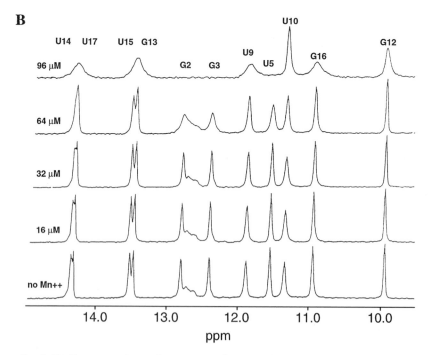

FIG. 5. (A) Sequence and secondary structure of the P1 helix from the *T. thermophila* group I intron. Nucleotides with strongly broadened imino protons by Mn^{2+} are boxed and those with slightly broadened imino protons are shown in outline form. (B) Mn^{2+} titration of the P1 helix from 1D spectra acquired in H_2O at 275 K at increasing concentrations of Mn^{2+}, as indicated next to each spectrum. Assignments are indicated at the top of the figure. (Reprinted from Ref. 40.) (C) Comparison of the $^1H–^{13}C$ correlated (HSQC) spectra of the P1 helix in the presence (top) and absence (bottom) of manganese impurities. The region shown on the right-hand side contains correlations between base resonances (H2, H8, and H6) and their attached carbons. The region on the left-hand side contains correlations between pyrimidine H5 and all H1′ resonances and their attached carbons. [Reprinted with permission from F. H.-T. Allain and G. Varani, *Nucl. Acids Res.* **23,** 341 (1995).]

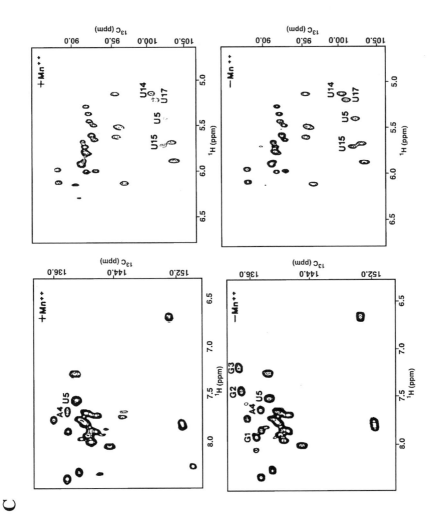

paramagnetic Ni^{2+} has been used to identify a Ni^{2+} ion binding site in an *in vitro* selected Ni^{2+} binding RNA aptamer.[33]

In a typical experiment,[26] a one-dimensional experiment in 90% H_2O/10% D_2O is recorded using the jump-return water suppression technique on a 1 mM RNA sample in the absence of divalent metal ions. $MnCl_2$ is then added directly to the sample in small increments (of approximately 10–20 μM) from a 2 mM $MnCl_2$ stock solution to cover a range of 4–200 μM Mn^{2+} concentration. This allows analysis of the imino proton region of the RNA spectrum (Fig. 5B). The titration can also be performed using a one-dimensional experiment or two-dimensional COSY, TOCSY, or NOESY type experiments in 99.9% D_2O with a presaturation pulse during the relaxation delay to suppress the residual HDO peak, in order to observe effects on nonexchangeable protons. Natural abundance ^1H–^{13}C or, in an isotopically labeled RNA molecule, ^1H–^{15}N or ^1H–^{13}C HSQC[19,26] or HMQC experiments can be used to observe line broadening of ^1H and ^{13}C or ^{15}N (Fig. 5C). ^{31}P line broadening can be observed in one-dimensional phosphorus experiments[18] for unusually well resolved ^{31}P resonances, or two-dimensional ^1H-^{31}P COSY experiments can be used to measure ^1H and ^{31}P line broadening for less well resolved ^{31}P resonances.

Contact and Dipolar (Pseudocontact) Shifts by Paramagnetic Metal Ions

The chemical shift changes and relaxation rate enhancements observed for magnetic nuclei in the vicinity of paramagnetic metal ions are due to two types of interactions between the unpaired electron spin and the nuclear spin. The first of these, the Fermi contact interaction, is a through-bond, scalar interaction that involves the direct transfer of unpaired electron spin density to the nuclear spin.[23,34,35] Although the contact interaction is useful in identifying nuclei near the paramagnetic metal ion, no quantitative distance information can be derived from the magnitude of the interaction.

The second type of interaction between an unpaired electron spin and a nuclear spin is a through-space dipolar interaction also known as the pseudocontact interaction. Chemical shift changes caused by this mechanism are called pseudocontact shifts and depend primarily on two factors: the angular dependence of the magnetic susceptibility tensor centered on the paramagnetic metal ion center, and a long-range, r^{-3} distance dependence where r is the distance between the paramagnetic metal ion center and the nucleus. Pseudocontact shifts have been measured in cytochromes where chemical shifts are measured for the diamagnetic, reduced

[33] H.-P. Hoffman, S. Limmer, V. Hornung, and M. Sprinzl, *RNA* **3**, 1289 (1997).
[34] M. Gochin and H. Roder, *Protein Sci.* **4**, 296 (1995).
[35] M. Gochin and H. Roder, *Bull. Magn. Reson.* **17**, 1 (1995).

[Fe(II)] heme state and for the paramagnetic, oxidized [Fe(III)] heme state.[34–38] Pseudocontact shifts have also been measured in the N-terminal domain of calmodulin where chemical shifts in the diamagnetic reference state were measured for the Ca^{2+}-bound form and the chemical shifts in the paramagnetic state were measured by replacing the Ca^{2+} ions with paramagnetic Ce^{2+} ions.[28] Most recently, pseudocontact shifts have been measured in a DNA–chromomycin–Co(II) complex.[29,30] To make use of the measured pseudocontact shifts, the relationship between the magnetic susceptibility tensor centered on the paramagnetic ion and the molecular coordinate frame must be known.[28–30,34–38]

Although pseudocontact shifts from paramagnetic metal ions have not yet been applied to structural determination of RNA metal ion binding sites, their use clearly seems advantageous. The distance constraints derived from the shifts would be useful not only in very accurately locating the metal ion, but also in providing long-range distance constraints that would better define global features of the RNA structure. In fact, work on the structure of a DNA–chromomycin–Co(II) complex refined against pseudocontact shifts has provided a structure determined to 0.7 Å resolution.[29,30] Global features and fine details of this structure, including differences in the precise location of the drug, are defined to much higher precision in this structure compared to previous NMR structural studies of this complex.[29,30] Co(II),[39–41] Ni(II),[42] Fe(II),[37,38] and the lanthanide(III) metal ions [except for Gd(III)][28] are all paramagnetic metal ions with asymmetric magnetic susceptibility tensors and fast electron spin relaxation; they are ideal for measurement of pseudocontact shifts. Tb^{3+} has been shown to block activity of a hammerhead ribozyme by substituting one catalytically important Mg^{2+} ion,[43] and both Tb^{3+} and Eu^{3+} have been used in luminescence spectroscopy studies of metal ion binding sites in the hammerhead ribozyme.[44] These results demonstrate that substitution of Mg^{2+} by lanthanide(III) ions in RNA is possible and perhaps could be used to measure pseudocontact shifts. Studies are currently underway to

[36] R. D. Guiles, S. Sarma, R. J. Digate, D. Banville, V. J. Basus, I. D. Kuntz, and L. Waskell, *Nature Struct. Biol.* **3**, 333 (1996).

[37] L. Banci, I. Bertini, H. B. Gray, C. Luchinat, T. Reddig, A. Rosato, and P. Turano, *Biochemistry* **36**, 9867 (1997).

[38] L. Banci, I. Bertini, K. L. Bren, H. B. Gray, P. Sompornpisut, and P. Turano, *Biochemistry* **36**, 8992 (1997).

[39] I. Bertini, L. Banci, and C. Luchinat, *Methods Enzymol.* **177**, 246 (1989).

[40] I. Bertini, C. Luchinat, and M. Piccioli, *Prog. Nucl. Magn. Reson. Spectrosc.* **26**, 91 (1994).

[41] J. M. Mortal, J. Salgado, A. Donaire, H. R. Jiménez, and J. Castells, *Inorg. Chem.* **32**, 3578 (1993).

[42] J. M. Mortal, J. Salgado, A. Donaire, H. R. Jiménez, and J. Castells, *J. Chem. Soc., Chem. Commun.* **110**, (1993).

[43] A. L. Feig, W. G. Scott, and O. C. Uhlenbeck, *Science* **279**, 81 (1998).

[44] A. L. Feig, M. Panek, W. D. Horrocks, and O. C. Uhlenbeck, *Chem. Biol.* **6**, 801 (1999).

use Yb^{3+} to substitute Mg^{2+} binding sites in order to measure pseudocontact shifts in various RNA molecules (G. Pintacuda, personal communication, 2000).

NMR-Active Metal Ion Probes

The direct observation of NMR-active metal ions such as ^{199}Hg[45,46] and ^{113}Cd[46,47] has provided much information about metal ion binding sites in proteins. ^{59}Co NMR and ^{23}Na have been used to study $Co(NH_3)_6^{3+}$ and Na^+ binding to B-DNA,[48] and ^{113}Cd has been used to study a metal ion binding site in a small ribozyme.[49] Direct NMR spectroscopy of the metal ion provides unique information from analysis of NMR parameters such as chemical shifts, scalar couplings, and resonance line widths in one-dimensional experiments. It is possible to record spectra with reasonable signal-to-noise on 0.5 ml of a 96% isotopically enriched ^{113}Cd sample at a concentration of 1–5 mM in a 5 mm NMR tube and a field strength of 11.5 T (500 MHz for ^1H).[47] Similarly, a 91% isotopically enriched, 0.4 ml sample of 1–2 mM ^{199}Hg in a 5 mm NMR tube at a field strength of 14.09 T (600 MHz for ^1H) is adequate to detect ^{199}Hg with a reasonable signal-to-noise ratio.[45]

Chemical Shift

The NMR spectrum of an NMR-active metal nucleus is very simple, as only one resonance is observed for each metal ion bound. In most cases, the chemical shift range of the metal ion is very large. For example, ^{113}Cd has a chemical shift range of 800–900 ppm[47] and ^{199}Hg has a chemical shift range of more than 5000 ppm.[45] Because of these very large chemical shift ranges, the chemical shift of an NMR-active metal nucleus is exquisitely sensitive to the complexation state of the ion.[49,50] The chemical shift can be used to determine the number of ligands,[45] identity of the ligand donor atoms,[45,47] and the complexation geometry in some cases.[45] Currently, such information about metal ion binding sites in RNA is generally only available from crystallographic studies. A ^{113}Cd NMR study of a small ribozyme revealed a 9 ppm chemical shift change and a 40 Hz increase in line width in the single ^{113}Cd resonance upon addition of the ribozyme. This effect on the ^{113}Cd chemical shift and line width indicated specific binding of the metal ion by the RNA, but no details of the interactions were elucidated.[49]

[45] L. M. Utschig, J. W. Bryson, and T. V. O'Halloran, *Science* **268**, 380 (1995).
[46] P. R. Blake, B. Lee, M. F. Summers, M. W. W. Adams, J.-B. Park, Z. H. Zhou, and A. Bax, *J. Biomolec. NMR* **2**, 527 (1992).
[47] J. E. Coleman, *Methods Enzymol.* **227**, 16 (1993).
[48] W. H. Braunlin, C. F. Anderson, and M. T. Record, Jr., *Biochemistry* **26**, 7724 (1987).
[49] M. Vogtherr and S. Limmer, *FEBS Lett.* **433**, 301 (1998).
[50] R. J. Goodfellow, *in* "Multinuclear NMR" (J. Mason, ed.), p. 563. Plenum Press, New York, 1987.

Scalar Coupling

In proteins, scalar coupling has been observed between ^{199}Hg or ^{113}Cd metal ions and ^{1}H, ^{13}C, ^{15}N, or ^{31}P resonances in the protein in cases of direct coordination of the metal ion.[45-47] These scalar couplings can be observed indirectly by comparing cross-peak patterns in ^{1}H–^{1}H COSY or ^{1}H–^{13}C/^{15}N HMQC experiments recorded in the presence of a spin-0 metal ion and in the presence of a spin-1/2 (^{199}Hg or ^{113}Cd) metal ion without decoupling of the metal ion. Scalar couplings can also be detected and their values measured directly by analyzing fine structure in the one-dimensional ^{113}Cd or ^{199}Hg spectrum[47] or by recording ^{1}H–^{199}Hg or ^{1}H–^{113}Cd HMQC experiments.[45,47] Observation of scalar coupling can provide much structural data about the metal ion binding site. It can help identify which atoms on which side chains are directly coordinated to the metal ion, information that sometimes cannot be unambiguously determined by X-ray crystallographic analysis. In addition, two-bond scalar coupling values can potentially provide information about bond angles, and three-bond scalar coupling values will vary in a Karplus fashion and provide dihedral angle information. Further development of these methods for studying RNA–metal ion interactions should provide new and unique structural information.

Structure Determination of Metal Ion Binding Sites in RNA

Distance and/or angular constraints generated from intermolecular NOEs, paramagnetic T_2 relaxation rate enhancements, or paramagnetic pseudocontact shifts can be used in a structure calculation protocol in order to solve the structure of the metal ion binding site. Intermolecular NOEs between Co(NH$_3$)$_6$$^{3+}$ protons and RNA protons have been used to solve the structures of four metal ion binding sites in RNA.[5-8] If the metal ion complex is observed to sample all possible orientations within the binding pocket, the metal atom can be used as a pseudoatom for the entire metal ion complex with a correction to the intermolecular distance to account for the radius of the ion complex [2.5 Å for Co(NH$_3$)$_6$$^{3+}$].[6,51,52] Direct coordination of the metal ion complex will exhibit a preferred orientation within the binding pocket, and intermolecular distances can be referenced to a unique proton type in the metal ion complex. NOEs from the Co(NH$_3$)$_6$$^{3+}$ protons to RNA imino and amino protons are assigned conservative distance constraints of 0–5 Å (not including the 2.5 Å ionic radius) due to the nonuniform excitation profile of the jump-return water suppression technique and chemical exchange of the protons with solvent. Since the Co(NH$_3$)$_6$$^{3+}$ protons exchange slowly on the time scale of

[51] G. J. Kruger and E. C. Reynhardt, *Acta Crystallogr.* **34**, 915 (1978).

[52] J. S. Kieft, "Structure and Thermodynamics of a Metal Ion Binding Site in the RNA Major Groove: Cobalt(III) Hexammine as a Probe." Ph.D. Thesis, University of California, Berkeley, 1997.

one t_1 acquisition block during the NMR experiment,[53] these can be essentially treated as nonexchangeable. Therefore, intermolecular NOEs to nonexchangeable RNA protons can be categorized into strong, medium, and weak categories based on NOE cross-peak intensities. There are two general methods for carrying out the structure calculation as outlined below.

Docking of Metal Ion onto Static NMR Structure

In this approach the structure of the RNA molecule is determined initially using well-established structure calculation protocols.[54,55] The lowest energy structure or the average structure can then be used to dock the metal ion complex. The RNA coordinates are held static throughout the rest of the calculation and the metal ion complex is docked using the experimentally determined intermolecular distance restraints. There are several structure calculation programs available to do the docking. Both Discover 3[6,52] (Molecular Simulations, Inc., San Diego, CA) and XPLOR[7] scripts have been utilized to dock $Co(NH_3)_6^{3+}$ to two RNA hairpins containing tandem G · U wobble base pairs. The protocol in both cases is similar. The $Co(NH_3)_6^{3+}$ metal ion coordinates are generated using the Builder Module within InsightII (Molecular Simulations, Inc.). The bond lengths and bond angles can be adjusted by energy minimization with Discover 3 (Molecular Simulations, Inc.) using the extensible systematic force field (ESFF). Coordinates derived in this manner are consistent with the published crystal structure of $Co(NH_3)_6^{3+}$.[6,51,52] Alternatively, the equilibrium bond lengths and bond angles can be taken directly from the published crystal structure[51] and input into X-PLOR.[7] The Lennard–Jones terms and atomic partial charges in both cases are derived from the ESFF forcefield. The $Co(NH_3)_6^{3+}$ metal ion complex is placed in a random translational and rotational position relative to the RNA molecule, and intermolecular distance constraints are then applied during an energy minimization protocol in Discover 3[6,51] or a restrained molecular dynamics protocol using XPLOR[7] until the intermolecular distance constraints are satisfied without the Lennard–Jones or electrostatics force field terms turned on. In a second stage an energy minimization is carried out using the previous constraints, as well as full Lennard–Jones and electrostatic force field terms in order to find the lowest energy orientation(s) of the $Co(NH_3)_6^{3+}$ metal ion complex within the binding pocket. This procedure is repeated with random initial positioning of the $Co(NH_3)_6^{3+}$ ion in order to obtain a family of structures displaying the metal ion binding site. The final structures can be analyzed to determine the RMSD of the metal ion complex within the binding

[53] J. S. Anderson, H. V. A. Briscoe, and N. L. Spoor, *J. Chem. Soc.,* 361 (1943).
[54] B. T. Wimberley, "NMR Derived Structures of RNA Loops: The Conformation of Eukaryotic 5S Ribosomal Loop E." Ph.D. Thesis, University of California, Berkeley, 1992.
[55] G. Varani, F. Aboul-ela, and F. H.-T. Allain, *Prog. Nucl. Magn. Reson. Spect.* **29,** 51 (1996).

pocket as well as to determine the range of possible orientations of the metal ion complex that are consistent with the NMR data.

Complete Structure Calculation Including Metal Ions

A different approach was used in solving the structure of the metal ion binding sites in the VPK frameshifting viral RNA pseudoknot[8] and the GAAA tetraloop[5] in which the metal ion complex was included in all stages of the structure calculations and was used as a structural constraint in itself by not holding the RNA coordinates static at any time during the calculation. Starting structures with random torsion angles were generated, including random positioning of a $Co(NH_3)_6^{3+}$ ion complex. Loose intermolecular distance constraints of 0–10 Å (not including the pseudoatom correction) were used during the global fold stage[54,55] of the VPK pseudoknot structure calculation. During the refinement stage[54,55] of the structure calculation, intermolecular distance constraints were tightened to 0–5 Å (not including the pseudoatom correction) for distance constraints involving exchangeable RNA protons, and categories of strong, medium, and weak for distance constraints involving nonexchangeable RNA protons. In the final energy minimization stage,[54,55] full Lennard–Jones and electrostatic potential force field terms were turned on. In the GAAA tetraloop structure calculation, the same intermolecular distance ranges were used in the global fold, refinement, and minimization stages. $Co(NH_3)_6^{3+}$ proton to RNA exchangeable proton cross peaks were put into a loose 1.8–5.0 Å (without pseudoatom correction) category. Intermolecular cross peaks involving nonexchangeable RNA cross peaks were categorized as strong, medium, weak, or very weak. The rest of the structure calculation was carried out in a manner analogous to that for the VPK pseudoknot. This approach was particularly important in the VPK pseudoknot structure calculation where intermolecular NOEs were observed between the $Co(NH_3)_6^{3+}$ and RNA protons located in stem 2 of and loop 1—a unique tertiary structural feature—of the pseudoknot. The short two-nucleotide loop 1 crosses the major groove face of stem 2 and could certainly impede docking of the metal ion complex if the RNA coordinates were held static. In addition, this provided indirect loop-to-stem connectivities, connectivities that previously have not been observed for this pseudoknot and clearly would be important in determining the structural relationship between loop 1 and stem 2. In fact, previous work on the VPK pseudoknot had required the use of additional, nonexperimental, long-range constraints to prevent the starting structures from forming knotted structures during the initial global folding stage of structure calculation. The inclusion of the $Co(NH_3)_6^{3+}$ metal ion during the global fold eliminated the need for these constraints.[8,56]

[56] L. X. Shen and I. Tinoco, Jr., *J. Mol. Biol.* **247**, 963 (1995).

Thermodynamic and Kinetic Characterization of Metal Ion Binding Sites

An RNA nucleus in close proximity of a metal ion binding site, or in the vicinity of a metal-induced conformational change, will experience two unique chemical environments. The analogous situation exists for a metal ligand resonance. Analysis of the chemical exchange properties between the metal-free and metal-bound forms of the RNA–metal ion complex can provide thermodynamic and kinetic information about the interaction between the metal ions and the RNA molecule. In this way, NMR provides a unique opportunity to study the kinetics and thermodynamics of the interaction, as well as determine the structure of the RNA–metal ion complex. Diamagnetic or NMR-active metal ion probes are the most useful for thermodynamic and kinetic characterization since the weak binding and the fast exchange condition typical of RNA–metal ion complexes mean that even trace amounts of paramagnetic ions will affect many RNA sites on many RNA molecules.[18] The methods used to obtain thermodynamic and kinetic parameters depend on whether the kinetics of binding are in fast, slow, or intermediate exchange. The following discussion will consist of chemical exchange on the chemical shift time scale, since chemical exchange on the scalar coupling and relaxation rate time scales is generally found in the fast exchange regime and are therefore measured as averages. The fast exchange regime occurs when the rate of exchange between the bound and free states is significantly faster (a factor of 5) than the difference between chemical shifts (in hertz) of the bound and free states. Resonances in the fast exchange regime will display a single resonance at a chemical shift determined by the population weighted average of the individual chemical shifts of the bound and free states. The slow exchange regime occurs when the rate of exchange between the bound and free state is significantly slower (a factor of 1/5) than the difference between the chemical shift of the bound and free states. Resonances in the slow exchange regime will display two distinct resonances at the characteristic chemical shift of the bound and free states with areas proportional to the population of each state. Intermediate exchange occurs when the rate of exchange is approximately equal to the difference in chemical shifts between the bound and free states. Resonances in intermediate exchange will exhibit complicated spectral behavior between that observed in the fast and slow exchange cases.[57]

The most general way to obtain thermodynamic and kinetic information of a chemical exchange process, including metal ion binding to RNA, is line-shape analysis. Line-shape analysis is covered in detail in several reviews.[23,57,58] This technique involves the fitting of computer-simulated line shapes to experimental

---

[57] L.-Y. Lian and G. C. K. Roberts, in "NMR of Macromolecules: A Practical Approach" (G. C. K. Roberts, ed.), p. 153. Oxford University Press, New York, 1993.

[58] H. S. Gutowsky and C. H. Holm, *J. Chem. Phys.* **25**, 1228 (1956).

line shapes. In general, the imaginary component of the McConnell-modified Bloch equations[59] is used to simulate line shapes for a two-site exchange process in the absence of scalar coupling. These line shapes depend on the population, resonance frequency, T_2 relaxation rate, and exchange lifetime of the metal free and metal bound states. Usually the resonance frequencies and the T_2 relaxation rates for the two states can be experimentally measured or estimated, and the populations and exchange lifetime can then be fitted to the observed line shape using a nonlinear regression program.

Line-shape analysis of RNA (or metal ion) resonances can be done as a function of increasing metal ion (or RNA) concentration. Assuming that a particular resonance is affected by a single metal ion binding at this site, line-shape analysis can provide the on-rate (k_{on}) and off-rate (k_{off}) of metal ion binding, as well as the corresponding equilibrium dissociation constant (K_d) at a given temperature. The fast and slow exchange regimes allow a number of simplifications to line-shape analysis, and thermodynamic and kinetic measurements are straightforward in these two extremes.

Thermodynamic and Kinetic Analysis in Fast Exchange Case

When metal ion binding to RNA is relatively weak ($K_d \approx 10^{-3} M$), the exchange rate is usually in the fast exchange limit. An RNA (or metal ion) resonance that undergoes a change in chemical shift, with no accompanying change in intensity or line width, is in the fast exchange limit.

Changes in chemical shift of RNA protons on titration with $Co(NH_3)_6^{3+}$ or $Mg(H_2O)_6^{2+}$ have been used to study four RNA–metal ion complexes.[5-8] In each case, aliquots of a concentrated (1 M) stock solution of $Co(NH_3)_6^{3+}$ or $Mg(H_2O)_6^{2+}$ are added directly to an NMR sample containing the RNA of interest at a known total RNA concentration in the range of 0.2–0.8 mM in 10 mM sodium phosphate (pH 6.4), 200 mM NaCl, 100 μM EDTA dissolved in 90% H_2O/10% D_2O. A one-dimensional experiment utilizing the jump-return method of water suppression is done at each metal ion concentration and the chemical shift and line width of each imino proton resonance are recorded. The good dispersion of the RNA imino proton spectral region makes it ideal for these titrations. The sample volume change in each titration point is accounted for and the RNA concentration is corrected.

$Co(NH_3)_6^{3+}$ or Mg^{2+} have been found to be in fast exchange between the free and bound states as well as in fast exchange within the RNA binding pocket in all four RNA–$Co(NH_3)_6^{3+}$ complexes studied thus far. During titration of RNA with $Co(NH_3)_6^{3+}$, all 18 $Co(NH_3)_6^{3+}$ protons resonate in a single, broad ($\sim$15 Hz) resonance at 3.65 ppm with no significant chemical shift or line-width changes with increasing amounts of $Co(NH_3)_6^{3+}$. The chemical shift of free $Co(NH_3)_6^{3+}$

[59] H. M. McConnell, *J. Chem. Phys.* **28**, 430 (1958).

protons, with no RNA present, displays similar line width and chemical shift as that of $Co(NH_3)_6^{3+}$ protons in the presence of RNA. The lack of a dramatic chemical shift or line-width change of the $Co(NH_3)_6^{3+}$ proton resonance and the close agreement of these NMR parameters with the same parameters for $Co(NH_3)_6^{3+}$ free in solution is characteristic of a fast exchange process with a large population of free $Co(NH_3)_6^{3+}$ vs bound $Co(NH_3)_6^{3+}$, or a very small chemical shift difference between the bound and free state of the $Co(NH_3)_6^{3+}$ protons, or a combination of the two.

Analysis of RNA imino protons during the titration also indicate fast exchange between bound and free state of the RNA molecule. During the titration, no resonances disappear and no new resonances are observed. Several peaks undergo changes in chemical shift value with no significant changes in line width (Fig. 6). All of this is consistent with a fast exchange process for the binding of the metal ion complex. Some imino protons do display minor amounts of line broadening at higher metal ion concentrations, but this may be due to changes in base pair stability and imino proton exchange with bulk solvent water. Titrations with Mg^{2+} or with $Co(NH_3)_6^{3+}$ have similar effects on RNA proton resonances.

In the fast exchange limit, the rate of exchange and the corresponding exchange lifetime cannot be calculated. Calculation of the equilibrium binding constant (K_d), however, is very straightforward in the limit of fast exchange. Since there are no changes in line width, the experimental data consist of the chemical shift for the free state (measured in the absence of metal ion) and the observed chemical shift at various increasing concentrations of metal ion. Ideally, the NMR sample is dialyzed against buffer containing the desired metal ion concentration at each titration point. In this way the free metal ion concentration at each titration point is known, and a binding curve of observed chemical shift vs free metal ion concentration can be constructed. A Scatchard plot can provide the number of binding sites and their K_d. Alternatively, metal ion can be directly added to the NMR sample, and a binding curve of observed chemical shift vs total metal ion concentration can then be plotted (Fig. 6). In favorable cases, the NMR properties of an observed RNA

FIG. 6. Binding curve of the P5b U15 imino proton vs (A) Mg^{2+} or (B) $Co(NH_3)_6^{3+}$ concentration. A change in chemical shift of the U15 imino proton resonance is observed with increasing concentration of (A) Mg^{2+} or (B) $Co(NH_3)_6^{3+}$ with no significant change in line width, indicating a fast exchange process on the chemical shift time scale. The equilibrium dissociation constant (K_d) for (A) Mg^{2+} or (B) $Co(NH_3)_6^{3+}$ binding to the U15 imino proton is calculated by fitting the data to an equation for a one-site binding mode in fast exchange. In (A) and (B) the data are plotted as points and the calculated curve is shown as a solid line. An average K_d for the metal ion binding site can be determined by performing similar calculations for other RNA protons in the vicinity of the binding site that undergo chemical shift changes as a function of metal ion concentration.

A

Chemical Shift vs. Mg²⁺ Concentration
U15 Imino Proton of P5b Helix

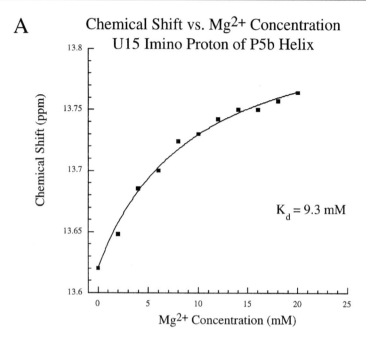

B

Chemical Shift vs. Co(NH₃)₆³⁺ Concentration
U15 Imino Proton of P5b Helix

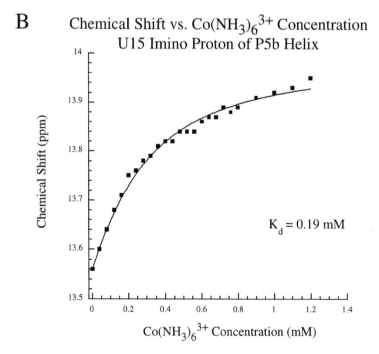

resonance will be affected chiefly by the direct binding of only one metal ion. The plot of observed chemical shift vs total metal ion concentration can be fitted to an equation for a single metal ion binding site in order to determine K_d (Fig. 8).[6,8,52,57] This equation has been applied to metal ion titrations in order to determine K_d for $Co(NH_3)_6^{3+}$ or $Mg(H_2O)_6^{2+}$ in the four metal ion–RNA complexes. In principle, a van't Hoff analysis of the temperature dependence of the binding can provide the enthalpy, entropy, and free energy of binding.

Thermodynamic and Kinetic Analysis in Slow Exchange Case

Slow exchange on the chemical shift time scale usually occurs in cases of tight binding $(K_d \approx 10^{-6} M)$[23] and is not commonly observed in metal ion interactions with RNA. In cases of slow exchange on the chemical shift time scale, individual resonances are observed for the metal free state and the metal bound state. For a well-resolved RNA resonance, the area of each peak will correspond to the population of each state and allow calculation of the free RNA and the RNA–metal ion complex concentrations. Assuming a one metal ion binding site model, the free metal ion concentration and the K_d can be calculated directly from the population measurements. In cases where both an RNA resonance and a metal ion complex resonance are well resolved, the free RNA, free metal ion, and RNA–metal ion complex concentrations can all be determined from population measurements as a function of metal ion concentration. This again would allow Scatchard analysis of the data. Furthermore, the metal ion concentration dependence of the line widths of the free state resonance from the RNA molecule (or the RNA concentration dependence on a metal ion complex free state resonance) can be used to determine the exchange rates.[57]

Thermodynamic and Kinetic Analysis in Intermediate Exchange Case

When metal ion binding produces resonances in intermediate exchange,[16] or in slow exchange where only the resonance of one of the two states is well resolved,[18] the simplifications of the fast exchange or the slow exchange limits do not apply and the thermodynamics and kinetics of binding must be determined from a full line-shape analysis.[18,23,57] Line-shape analysis was used to measure the binding affinity of a Mg^{2+} ion to the catalytic core of a hammerhead ribozyme[18]—a metal ion that despite its unusually tight binding has not been previously observed by crystallographic or biochemical methods. In this case an unusually well resolved phosphorus resonance was observed to undergo chemical exchange on the chemical shift time scale between the metal bound and the metal free state as a function of increasing Mg^{2+} concentration. The chemical shift of the free state was well resolved and was measured in the absence of metal ion. The bound chemical shift, on the other hand, resides in the overcrowded and unresolved region of the RNA phosphorus spectrum. Therefore, several values for the bound state chemical shift were used, which leads to the uncertainty reported in the measurement. Using the experimentally determined free state chemical shift and a range of bound state chemical shifts, the population and exchange lifetime of the free and bound states were fitted to the experimental data. An equilibrium binding constant of 250–570 μM for the binding of a single Mg^{2+} to a specific phosphate group was obtained.

Conclusion

The interactions of metal ions with RNA are clearly important to RNA structure and function. NMR spectroscopy provides a powerful methodology for the study of these interactions. NMR spectroscopy can provide distance constraints between a metal ion and a binding site in an RNA molecule generated from analysis of paramagnetic relaxation rate enhancements, or intermolecular NOEs. The potential exists for further geometric constraints from pseudocontact shift distance estimates and observation of scalar coupling of RNA atoms to tightly bound, directly coordinated metal ions. In addition to the detailed structural information that is available from NMR spectroscopic techniques, the kinetics and thermodynamics of the interaction can also be determined. Exchange rate constants and equilibrium dissociation constants can be estimated from relatively simple NMR experiments.

Acknowledgments

This research was supported in part by the National Institutes of Health Grant GM 10840, by the Department of Energy Grant DE-FG03-86ER60406, and through instrumentation grants from the Department of Energy (DE-FG05-86ER75281) and from the National Science Foundation (DMB 86-09305). We thank Ms. Barbara Dengler for managing the laboratory, Mr. David Koh for synthesizing DNA oligonucleotides, and Dr. Jeffrey Pelton for valuable NMR advice.

Author Index

A

Abdul-Manan, N., 202, 245
Abildgaard, F., 4
Aboul-ela, F., 372, 373(5), 378, 382(5), 383(5), 384(5), 388(5), 436, 437(55)
Abragam, A., 82, 128, 187
Abraham, R. J., 14
Acevedo, O. L., 405
Ackermann, J. L., 36, 62(3)
Acree, S. F., 424
Adamo, C., 10
Adams, J. L., 107(59), 108
Adams, M. W., 114(54; 74), 121, 122(54), 131, 434, 435(46)
Adams, P. D., 22, 24
Agback, P., 278
Agris, P. F., 429
Agrofoglio, L. A., 264, 273
Ahluwalia, M., 286
Ajay, 205, 207(8), 216(8), 217(8), 218(8), 221(8), 226(8), 228(8), 231, 239, 239(25), 241(25), 244(25)
Akasaka, K., 134, 135, 136, 137(28), 144, 145, 145(35), 146, 146(39), 147, 147(35; 39), 148, 149, 149(38), 150, 150(35; 50), 151(50), 152(56), 153(28; 35; 49; 50), 155(27; 50), 156(27), 157, 158, 158(56)
Alajarin, R., 311
Alam, T. M., 278
Alberty, R. A., 424
Albrand, J. P., 205
Alexander, S., 205
Al-Hashimi, H. M., 316, 317(52)
Allain, F. H.-T., 372, 373(5), 378, 382(5), 383, 383(5), 384(5), 385, 388(5), 390, 391(74), 400, 417, 417(11), 418, 418(11), 424, 428(19), 429, 429(19), 430, 432(26), 436, 437(55)
Allard, P., 27
Allen, M., 262
Altman, L. J., 112

Altona, C., 69, 70(43), 349, 355, 382, 386, 388(90), 393
Amos, R. D., 10, 11(47)
Anber, M., 269
Anderson, C. F., 401, 424
Anderson, J. S., 436
Anderson, W. T., 159
Andersson, T., 186
Ando, I., 19, 112
Andrec, M., 29
Anet, F. A. L., 402
Anglister, J., 231, 320
Angwin, D. T., 230
Antzutkin, O. N., 319
Aoki, S., 317
Aoyama, Y., 266, 267(20a; 20b; 20c), 273
Appella, E., 355, 391
Arata, Y., 4, 69, 317
Ariza, X., 269
Arnau, M. J., 24, 25(115)
Arnold, L., 417
Arnold, W. D., 14
Arrowsmith, C. H., 320
Asakawa, N., 19, 112
Asakura, T., 3, 4, 13, 15, 15(15; 78), 16(7; 92), 17(78; 92), 18(78), 19(15), 22, 24(78; 92), 29(78)
Ascheim, K., 33
Assfalg, M., 25
Aubin, Y., 308, 316(17–19)
Augeri, D. J., 219
Au-Yeung, S. C. F., 12
Avizonis, D., 247
Aygen, S., 134
Azhderian, E., 428(22), 429
Azorín, F., 349, 377

B

Baber, J. L., 385, 386(83), 390(83)
Bachmann, P., 213

Subject Index